DETERMINATION OF METALS AND ANIONS IN SOILS, SEDIMENTS AND SLUDGES

T0199529

DETERMINATION OF METALS AND ANIONS IN SOILS, SEDIMENTS AND SLUDGES

T. R. Crompton

CRC Press
Taylor & Francis Group
Boca Raton London New York

CRC Press is an imprint of the
Taylor & Francis Group, an **informa** business

A SPON PRESS BOOK

CRC Press
Taylor & Francis Group
6000 Broken Sound Parkway NW, Suite 300
Boca Raton, FL 33487-2742

First issued in paperback 2019

ISBN-13: 978-0-415-23882-3 (hbk)
ISBN-13: 978-0-367-87197-0 (pbk)

Publisher's Note

This book has been prepared from camera-ready copy supplied by the author.

British Library Cataloguing in Publication Data
A catalogue record for this book is available from the British Library

Library of Congress Cataloging in Publication Data
A catalogue record for this book has been requested

Visit the Taylor & Francis Web site at
http://www.taylorandfrancis.com

and the CRC Press Web site at
http://www.crcpress.com

Determination of Metals and Anions in Soils, Sediments and Sludges

The publisher would like to bring to the attention of the reader, the following textual errors:

Tamerium should be replaced by Thulium on pages xii, xx (contents), 233, 276, 341, 359 and the index (pages 719, 722, 723, 725).

Contents

10 Sampling procedures 631

11 Accumulation processes in sediments 660

Preface

This book is concerned with a discussion of methods currently available in the world literature up to 1999 for the determination of metals and anions in soils, river and marine sediments and industrial sludges.

No books have previously been published on the determination of cations and anions in these solids. Yet the occurrence of metals and anions, many of which are toxic, can have profound effects on the ecosystem.

In the case of soils the presence of deliberately added or adventitious metallic compounds can cause contamination of the tissues of crops grown on the land or animals feeding on the land and, consequently, can cause adverse toxic effects on man, animals, birds and insects. Also drainage of these substances from the soil can cause pollution of adjacent streams, rivers and eventually the oceans. Some of the substances included in this category are fertilizers, crop sprays, sheep dips, etc. In the case of river and marine sediments, a major input of metals is a consequence of industrial activity. Many industries discharge metal-containing effluents to rivers or directly to sea or discharge of such wastes by ship dumping.

Surface water drainage is another input, particularly in the case of lead originating in automobile exhausts. The presence of metallic compounds in river and oceanic sediments is due, in part, to manmade pollution and monitoring the levels of these substances in the sediment and sediment cores provides an indication of the time dependence of their concentration over large time spans. Contamination of sediments is found not only in rivers but also in estuarine and oceanic sediments and thus sediment analysis provides a means of tracking metallic compounds from their source through the ecosystem.

Another consideration is that fish, particularly bottom feeders and crustacea, pick up contaminants when sediments enter their gills and the contamination of these creatures has definite toxicological implications both for the creatures themselves, for man who eats them and, in the case of fish meal, for animals.

Sediments have the property of absorbing such contaminants from water within their bulk (accumulation) and, indeed, it has been shown that the concentration, for example, of some types of metallic contaminants in river sediments can be up to one million times greater than occurs in the surrounding water.

Most of the metallic elements in the periodic table have been found to be present in soils, sediments and sludges, some naturally occurring and others as a result of industrial activity of one kind or another. To date, insufficient attention has been given to the analysis of sediments and one of the objects of this book is to draw the attention of analysts and others concerned to the methods available and their sensitivity and limitations.

The purpose of this book is to draw together worldwide literature, up to early 1999, on the occurrence and determination of metals and anions in solid samples. In this way reference to a very scattered literature can be avoided.

This is not a recipe book, ie methods are not presented in detail. Space considerations alone would not permit this. Instead, the chemist is presented with details of methods available for the determinations of all metals and anions in soils, sediments and sludges. Methods are described in broad outline giving enough information for the chemist to decide whether he or she wishes to refer to the original paper. To this end, information is supplied on applicability of methods, advantages and disadvantages of one method against another, interferences, sensitivity and detection limits.

Chapter 1 discusses the principles of the various techniques now being employed in the analysis of soils, sludges and sediments and the types of determinations to which these methods can be applied. This chapter also contains a useful key system which enables the reader to quickly locate in the book sections in which are discussed the determination by various techniques of particular metals and anions in particular types of sample.

The contents of the book are presented in as logical a manner as possible starting in Chapter 2 with a discussion of the determination of 59 different metals in soils.

Chapters 3 to 5 similarly deal with the determination of up to 64 metals in non-saline sediments, marine and estuarine sediments. Chapter 6 deals with the determination of 52 metals in sludges. In each of these chapters, methods for the extraction of metals from soil sediments and sludges prior to analysis are discussed. Featured in each of Chapters 3 to 5 is a discussion of methods of multi-metal analysis whereby a range of metals is determined in a single analysis by techniques such as atomic absorption spectrometry, inductively coupled plasma atomic emission spectrometry, inductively coupled plasma mass spectrometry, neutron activation analysis, photon activation analysis, polarographic and potentiometric stripping methods of analysis, emission spectrometry including X-ray secondary emission spectrometry, γ-ray spectrometry and gas chromatography. Chapters 7 to 9 respectively deal with the determination of anions in soils, sediments and sludges.

The use of correct sampling procedures is mandatory in the analysis of solid materials and this is discussed fully in Chapter 10 which discusses aspects of sampling including sample homogeneity, comminution and grinding of samples, sample digestion procedures. The application of non-destructive methods of analysis to solid samples is also discussed.

Chapter 11 and 13 respectively deal with the important subjects of accumulation of metals in sediments and the relationship between metal contaminant levels in soil and in crops. A growing practice is the application of sewage sludge as a fertiliser to land. Such sludge is often contaminated with metals especially if it originates in sewage processed in an industrial area. The implications of this are discussed in Chapter 12.

Examination of solid samples for metals combines all the exciting features of analytical chemistry. First, the analysis must be successful and in many cases must be completed quickly. Often the nature of the substances to be analysed for is unknown, might occur at exceeding low concentrations and might, indeed, be a complex mixture. To be successful in such an area requires analytical skills of a high order and the availability of sophisticated instrumentation.

The work has been written with the interests of the following groups of people in mind: management and scientists in all aspects of the water industry, river management, fishery industries, sewage effluent treatment and disposal, land drainage and water supply; also management and scientists in all branches of industry. It will also be of interest to agricultural chemists, agriculturalists concerned with the ways in which metals used in crop or soil treatment permeate through the ecosystem, the biologists and scientists involved in fish, plant, insect and plant life, and also to the medical profession, toxicologists and public health workers and public analysts. Other groups or workers to whom the work will be of interest include oceanographers, environmentalists and, not least, members of the public who are concerned with the protection of our environment.

Finally, it is hoped that the work will act as a spur to students of all subjects mentioned and assist them in the challenge that awaits them in ensuring that the pollution of the environment is controlled so as to ensure that into the new millennium we are left with a worthwhile environment to protect.

Chapter I

Introduction

I.I Summary of methodologies for determining cations

I.I.I Polarographic, potentiometric stripping and differential pulse anodic stripping voltammetric methods

Direct polarography has been used to determine lead in soil. Potentiometric stripping methods have been used to determine copper and lead in non-saline sediments, lead in saline sediments and copper and lead in sewages. The applications of differential pulse anodic stripping voltammetry are more numerous.

In soils: the full range of heavy metals *viz* cadmium, chromium, cobalt, copper, iron, lead, manganese, mercury, nickel and zinc.

In non-saline sediments: mercury, lead and vanadium (also selenium by cathodic stripping voltammetry).

In saline sediments: copper, lead and thallium.

In sludges: heavy metals, *viz* cadmium, copper, chromium, iron, lead, manganese and zinc.

Three basic techniques of polarography are of interest and the basic principles of these are outlined below.

Universal: Differential Pulse (DPN, DPI, DPR)

In this technique a voltage pulse is superimposed on the voltage ramp during the last 40 ms of controlled drop growth with the standard dropping mercury electrode; the drop surface is then constant. The pulse amplitude can be preselected. The current is measured by integration over a 20 ms period immediately before the start of the pulse and again for 20 ms as the pulse nears completion. The difference between the two current integrals (1_2–1) is recorded and this gives a peak-shaped curve. If the pulse amplitude is increased, the peak current value is raised but the peak is broadened at the same time.

Classical Direct Current (DCT)

In this direct current method, integration is performed over the last 20 ms of the controlled drop growth (Tast procedure): during this time, the drop surface is constant in the case of the dropping mercury electrode. The resulting polarogram is step-shaped. Compared with classical DC polarography according to Heyrovsky, i.e. with the free-dropping mercury electrode, the DCT method offers great advantages; considerably shorter analysis times, no disturbance due to current oscillations, simpler evaluation and larger diffusion-controlled limiting current.

Rapid Square Wave (SQW)

Five square-wave oscillations of frequency around 125 Hz are superimposed on the voltage ramp during the last 40 ms of controlled drop growth; with the dropping mercury electrode the drop surface is then constant. The oscillation amplitude can be pre-selected. Measurements are performed in the second, third and fourth square-wave oscillation; the current is integrated over 2 ms at the end of the first and at the end of the second half of each oscillation. The three differences of the six integrals (1_1–1_2, 1_3–1_4, 1_5–1_6) are averaged arithmetically and recorded as one current value. The resulting polarogram is peak shaped.

Metrohm are leading suppliers of polarographic equipment. They supply three main pieces of equipment: the Metrohm 646 VA processor, the 647 VA stand (for single determinations) and the 675 VA sample changer (for a series of determinations). Some features of the 646 VA processor are listed below:

- Optimized data acquisition and data processing
- High-grade electronics for a better signal-to-noise ratio
- Automatic curve evaluation as well as automated standard addition for greater accuracy and smaller standard deviation
- Large, non-volatile methods memory for the library of fully developed analytical procedures
- Connection of the 675 VA sample changer for greater sample throughout
- Connection of an electronic balance
- Simple perfectly clear operation principle via guidance in the dialogue mode yet at the same time high application flexibility thanks to the visual display and alphanumeric keyboard
- Complete and convenient result recording with built-in thermal recorder/printer

The 675 VA sample changer is controlled by the 646 VA processor on which the user enters the few control commands necessary. The 646 VA processor also controls the 677 drive unit and the 683 pumps. With these auxiliary units, the instrument combination becomes a polarographic analysis station which can be used to carry out on-line measurements.

The 646 VA processor is conceived as a central compact component for automated polarographic and voltammetric systems. Thus two independent 647 VA stands or a 675 VA sample changer can be added. Up to 4 multidosimats of the 665 type for automated standard additions and/or addition of auxiliary solutions can be connected to each of these wet-chemical workstations. Connection of an electronic balance for direct transfer of data is also possible.

Program-controlled automatic switching and mixing of these three electrode configurations during a single analysis via software commands occur. The complete electrode is pneumatically controlled. A hermetically sealed mercury reservoir of only a few millilitres suffices for approximately 200,000 drops. The mercury drops are small and stable, consequently there is a good signal-to-noise ratio. Mercury comes into contact only with the purest inert gas and plastic free of metal traces. Filling is seldom required and very simple to carry out. The system uses glass capillaries which can be exchanged simply and rapidly.

Up to 30 complete analytical methods (including all detailed information and instructions) can be filled in a non-volatile memory and called up. Consequently, a large extensive and correspondingly efficient library of analytical methods can be built up, comprehensive enough to carry out all routine determinations conveniently via call-up of a stored method.

The standard addition method (SAM) is the procedure generally employed to calculate the analyte content from the signal of the sample solution; electric current SAM amount of substance/mass concentration. The SAM is coupled directly to the determination of the sample solution so that all factors which influence the measurement remain constant. There would be no doubt that the SAM provides results that have proved to be accurate and precise in virtually every case.

The addition of standard solutions can be performed several times if need be (multiple standard addition) to raise the level of quality of the results still further.

Normally, a real sample solution contains the substances to be analysed in widely different concentrations. In a single multi-element analysis, however, all components must be determined simultaneously. The superiority of the facilities offered by segmented data acquisition in this respect is clear when a comparison is made with previous solutions. The analytical conditions were inevitably a compromise; no matter what type of analytical conditions were selected, such large differences could rarely be reconciled. In the recording, either the peaks of some of the components were shown meaningfully – each of the other two were either no longer recognisable or led to gigantic signals with cut-off peak tips.

And all too often the differences were still too large even within the two concentration ranges. Since the recorder sensitivity and also all other instrument and electrode functions could only be set and adjusted for a single substance even automatic range switching of the recorder was of very little use.

The dilemma is solved with the 646 VA processor: the freedom to divide the voltage sweep into substance-specific segments and to adjust all conditions individually and independently of one and another within these segments opens up quite a new and to date unknown analytical possibility. Furthermore, it allows optimum evaluation of the experimental data.

Various suppliers of polarographs are summarised in Table 1.1.

Metrohm, in addition to the 646 VA processor and 647 VA stand or 675 VA sample changer which can carry out differential pulse direct current and square-wave measurements, also supply two other instruments capable of carrying out different kinds of measurements.

The SO6 Polarecord

- direct current
- normal pulse
- differential pulse
- 1st harmonic ac
- 2nd harmonic ac
- Kalousek

The 626 Polarecord

- direct current sampled dc
- differential pulse

The latter is a basic instrument intended for routine analysis and teaching applications. It does not have sensitivity of the 646 VA but has, nevertheless, been used for the determination of down to 200 µg L $^{-1}$ levels of metals in tap water. The stripping differential pulse voltammetry at a hanging mercury drop electrode in a pH 4.7 ammonium citrate buffered medium.

1.1.2 Spectrophotometric methods

1.1.2.1 Visible spectrophotometry

This technique is only of value when the identity of the element to be determined is known. There are also limitations on the sensitivity that can be achieved usually mg L $^{-1}$ or occasionally ug L $^{-1}$ i.e. is inadequate for the

Table 1.1 Polarographic differential pulse direct current square wave and anodic scanning

Supplier	Type	Model No.	Detection limits
Metrohm	Differential Pulse Direct current Square wave	646 VA processor 647 VA stand 675 sample changer 665 Dosimat (motor driven piston burettes for standard additions)	0.05 µg L^{-1} quoted for basic metals 2–10 µg L^{-1} for nitriloacetic acid
	Direct current normal pulse differential pulse 1st harmonic ac. 2nd harmonic ac. Kalousek		
	Direct current sampled differential pulse	DC 626 Polarecord	
Chemtronics Ltd	On-line voltammetric analyser for metals in effluents and field work	PDV 2000	~0.1 µg L^{-1}
RDT Analytical Ltd	Differential pulse anodic stripping on-line voltammetric analyser for metals in effluents and field work	ECP 100 plus ECP 104 programmes ECP 140 PDV 200	–
	On-line voltametric analyser for continuous measurement of metals in effluents and water	OVA 2000	–
EDT Analytical Ltd	Cyclic voltammetry differential pulse voltammetry linear scam voltammetry, square-wave voltammetry, single- and double-step chronopotentiometry and chronocoulometry	Cipress Model CYSY–1B (basic system) CY57–1H–(high-sensitivity system)	

Source: Own files

determination of the extremely low concentrations of some metals that can occur in solid samples.

This technique has been used to determine a range of elements in soils, sediments and sludges.

Soils

Aluminium, ammonium, arsenic, boron, calcium, mercury, molybdenum, selenium, titanium, tungsten, uranium and vanadium. Also the following heavy metals: chromium, cobalt, iron, lead, manganese and nickel.

Non-saline sediments

Antimony, arsenic, tungsten, lead, vanadium, mercury.

Saline sediments

Arsenic, scandium, cobalt, copper, molybdenum, selenium and zirconium.

Sludges

Molybdenum, silicon and zinc

Some commercially available instruments, in addition to visible spectrophotometers, can also perform measurements in the UV and near IR regions of the spectrum. These have not yet found extensive applications in the field of water analysis.

Suppliers of visible spectrophotometers are reviewed in Table 1.2.

1.1.2.2 Spectrofluorimetric methods

Spectrofluorimetric methods are applicable to the determination of lead and selenium in soils and non-saline sediments, thorium and uranium in non-saline sediments, mercury in marine sediments and cadmium in sludges.

Generally speaking, concentrations down to the picogram (0.001 ug L^{-1}) can be determined by this technique with recovery efficiencies near 100%.

Potentially, fluorimetry is valuable in every laboratory performing chemical analysis where the prime requirements are selectivity and sensitivity. While only 5–10% of all molecules possess a native fluorescence, many can be induced to fluoresce by chemical modification or tagged with a fluorescent module.

Luminescence is the generic name used to cover all forms of light emission other than that arising from elevated temperature (thermoluminescence). The emission of light through the absorption of UV or visible energy is called photoluminescence, and that caused by chemical reactions is called chemiluminescence. Light emission through the use of enzymes in living systems is called bioluminescence.

Photoluminescence may be further subdivided into fluorescence which is the immediate release (10^{-8} s) of absorbed light energy as opposed to phosphorescence which is delayed release (10^{-6}–10^{2} s) of absorbed light energy.

Table 1.2 Visible-ultraviolet-near infrared spectrophotometers

Spectral region	Range (nm)	Manufacturer	Model	Single or double beam	Cost range range
UV/visible	–	Philips	PU 8620 (optional PU 8620 scanner)	Single	Low
Visible	325–900	Celcil Instruments	CE 2343 Optical Flowcell	Single	Low
Visible	280–900	Celcil Instruments	CE 2393 (grating, digital)	Single	High
Visible	280–900	Celcil Instruments	CE 2303 (grating, non-digital)	Single	Low
Visible	280–900	Celcil Instruments	CE 2373 (grating, linear)	Single	High
UV/visible	190–900	Celcil Instruments	CE 2292 (digital)	Single	High
UV/visible	190–900	Celcil Instruments	CE 2202 (non-digital)	Single	Low
UV/visible	190–900	Celcil Instruments	CE 2272 (linear)	Single	High
UV/visible	200–750	Celcil Instruments	CE 594 (microcomputer controlled)	Double	High
UV/visible	190–800	Celcil Instruments	CE 6000 (with CE 6606 graphic plotter option)	Double	High
UV/visible	190–800	Celcil Instruments	5000 series (computerized and data station)	Double	High
UV/visible	–	Philip	PU 8800		High
UV/visible	–	Kontron	Unikon 860 (computerized with screen)	Double	High
UV/visible	–	Kontron	Unikon 930 (computerized with screen)	Double	High
UV/visible	190–1100	Perkin–Elmer	Lambda 2 (microcomputer electronics screen)	Double	High
UV/visible	190–750 or 190–900	Perkin–Elmer	Lambda 3 (microcomputer electronics)	Double	Low to High
UV/visible	190–900	Perkin–Elmer	Lambda 5 and Lambda 7 (computerized with screen)	Double	High
UV/visible	185–900 &	Perkin–Elmer	Lambda 9 (computerized with screen)	UV/vis/NIR	High
UV/visible	190–900	Perkin–Elmer	Lambda Array 3840 (computerized with screen)	Photodiode	High

The excitation spectrum of a molecule is similar to its absorption spectrum while the fluorescence and phosphorescence emission occur at longer wavelengths than the absorbed light. The intensity of the emitted light allows quantitative measurement since, for dilute solutions, the emitted intensity is proportional to concentration. The excitation and emission spectra are characteristic of the molecule and allow qualitative measurements to be made. The inherent advantages of the technique, particularly fluorescence, are:

1   Sensitivity, picogram quantities of luminescent material are frequently studied.
2   Selectivity, derived from the two characteristic wavelengths.
3   The variety of sampling methods that are available, i.e. dilute and concentrated samples, suspensions, solids, surfaces and combination with chromatographic methods.

Fluorescence spectrometry forms the majority of luminescence analysis. However, the recent developments in instrumentation and room-temperature phosphorescence techniques have given rise to practical and fundamental advances which should increase the use of phosphorescence spectrometry. The sensitivity of phosphorescence is comparable to that of fluorescence and complements the latter by offering a wider range of molecules of study.

The pulsed xenon lamp forms the basis for both fluorescence and phosphorescence measurement. The lamp has a pulse duration at half peak height of 10 µs. Fluorescence is measured at the instant of the flash. Phosphorescence is measured by delaying the time of measurement until the pulse has decayed to zero.

Several methods are employed to allow the observation of phosphorescence. One of the most common techniques is to supercool solutions to a rigid glass state, usually at the temperature of liquid nitrogen (77 K). At these temperatures molecular collisions are greatly reduced and strong phosphorescence signals are observed.

Under certain conditions phosophorescence can be observed at room temperature from organic molecules adsorbed on solid supports such as filter paper, silica and other chromatographic supports.

Phosphorescence can also be detected when the phosphor is incorporated into an ionic micelle. Deoxygenation is still required either by degassing with nitrogen or by the addition of sodium sulphite. Micelle-stabilised room-temperature phosphorescence (MS RTP) promises to be a useful analytical tool for determining a wide variety of compounds.

Perkin-Elmer and Hamilton both supply luminescence instruments.

Perkin-Elmer LS-3B and LS-5B luminescence spectrometers

The LS–3B is a fluorescence spectrometer with separate scanning monochromators for excitation and emission, and digital displays of both monochromator wavelengths and signal intensity. The LS–5B is a ratioing luminescence spectrometer with the capability of measuring fluorescence, phosphorescence and bio– and chemiluminescence. Delay time (t_d) and gate width (t_g) are variable via the keypad in 10 μs intervals. It corrects excitation and emission spectra.

Both instruments are equipped with a xenon discharge lamp source and have an excitation wavelength range of 230–720 nm and an emission wavelength range of 250–800 nm.

These instruments feature keyboard entry of instrument parameters which combined with digital displays, simplify instrument operation. A high–output pulsed xenon lamp, having low power consumption and minimal ozone production, is incorporated within the optical module.

Through the use of an RS 232C interface, both instruments may be connected to Perkin–Elmer computers for instrument control and external data manipulation.

With the LS–5B instrument, the printing of the sample photomultiplier can be delayed so that it no longer coincides with the flash. When used in this mode, the instrument measures phosphorescence signals. Both the delay of the start of the gate (t_d) and the duration of the gate (t_g) can be selected in multiples of 10 μs from the keyboard. Delay times may be accurately measured, by varying the delay time and noting the intensity at each value.

Specificity in luminescence spectrometry is achieved because each compound is characterised by an excitation and emission wavelength. The identification of individual compounds is made difficult in complex mixtures because of the lack of structure from conventional excitation or emission spectra. However, by collecting emission an excitation spectra for each increment of the other, a fingerprint of the mixture can be obtained. This is visualised in the form of a time–dimensional contour plot on a three–dimensional isometric plot.

Fluorescence spectrometers are equivalent in their performance to single-beam UV-visible spectrometers in that the spectra they produce are affected by solvent background and the optical characteristics of the instrument. These effects can be overcome by using software built into the Perkin–Elmer LS–5B instrument or by using application software for use with the Perkin–Elmer models 3700 and 7700 computers.

Perkin–Elmer LS–2B microfilter fluorimeter

The model LS–2B is a low-cost easy-to-operate filter fluorimeter that scans emission spectra over the wavelength range 390–700 nm (scanning) or 220–650 nm (individual interferences filters).

The essentials of a filter fluorimeter are as follows:

- a source of UV/visible energy (pulsed Xenon)
- a method of isolating the excitation wavelength
- a means of discriminating between fluorescence emission and excitation energy
- a sensitive detector and a display of the fluorescence intensity.

The model LS–2B has all these features arranged to optimise sensitivity for microsamples. It can also be connected to a highly sensitive 7 μL liquid chromatographic detector for detecting the constituents in the column effluent. It has the capability of measuring fluorescence, time-resolved fluorescence and bio- and chemiluminescent signals. A 40-portion autosampler is provided. An excitation filter kit containing six filters – 310, 340, 375, 400, 450 and 480 nm – is available.

1.1.2.3 Flow injection analysis

This technique which, in effect, automates spectrophotometric methods, has found limited applications. Its reported applications in soil analysis include the determination of aluminium and ammonium. Silica, cadmium lead and copper have been determined in non-saline sediments. The sensitivity achieved in this method of analysis is often inadequate for the determination of low levels of anions and cations.

Flow injection analysis (FIA) is a rapidly growing analytical technique. Since the introduction of the original concept by Ruzicka and Hanson [1] in 1975, about 1,000 papers have been published.

Flow injection analysis is based on the introduction of a defined volume of sample into a carrier (or reagent) stream. This results in a sample plug bracketed by carrier (Fig. 1.1(a)).

The carrier stream is merged with a reagent stream to obtain a chemical reaction between the sample and the reagent. The total stream then flows through a detector (Fig. 1.1(b)). Although spectrophotometry is the commonly used detector system in this application, other types of detectors have been used, namely fluorimetric, atomic absorption emission spectrometry and electrochemical (e.g. ion selective electrodes).

The pump provides constant flow and no compressible air segments are present in the system. As a result the residence time of the sample in the system is absolutely constant. As it moves towards the detector the sample is mixed with both carrier and reagent. The degree of dispersion (or dilution) of the sample can be controlled by varying a number of factors, such as sample volume, length and diameter of mixing coils and flow rates.

When the dispersed sample zone reaches the detector, neither the chemical reaction nor the dispersion process has reached a steady state.

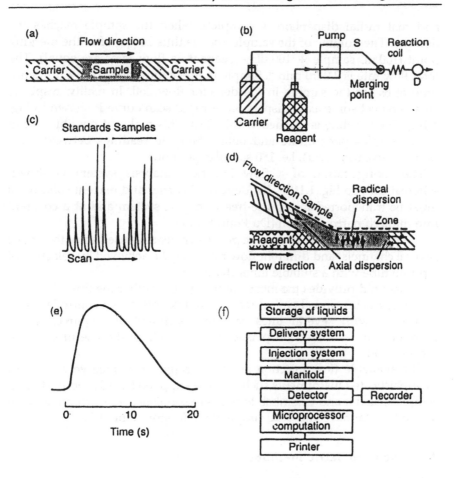

Fig. 1.1 Configuration of flow injection analysis system
Source: Own files

However, experimental conditions are held identical for both samples and standards in terms of constant residence time, constant temperature and constant dispersion. The sample concentration can thus be evaluated against appropriate standards injected in the same manner as samples (Fig. 1.1(c)).

The short distance between the injection site and the merging point ensures negligible dispersion of the sample in this part of the system. This means that sample and reagent are mixed in equal proportions at the merging point.

The mixing technique can be best understood by having a closer look at the hydrodynamic conditions in and around the merging point (Fig. 1.1(d)).

In Fig. 1.1(d) the hydrodynamic behaviour is simplified in order to explain the mixing process. Let us assume that there is no axial dispersion and that radial dispersion is complete when the sample reaches the detector. The volume of the sample zone is thus 200 µg after the merging point (100 µL sample + 100 µL – reagent as flow rates are equal). The total flow rate is 2.0 ml min $^{-1}$. Simple mathematics then gives a residence time of 6s for the sample in the detector flow cell. In reality, response curves reflect some axial dispersion. A rapid scan curve is shown in Fig. 1.1(e). The baseline is reached within 20 s. This makes it possible to run three samples per minute and obtain baseline readings between each sample (no carry–over), i.e. 180 samples per hour.

The configuration of a flow injection analysis system is shown schematically in Fig. 1.1(f). The (degassed) carrier and reagent solution(s) must be transported in a pulse-free transport system and at a constant rate through narrow Teflon (Du Pont) tubing.

In a practical system, peristaltic pumps are usually used since they have several channels, and different flow rates may be achieved by selection of a pump tube with a suitable inner diameter.

A manifold provides the means of bringing together the fluid lines and allowing rinsing and chemical reaction to take place in a controlled way. Manifolds with several lines can be assembled as required. These manifolds are mounted on plastic trays and allow the use of different reaction coils.

The available flow injection analysers range from relatively low-cost unsophisticated instruments such as those supplied by Advanced Medical Supplies, Skalar and ChemLab to the very sophisticated instruments such as the FIA star 5010 and 5020 supplied by Tecator (Table 1.3).

1.1.3 Spectrometric methods

1.1.3.1 Atomic absorption spectrometry

Since shortly after its inception in 1955 atomic absorption spectrometry has been the standard tool employed by analysts for the determination of trace levels of metals. In this technique a fine spray of the analyte is passed into a suitable flame, frequently oxygen acetylene, or nitrous oxide acetylene, which converts the elements to an atomic vapour. Through this vapour is passed radiation at the right wavelength to excite the ground state atoms to the first excited electronic level. The amount of radiation absorbed can then be measured and directly related to the atom concentration: a hollow cathode lamp is used to emit light with the characteristic narrow line spectrum of the analyte element. The detection system consists of a monochromator (to reject other lines produced by the lamp and background flame radiation) and a photomultiplier. Another key feature of the technique involves modulation of the source radiation so that it can be detected against the strong flame and sample emission radiation.

Table 1.3 Equipment for flow-injector analysis

Supplier	Model	Features	Detectors available
Advanced Medical Supplies	LCG 1	Relatively low-cost instrument, recorder output. No computerization on data processing (8 channels)	Colorimeter (other detectors can be used but are not linked in e.g. atomic absorption, fluorometer ion-selective electrodes
Chemlab	–	Relatively low-cost, recorder output or data analysis by microprocessor (3 channels)	Colorimeter
Skalar	–	Relatively low-cost, recorder output on data analysis by microprocessor also carries out segmented flow analysis	Colorimeter, flow cells for fluorometer and ion-selective electrodes available
Fiatron	Finlite 600	Laboratory process control and pilot plant instrument computerized	Colorimeter
	Fiatrode 400 Fiatrode 410 Fiatrode 430	Flow through analyser/controller, process control analyser	pH and ion-selective electrode
Tecator	FIA star 5025	Relatively low cost manual instrument specifically designed for fluoride, cyanide, potassium, iodide, etc.	Specially designed for use with ion-selective electrodes
	FIA star 5032	Relatively low-cost manual instrument (400–700 nm)	Spectrophotometer and/or photometer detectors
	Aquatec	Modular, semi-or fully automatic operation. Microprocessor controlled. A dedicated instrument designed for water analysis, i.e. dedicated method cassettes for phosphate and chloride, 600–100 samples h^{-1}	Flow through spectrophotometer (400–700nm)
	FIA star 5010	Modular, semi- or fully automatic operation. May be operated with process controller micro-processor. Can be set up in various combinations with 5017 sampler and superflow software which is designed to run on IBM PC/XT computer; 60–180 samples h^{-1}. Dialysis for in-line sample preparation and in-line solvent extraction. Thermostat to speed up reactions.	Spectrophotometer (400–700nm) or photometer can be connected to any flow through detector, e.g. UV/visible, inductively coupled plasma, atomic absorption spectrometer and ion-selective electrodes

Source: Own files

The technique can determine a particular element with little interference from other elements. It does however have two major limitations. One of these is that the technique does not have the highest sensitivity, and the other is that only one element at a time can be determined. This has reduced the extent to which it is currently used.

It is seen in Tables 1.4(a)–(e) that this technique has been applied to the determination of a wide range of elements in soils (Table 1.4(a)), sediments (Tables 1.4(b)–1.4(d)) and sludges (Table 1.4(e)). In particular, these include the heavy metals, group V elements and some alkali and alkali earth metals, and selenium, tin and mercury. Cresser [57] has reviewed the analyses of polluted soils by atomic absorption spectrometry.

1.1.3.1(a) Graphite furnace atomic absorption spectrometers

The graphite furnace atomic absorption technique, first developed in 1961 by L'vov, is an attempt to improve the detection limits achievable in this method of analysis. In this technique, instead of being sprayed as a fine mist into the flame, a measured portion of the sample is injected into an electrically heated graphite boat or tube, allowing a larger volume of sample to be handled. Furthermore, by placing the sample on a small platform inside the furnace tube, atomization is delayed until the surrounding gas within the tube has heated sufficiently to minimise vapour phase interferences, which would otherwise occur in a cooler gas atmosphere.

The sample is heated to a temperature slightly above 100 °C to remove free water, then to a temperature of several hundred degrees centigrade to remove water of fusion and other volatiles. Finally the sample is heated to a temperature near to 1000 °C to atomise it and the signals thereby produced are measured by the instrument.

The problem of background absorption in this technique is solved by using a broad-band source, usually a deuterium arc or a hollow cathode lamp, to measure background independently and subsequently to subtract it from the combined atomic and background signal produced by the analyte hollow cathode lamp. By interspersing the modulation of the hollow cathode lamp and 'background corrector' sources, the measurements are performed apparently simultaneously. Specific advances have been achieved such as optical control of furnace temperatures to improve heating rates and hence measurement sensitivities. Better furnace tube geometries and gas–flow systems have helped to reduce interference effects. Methods have been devised of mass producing pyrolytic graphite coatings on electrographite tubes and this has reduced the porosity of atom containment and stable carbide formation.

Besides background attenuation chemical interferences can be a major limitation in the practical use of graphite furnace atomic absorption

Table I.4(a) Metallic and metalloid elements in Periodic Table determined by atomic absorption spectrometry in soil

Group / Period	I	II											III	IV	V	VI	VIII	0
1	^{1}H																	^{2}He
2	^{2}Li	^{4}Be											^{5}B	^{6}C	^{7}N	^{8}O	^{9}F	^{10}Ne
3	^{11}Na√	^{12}Mg√											^{13}Al√	^{14}Si√	^{15}P	^{16}S	^{17}Cl	^{18}Ar
4	^{19}K	^{20}Ca	^{21}Sc	^{22}Ti	^{23}V	^{24}Cr√	^{25}Mn√	^{26}Fe√	^{27}Co√	^{28}Ni√	^{29}Cu√	^{30}Zn√	^{31}Ga	^{32}Ge	^{33}As√	^{34}Se√	^{35}Br	^{36}Kr
5	^{37}Rb	^{38}Sr	^{39}Y	^{40}Zr	^{41}Nb	^{42}Mo√	^{43}Tc	^{44}Ru	^{45}Rh	^{46}Pd	^{47}Ag√	^{48}Cd√	^{49}In√	^{50}Sn	^{51}Sb√	^{52}Te	^{53}I	^{54}Xe
6	^{55}Cs	^{56}Ba	^{57}La	^{72}Hf	^{73}Ta	^{74}W	^{75}Re	^{76}Os	^{77}Ir	^{78}Pt	^{79}Au	^{80}Hg√	^{81}Tl√	^{82}Pb√	^{83}Bi√	^{84}Po	^{85}At	^{86}Rn
7	^{87}Fr	^{88}Ra	^{89}Ac															

Lanthanides	^{58}Ce	^{59}Pr	^{60}Nd	^{61}Pm	^{62}Sm	^{63}Eu	^{64}Gd	^{65}Tb	^{66}Dy	^{67}Ho	^{68}Er	^{69}Tm	^{70}Yb	^{71}Lu
Actinides	^{90}Th	^{91}Pa	^{92}U	^{93}Np	^{94}Pu	^{95}Am	^{96}Cm	^{97}Bk	^{98}Cf	^{99}Es	^{100}Fm	^{101}Md	^{102}No	^{103}Lr

Total number of metals and metalloids 86 (include actinides and lanthanides).
Number determinable by atomic absorption spectrometry [22] (indicated by tick)

Source: Own files

Table 1.4(b) Metallic and metalloid elements in Periodic Table determined by atomic absorption spectrometry in non-saline sediments

Group / Period	I	II										III	IV	V	VI	VIII	0
1	^{1}H																^{2}He
2	^{2}Li√	^{4}Be√										^{5}B	^{6}C	^{7}N	^{8}O	^{9}F	^{10}Ne
3	^{11}Na	^{12}Mg√										^{13}Al√	^{14}Si	^{15}P	^{16}S	^{17}Cl	^{18}Ar
4	^{19}K	^{20}Ca√	^{21}Sc√ ^{22}Ti ^{23}V ^{24}Cr√ ^{25}Mn√ ^{26}Fe√ ^{27}Co√ ^{28}Ni√ ^{29}Cu√ ^{30}Zn√									^{31}Ga√	^{32}Ge	^{33}As√	^{34}Se√	^{35}Br	^{36}Kr
5	^{37}Rb	^{38}Sr	^{39}Y ^{40}Zr√ ^{41}Nb ^{42}Mo√ ^{43}Tc ^{44}Ru ^{45}Rh ^{46}Pd ^{47}Ag√ ^{48}Cd√									^{49}In√	^{50}Sn√	^{51}Sb	^{52}Te	^{53}I	^{54}Xe
6	^{55}Cs	^{56}Ba√	^{57}La ^{72}Hf ^{73}Ta ^{74}W ^{75}Re ^{76}Os ^{77}Ir ^{78}Pt ^{79}Au ^{80}Hg√									^{81}Tl√	^{82}Pb√	^{83}Bi√	^{84}Po	^{85}At	^{86}Rn
7	^{87}Fr	^{88}Ra	^{89}Ac														

Lanthanides	^{58}Ce	^{59}Pr	^{60}Nd	^{61}Pm	^{62}Sm	^{63}Eu	^{64}Gd	^{65}Tb	^{66}Dy	^{67}Ho	^{68}Er	^{69}Tm	^{70}Yb	^{71}Lu
Actinides	^{90}Th	^{91}Pa	^{92}U	^{93}Np	^{94}Pu	^{95}Am	^{96}Cm	^{97}Bk	^{98}Cf	^{99}Es	^{100}Fm	^{101}Md	^{102}No	^{103}Lr

Total number of metals and metalloids 86 (include actinides and lanthanides).
Number determinable by atomic absorption spectrometery [24] (indicated by tick)

Source: Own files

Table 1.4(c) Metallic and metalloid elements in Periodic Table determined by atomic absorption spectrometry in marine sediments

Group Period	I	II											III	IV	V	VI	VIII	0
1	1H																	2He
2	2Li	4Be√											5B	6C	7N	8O	9F	10Ne
3	11Na	12Mg											13Al	14Si	15P	16S	17Cl	18Ar
4	19K	20Ca√	21Sc	22Ti√	23V	24Cr√	25Mn√	26Fe√	27Co√	28Ni√	29Cu√	30Zn√	31Ga	32Ge	33As√	34Se√	35Br	36Kr
5	37Rb	38Sr	39Y	40Zr	41Nb	42Mo	43Tc	44Ru	45Rh	46Pd	47Ag√	48Cd√	49In	50Sn√	51Sb√	52Te	53I	54Xe
6	55Cs	56Ba√	57La	72Hf	73Ta	74W	75Re√	76Os	77Ir√	78Pt√	79Au	80Hg√	81Tl√	82Pb√	83Bi√	84Po	85At	86Rn
7	87Fr	88Ra	89Ac															

Lanthanides	58Ce	59Pr	60Nd	61Pm	62Sm	63Eu	64Gd	65Tb	66Dy	67Ho	68Er	69Tm	70Yb	71Lu
Actinides	90Th	91Pa	92U	93Np	94Pu	95Am	96Cm	97Bk	98Cf	99Es	100Fm	101Md	102No	103Lr

Total number of metals and metalloids 86 (include actinides and lanthanides).
Number determinable by atomic absorption spectrometery [24] (indicated by tick)

Source: Own files

Table 1.4(d) Metallic and metalloid elements in Periodic Table determined by atomic absorption spectrometry in estuarine sediments

Group Period	I	II												III	IV	V	VI	VIII	0
1	^{1}H																	^{1}H	^{2}He
2	^{2}Li	^{4}Be√												^{5}B	^{6}C	^{7}N	^{8}O	^{9}F	^{10}Ne
3	^{11}Na	^{12}Mg												^{13}Al	^{14}Si	^{15}P	^{16}S	^{17}Cl	^{18}Ar
4	^{19}K	^{20}Ca	^{21}Sc	^{22}Ti	^{23}V	^{24}Cr√	^{25}Mn√	^{26}Fe√	^{27}Co√	^{28}Ni√	^{29}Cu√	^{30}Zn√		^{31}Ga	^{32}Ge	^{33}As√	^{34}Se√	^{35}Br	^{36}Kr
5	^{37}Rb	^{38}Sr	^{39}Y	^{40}Zr	^{41}Nb	^{42}Mo	^{43}Tc	^{44}Ru	^{45}Rh	^{46}Pd	^{47}Ag	^{48}Cd√		^{49}In	^{50}Sn	^{51}Sb	^{52}Te	^{53}I	^{54}Xe
6	^{55}Cs	^{56}Ba	^{57}La	^{72}Hf	^{73}Ta	^{74}W	^{75}Re√	^{76}Os	^{77}Ir	^{78}Pt	^{79}Au	^{80}Hg		^{81}Tl	^{82}Pb√	^{83}Bi	^{84}Po	^{85}At	^{86}Rn
7	^{87}Fr	^{88}Ra	^{89}Ac																

Lanthanides	^{58}Ce	^{59}Pr	^{60}Nd	^{61}Pm	^{62}Sm	^{63}Eu	^{64}Gd	^{65}Tb	^{66}Dy	^{67}Ho	^{68}Er	^{69}Tm	^{70}Yb	^{71}Lu
Actinides	^{90}Th	^{91}Pa	^{92}U	^{93}Np	^{94}Pu	^{95}Am	^{96}Cm	^{97}Bk	^{98}Cf	^{99}Es	^{100}Fm	^{101}Md	^{102}No	^{103}Lr

Total number of metals and metalloids 86 (include actinides and lanthanides).
Number determinable by atomic absorption spectrometery [12] (indicated by tick)

Source: Own files

Table 1.4(e) Metallic and metalloid elements in Periodic Table determined by atomic absorption spectrometry in sludges

Group / Period	I	II											III	IV	V	VI	VII	0
1	^{1}H																^{1}H	^{2}He
2	^{2}Li	^{4}Be											^{5}B	^{6}C	^{7}N	^{8}O	^{9}F	^{10}Ne
3	^{11}Na√	^{12}Mg√											^{13}Al√	^{14}Si	^{15}P	^{16}S	^{17}Cl	^{18}Ar
4	^{19}K√	^{20}Ca√	^{21}Sc	^{22}Ti	^{23}V√	^{24}Cr√	^{25}Mn√	^{26}Fe√	^{27}Co√	^{28}Ni√	^{29}Cu√	^{30}Zn√	^{31}Ga	^{32}Ge	^{33}As√	^{34}Se√	^{35}Br	^{36}Kr
5	^{37}Rb	^{38}Sr	^{39}Y	^{40}Zr	^{41}Nb	^{42}Mo√	^{43}Tc	^{44}Ru	^{45}Rh	^{46}Pd	^{47}Ag√	^{48}Cd√	^{49}In	^{50}Sn√	^{51}Sb√	^{52}Te√	^{53}I	^{54}Xe
6	^{55}Cs	^{56}Ba	^{57}La	^{72}Hf	^{73}Ta	^{74}W	^{75}Re	^{76}Os	^{77}Ir	^{78}Pt	^{79}Au	^{80}Hg√	^{81}Tl√	^{82}Pb√	^{83}Bi√	^{84}Po	^{85}At	^{86}Rn
7	^{87}Fr	^{88}Ra	^{89}Ac															

Lanthanides	^{58}Ce	^{59}Pr	^{60}Nd	^{61}Pm	^{62}Sm	^{63}Eu	^{64}Gd	^{65}Tb	^{66}Dy	^{67}Ho	^{68}Er	^{69}Tm	^{70}Yb	^{71}Lu
Actinides	^{90}Th	^{91}Pa	^{92}U	^{93}Np	^{94}Pu	^{95}Am	^{96}Cm	^{97}Bk	^{98}Cf	^{99}Es	^{100}Fm	^{101}Md	^{102}No	^{103}Lr

Total number of metals and metalloids 86 (include actinides and lanthanides).
Number determinable by atomic absorption spectrometry [24] (indicated by tick)

Source: Own files

spectrometry. Chemical interferences in furnace atomic absorption spectrometry often lead to strong signal suppression. This cannot be removed by any type of background correction. The way to reduce or completely eliminate these matrix interferences is to atomise samples into the environment that has almost reached thermal equilibrium. Under such conditions the formation of free atoms is optimum and recombination of atoms to molecules or loss of atoms is effectively avoided.

The stabilised temperature platform furnace eliminates chemical interferences to such an extent that in most cases personnel and cost–intensive sample preparation steps, such as solvent extractions, as well as the time–consuming method of additions are no longer required. Graphite furnace techniques are about one order of magnitude more sensitive than direct injection techniques. Thus lead can be determined down to 50 μg L $^{-1}$ by direct AAS and down to μg L $^{-1}$ by graphite furnace AAS.

In recent years, even greater demands for increases in sensitivity have been placed on chemists. This has raised the popularity of the furnace technique. In turn, it has placed even greater demands on background correction systems. Great improvements have been made in continuum sources methods, and additionally two alternative techniques have reached commercial fruition. The first is based on measuring the background during a high current pulse of the source hollow cathode lamp, thereby broadening the lamp–emitted line profile. The second is by using Zeeman (see section 1.1.3.1(b)) or magnetically induced splitting of either the hollow cathode lamp emission profile or the sample adsorption profile. Both techniques thus allow the measurement of background levels without using a second source.

When considering the purchase of a flame or graphite furnace spectrometer the following are some of the important parameters that should be considered in addition to the all important question of cost and reliability:

1   Monochromator design
2   Monochromator dual grating option
3   Automatic wavelength drive option
4   Photometer type and design
5   Background correction option
6   Lamp turret option
7   Printer option
8   Readout facilities
9   Graphics option
10 Burner-atomiser design
11 Automation of flame gas controls
12 Microcomputer facilities

13 Availability of training courses
14 Repair and servicing facilities

Instrumentation

Increasingly, due to their superior intrinsic sensitivity, atomic absorption spectrometers available are capable of implementing the graphite furnace techniques. Some available instrumentation in flame and graphite furnace atomic absorption spectrometry is listed in Table 1.5.

In Fig. 1.2(a) and (b) are show the optics of a single-beam flame spectrometer (Perkin–Elmer 2280) and a double-beam instrument (Perkin–Elmer Model 2380).

The model 2280 has a very simple optical system. Energy from the primary source is focused through a lens into the sample compartment and then to the monochromator. As only one lens is used, compensation for the wavelength influence on energy throughput is easily done by positioning the hollow cathode lamp along the horizontal axis.

In the double-beam Model 2380 light from the primary source is divided into two beams – a sample beam and a reference beam. The sample beam travels through the sample compartment, while the reference beam travels around it. The beams are recombined before entering the monochromator. The double–beam system compensates for any changes that may occur in lamp intensity during an analysis. The signal produced is actually a ratio between the two beams. Therefore, any fluctuations in the light output will affect both beams equally and will be compensated for automatically. This results in a more stable baseline and ultimately in better detection limits.

The light beam is transmitted through the system by front–surfaced, quartz-coated mirrors. The advantages of using reflecting optics are that the efficiency in energy throughout is unaffected by the wavelength being considered and no additional focusing optics are required. All mirrors used in the system are also specially coated with silicon oxide, which protects the surface if the instrument is operated in a corrosive laboratory atmosphere.

The monochromator uses a very finely ruled grating, the efficiency of which is dependent upon its area, its dispersion and its blaze angle. A dual-blazed grating is used in the Perkin–Elmer Model 2380 with two blaze angles, one in the ultraviolet at 236 nm and the other in the visible range at 597 nm. This distributes energy more evenly throughout the wavelength range (190–870 nm). The model 2280 uses a blazed grating with one blaze angle at 255 nm.

The burner assembly of the Perkin–Elmer models 2280 and 2380 instruments incorporates an impact bead which can lead to an appreciable improvement in detection limits achieved to those attainable with a conventional flow spoiler design.

Table 1.5 Available flame and graphite furnace atomic absorption spectrometers

Type instrument	Supplier	Model no and type
Flame (direct injection)	Thermo-electron	IL 157 single channel
		IL 357 single beam
		IL 457 single channel double beam
		Video 11 single channel single beam
		Video 12 single channel double beam
		Video 22 two double channels
Graphite furnace Direct injection	Thermo-electron Perkin–Elmer	IL 655 CTF
		2280 single beam
		2380 double beam
Graphite furnace	Perkin–Elmer	100 single beam
		2100 single path double beam
Graphite furnace	Varian Associates	SpectrA A30 + 40 multi-element analysis Method storage
		SpectrA A10 (low cost, single beam)
		SpectrA A20 (medium cost, double beam)
Flame graphite furnace		SpectrA A300/400 multi-element analysis, centralized instrument control
		STA 95 and GTA 96 graphite tube atomizer units – compatible with all SpectrA A instruments
Graphite furnace	GBC Scientific Pty Ltd	903 single beam
		902 double beam
		(both with impact head option)
Flame (direct injection) graphite furnace	Shimadzu	AA670 double beam
		AA670 G Double beam

Source: Own file

The impact bead cannot be used for all applications. It is not recommended for the nitrous oxide–acetylene flame; the sensitivity improvement is not noticed here. Furthermore, due to minor flame instability, poorer precision and somewhat poorer detection limits will result. The bead also should not be used when analysing solutions containing high concentrations of dissolved salts as those solutions may clog the burner system.

The flow spoiler is selected when determining elements requiring the nitrous oxide–acetylene flame for best precision or when analysing solutions containing high concentrations of dissolved solids. In addition the flow spoiler is used when working with solutions which may corrode the impact bead and cause contamination problems.

Microprocessor	Hydride and mercury attachment	Autosampler	Wavelength range
} –	Yes	Yes	
with graphics			
with graphics			
with graphics			
computer interference			
–	–	Yes	
Yes			
Yes (with automatic background correction)	–	–	190–870
Yes	–	–	190–870
Yes	Yes	–	190–870
Yes	Yes	Yes	190–900
Yes (built in VDU)	Yes	Yes	190–900
Yes (built in VDU)	Yes	Yes	190–900
Yes (with colour graphics and 90 elements on disk)	Yes	Yes	190–900
Furnace and programmable sample dispenser operated from SpectrA A keyboard. Rapid interchange between flame and furnace operation			
Yes	Yes	Yes	176–900
Yes	Yes	Yes	170–900
Yes	Yes	Yes	190–900
Yes	Yes	Yes	190–900

The Perkin–Elmer burner system has the ability to use an auxiliary oxidant flow. This capability contributes to improved flame stability and precision and it provides a convenient and simple means of obtaining proper flame conditions when using combustible organic solvents.

The Perkin–Elmer models 2280 and 2380 spectrometers are available with several types of burner control systems. They differ in degree of automation and provide full operational safety with the air–acetylene or nitrous oxide–acetylene flames.

The models 2280 and 2380 are microcomputer–controlled atomic absorption spectrophotometers. The use of the microcomputer greatly simplifies the operation of the instrument and makes instrument calibration a simple operation, at the same time providing versatility for the analyst.

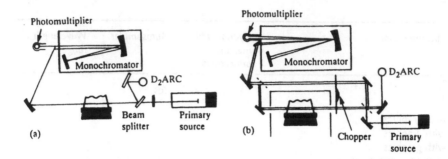

Fig. 1.2 (a) Optics Perkin–Elmer Model 2280 single beam atomic absorption spectrometer; (b) Optics Perkin–Elmer 2380 double beam atomic absorption spectrometer
Source: Own files

Another advantage of the microcomputer electronics in the models 2280 and 2380 is the instrument's ability to alert the analyst to any error–producing conditions. Error codes are displayed on the readout. For example, if the standards have been entered out of sequence, an error code (E–11) will be displayed on the readout. A quick referral to the instrument manual will provide the cause and remedy for the error message.

Background absorption is a problem for analysts to consider when utilising the atomic absorption technique. Background absorption is a collective term used to describe the combined effects of such phenomena as flame absorption, molecular absorption and light scattering on atomic absorption determinations. If no correction is made for background absorption, sample absorbance may appear to be more significant than it is and the analytical result can be erroneously high. Specific applications where background correction is necessary are:

1 HGA graphite furnace analyses
2 Flame determinations of low concentrations of an element in the presence of high concentrations of dissolved salts
3 Flame analyses where sample matrix may show molecular absorption at the wavelength of the resonance line and
4 Flame determination of an element at a wavelength where flame absorption is high

Both the models 2280 and 2380 offer an optional background corrector with a deuterium arc lamp.

Some flame instruments, e.g. the Varian Spectra AA–40 has a fully automated lamp turret, providing the mounting for and automatic selection of eight lamps. When an element is selected, the appropriate lamp is brought into the operating position automatically. During automated analysis, the next lamp needed is being warmed up for maximum stability.

With many flame–graphite furnace instruments now available, the flame-to-furnace changeover can be brought about in a few seconds.

Graphite tube design

The GTA–96 graphite tube atomiser used by Varian is contained in a cell fitted at each end with removable quartz windows. A stream of insert gas such as nitrogen or argon flows through the cell, protecting the graphite tube from oxidation. This gas emerges from the sampling hole carrying with it the products of drying and ashing. The graphite tube is electrically heated by current passing along its length.

Flame and graphite furnace atomic absorption spectrometry have adequate sensitivity for the determination of metals in sediment samples, In this technique up to 1 g of dry sample were digested in a microwave oven for a few minutes with 5 ml aqua regia in a small PTFE lined bomb and then the bomb washings transferred to a 50 ml volumetric flask prior to analysis by flame atomic absorption spectrometry (Table 1.6). Detection limits (mg kg $^{-1}$) achieved by this technique were: 0.25 (cadmium, zinc); 0.5 (chromium, manganese); 1 (copper, nickel and iron); and 2.5 (lead). Application of this technique to an NBS centrified sediment gave recoveries ranging between 85% (cadmium) and 101% (lead, nickel and iron) with an overall recovery of 95%.

1.1.3.1(b) Zeeman atomic absorption spectrometry techniques

The Zeeman technique, though difficult to establish, has an intrinsic sensitivity perhaps five times greater than that of the graphite furnace technique, say 1 µg L $^{-1}$ detection limit for lead.

Non-specific background attenuation has always been the most common type of interference in graphite furnace atomic absorption spectrometry. It is caused by higher concentrations of molecular species, small droplets, salt particles or smoke, which may absorb or scatter the light emitted from the primary light source.

Most atomic absorption spectrometers for use with the graphite furnace technique are therefore equipped with continuum source background correctors which automatically compensate for broad band absorption interferences. These are sufficient for many routine applications. Background correction with a continuum light source is however limited to background absorption which is 'uniform' within the spectral bandwidth. In addition, background signals can be corrected only up to 0.7 absorbance units.

The Zeeman effect is exhibited when the intensity of an atomic spectral line, emission or absorption, is reduced when the atoms responsible are

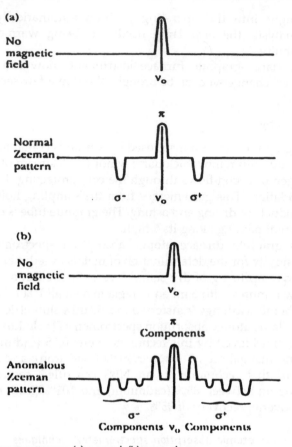

Fig. 1.3 Zeeman patterns: (a) normal; (b) anomalous
Source: Own files

Table 1.6 Analysis 1 NBS 1645 standard sediment

	Certif analysis mg kg^{-1}	Certif range mg kg^{-1}	Aqua regia digestion mg kg^{-1}	Recovery aqua regia digestion (%)
Cd	10.2 ± 1.5	8.7–11.7 Mean 8.67	8.57, 8.77	85.0
Cu	109±19	90–128	104.6	96.0
Pb	714±28	686–742	721	101.0
Ni	45.8±2.9	42.9–48.7	46.2	101.0
Zn	1750±169	1551–1889 Mean 1582	1514, 1651	92.0
Fe	113,000±12,000	101,000–125,000	114, 130	101.0
Mn	785±97	688–882 Mean 682	651, 714	86.9

Source: Own files

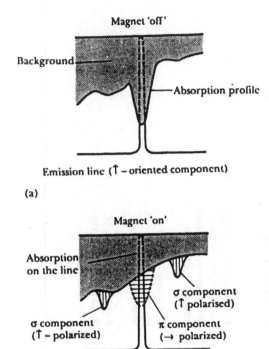

Fig. 1.4 Zeeman patterns: (a) analyte signal plus background; (b) background only
Source: Own files

subjected to a magnetic field, nearby lines arising instead (Fig. 1.3). This makes a powerful tool for correction of background attenuation caused by molecules or particles, which does not normally show such an effect. The technique is to subtract from a 'field–off' measurement the average of 'field-on' measurements made just beforehand and just afterwards (see Fig. 1.4).

The simultaneous highly resolved graphic display of the analyte and the background signals on the video screen provides a means of reliable monitoring of the determination and simplifies methods development.

The advantages of Zeeman background correction are:

1   correction over the complete wavelength range
2   correction for structural background
3   correction for spectral interferences
4   correction for high background absorbances
5   single element light source with no possibility of misalignment

Table 1.7 Available Zeeman atomic absorption spectrometers

Supplier	Model	Microprocessor	Type	Hydride and mercury attachment	Autosampler
Perkin–Elmer	Zeeman 3030	Yes (method storage on floppy disk)	Integral flame/graphite	–	Yes
	Zeeman 5000	Yes, with programmer	Fully automated integral flame/graphite furnace double-beam operation roll-over protection	Yes	Yes
Varian	SpectrA A30/40	Yes, method storage on floppy disk	Automated analysis of up to 12 elements; roll-over protection	–	Yes
	SpectrA A300/400	Yes, total system control and colour graphics; 90 methods stored on floppy disk	Automated analysis of up to 12 elements; roll-over protection	–	Yes

Source: Own files

Instrumentation

Some available instrumentation for Zeeman atomic absorption spectrometry is given in Table 1.7.

The Perkin–Elmer 5000 system

The Zeeman 5000 system is a further approach to atomic absorption spectroscopy. This system offers superb features for the determination of very low elements concentrations in complex matrices with high solids contents, e.g. seawater, and provides an excellent capability for accurately correcting for very high levels of non–specific absorption (background).

Background absorption has always been the most common type of interference in electrothermal atomic absorption spectroscopy with graphite furnace spectrometers. Most atomic absorption spectrometers for use with the graphite furnace technique are therefore equipped with continuum source background correctors which automatically compensate for broadband absorption interferences. These are sufficient for many routine applications.

Background correction with a continuum light source is, however, limited, by definition, to background absorption which is uniform within the observed spectral bandwidth. In addition background signals can be corrected only up to certain limits which are between 0.5 and 1.0 absorbance. Background absorbance higher than 1 absorbance cannot usually be fully compensated by a continuum source background corrector and may lead to erroneous results.

Non–uniform background absorption may be caused by atomic lines of matrix elements accompanying the analyte element or by a rotational fine structure of molecular spectra. Structured background absorption may lead to over- or under-compensation when using a continuum source background corrector.

With the transverse AC Zeeman system used in the Zeeman 5000 system, it becomes possible to correct easily and accurately for background absorption up to 2.0 absorbance, even when the background exhibits fine structure.

The 5000 system includes the Zeeman furnace model, the furnace programmes and the model 5000 atomic absorption spectrometer. The Zeeman module is located to the right of the model 5000 and is interfaced to the model 5000 with transfer optics.

Naturally the model 5000 can be used for double-beam flame operation as well as with other flameless devices such as the MH 510 or the MHS–20 mercury hydride systems. Changeover from the Zeeman effect furnace operation to an alternative sampling technique can be done in seconds by simply turning a control.

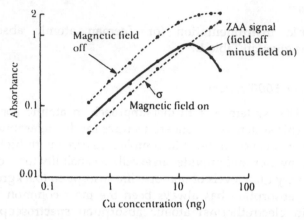

Fig. 1.5 Copper calibration curves (324.8 nm) measured with the Zeeman 5000
Source: Own files

Fig. 1.4 illustrates the operating principle of the Zeeman 5000 system. For Zeeman operation, the source lamps are pulsed at 100 Hz (120 Hz) while the current to the magnet is modulated at 50 Hz (60 Hz). When the field is off, both analyte and background absorption are measured at the unshifted resonance line. This measurement directly compares with a 'conventional' atomic absorption measurement without background correction. However, when the field is on, only background is measured since the σ absorption line profiles are shifted away from the emission line while the static polariser, constructed from synthetic crystalline quartz, rejects the signal from the π components. Background correction is achieved by subtraction of the field-on signal from the field-off signal. With this principle of operation, background absorption of up to 2 absorbance can be corrected most accurately even when the background shows a fine structure.

In assessing overall performance with a Zeeman effect instrument the subject of analytical range must also be considered. For most normal class transitions, σ components will be completely separated at sufficiently high magnetic fields. Consequently, the analytical curves will generally be similar to those obtained by standard atomic absorption spectrometry. However, for certain anomalous transitions some overlap may occur. In these cases, curvature will be greater and may be so severe as to produce double–valued analytical curves. Fig. 1.5, which shows calibration curves for copper, illustrates the reason for this behaviour. The Zeeman pattern for copper (324.8 nm) is particularly complex due to the presence of hyperfine structure. The dashed lines represent the standard field-off and field-on measurements. As sample concentration increases, field-off absorbance begins to saturate as in standard atomic absorption spectrometry. The σ absorbance measured with the field-on saturates at higher concentrations because of the greater separation from the emission line. When the increase in σ absorbance exceeds the incremental change in the field-off absorbance, the analytical curve, shown as the solid line, rolls over back towards the concentration axis. This behaviour can be observed with all Zeeman designs regardless of how the magnet is positioned or operated. The existence of roll-over does introduce the possibility of ambiguous results, particularly when peak area is being measured.

1.1.3.1(c) Vapour generation atomic absorption spectrometry

In the past certain elements, e.g. arsenic, antimony, bismuth, selenium, tellurium, mercury, germanium, tin, lead were difficult to measure by direct atomic absorption spectrometry. In the past 10 years or so, a novel technique of atomisation, known as vapour generation via generation of the metal hydride, has been evolved, which has increased sensitivity and specificity enormously for these elements. In these methods the hydride generator is linked to an atomic absorption spectrometer (flame graphite furnace) or inductively coupled plasma atomic emission spectrometer (ICP–AES) or an inductively coupled plasma mass spectrometer (ICP–MS). Some typical detection limits achievable by these techniques for the determination of elements in seawater are listed in Table 1.8.

This technique makes use of the property that the above elements exhibit, i.e. the formation of covalent, gaseous hydrides which are not stable at high temperatures. Arsenic, selenium, tellurium, antimony, bismuth and tin (and to a lesser degree lead and germanium) are volatilized by the addition of a reducing agent like sodium tetrahydroborate (III) to an acidified solution. Mercury is reduced to the atomic form in a similar manner. Other systems have used titanium (II) chloride/magnesium powder and tin (II) chloride/potassium iodide/zinc powder as reducing agents. As a general principle sodium tetrahydroborate (III) is the preferred method because it gives faster hydride formation, higher conversion efficiency, lower blank levels and is more simple to use.

After the gaseous hydride is formed it is passed, together with any excess hydrogen gas, directly to the inductively coupled plasma (ICP) torch, directly coupled plasma (DCP) system or the atomizer mounted in the light beam of the atomic absorption spectrophotometer. Flames and tubes (which can be flame or electrically heated) have been used to atomize the elements from their hydrides. When the hydride is injected into an air–acetylene flame, extremely poor signal-to-noise (S/N) ratios occur because the resonance lines of arsenic and selenium, for instance, occur in the UV end of the spectrum. The air/acetylene flame absorbs 62% of the available light energy at 193.7 nm, the arsenic wavelength. The introduction of the argon or nitrogen/hydrogen/entrained air flame in 1967 gave much better S/N ratios, because it absorbs only 15% of the light. However, the use of the cooler flames gives rise to interference due to molecular absorption and incomplete salt dissociation.

Table 1.8

Element	Detection limit (µg L $^{-1}$)
Arsenic	3
Antimony	0.3
Bismuth	0.02
Selenium	0.09

Source: Own files

These problems have largely been overcome by the use of heated tubes. The light beam of the spectrophotometer is focused through the tube and is never affected by the flame which lies outside the tube; hence flame absorption is eliminated. When linking to directly coupled plasma (DCP) the hydrides must be sheathed in argon to direct flow into the central arc of the DCP. Linking to the ICP may require a change in the automatic power control settings but is dependent on a continuous generation of the hydride and hydrogen or the plasma will be extinguished.

When the chemical composition of a material is analysed, either by atomic absorption inductively coupled plasma(ICP)–mass spectrometry or fluorescence, the sample must normally be in a liquid form. This is then aspirated or nebulized by a high-pressure gas flow. This gas, argon in the case of ICPs and other support gas such as oxygen or air in the case of atomic absorption, produces a very fine aerosol (very small liquid droplets suspended in a gas) which is carried to the 'analytical region' by the gas. In the case of ICP, the analytical region is the high temperature (7000 °C) glowing electrical discharge, and in the case of atomic absorption the analytical region is usually a flame. Unfortunately, the nebulization process is very inefficient. Typically the nebulizer converts only about 3% of the sample volume into useful aerosol. For a flow rate of 2 ml min $^{-1}$ into the nebulizer, only 0.06 ml min $^{-1}$ of sample actually reaches the analytical region. However, when the hydride generator is used, the nebulizer is not needed and the rate of sample presentation to the analytical region is changed. Sample typically is pumped to the hydride reaction zone at a rate of 10 ml min $^{-1}$ in this zone. Those elements capable of forming gaseous hydrides react and these hydrides are subsequently released from the liquid phase into the argon gas flowing through the interface vessel. This argon gas, in the case of an ICP, is then directed into the plasma analytical region for analysis. The gaseous conversion is approximately 100% so that approximately 10 ml min $^{-1}$ of sample are directed to the analytical region in contrast to the 0.06 ml mn $^{-1}$ as in the case of nebulisation. This produces an increase, by a factor of 160, in the rate of sample presentation to the analytical region, which obviously increases the analytical sensitivity by a similar factor.

The vast majority of publications refer to batch devices which have been automated to a greater or lesser extent. The device described by Rence [46] and others [41–45, 47, 48] presents a stateof-the-art batch system. However, these systems are subject to one major disadvantage in that the hydrogen and hydride formed in the reaction chamber are forced into the atom cell simultaneously and disturb the equilibrium conditions, the measured signal being the sum of disturbances. In many batch systems, the apparatus has to be disassembled after one analysis. This approach can be tolerated for AA systems but has no application in ICP/OPS, ICP/MS or DCP systems.

In the batch system, sodium tetrahydroborate (III) is added to a solution of the sample and hydrochloric acid, the hydride formed is then stripped from the solution with an inert gas. Continuous-flow methods can also be used; here, the hydride is stripped in a gas–liquid separator.

The batch system gives better sensitivity but the continuous system is more suitable for automation. It gives a continuous flame–like signal, the absorbance depending mainly upon the analyte concentration and the sample flow rate. A batch system gives a peak type signal and the absorbance maximum depends upon the analyte mass, not its concentration. The sensitivity can be increased by simply adding larger volumes of sample solution and volumes of up to 50 ml can usually be handled without problems. When large sample volumes are used the sensitivity of a batch system may be as much as an order of magnitude better than that of a continuous-flow system.

In contrast, the continuous approach relies on the continuous generation of hydrogen and a steady flow into the analyses device. Despite some objections, the inductively coupled plasma system can accommodate a reasonable level of hydrogen input without a deleterious effect on the system. Goulden and Brooksbank [2] described an analytical system based on Technicon autoanalyser technology; however, this pumped an aluminium slurry and sulphuric acid streams – the latter to dry the hydride prior to entry into an atom cell. Stockwell [3] described the inherent disadvantages of the Goulden system, and outlines the advantages in that developed by Dennis and Porter [4], again using the aluminium slurry approach.

Instrumentation

Automating the sodium tetrahydroborate system based on continuous-flow principles represents the most reliable approach in the design of commercial instrumentation. Thompson *et al.* [5] described a simple system for multi-element analysis using an ICP spectrometer, based on the sodium tetrahydroborate approach. PS Analytical Ltd developed a reliable and robust commercial analytical hydride generator system, along similar lines, but using different pumping principles to those discussed by Thompson *et al.* [5]. A further major advantage of this range of instruments is that different chemical procedures can be operated in the instrument with little, if any, modification. Thus, in addition to sodium tetrahydroborate as a reductant, stannous chloride can be used for the determination of mercury at very low levels.

The main advantage of hydride generation atomic absorption spectrometry for the determination of antimony, arsenic, selenium, etc. is its superior sensitivity. The limits of determination which can be routinely obtained with a batch system (0.01–0.04 µg L^{-1}) compares well with those of graphite-furnace atomic absorption spectrometry (0.2–1 µg L^{-1}), another well-known technique for ultra-trace element determination. A batch system gives better sensitivity than a continuous-flow system while a quartz tube atomizer is better than a flame one.

Hydride generation techniques are not completely free from problems of interference. Interferences can arise in the liquid phase caused by changes in the hydride generation rate and/or by a decreased fraction of the analyte reduced or released from the sample solution. Gas phase interferences can occur during transport of the hydride due to a delay or loss of the analyte en route or inside the atomiser.

Continuous-flow hydride generator produced by PSA Analytical Ltd

The PSA 10.002 hydride generator has, since its introduction, been widely used as a continuous source of hydride generation in ICP, DCP and most recently atomic absorption spectroscopy. Its inherent design features allow the user of each of these techniques to provide a stable baseline signal with low noise characteristics. In addition, the gas–liquid separator system provides signal rise times and also minimal memory effects. In both of these areas, which are fundamental to obtaining excellent detection levels and precision, the pumping action is critical.

The PSA 10.002 has been significantly improved by the addition of a Watson Marlow 304D pump head arrangement. This avoids deterioration of pump tubes, prolongs the tube life because the tension can be released from the head and increases the new acceptability of performance by improving the methods of connecting and disconnecting the tubing itself and also the head mechanism to the motor.

For ICP applications it is important to force the hydrides into the plasma, whereas for AA applications a compromise must be reached to allow sufficient residence time to provide a measurement while retaining the hydrides in a stable form.

To complement this hydride generator which couples easily to all makes of AA, ICPs and DCPs, PSA has designed a complete range of mechanical interfaces. These are designed to allow the user to switch quickly between conventional analysis through a nebuliser, to the use of the hydride generator. Change-over takes less than 30 seconds, thereby removing the user resistance to making changes to hydride analysis. Autosamplers are also available. An interface system is available of suitable IBM-compatible software to operate both hydride generator and autosamplers.

Detection limits achieved for some of the above elements by the PS Analytical Model 10.002 system coupled with various types of detector are between 0.15 µg L $^{-1}$ (bismuth) and 0.4 µgL $^{-1}$ (selenium).

Continuous-flow and discrete analyser to determine down to 0.05 µg L $^{-1}$ of mercury supplied by PS Analytical Ltd

Fluorescence method

The fluorescence technique utilises the PSA 20.020 autosampler linked to the PSA 10.002 hydride generator; the vapour produced is then transferred into a PSA 10.022 fluorescence monitor. The interface between the hydride and fluorescence unit provides a flow of mercury vapour in argon, which is sheathed in an additional argon stream to reduce any quenching effects. Two argon flows are therefore provided, one to pick up mercury from the separator and the other as the sheath gas. The PSA 10.002 is used in its normal configuration, although the stannous chloride reaction scheme is preferred. In this chemical scheme there is no direct agitation produced; as with the sodium between borohydride method, a modified gas–liquid mercury separator is used (PSA 10.101B). With this separator at a high rate (21 min $^{-1}$) the reagents are fed into the separator at a different level, the transfer of the mercury into the measurement cell is improved and stable readings are obtained. The mercury separator fits into the standard fixtures in the PSA 10.002.

The fluorescence system has extremely low levels of detection and is capable of rapid analysis. Typically the mercury signal will reach a peak in about 30 s and also return to the baseline in the same time. Sampling rates in the region of 40/hour with detection levels better than 0.0025 p.p.b. are obtainable at maximum sensitivity.

Using the PSA 10.002 hydride generator in the cold vapour mode, mercury in the samples is continuously generated as the metallic vapour. This is then trapped on a Pt/Au foil as the amalgam. The mercury can be concentrated in this way as sample is continuously pumped. On completion of the mercury timing cycle the PSA 10.002 triggers the initiation of the amalgam system. After a delay period the Pt/Au foil is rapidly heated to force off the trapped mercury as a transient pulse into the atomic cell of an atomic absorption system. Using this method the detection levels for mercury can be substantially reduced. Complete automation can be achieved using the PSA 20.020 large–volume autosampler.

The VGA–76 vapour generator accessory supplied by Varian AG

The VGA–76 vapour generation assembly is fully compatible with both the SpectrAA 10 single-beam and SpectrAA 20 double-beam atomic absorption spectrometers.

The VGA–76 system pumps the sample through a manifold where it is automatically acidified and mixed with sodium borohydride. The resulting vapour is swept by a stream of nitrogen into a quartz atomization cell (heated with an air-acetylene flame) in the atomic absorption spectrometer.

1.1.3.2 Inductively coupled plasma atomic emission spectrometry

It is seen in Tables 1.9(a) to (e) that this technique has been applied to the determination of a wide range of elements in soils (Table 1.9(a)), sediments (Tables 1.9(b) to (d)) and sludges (Table 1.9(e)).

An inductively coupled plasma is formed by coupling the energy from a radiofrequency (1–3k W or 27–50 MHz) magnetic field to free electrons in a suitable gas. The magnetic field is produced by a two- or three turn water-cooled coil and the electrons are accelerated in circular paths around the magnetic field lines that run axially through the coil. The initial electron 'seeding' is produced by a spark discharge but, once the electrons reach the ionisation potential of the support gas, further ionisation occurs and a stable plasma is formed.

The neutral particles are heated indirectly by collisions with the charged particles upon which the field acts. Macroscopically the process is equivalent to heating a conductor by a radio-frequency field, the resistance to eddy-current flow producing Joule heating. The field does not penetrate the conductor uniformly and therefore the largest current flow is at the periphery of the plasma. This is the so-called 'skin' effect and, coupled with a suitable gas-flow geometry, it produces an annular or doughnut-shaped plasma. Electrically, the coil and plasma form a transformer with the plasma acting as a one-turn coil of finite resistance.

The properties of an inductively coupled plasma closely approach those of an ideal source for the following reasons:

- The source must be able to accept a reasonable input flux of the sample and it should be able to accommodate samples in the gas, liquid or solid phases;
- the introduction of the sample should not radically alter the internal energy generation process or affect the coupling of energy to the source from external supplies;
- the source should be operable on commonly available gases and should be available at a price that will give cost-effective analysis;

Table 1.9(a) Metals and metalloid element in Periodic Table determined by inductively coupled plasma atomic emission spectrometry (ICPAES) in soils

Group / Period	I	II											III	IV	V	VI	VIII	0
1	^{1}H																^{1}H	^{2}He
2	^{2}Li	^{4}Be											^{5}B√	^{6}C	^{7}N	^{8}O	^{9}F	^{10}Ne
3	^{11}Na√	^{12}Mg√											^{13}Al√	^{14}Si√	^{15}P	^{16}S	^{17}Cl	^{18}Ar
4	^{19}K√	^{20}Ca√	^{21}Sc	^{22}Ti√	^{23}V√	^{24}Cr√	^{25}Mn√	^{26}Fe√	^{27}Co√	^{28}Ni√	^{29}Cu√	^{30}Zn√	^{31}Ga	^{32}Ge	^{33}As√	^{34}Se√	^{35}Br√	^{36}Kr
5	^{37}Rb	^{38}Sr√	^{39}Y	^{40}Zr√	^{41}Nb	^{42}Mo√	^{43}Tc	^{44}Ru	^{45}Rh	^{46}Pd	^{47}Ag	^{48}Cd√	^{49}In	^{50}Sn	^{51}Sb√	^{52}Te	^{53}I	^{54}Xe
6	^{55}Cs	^{56}Ba√	^{57}La√	^{72}Hf	^{73}Ta	^{74}W	^{75}Re	^{76}Os	^{77}Ir	^{78}Pt√	^{79}Au	^{80}Hg	^{81}Tl	^{82}Pb√	^{83}Bi√	^{84}Po	^{85}At	^{86}Rn
7	^{87}Fr	^{88}Ra	^{89}Ac															

Lanthanides	^{58}Ce	^{59}Pr	^{60}Nd	^{61}Pm	^{62}Sm	^{63}Eu	^{64}Gd	^{65}Tb	^{66}Dy	^{67}Ho	^{68}Er	^{69}Tm	^{70}Yb	^{71}Lu
Actinides	^{90}Th	^{91}Pa	^{92}U	^{93}Np	^{94}Pu√	^{95}Am	^{96}Cm	^{97}Bk	^{98}Cf	^{99}Es	^{100}Fm	^{101}Md	^{102}No	^{103}Lr

Total number of metals and metalloids 86 (include actinides and lanthanides).
Number determinable by ICPAES [27] (indicated by tick)

Source: Own files

Table 1.9(b) Metallic and metalloid elements in Periodic Table determined by inductively coupled plasma atomic emission spectrometry (ICPAES) in non-saline sediments

Group / Period	I	II											III	IV	V	VI	VIII	0
1	^{1}H																^{1}H	^{2}He
2	^{2}Li	^{4}Be											^{5}B	^{6}C	^{7}N	^{8}O	^{9}F	^{10}Ne
3	^{11}Na√	^{12}Mg√											^{13}Al√	^{14}Si√	^{15}P	^{16}S	^{17}Cl	^{18}Ar
4	^{19}K√	^{20}Ca√	^{21}Sc	^{22}Ti√	^{23}V√	^{24}Cr√	^{25}Mn√	^{26}Fe√	^{27}Co	^{28}Ni	^{29}Cu	^{30}Zn√	^{31}Ga√	^{32}Ge	^{33}As√	^{34}Se√	^{35}Br√	^{36}Kr
5	^{37}Rb	^{38}Sr√	^{39}Y	^{40}Zr	^{41}Nb	^{42}Mo	^{43}Tc	^{44}Ru	^{45}Rh	^{46}Pd	^{47}Ag	^{48}Cd√	^{49}In√	^{50}Sn√	^{51}Sb√	^{52}Te	^{53}I	^{54}Xe
6	^{55}Cs	^{56}Ba√	^{57}La	^{72}Hf	^{73}Ta	^{74}W	^{75}Re	^{76}Os	^{77}Ir	^{78}Pt	^{79}Au	^{80}Hg√	^{81}Tl	^{82}Pb√	^{83}Bi√	^{84}Po√	^{85}At	^{86}Rn
7	^{87}Fr	^{88}Ra	^{89}Ac															

Lanthanides	^{58}Ce	^{59}Pr	^{60}Nd	^{61}Pm	^{62}Sm	^{63}Eu	^{64}Gd	^{65}Tb	^{66}Dy	^{67}Ho	^{68}Er	^{69}Tm	^{70}Yb	^{71}Lu
Actinides	^{90}Th	^{91}Pa	^{92}U	^{93}Np	^{94}Pu	^{95}Am	^{96}Cm	^{97}Bk	^{98}Cf	^{99}Es	^{100}Fm	^{101}Md	^{102}No	^{103}Lr

Total number of metals and metalloids 86 (include actinides and lanthanides).
Number determinable by ICPAES [24] (indicated by tick)

Source: Own files

Table 1.9(c) Metallic and metalloid elements in Periodic Table determined by inductively coupled plasma atomic emission spectrometry (ICPAES) in marine sediments

Group Period	I	II	III	IV	V	VI	VIII	0
1	^{1}H						^{1}H	^{2}He
2	^{2}Li√	^{4}Be√	^{5}B	^{6}C	^{7}N	^{8}O	^{9}F	^{10}Ne
3	^{11}Na√	^{12}Mg√	^{13}Al√	^{14}Si√	^{15}P	^{16}S	^{17}Cl	^{18}Ar
4	^{19}K	^{20}Ca√ ^{21}Sc√ ^{22}Ti√ ^{23}V√ ^{24}Cr√ ^{25}Mn√ ^{26}Fe√ ^{27}Co√ ^{28}Ni√ ^{29}Cu√ ^{30}Zn√	^{31}Ga√	^{32}Ge	^{33}As√	^{34}Se√	^{35}Br	^{36}Kr
5	^{37}Rb	^{35}Sr ^{39}Y√ ^{40}Zr√ ^{41}Nb√ ^{42}Mo√ ^{43}Tc ^{44}Ru ^{45}Rh ^{46}Pd ^{47}Ag ^{48}Cd√	^{49}In√	^{50}Sn	^{51}Sb√	^{52}Te	^{53}I	^{54}Xe
6	^{55}Cs	^{50}Ba ^{57}La√ ^{72}Hf ^{73}Ta ^{74}W ^{75}Re ^{76}Os ^{77}Ir ^{78}Pt ^{79}Au ^{80}Hg√	^{81}Tl	^{82}Pb√	^{83}Bi√	^{84}Po	^{85}At	^{86}Rn
7	^{87}Fr	^{88}Ra ^{89}Ac						

Lanthanides ^{58}Ce√ ^{59}Pr ^{60}Nd ^{61}Pm ^{62}Sm ^{63}Eu ^{64}Gd ^{65}Tb ^{66}Dy√ ^{67}Ho ^{68}Er ^{69}Tm ^{70}Yb√ ^{71}Lu

Actinides ^{90}Th ^{91}Pa ^{92}U ^{93}Np ^{94}Pu ^{95}Am ^{96}Cm ^{97}Bk ^{98}Cf ^{99}Es ^{100}Fm ^{101}Md ^{102}No ^{103}Lr

Total number of metals and metalloids 86 (include actinides and lanthanides).
Number determinable by ICPAES [30] (indicated by tick)

Source: Own files

Table 1.9(d) Metallic and metalloid elements in Periodic Table determined by inductively coupled plasma atomic emission spectrometry (ICPAES) in estaurine sediments

Group / Period	I	II											III	IV	V	VI	VIII	0
1	^{1}H																^{1}H	^{2}He
2	^{2}Li	^{4}Be											^{5}B	^{6}C	^{7}N	^{8}O	^{9}F	^{10}Ne
3	^{11}Na	^{12}Mg											^{13}Al√	^{14}Si√	^{15}P	^{16}S	^{17}Cl	^{18}Ar
4	^{19}K	^{20}Ca	^{21}Sc	^{22}Ti√	^{23}V√	^{24}Cr√	^{25}Mn√	^{26}Fe√	^{27}Co√	^{28}Ni√	^{29}Cu√	^{30}Zn√	^{31}Ga√	^{32}Ge	^{33}As	^{34}Se	^{35}Br	^{36}Kr
5	^{37}Rb	^{38}Sr	^{39}Y√	^{40}Zr√	^{41}Nb	^{42}Mo	^{43}Tc	^{44}Ru	^{45}Rh	^{46}Pd	^{47}Ag	^{48}Cd	^{49}In	^{50}Sn	^{51}Sb	^{52}Te	^{53}I	^{54}Xe
6	^{55}Cs	^{56}Ba	^{57}La√	^{72}Hf	^{73}Ta	^{74}W	^{75}Re	^{76}Os	^{77}Ir	^{78}Pt	^{79}Au	^{80}Hg	^{81}Tl	^{82}Pb	^{83}Bi	^{84}Po	^{85}At	^{86}Rn
7	^{87}Fr	^{88}Ra	^{89}Ac															

Lanthanides	^{58}Ce√	^{59}Pr	^{60}Nd	^{61}Pm	^{62}Sm	^{63}Eu	^{64}Gd	^{65}Tb	^{66}Dy√	^{67}Ho	^{68}Er	^{69}Tm	^{70}Yb√	^{71}Lu
Actinides	^{90}Th	^{91}Pa	^{92}U	^{93}Np	^{94}Pu	^{95}Am	^{96}Cm	^{97}Bk	^{98}Cf	^{99}Es	^{100}Fm	^{101}Md	^{102}No	^{103}Lr

Total number of metals and metalloids 86 (include actinides and lanthanides).
Number determinable byICPAES [17] (indicated by tick)

Source: Own files

Table 1.9(e) Metallic and metalloid elements in Periodic Table determined by inductively coupled plasma atomic emission spetrometry (ICPAES) in sludges

Group / Period	I	II											III	IV	V	VI	VIII	0
1	^{1}H																^{1}H	^{2}Hc
2	^{3}Li	^{4}Be											^{5}B	^{6}C	^{7}N	^{8}O	^{9}F	^{10}Ne
3	^{11}Na	^{12}Mg											^{13}Al√	^{14}Si	^{15}P	^{16}S	^{17}Cl	^{18}Ar
4	^{19}K	^{20}Ca	^{21}Sc	^{22}Ti	^{23}V√	^{24}Cr√	^{25}Mn	^{26}Fe	^{27}Co	^{28}Ni√	^{29}Cu√	^{30}Zn√	^{31}Ga√	^{32}Ge	^{33}As√	^{34}Se	^{35}Br	^{36}Kr
5	^{37}Rb	^{38}Sr	^{39}Y	^{40}Zr	^{41}Nb	^{42}Mo	^{43}Tc	^{44}Ru	^{45}Rh	^{46}Pd	^{47}Ag	^{48}Cd√	^{49}In	^{50}Sn	^{51}Sb	^{52}Te	^{53}I	^{54}Xe
6	^{55}Cs	^{56}Ba	^{57}La	^{72}Hf	^{73}Ta	^{74}W	^{75}Re	^{76}Os	^{77}Ir	^{78}Pt	^{79}Au	^{80}Hg	^{81}Tl	^{82}Pb√	^{83}Bi	^{84}Po	^{85}At	^{86}Rn
7	^{87}Fr	^{88}Ra	^{89}Ac															

Lanthanides	^{58}Ce	^{59}Pr	^{60}Nd	^{61}Pm	^{62}Sm	^{63}Eu	^{64}Gd	^{65}Tb	^{66}Dy	^{67}Ho	^{68}Er	^{69}Tm	^{70}Yb	^{71}Lu
Actinides	^{90}Th	^{91}Pa	^{92}U	^{93}Np	^{94}Pu	^{95}Am	^{96}Cm	^{97}Bk	^{98}Cf	^{99}Es	^{100}Fm	^{101}Md	^{102}No	^{103}Lr

Total number of metals and metalloids 86 (include actinides and lanthanides).
Number determinable by ICPAES [7] (indicated by tick)

Source: Own files

- the temperature and residence time of the sample within the source should be such that all the sample material is converted to free atoms irrespective of its initial phase or chemical composition; such a source should be suitable for atomic absorption or atomic fluorescence spectrometry;
- if the source is to be used for emission spectrometry, then the temperature should be sufficient to provide efficient excitation of a majority of elements in the periodic table;
- the continuum emission from the source should be of a low intensity to enable the detection and measurement of weak spectral lines superimposed upon it, the sample should experience a uniform temperature field and the optical density of the source should be low so that a linear relationship between the spectral line intensity and the analyte concentration can be obtained over a wide concentration rate.

Greenfield *et al.* [6] were the first to recognise the analytical potential of the annular inductively coupled plasma.

Wendt and Fassel [7] reported early experiments with a 'tear drop' shaped inductively coupled plasma but later described the medium power 1–3 kW, 18 mm annular plasma now favoured in modern analytical instruments [8].

The current generation of inductively coupled plasma emission spectrometers provide limits of detection in the range of 0.1–500 µg L $^{-1}$ in solution, a substantial degree of freedom from interference and a capability for simultaneous multi-element determination facilitated by a directly proportional response between the signal and the concentration of the analyte over a range of about five orders of magnitude.

The most common method of introducing liquid samples into the inductively coupled plasma is by using pneumatic nebulisation [9] in which the liquid is dispensed into a fine aerosol by the action of a high-velocity gas stream. To allow the correct penetration of the central channel of the inductively coupled plasma by the sample aerosol, an injection velocity of about 7 ms $^{-1}$ is required. This is achieved using a gas injection with a flow rate of about 0.5–11 min $^{-1}$ through an injector tube of 1.5–2.0 mm internal diameter. Given that the normal sample uptake is 1–2 mL min $^{-1}$ this is an insufficient quantity of gas to produce efficient nebulization and aerosol transport. Indeed, only 2% of the sample reaches the plasma. The fine gas jets and liquid capillaries used in inductively coupled plasma nebulisers may cause inconsistent operation and even blockage when solutions containing high levels of dissolved solids and particulate matter are used. Such problems have led to the development of a new type of nebulizer, the most successful being based on a principle originally described by Babington (US Patents). In these, the liquid is pumped from a wide bore tube and thence conducted to the nebulizing orifice by a V-shaped groove [10] or by divergent wall of an over-expanded nozzle [11]. Such devices handle most liquids and even slurries without difficulty.

Nebulisation is inefficient and therefore not appropriate for very small liquid samples. Introducing samples into the plasma in liquid form reduces the potential sensitivity because the analyte flux is limited by the amount of solvent that the plasma will tolerate. To circumvent these problems a variety of thermal and electrothermal vaporisation devices have been investigated. Two basic approaches are in use. The first involves indirect vaporisation of the sample in an electrothermal vaporiser, e.g. a carbon rod or tube furnace or heated metal filament as commonly used in atomic absorption spectrometry [12–14]. The second involves inserting the sample into the base of the inductively coupled plasma on a carbon rod or metal filament support [15–16]. Atomic absorption has, for the past fifteen years, proved to be the most generally useful technique for the determination of metallic elements. As the user of atomic absorption reads the current literature, however, he finds the technique of inductively coupled plasma emission is finding favour in some applications as the method of choice for metallic determinations. Certainly the high sample throughput possible with direct reading ICP spectrometers appears to be attractive. The superior detection limits for those refractory elements that are relatively insensitive by AA, such as zirconium and tungsten, are an obvious favourable characteristic. Claims have been made as to reduction in chemical and ionisation interferences with the ICP and for an extended linear analytical range. A minor but still significant factor is the elimination of the instrumental manipulations required by the use of specific source lamps for atomic absorption.

Workers at Thermo Electron Ltd, a subsidiary of Jarrell Ash, have published the following comparison of the two techniques:

1   *Detection limits* Inductively coupled plasma spectrometry has similar detection limits to flame atomic absorption for most elements but is vastly superior for those elements that require a nitrous oxide/acetylene flame in atomic absorption spectrometry, e.g. barium, aluminium, tungsten, boron, titanium and zirconium, rare earths etc., plus inductively coupled plasma spectrometry can measure non-metals such as phosphorus, sulphur, iodine and bromine and high-level chlorine. Inductively coupled plasma spectrometry has inferior detection limits for zinc, lead and alkali metals, although detection limits of $2\,\mu g\,L^{-1}$ and $25\,\mu g\,L^{-1}$ for zinc and lead are still acceptable. Detection limits for ICP are matrix dependent.

2   *Speed* For multi-element analysis, inductively coupled plasma spectrometry is much faster than atomic absorption spectrometry, the greater the number of elements required the greater is the advantage; for example, a typical 13-element programme for major and trace

elements takes approximately five minutes on a dual channel instrument.

3   *Linear dynamic range* Inductively coupled plasma spectrometry has a linear dynamic range of approximately 10^6, flame atomic absorption spectrometry is approximately 10^3 at best.

4   *Interferences* Inductively coupled plasma spectrometry has virtually no chemical interferences but may have spectral interferences in a given matrix on a given elemental wavelength. Flame atomic absorption spectrometry has many chemical interferences but few spectral (ionisation) interferences.

5   *Precision* Short-term precision for flame atomic absorption spectrometry and inductively coupled plasma spectrometry are similar, 0.3–2% RSD, but in the long-term precision inductively coupled plasma spectrometry is vastly superior with precisions of better than 5% over a full eight hour day.

6   *Overnight running* Due to the fact that the plasma is not a flame there are no combustible gases, the plasma can run unattended overnight on aqueous matrices and in fact many existing plasma users have been doing this for some years, i.e. operating a 24-hour working day instead of eight hours. No matter how automated the atomic absorption spectrometry and the safety features present, one cannot leave this instrument unattended.

7   *Simultaneous multi-element determinations* Unlike atomic absorption spectrometry the inductively coupled plasma technique is capable of simultaneous multi-element determinations.

Both techniques have their advantages; what many chemists are finding is that it is distinctly advantageous to have both techniques in the laboratory.

In Table 1.10 are compared detection limits claimed for atomic absorption spectrometry, graphite furnace atomic absorption spectrometry and inductively coupled plasma optical emission spectrometry.

Simultaneous versus sequential inductively coupled plasma techniques

There are two main types of inductively coupled plasma spectrometer systems.

Monochromator system for sequential scanning
This consists of a high-speed, high-resolution scanning monochromator which views one element wavelength at a time. Typical layouts are shown in Fig. 1.6. Fig. 1.6(a) represents a one-channel air path double monochromator design with a pre-monochromator for order sorting and stray light rejection and a main monochromator to provide resolution of up to 0.20 nm. The air-path design is capable of measuring wavelengths

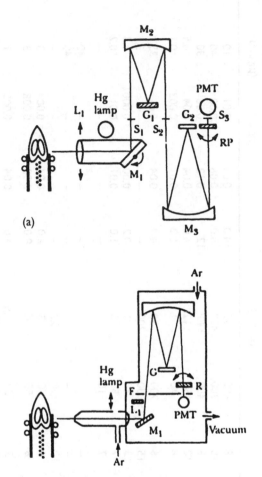

Fig. 1.6 (a) A double monochromator consisting of an air-path monochromator with a pre-monochromator for order sorting and stray light rejection to determine elements in the 190–900 nm range; (b) the vacuum UV monochromator – an evacuated and argon-purged monochromator to routinely determine elements in the 160 to 500 μm range
Source: Own files

in the range of 190–900 nm. The wide wavelength range enables measurements to be performed in the ultraviolet, visible and near infrared regions of the spectrum (allowing determinations of elements from arsenic at 193/70 nm to caesium at 852.1 nm).

The second channel (Fig. 1.6(b)) is a vacuum monochromator design allowing measurements in the 160–500 nm wavelength range. The exceptionally low wavelength range gives the capability of determining trace levels of metals at low UV wavelengths such as the extremely sensitive aluminium emission line at 167.08 nm.

Table 1.10 Guide to analytical values for IL spectrometers (IL 157/357/457/451/551/951/Video 11/12/22/S11/S12 Atomic Absorption Spectrometers IL Plasma-100/-200/-300 ICP Emission Spectrometers)

| Element | Wavelength (nm) | | AA Lamp current (mA) | Flame AA | | Furnace AA (IL755CTF Atomizer) | | | ICP |
	AA	ICP		Sensitivity[2] (µg L⁻¹)	Detection limit (µg L⁻¹)	Sensitivity[2] (µg L⁻¹)	Detection limit (µg L⁻¹)	Detection limit (µg L⁻¹)	Detection limit (µg L⁻¹)
Aluminium (Al)[1]	309.3	396.15	8	400	25	4.0	0.04	0.01	10
Antimony (Sb)	217.6	206.83	10	200	400	8.0	0.08	0.08	40
Arsenic (As)	193.7	193.70	8	400	140[3]	12	0.12	0.08	30
Barium (Ba)[1]	553.5	455.40	10	150	12	4.0	0.04	0.04	0.5
Beryllium (Be)[1]	234.9	313.04	8	10	1	1.0	0.01	0.003	0.1
Bismuth (Bi)	223.1	223.06	6	200	30	4.0	0.04	0.01	35
Boron (B)[1]	249.7	249.77	15	9,000	700	–	–	–	3
Cadmium (Cd)	228.8	214.44	3	10	1	0.2	0.002	0.0002	1.5
Calcium (Ca)	422.7	393.37	7	50	2	1.0	0.01	0.01	0.2
Calcium[1]	422.7	–	7	10	1	–	–	–	–
Carbon (C)	–	193.09	–	–	–	–	–	–	40
Cerium (Ce)	–	413.77	–	–	–	–	–	–	40
Caesium (Cs)	852.1	455.53	10	150	20	–	–	–	–
Chromium (Cr)	357.9	205.55	6	40	3	4.0	0.04	0.004	3
Cobalt (Co)	240.7	238.89	8	50	4	8.0	0.08	0.008	3
Copper (Cu)	324.7	324.75	5	30	1.8	4.0	0.04	0.005	1
Dysprosium (Dy)[1]	421.2	353.17	8	600	50	–	–	–	4
Erbium (Er)[1]	400.8	337.27	8	400	40	50	0.5	0.3	3
Europium (Eu)	–	381.97	–	–	–	–	–	–	–
Gadolinium (Gd)[1]	368.4	342.25	9	13,000	2,000	1,600	16	8	4

Element	Wavelength (nm) AA	Wavelength (nm) ICP	AA Lamp current (mA)	Flame AA Sensitivity² (µg L⁻¹)	Flame AA Detection limit (µg L⁻¹)	Furnace AA (IL755CTF Atomizer) Sensitivity²	Furnace AA (IL755CTF Atomizer) (µg L⁻¹)	Furnace AA (IL755CTF Atomizer) Detection limit (µg L⁻¹)	ICP Detection limit (µg L⁻¹)
Gallium (Ga)	287.4	294.36	5	400	50	5.2	0.05	0.01	15
Germanium (Ge)¹	265.1	209.43	5	800	50	40	0.4	0.1	20
Gold (Au)	242.8	242.80	5	100	6	5.0	0.05	0.01	10
Hafnium (Hf)¹	307.3	339.98	10	14,000	2,000	–	–	–	5
Holmium (Ho)¹	410.4	345.60	12	660	60	90	0.9	0.7	1
Indium (In)	303.9	325.61	5	180	30	11	0.11	0.02	15
Iridium (Ir)¹	208.8	224.27	15	1,500	500	170	1.7	0.5	8
Iron (Fe)	248.3	238.20	8	40	5	3.0	0.03	0.01	2
Lanthanum (La)¹	550.1	333.75	10	22,000	2,000	58	0.58	0.5	2
Lead (Pb)	217.0	220.35	5	100	9	4.0	0.04	0.007	25
Lithium (Li)	670.8	670.78	8	16	1	4.0	0.04	0.01	2⁵
Lutetium (Lu)	–	261.54	–	–	–	–	–	–	0.2
Magnesium (Mg)	285.2	279.55	3	3	0.3	0.07	0.007	0.002	0.1
Manganese (Mn)	279.5	257.61	5	20	1.8	1.0	0.01	0.0005	–
Mercury (Hg)	253.7	253.65	3	2,500	140	40	0.4	0.2	12
Molybdenum (Mo)¹	313.3	202.03	6	200	25	12	0.12	0.03	4
Neodynium (Nd)¹	492.5	401.23	10	5,000	700	–	–	–	8
Nickel (Ni)	232.0	221.65	10	60	5	20	0.2	0.05	4
Niobium (Nb)¹	334.9	309.42	15	12,000	2,000	–	–	–	6
Osmium (Os)¹	290.9	225.59	15	1,000	90	270	2.7	2	0.6
Palladium (Pd)	247.6	340.46	5	140	20	13	0.13	0.05	13
Phosphorus (P)	213.6	213.62	8	125,000	30,000	4,900	49	20	16
Platinum (Pt)	265.9	214.42	10	1,000	50	80	0.8	0.2	16

Element	Wavelength (nm)		AA Lamp current (mA)	Flame AA		Furnace AA (IL755CTF Atomizer)			ICP
	AA	ICP		Sensitivity[2] (µg L⁻¹)	Detection limit (µg L⁻¹)	Sensitivity[2] (µg L⁻¹)	(µg L⁻¹)	Detection limit (µg L⁻¹)	Detection limit (µg L⁻¹)
Potassium (K)	766.5	766.49	7	10	1	0.4	0.004	0.004	30[5]
Praesodymium (Pr)[1]	495.1	390.84	15	20,000	2,000	–	–	–	20
Rhenium (Re)	346.1	221.43	15	8,000	800	1,000	10	10	6
Rhdoium (Rh)	43.5	343.49	5	200	2	20	0.20	0.	8
Rubidium (Rb)	780.0	–	10	30	2	–	–	–	–
Ruthenium (Ru)[1]	349.9	240.27	10	800	400	–	–	–	8
Samarium (Sm)[1]	429.7	359.26	10	3,000	500	–	–	–	8
Scandium (Sc)[1]	391.2	361.38	10	100	20	–	–	–	–/5
Selenium (Se)	196.0	196.03	12	300	80[3]	8.0	0.08	0.05	30
Silicon (Si)[1]	251.6	251.61	12	800	60	60	0.60	0.6	6
Silver (Ag)[1]	328.1	328.07	3	30	1.2	0.5	0.005	0.001	3
Sodium (Na)	589.0	589.59	8	3	0.4	0.4	0.004	0.004	7
Strontium (Sr)	460.7	407.77	12	80	6	1.8	0.018	0.01	0.2
Tantalum (Ta)[1]	271.5	240.06	15	10,000	800	–	–	–	13
Tellarium (Te)	214.3	214.28	7	200	300	7.0	0.07	0.03	20
Terbium (Tb)[1]	432.7	350.92	8	3,300	1,000	–	–	–	3
Thallium (Tl)	276.8	276.79	8	100	30	4.0	0.04	0.01	27
Thorium (Th)	–	283.73	–	–	–	–	–	–	8
Thulium (Tm)	–	313.13	–	–	–	–	–	–	0.9
Tin (Sn)[1]	235.5	189.99	6	1,200	90	7.0	0.7	0.03	30
Titanium (Ti)[1]	364.3	334.94	8	900	60	50	0.50	0.3	–
Tungsten (W)[1]	255.1	207.91	15	5,000	500	–	–	–	14

Element	Wavelength (nm)		AA Lamp current (mA)	Flame AA		Furnace AA (IL755CTF Atomizer)			ICP
	AA	ICP		Sensitivity[2] (µg L⁻¹)	Detection limit (µg L⁻¹)	Sensitivity[2]	(µg L⁻¹)	Detection limit (µg L⁻¹)	Detection limit (µg L⁻¹)
Uranium (U)[1]	358.5	263.55	15	100,000	7,000	3,100	31	30	70
Vanadium (V)[1]	318.5	309.31	8	600	25	40	0.40	0.1	3
Ytterbium (Yb)	398.8	328.94	5	80	–	1.3	0.01	0.01	–
Yttrium (Y)[1]	410.2	371.03	6	1,800	200	1,300	13	10	0.7
Zinc (Zn)	213.9	213.86	3	8	1.2[3]	0.3	0.003	0.001	2
Zirconium (Zr)[1]	360.1	343.82	10	10,000	2,000	–	–	–	2

[1] Nitrous oxide/acetylene flame (AA).
[2] Sensitivity is concentration (or mass) yielding 1% absorption (0.0044 absorbance units)
[3] With background correction
[4] Furnace AA concentration values are based on cuvette capacity of 100 µl
[5] Requires use of red-sensitive PMT

Source: Own files

The sequential instrument, equipped with either or both monochromators facilitates the sequential determination in a sample of up to 63 elements in turn at a speed as fast as 18 elements per minute. Having completed the analysis of the first sample, usually in less than a minute, it proceeds to the second sample, etc.

Polychromator system for simultaneous scanning
The polychromator systems scan many wavelengths simultaneously, i.e. several elements are determined simultaneously at higher speeds than are possible with monochromator systems. It then moves onto the next sample. A typical system is shown in Fig. 1.7.

It is possible to obtain instruments which are equipped for both sequential and simultaneous scanning, e.g. the Labtam 8410.

For the chemist the decision whether to purchase a sequential or a simultaneous system depends on obtaining answers to the following questions:

1   What matrices do you want to analyse? The more diverse the matrices the greater the need for a sequential instrument. If the matrix is very well defined, the case for a simultaneous system becomes stronger.
2   How many elements are required and could this number increase? The simultaneous plasma is custom built so a decision has to be made at time of purchase on how many analytical channels are put on the focal curve and also on the best wavelengths for the defined matrices and concentration levels expected.
3   How many samples need to be analysed per day? The great advantage of simultaneous inductively coupled plasma is speed. In many cases full analysis can be done within one minute, whether it is one element or 61 elements. With sequential inductively coupled plasma the larger the number of elements, the longer the analysis time.
4   What are the relevant advantages of either technique to achieve accurate results? Simultaneous inductively coupled plasma has fixed positions by which background correction can be applied leading to possible errors on some elements in some matrices. Inter-element corrections are routinely required on simultaneous inductively coupled plasma but a bad wavelength for a given element in a given matrix may not be possible even if inter-element correction is applied. A sequential inductively coupled plasma can be programmed to avoid most or all of any encountered spectral interferences.
5   What are the relevant precisions of the two inductively coupled plasma versions? Sitting on a peak (i.e. sequential) will always give better precision and reproducibility than scanning past the peak (i.e. simultaneous) (as most scanning inductively coupled plasma systems do). Low-level detection rates may be superior on the simultaneous inductively coupled plasma in terms of reproducibility over a period of time.

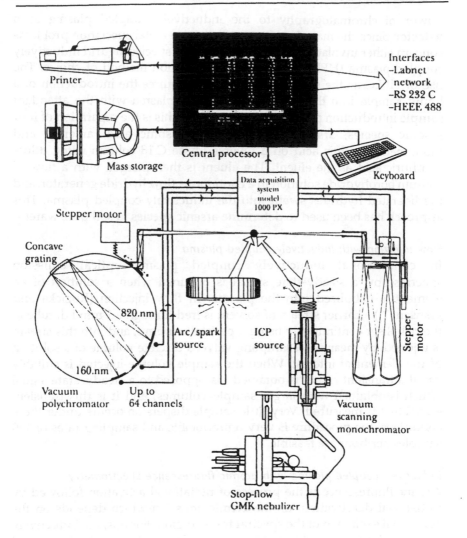

Fig 1.7 Polychromator system for inductively coupled plasma atomic emission spectrometer
Source: Own files

Hybrid inductively coupled plasma techniques

Chromatography-inductively coupled plasma

Direct introduction of the sample into an inductively coupled plasma produces information only on the total element content. It is now recognised that information on the form of the element present, or trace element speciation, is important in a variety of applications including sediment and soil water analysis. One way of obtaining quantitative measurement of trace element speciation is to couple the separation power of chromatography to the inductively coupled plasma as a detector. Since the majority of interesting trace metal speciation problems concern either involatile or thermally unstable species, hydrid inductively coupled plasma (HPLC) becomes the separation method of choice. The use of HPLC as the separation technique requires the introduction of a liquid sample into the inductively coupled plasma with the attendant sample introduction problem. An example of this is the separation of four arsenic species, arsenate, monomethyl arsonic acid, arsenite and monomethyl arsinic acid on a reverse phase C 18 column using dilute sulphuric acid as the eluent. The eluent is then merged with a flow of sodium borohydride solution in a continuous flow hydride generator, and the liberated hydrides swept into the inductively coupled plasma. This approach has been used to determine arsenic species in soil pore waters.

Flow injection with inductively coupled plasma
In conventional inductively coupled plasma optical emission spectroscopy, a steady-state signal is obtained when a solution of an element is nebulized into the plasma. In flow injection (Ruzicka and Hansen [1]) a carrier stream of solvent is fred through a 1 mm i.d. tube to the nebuliser continuously using a peristaltic pump, and into this stream is injected, by means of a sampling valve, a discrete volume of a solution of the element of interest. When the sample volume injected is suitably small a transient signal is obtained (as opposed to a steady–state signal which is obtained with larger sample volumes) and it is this transient signal that is measured. Very little sample dispersion occurs under these conditions, the procedure is very reproducible and sampling rates of 180 samples per hour are feasible.

Inductively coupled plasma with atomic fluorescence spectrometry
Atomic fluorescence is the process of radiational activation followed by radiational deactivation, unlike atomic emission which depends on the collisional excitation of the spectral transmission. For this, the inductively coupled plasma is used to produce a population of atoms in the ground state and a light source is required to provide excitation of the spectral transitions. Whereas a multitude of spectral lines from all the accompanying elements are emitted by the atomic emission process, the fluorescence spectrum is relatively simple, being confined principally to the resonance lines of the element used in the excitation source.

The inductively coupled plasma (ICP) is a highly effective line source with a low background continuum. It is optically thin – it obeys Beer's law – and therefore exhibits little self-absorption. It is also a very good atomiser and the long tail flame issuing from the plasma has such a range of temperatures that conditions favourable to the production of atoms in the ground state for most elements are attainable. It is therefore possible to use two plasmas in one system, a source plasma to supply the radiation to activate the ground rate atoms in another, the atomiser. This atomic fluorescence (AFS) mode of detection is relatively free from spectral interference, the main drawback of inductively coupled plasma optical emission spectroscopy.

Good results have been obtained using a high power (6 kW) ICP as a source and a low power (< 1 kW) plasma as an atomizer.

Schramel [55] has demonstrated the advantages of using a high resolution inductively coupled plasma for the determination of trace metals in soil.

In-situ laser ablation inductively coupled plasma atomic emission spectrometry

Field analyses of soils by in-situ laser ablation – inductively coupled plasmal/atomic emission spectrometry techniques are reported to be generally in good agreement with results obtained in laboratory determinations [36, 49].

Instrumentation

The chemist interested in purchasing an inductively coupled plasma optical emission spectrometer is faced with an embarrassingly large variety of suppliers (see Table 1.11). The eventual choice will be made on the basis of selecting a sequential or a simultaneous instrument for general requirements (e.g. degree of automation and computerisation) and cost (price range £40,000–£90,000).

The Perkin–Elmer ICP 5000 combined graphite furnace spectrometer inductively coupled plasma emission spectrometer is a fairly sophisticated instrument with the following features:

- Every instrumental parameter is controlled by the built–in microprocessor or the external data system
- High-energy monochromator has both an 84 × 84 mm holographic grating blazed at 210 nm and a ruled grating blazed at 580 nm, covering a spectral range of 170 to 900 nm in the first order
- Storage and recall of six complete AA programs by the microprocessor
- Correction of non-specific atomic absorption over the complete wavelength range available for the AA mode
- Optical interface:
  - Transmits either ICP emission or radiation from the AA source into the monochromator
  - Allows change from AA to ICP in two seconds with the turn of a knob
  - Both optical interface and monochromator can be purged for working in the far UV at wavelengths down to 170 nm

Table 1.11 Inductively coupled plasma optical emission spectrometers available on the market

Supplier	Model	System	Number of elements claimed	Maximum analysis rate (elements min^{-1})	Microprocessor	Auto-sampler	Range (nm)
Perkin–Elmer	Plasma II	Optimised sequential system	70	Up to 50	Yes	Yes	160–800
Perkin–Elmer	ICP 5500	Sequential		15	Yes	Yes	170–900
Perkin–Elmer	ICP5500B	Sequential		20	Yes	Yes	170–900
Perkin–Elmer	ICP6500	Sequential		20	Yes	–	170–900
Perkin–Elmer	ICp500	Can be used for flame and graphite furnace ASS and inductively coupled plasma atomic emission spectrometry (sequential)		–	Yes	Yes	175–900
Perkin–Elmer	Plasma 40	Lower-cost sequential			Personal computer control	Yes	160–800
Labtam	Plasma scan 8440	Simultaneous (poly-chromator or with optional mono-chromator for sequential)	60–70	Up to 64	Yes	Yes	170–820
Labtam	Plasma 8410	Sequential	More than 70	–		Yes	170–820
Thermo-electron	Plasma 300 (replacing the Plasma 200) (single (air) or double (air/vacuum) options available	Sequential	Up to 63	Up to 18 (single channel air); up to 24 (double channel air/vacuum)	Yes	Yes	160–900

Supplier	Model	System	Number of elements claimed	Maximum analysis rate (elements min^{-1})	Microprocessor	Auto-sampler	Range (nm)
Philips	PV 8050 series PV 8055 PV 8060 PV 8065	Simultaneous	56	–	Yes	Yes	165–485 and 530–860
Philips	PU 7450	Sequential	70	–	Yes	Yes	190–800
Baird	Spectrovac PS3/4 plasma hydride device option	Simultaneous and sequential	Up to 60	Up to 80 samples per hour each up to 60 elements	Yes	Yes	162–766 and 162–800
Baird	Plasmatest system 75	Simultaneous and sequential	Up to 64	–	Yes	Yes	175–768 and 168–800
Spectro Analytical Ltd	Spectroflame plasma hydride device option	Simultaneous and sequential	Up to 64	–	Yes	Yes	165–800

Source: Own files

- provides the adjustment of viewing height of the ICP torch ICP emission source and power supply
- High analytical reproducibility by automatic tuning of the plasma to maintain its most efficient operating level
- Power adjustable to 2500 watts for analysis of all types of aqueous and organic sample solutions
- Sensors for argon and water flow prevent accidental damage to the torch components

The first enthusiastic reports on analytical applications of ICP gave the impression that the ultimate technique for the determination of metallic and semi-metallic elements had been discovered. However, as time passed it became clear that while ICP emission substantially enhances the capability of atomic spectroscopy as an analytical technique ICP must be considered a complement to, rather than a replacement for, atomic absorption. And indeed, each technique offers advantages and disadvantages of its own, as described below.

ICP emission
- Excellent detection limits for refractory elements that are relatively insensitive to determination by flame AA including U, B, P, Ta, Ti and W. They can be determined at submilligram/litre levels with ICP.
- Multi-element analysis is more rapid by ICP if more than eight elements are to be determined in the same sample.
- Matrix dependence is less of a problem with ICP which is practically free of ionization and chemical interferences.
- Although ICP does not offer the intrinsic specificity of atomic absorption, the spectral interferences which are observed may be eliminated by background correction techniques or the use of alternate wavelengths.

Flame atomic absorption
- The analytical precision of AA is usually better than 0.3% for many analyses and cannot currently be matched by ICP. This is particularly important for applications where high precision is required.
- The sample throughput of flame AA is higher for single-element determinations of the automated determination of up to six to eight elements per sample. This is due to the rapid equilibration of the AA signal after sampling begins.

Graphite furnace atomic absorption
- The HGA graphite furnace offers outstanding detection limits and is the method of choice for trace analysis of the majority of elements.
- If the sample quantity is limited, the HGA graphite furnace should be selected for its microsampling capability. Microlitre quantities of sample are sufficient to perform an analysis.

- Direct analysis of solid samples (e.g. sediments) is possibly only with the HGA graphite furnace technique.

Sequential ICP analysis
The sequential ICP multi-element determines each element under its own optimum, uncompromised set of analytical conditions. By necessity, simultaneous multi-element instruments always operate under compromised conditions. However, sequential multi–element instruments can be programmed to perform each elemental analysis under specific optimized conditions.

Some detection limits achieved by the inductively coupled plasma atomic emission systems are given in Table 1.10.

Autosampler for ICP
The Labtam autosampler has been specifically designed for inductively coupled plasma applications and provides total flexibility for both simultaneous and sequential operations. The design and construction employed in this autosampler has simplified difficult and arduous applications commonly experienced in laboratories with high sample throughput.

1.1.3.3 Inductively coupled plasma atomic emission spectrometry–mass spectrometry

This technique is of growing importance. Its applications to soil analysis are so far limited to the determination of arsenic, cadmium, lead, zinc, technecium (astatine) and thorium. However extensive applications have recently been made to the analysis of non-saline and saline sediments, viz: non-saline sediments, heavy metals (cadmium, lead and zinc), arsenic, antimony, thallium, thorium, calcium, uranium, mercury, selenium and plutonium, also rare earths (cerium, dysprosium, europium, gadolinium, tamerium, terbium, ytterbium, lanthanum, lutecium, neodynium and samerium).

Saline sediments
Heavy metals (chromium, nickel, zinc, cadmium and lead) and mercury, strontium, molybdenum, tin, antimony, sodium, thallium and uranium have been determined.

Inductively coupled plasma mass spectrometry has been widely recognised as a suitable technique for the determination of trace elements, with particular advantages of its high sensitivity, large dynamic range and low background. However, it has been reported that many problems related to mutual effects involving matrix interferences exist in this technique. In particular, it has been noticed that determination of trace elements in natural materials by inductively coupled plasma atomic emission spectrometry is strongly affected by two problems: suppression or enhancement of ion intensity by the matrix and gradual drift in sensitivity by deposition of matrix elements on the sampling cone.

In order to reduce the matrix–associated problems mentioned above, a number of approaches for elemental analyses have been suggested. These include matrix matching, standard additions and internal standardisation. However, these compensation methods have not always been successful since the matrix problems due to concomitant elements are very complicated. In addition, because of the low concentrations of many elements of interest in natural materials, the dissolution procedure is also an especially critical step. It has been reported that in some types of solid samples the metals are contained in mineral phases which are resistant to acid attack. In such materials, samples should be decomposed by the alkali fusion method. This method of decomposition, however, results in a solution containing a high level of dissolved solids. Therefore, direct introduction of such a solution to the sampling system of inductively coupled plasma atomic emission spectrometry is restricted.

Pretreatment prior to instrumental analysis solves important problems related to matrix and sensitivity in a large number of cases of determination and extends the range of application of the instrument in a substantial manner. Solvent extraction has been proved to be an effective mean of increasing sensitivity as well as a means of removing matrix elements.

Although inductively coupled plasma atomic emission spectrometry instrumental determination is rapid and can handle a large number of samples, the manual solvent extraction preconcentration is tedious and time-consuming and incompatible with the final rapid determination step. Recently, advances have been made in developing flow injection systems for on-line solvent extraction with the aim of speeding up and simplifying the preconcentration step.

Inductively coupled plasma mass spectrometry is a multi-element technique with sub-part-per-billion detection limits for many elements. Quantitative or semi–quantitative information on over 50 elements is attainable by scanning over the mass-to-charge range 6 to 238, but studies to date have not yet fully utilized this capability of inductively coupled plasma mass spectrometry.

The detection power of inductively coupled plasma mass spectrometry makes possible the determination in geological reference materials of many trace elements for which relatively few reliable values have been previously established. This lack of data in many cases prevents a full assessment of the accuracy of inductively coupled plasma mass spectrometry results. A partial solution to this problem is the use of stable isotope dilution techniques which are immune to many of the sources of error which can adversely affect inductively coupled plasma mass spectrometry results obtained by other calibration strategies. This approach would, for example, be an effective means to compensate for the suppression of ion sensitivities by concomitant elements observed in many of the early applications of inductively coupled plasma mass spectrometry. Also, more calibration drift can be tolerated in an isotope dilution analysis because an isotope ratio, rather than an absolute intensity measurement, is used in the calculation of the analyte concentration. This suggests that it may be easier to obtain accurate and precise inductively coupled plasma mass spectrometry results for solutions with appreciable dissolved solids concentrations if isotope dilution techniques are used.

Inductively coupled mass spectrometry combines the established inductively coupled plasma to break the sample into a stream of positively charged ions which are subsequently analysed on the basis of their mass. Inductively coupled mass spectrometry does not depend on indirect measurements of the physical properties of the sample. The elemental concentrations are measured directly – individual atoms are counted giving the key attribute of high sensitivity. The technique has the additional benefit of unambiguous spectra and the ability to measure different isotopes of the same element directly.

The sample under investigation is introduced, most typically in solution, into the inductively coupled plasma at atmospheric pressure and a temperature of approximately 6000 K. The sample components are rapidly dissociated and ionised and the resulting atomic ions are introduced via a carefully designed interface into a high-performance quadrupole mass spectrometer at high vacuum.

A horizontally mounted ICP torch forms the basis of the ion source. Sample introduction is via a conventional nebulizer optimised for general-purpose solution analysis and suitable for use with both aqueous and organic solvents.

Nebulised samples enter the central channel of the plasma as a finely dispersed mist which is rapidly vaporized; dissociation is virtually complete during passage through the plasma core with most elements fully ionised.

Ions are extracted from the plasma through a water–cooled sampling aperture and a molecular beam is formed in the first vacuum stage and passes into the high–vacuum stages of the quadrupole mass analyser.

In an inductively coupled plasma mass spectrometry system a compact quadrupole mass analyser selects ions on the basis of their mass-to-charge ratio, m/e.

The quadrupole is a simple compact form of mass analyser which relies on a time-dependent electric field to filter the ions according to their mass-to-charge ratio.

Ions are transmitted sequentially in order of their m/e with constant resolution, across the entire mass range.

A multi-stage vacuum system provides the required low pressure in the analyser and is under computer control, enabling the control of the transition from atmospheric pressure at the sampling cone to the 2×10^{-4} torr within the quadrupole to be invisible to the operator.

In addition to the need for good resolution between peaks and high sensitivity, another instrument performance parameter, abundance sensitivity, is of particular importance to this technique.

Abundance sensitivity is a measure of the ability of the spectrometer to detect a small peak in the presence of a very large adjacent peak. Often the requirement is for detection limits of nanograms per litre or better in the presence of major elements at nearby masses (e.g. ultratrace levels of toxic impurities in the presence of large peaks of matrix elements).

A variety of sample introduction techniques which have been used successfully with ICP optical emission instruments depend on the measurement of transient signals. For a quadrupole with a scanning detector it is necessary to scan significantly faster than the analyte signal is changing and many discrete sample introduction systems have pulses which last a few seconds. With the plasma quad instrument manufactured by VG Isotopes Ltd, for example, it is possible to perform a sweep over all the elements in 100 ms.

ICP–MS has some significant advantages over ICP–AES when looking at pulses of this kind, such as greater sensitivity, simpler spectra and wider element coverage.

This technique is being adopted cautiously by several water laboratories. Its considerable cost has certainly been a deterrent to all but the most enterprising laboratories. It does seem to satisfy the seemingly conflicting demands made of chemists for high sample throughput and extremely low detection limits.

Automatic multi-element operation, with semi-quantitative analysis of the entire mass range in less than a minute (with subsequent quantitation by comparison with standards) and detection limits at least an order of magnitude better than any alternative techniques are most impressive credentials.

For many elements the most useful spectral lines arise from singly charged ions of which the inductively coupled plasma is an abundant source. These ions can be extracted for mass spectrometry. Gray [17] pioneered the use of plasmas fed by solutions and realised the potential of the ICP as a source for mass spectrometry. He was joined at Surrey University by Date and together with Fassel's group at Ames, Iowa, they established ICP–MS as a viable analytical technique [17–21].

In principle, the inductively coupled plasma offers many advantages as an ion source for mass spectrometry. Making it work, however, presents some formidable difficulties. The ICP reaches temperatures of 5000–9000 K at its core, operates with 10–20 Lmin $^{-1}$ of argon support gas and, of course, operates at atmospheric pressure. The mass spectrometer, on the other hand, operates at 10^{-5} torr or below and then obviously cannot be subjected to the heat of the plasma. Clearly, then, the critical aspect in the development of ICP–MS is in the design of a suitable interface between the two.

The interface consists essentially of two orifices; the first, known as the 'sampler' is a water–cooled metal cone containing a circular hole of 0.5–1.00 mm diam. A second metal cone, known as a 'skimmer', is placed behind the sampler and divides an initial expansion chamber (~1 torr) from the vacuum region that contains the ion optics and quadrupole mass filter. Expansion through the sampler orifice produces a supersonic beam of ions and neutral argon atoms which is further narrowed by the exclusion of peripheral ions at the succeeding skimmer nozzle. The sampler is immersed in the so-called 'normal analytical region' of the plasma, 5–15 mm above the radio frequency load coils used for plasma generation. Because the sample is able to accept gases from the axial channel of the plasma over a cross-section of around two to three times the orifice diameter a sizeable fraction of the central plasma channel is sampled.

Complementary aspects of inductively coupled plasma emission spectrometry and mass spectrometry

The combination of a mass spectrometer with an inductively coupled plasma has proved to be a harmonious relationship enabling the speed and convenience of sample introduction into the inductively coupled plasma to be combined with the high sensitivity and isotope-ratio capability of mass spectrometry. Some complementary features of the ICP–MS union are summarised in Table 1.12, demonstrating that many of the weaknesses of existing sample-introduction techniques for the inorganic mass spectrometer can be overcome when an inductively coupled plasma is used as an ion source.

Figs. 1.8(a) and (b) compare the conventional optical spectrum for a solution containing cerium with a mass spectrum for the same element.

ICP–MS system

The facilities in the ICP–MS experiments consisted of an argon–ICP system incorporating a 2.5 kW, 27.12 Mi-iz (Plasma Therm model HFP–2500F, Kresson, N,H.) plasma power supply and impedance-matching unit with a conventional-sized, water-cooled three-turn load coil and a plasma torch of optimised low-flow, low-power design. A laboratory-constructed concentratic pneumatic nebuliser with a Scott type double-pass spray chamber was used for sample introduction. In the spectra the plasma was operated at 1.25 kW with a nebuliser flow of 0.81 min $^{-1}$ and a coolant flow of 121 min $^{-1}$. The mass-spectral system consisted of an interface arrangement, a series of ion lenses to focus the ion beam, a quadrupole mass spectrometer (QMG 511 Balzers, Hudson, N,H.) and a detector (secondary electron multiplier) located off the optical axis to prevent entry of extraneous photons.

Table 1.12 Complementary aspects of the ICP–MS

ICP emission spectrometry	Mass spectrometry
Efficient but mild ionization source (produces mainly singly charged ions)	Ion source required
Sample introduction for solutions is rapid and convenient	Sample introductiono can be difficult for inorganic samples. Thermal spark-source, or secondary ion sources are generally restricted to solid samples and are time-consuming
Sample introduction is at atmospheric pressure	Often requires reduced-pressure sample introduction
Few matrix or interelement effects are observed and relatively large amounts of dissolved solids can be tolerated	Limited to small quantities of sample
Complicated spectra with frequent spectral overlaps	Relatively simple spectra
Detectability is limited by relatively high background continuum over much of the useful wavelength range	Very low background level throughout a large selection of the mass range
Moderate sensitivity	Excellent sensitivity
Isotope ratios cannot usually be determined	Isotope-ratio determinations are possible

Source: Own files

ICP–AES system

The ICP system in this case was operated with a 2.5 kW, 40.68 MHz power supply (Plasma Therm, model HFL–2000L) at 1.0 kW with a nebuliser flow of 1.01 min $^{-1}$ and a coolant flow of 131 min $^{-1}$, but was otherwise similar to that described above. An optical detection scheme consisting of a Heath monochromator (model EU–700) and an RCA 1P28 photo-multiplier was used to record the spectrum.

The optical spectrum contains dozens of strong lines and hundreds of weaker lines for cerium (Fig. 1.8), all of which are superimposed on top of an already complicated background emission spectrum resulting from emission lines of the argon-support gas and molecular bands from atmospheric contaminants (such as OH, NH, N_2). In addition, a relatively large continuous background is observed (Fig. 1.8(a)). The corresponding mass spectrum (Fig. 1.8(b)), on the other hand, is considerably simpler and contains fewer background peaks especially above $m/z = 40$. The

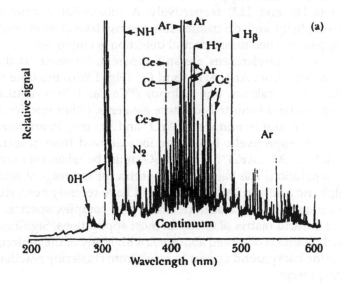

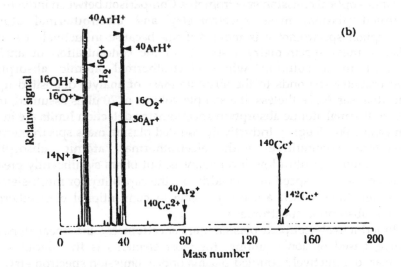

Fig. 1.8 Comparison between optical emission detection (a) for 100 p.p.m. Ce in the ICP and mass spectral detection; (b) for 10 p.p.m. of the same element. The optical spectrum contains dozens of strong Ce lines in the 350–450 nm region in addition to a number of strong background features due to OH, N_2, NH, etc. (only the most intense features are labelled). The corresponding mass spectrum is simpler and shows clearly resolved Ce isotopes at 140 and 142 mass units. (The minior isotopes at 136 and 138 are not visible at the detection sensitivity shown.) A small peak due to Ce $^{2+}$ is also apparent at *m/e* = 70
Source: Own files

response due to cerium appears at the m/z positions for cerium isotopes (mass units 140 and 142) respectively. A substantial improvement in spectral simplicity and discrimination against background response is readily apparent when mass spectral detection is employed.

Mass spectral interferences are still apparent, however, as the major background ions from Ar^+, ArH^+ and O^+ (Fig. 1.8(b)) interfere with the major isotopes of calcium and sulphur ($^{40}Ca^+$ and $^{32}S^+$) requiring that minor isotopes be substituted for these elements. Other interferences are less severe, for example, minor ions ArO^+ and Ar^+ may interfere with iron and selenium respectively. Molecular ions derived from reagent acids, such as ArN^+, ClO^+, $ArCl^+$, $S)^+$ and SO_2 must be taken into account in sample preparation. The background spectra for the reagent acids nitric acid, sulphuric acid and hydrochloric acid have recently been studied in detail, sulphuric and hydrochloric acids give complex spectra, making nitric acid the acid matrix of choice in most applications. Significantly, no major background is observed above $m/z = 82$. Many of the molecular ions present in the background appear to result from clustering reactions in the extraction process.

Inductively coupled plasma mass spectrometry also shows an advantage over flame atomic absorption spectrometry which has essentially similar detection limits to those for inductively coupled plasma – optical emission spectrometry. Comparison between inductively coupled plasma, mass spectrometry and electrothermal atomic absorption spectrometry is more difficult because inductively coupled plasma mass spectrometry responds to the concentration of analyte present in a solution while electrothermal atomic absorption spectrometry responds to the absolute mass of analyte deposited upon the atomiser. Nevertheless, if a sample volume of 20 µL is assumed, then electrothermal atomic absorption spectrometry detection limits are in the range of 0.005–5 µgL $^{-1}$. Inductively coupled plasma mass spectrometry is therefore competitive with electrothermal atomic absorption spectrometry for ultra low-level analysis, but offers significantly greater convenience and speed and, in addition, the capability for multi-element determinations which are not possible in conventional electrothermal atomic absorption spectrometry.

In inductively coupled plasma mass spectrometry, the linear dynamic range, based on ion counting, for most elements is five decades, i.e. similar to inductively coupled plasma optical emission spectrometry.

A major advantage of inductively coupled plasma mass spectrometry is the unique capacity for isotope ratio determination. Precision in these measurements is usually in the range 0.1–1% RSD.

Reports on interference effects in inductively coupled plasma mass spectrometry are mixed. Many workers report relative freedom from inter-element effects similar to that which had been found previously for inductively coupled plasma optical emission spectrometry. Little suppression of ionisation occurs in the presence of 1000 mg L $^{-1}$ sodium [20, 22] the classical interference of phosphate on calcium determinations is absent [23].

Conversely, other workers [24, 25] report severe and widespread inter-element effects in inductively coupled plasma mass spectrometry.

To sum up, the inductively coupled plasma mass spectrometry method offers remarkably high detection powers with the capability for isotope-ratio determination with speed and convenience. However, at this present stage of development a number of limitations are apparent; samples with highly dissolved solids cannot be optimally analysed unless they are diluted to at least 0.2% of total dissolved solute; instrument drift can be severe (5–10% per hour); interference effects appear to be more severe than in inductively coupled plasma optical emission spectrometry, accuracy of analysis is generally worse than with inductively coupled plasma optical mass spectrometry and capital costs are high (approximately US$2,000,000 for commercial instruments). Strenuous efforts are currently being made to minimise these limitations.

Taylor *et al.* [50] have discussed the use of inductively coupled plasma mass spectrometry as an element specific detector for field-flow fractionation particle separation. Dolan *et al.* [56] have demonstrated the advantages of using inductively coupled plasma mass spectrometry for the determination of trace metals in soil.

Workers at the Yorkshire Water Authority have evaluated the VG Isotopes Ltd plasmaquad inductively coupled plasma mass spectrometer during 1984 and 1986. The instrument failed to meet a specification they had laid down for its use in water and sediment analysis. Failure was on both reliability and performance grounds. They then went on to study sequential inductively coupled plasma optical emission spectrometry and indeed, based on their experience in these trials, ultimately purchased two of these instruments.

Instrumentation
Several manufacturers, including VG Isotopes and Perkin–Elmer can now supply equipment for inductively coupled plasma mass spectrometry.

The Perkin–Elmer Elan 500 instrument is discussed below. This spectrometer is designed for routine and rapid multi-element quantitative determinations of trace and ultratrace elements and isotopes. The Elan 500 can determine nearly all of the elements in the periodic table with exceptional sensitivity.

The entire Elan 500 Plasmalok system is designed for simplicity of operation. A typical daily start–up sequence from the standby mode includes turning on the plasma and changing to the operating mode. After a brief warm-up period for the plasma, routine sample analysis can begin.

Plasmalok eliminates undesirable secondary discharges in the interface region. The result is increased sampling orifice lifetime and minimal contamination from vaporised orifice materials. Plus, a sharply reduced presence of doubly charged ions simplifies the mass spectrum.

More importantly, Plasmalok helps maintain ion energies at constant low levels with minimal spreads, even when plasma conditions are changed. Even though the Elan 500's optics (the electrical devices that focus the stream of ions) are easily adjusted, Plasmalok ensures that such adjustments are rarely required for routine analysis.

An exclusive feature of the Perkin–Elmer Elan 500 instrument is the Omnirange system. The high sensitivity of inductively coupled plasma mass spectrometry makes it an ideal technique for trace and ultratrace analysis. However, the sensitivity can be a hindrance when determining sample components of widely diverse ion concentrations. Omnirange allows the analyst to selectively reduce the sensitivity for specific individual analyte masses directly from the system computer. Normally, an analyst would have to bring signals within the working range by sample dilution. With Omnirange, sample dilutions can be avoided. The user can bring off-scale peak mass values into the useful working range simply by indicating which mass values are to have reduced sensitivity. Omnirange extends the useful upper limits of the technique without sacrificing sensitivity and performance.

VG Isotopes Ltd are another leading manufacturer of inductively coupled plasma mass spectrometers. The special features of their VG Plasmaquad PQ2 include a multi-channel analyser which ensures rapid data acquisition over the whole mass range. The multi-channel analyser facilities include 4096 channels, 300 m facility for spectral analysis, user-definable number of measurements per peak in peak jumping mode and the ability to monitor data as it is acquired. A multi-channel analyser is imperative to acquire short-lived signals from accessories such as flow injection, electrothermal vaporisation, laser ablation, etc. or for fast multi-element survey scans (typically 1 min.).

A variant of the Plasmaquad PQ2 is the Plasmaquad PQ2 plus instrument. This latter instrument has improved detector technology which incorporates a multimode system which can measure higher concentrations of elements without compromising the inherent sensitivity of the instrument. This extended dynamic range system (Table 1.13) produces an improvement in effective linear dynamic range to eight orders of magnitude. Hence, traces at microgram per litre levels can be measured in the same analytical sequence as major constituents.

1.1.3.4 Atomic emission spectrometry

This technique has been used for the determination of aluminium, barium, molybdenum, strontium, tellurium, terbium, thallium, thorium beryllium, tamerium, cadmium, chromium, cobalt, copper, iron, lead, nickel and manganese in soils. No work appears to have been done on the analysis of sediments or sludges.

Table 1.13 Dynamic ranges of various techniques

Graphite furnace AAS	0.1 µg L^{-1} to 1 mg L^{-1}
Plasmaquad PQ2	0.1 µg L^{-1} to 10 mg L^{-1}
Inductively coupled plasma–atomic absorption spectrometry	10.1 µg L^{-1} to 1000 mg L^{-1}
Plasmaquad PQ2 plus	0.1 µg L^{-1} to 1000 mg L^{-1}

Source: Own files

1.1.3.5 Plasma emission spectrometry

Applications of this technique are limited to the determination of beryllium in soil and heavy metals, and aluminium in non-saline sediments.

1.1.3.6 Spark source mass spectrometry

Again, applications are limited to the determination in soil of antimony, caesium, rubidium, tin, chromium and nickel, and of radium in non-saline sediments.

1.1.3.7 Glow discharge mass spectrometry

Duckwork analysed soils by using glow discharge mass spectrometry [51]. The method produced variable results for some metals but was generally suitable for rapid–screening applications. Sample preparation only required grinding, drying, and mixing the sample with a conducting host material prior to electrode formation.

1.1.3.8 Laser-induced breakdown spectroscopy

Laser-induced breakdown spectroscopy has been used [52] for the identification of metals on soil surfaces. Detection limits ranged from 10–300 ppm.

1.1.4 Neutron activation analysis

Due to the complexity and cost of the technique no environmental laboratory in the UK has its own facility for carrying out neutron activation analysis. Instead, samples are sent to one of the organisations that possess the facilities, e.g. the Atomic Energy Research Establishment at Harwell, or the Joint Manchester–Liverpool University Reactor located at Risley.

As mentioned above, the technique is extremely sensitive and tends to be used when a referee analysis is required on a material which has become a standard for checking out other methods. Another advantage of the technique is that a foreknowledge of the elements present is not essential. It can be used to indicate the presence and concentration of entirely unexpected elements, even when present at very low concentrations.

In neutron activation analysis, the sample in a suitable container, often a pure polyethylene tube, is bombarded with slow neutrons for a fixed time together with standards. Transmutations convert analyte elements into radioactive elements, which are either different elements or isotopes of the original analyte.

After removal from the reactor the product is subject to various counting techniques and various forms of spectrometry to identify the elements present and their concentration.

This is a very sensitive technique with numerous applications in the analysis of soils (Table 1.14(a)), marine sediments (Tables 1.14(b) and (c)) and sludges (Table 1.14(d)).

1.1.5 Photon activation analysis

This is a sister technique to neutron activation analysis which has found applications in the determination of metals in soils (Table 1.15 (a)), marine sediments (Table 1.15(b)) and sludges (Table 1.15(c)).

1.1.6 X-ray methods

1.1.6.1 X-ray fluorescence spectroscopy

This technique has applications in the analysis of soils (chromium, manganese, iron, cobalt, nickel, copper, cadmium, lead and selenium) and non-saline sediments (lead, cadmium, arsenic and uranium).
Raab *et al.* [58] have described a rapid method for the determination of metals in soils based on a field-portable X-ray fluorescence spectrometer.

1.1.6.1(a) Energy-dispersive X-ray fluorescence spectrometry

Energy-dispersive X-ray fluorescence (EDXRF) spectrometry is an instrumental analytical technique for non-destructive multi-elemental analysis. The use of modern-day technologies coupled with the intrinsic simplicity of X-ray fluorescence spectra (as compared for instance with optical emission (OE) spectra) means that the powerful EDXRF technique can be used routinely. The EDXRF spectrum for iron is a clearly resolved doublet while the optical emission spectrum contains more than 4000 lines. This simplicity is a direct consequence of the fact that XRF spectra are a result of inner shell electron transitions which are possible only between a limited number of energy levels for the relatively few electrons. Optical emission spectra, on the other hand, arise from electron transitions in the outer, valence shells which are closer together in energy, more populated than the inner shells and from which it is easier to promote electron transitions.

In order to generate X–ray spectra, we may excite the elements in the specimen with any one of the following:

- X-ray photons
- High-energy electrons
- High-energy charged particles
- Gamma rays
- Synchrotron radiation.

The term XRF is generally applied when X-ray photons are used to generate characteristic X-rays from the elements in the specimen. The most commonly used source of such X–rays (in the 2–100 keV range) are radioisotopes and X-ray tubes. An EDXRF spectrometer such as the Link Analytical XR300 (see Appendix 1) uses a compact low power (10–100 W typical) X-ray tube capable of delivery of X-ray photons with a maximum energy of 30 or 50 keV.

Why is the technique referred to as 'energy-dispersive' XRF?
The classical XRF spectrometer which has been commercially available since the 1959s uses crystal structures to separate (resolve) the X-rays emanating from the fluorescence process in the irradiated specimen. These crystals diffract the characteristic X-rays from the elements in the specimen, allowing them to be separated and measured. The characteristic fluorescent X-rays are said to have been separated from each other by the process of 'wavelength disperser' (WDXRF). Each element emits characteristic lines which can be separated by WDXRF before being individually counted. For each line and diffracting crystal, we can set a detector at a particular angle (from the Bragg equation) and collect X-rays, which are primarily from the selected element.

The EDXRF system uses the Si(Li) (lithium-drifted silicon) detector to simultaneously collect all X-ray energies from the specimen. Each detected X-ray photon gives rise to a signal from the detector. The magnitude of this signal is proportional to the energy of the detected X-ray and when amplified and digitised can be passed to a multi-channel analyser which displays a histogram of number of X-rays (intensity) against energy. The incident photons, therefore, have been electronically separated (dispersed) according to their energy. The energy of each of the X-rays from all the elements is readily accessible from published tables.

Due to the simple spectra and the extensive element range (sodium upwards) which can be covered using the Si(Li) detector and a 50kV X-ray tube, EDXRF spectrometry is perhaps unparalleled for its quantitative element analysis power.

Table 1.14(a) Metallic and metalloid elements in Periodic Table determined by neutron activation analysis in soil

Group / Period	I	II											III	IV	V	VI	VII	O
1	^{1}H																	^{2}He
2	^{3}Li	^{4}Be											^{5}B	^{6}C	^{7}N	^{8}O	^{9}F	^{10}Ne
3	^{11}Na	^{12}Mg											^{13}Al√	^{14}Si	^{15}P	^{16}S	^{17}Cl	^{18}Ar
4	^{19}K	^{20}Ca	^{21}Sc√	^{22}Ti	^{23}V	^{24}Cr√	^{25}Mn	^{26}Fe√	^{27}Co√	^{28}Ni	^{29}Cu√	^{30}Zn	^{31}Ga	^{32}Ge	^{33}As√	^{34}Se√	^{35}Br√	^{36}Kr
5	^{37}Rb	^{38}Sr√	^{39}Y	^{40}Zr√	^{41}Nb	^{42}Mo√	^{43}Tc	^{44}Ru	^{45}Rh	^{46}Pd	^{47}Ag√	^{48}Cd	^{49}In√	^{50}Sn	^{51}Sb√	^{52}Te	^{53}I	^{54}Xe
6	^{55}Cs	^{56}Ba	^{57}La√	^{72}Hf√	^{73}Ta√	^{74}W	^{75}Re	^{76}Os	^{77}Ir√	^{78}Pt√	^{79}Au√	^{80}Hg√	^{81}Tl	^{82}Pb	^{83}Bi	^{84}Po	^{85}At	^{86}Rn
7	^{87}Fr	^{88}Ra	^{89}Ac															

Lanthanides	^{58}Ce√	^{59}Pr	^{60}Nd	^{61}Pm	^{62}Sm	^{63}Eu√	^{64}Gd√	^{65}Tb√	^{66}Dy√	^{67}Ho√	^{68}Er√	^{69}Tm√	^{70}Yb√	^{71}Lu√
Actinides	^{90}Th√	^{91}Pa	^{92}U√	^{93}Np	^{94}Pu	^{95}Am	^{96}Cm	^{97}Bk	^{98}Cf	^{99}Es	^{100}Fm	^{101}Md	^{102}No	^{103}Lr

Total number of metals and metalloids 86 (include actinides and lanthanides).
Number determinable by neutron activation analysis [21] (indicated by tick)

Source: Own files

Table 1.14(b) Metallic and metalloid elements in Periodic Table determined by neutron activation analysis in non-saline sediments

Group / Period	I	II											III	IV	V	VI	VII	0
1	^{1}H																^{1}H	^{2}He
2	^{2}Li	^{4}Be											^{5}B	^{6}C	^{7}N	^{8}O	^{9}F	^{10}Ne
3	^{11}Na√	^{12}Mg√											^{13}Al√	^{14}Si	^{15}P	^{16}S	^{17}Cl√	^{18}Ar
4	^{19}K√	^{20}Ca√	^{21}Sc√	^{22}Ti√	^{23}V√	^{24}Cr√	^{25}Mn√	^{26}Fe√	^{27}Co√	^{28}Ni√	^{29}Cu	^{30}Zn√	^{31}Ga√	^{32}Ge	^{33}As√	^{34}Se√	^{35}Br√	^{36}Kr
5	^{37}Rb	^{35}Sr√	^{39}Y	^{40}Zr√	^{41}Nb	^{42}Mo√	^{43}Tc	^{44}Ru√	^{45}Rh	^{46}Pd√	^{47}Ag√	^{48}Cd	^{49}In√	^{50}Sn√	^{51}Sb√	^{52}Te	^{53}I	^{54}Xe
6	^{55}Cs√	^{50}Ba√	^{57}La√	^{72}Hf√	^{73}Ta√	^{74}W√	^{76}Re	^{76}Os√	^{77}Ir√	^{78}Pt√	^{79}Au√	^{80}Hg√	^{81}Tl	^{82}Pb√	^{83}Bi	^{84}Po	^{85}At	^{86}Rn
7	^{87}Fr	^{88}Ra	^{89}Ac															

Lanthanides	^{58}Ce√	^{59}Pr	^{60}Nd√	^{61}Pm	^{62}Sm√	^{63}Eu√	^{64}Gd√	^{65}Tb√	^{66}Dy√	^{67}Ho√	^{68}Er	^{69}Tm√	^{70}Yb√	^{71}Lu√
Actinides	^{90}Th√	^{91}Pa	^{92}U√	^{93}Np	^{94}Pu	^{95}Am	^{96}Cm	^{97}Bk	^{98}Cf	^{99}Es	^{100}Fm	^{101}Md	^{102}No	^{103}Lr

Total number of metals and metalloids 86 (include actinides and lanthanides).
Number determinable by neutron activation analysis [48] (indicated by tick)

Source: Own files

Table 1.14(c) Metallic and metalloid elements in Periodic Table determined by neutron activation analysis in marine sediments

Group Period	I	II											III	IV	V	VI	VII	0
1	¹H																¹H	²He
2	²Li	⁴Be											⁵B	⁶C	⁷N	⁸O	⁹F	¹⁰Ne
3	¹¹Na	¹²Mg											¹³Al	¹⁴Si	¹⁵P	¹⁶S	¹⁷Cl	¹⁸Ar
4	¹⁹K	²⁰Ca	²¹Sc√	²²Ti√	²³V	²⁴Cr√	²⁵Mn	²⁶Fe√	²⁷Co√	²⁸Ni	²⁹Cu	³⁰Zn	³¹Ga	³²Ge	³³As	³⁴Se	³⁵Br√	³⁶Kr
5	³⁷Rb√	³⁸Sr√	³⁹Y	⁴⁰Zr	⁴¹Nb	⁴²Mo	⁴³Tc	⁴⁴Ru	⁴⁵Rh	⁴⁶Pd	⁴⁷Ag√	⁴⁸Cd	⁴⁹In	⁵⁰Sn√	⁵¹Sb√	⁵²Te	⁵³I	⁵⁴Xe
6	⁵⁵Cs√	⁵⁶Ba	⁵⁷La	⁷²Hf	⁷³Ta	⁷⁴W	⁷⁵Re	⁷⁶Os	⁷⁷Ir	⁷⁸Pt	⁷⁹Au	⁸⁰Hg	⁸¹Tl	⁸²Pb	⁸³Bi	⁸⁴Po	⁸⁵At	⁸⁶Rn
7	⁸⁷Fr	⁸⁸Ra	⁸⁹Ac															

Lanthanides	⁵⁸Ce	⁵⁹Pr	⁶⁰Nd	⁶¹Pm	⁶²Sm	⁶³Eu	⁶⁴Gd	⁶⁵Tb	⁶⁶Dy	⁶⁷Ho	⁶⁸Er	⁶⁹Tm	⁷⁰Yb	⁷¹Lu
Actinides	⁹⁰Th	⁹¹Pa	⁹²U	⁹³Np	⁹⁴Pu	⁹⁵Am	⁹⁶Cm	⁹⁷Bk	⁹⁸Cf	⁹⁹Es	¹⁰⁰Fm	¹⁰¹Md	¹⁰²No	¹⁰³Lr

Total number of metals and metalloids 86 (include actinides and lanthanides).
Number determinable by neutron activation analysis [9] (indicated by tick)

Source: Own files

Table 1.14(d) Metallic and metalloid elements in Periodic Table determined by neutron activation analysis in sludges

Group / Period	I	II											III	IV	V	VI	VIII	0
1	^{1}H																^{1}H	^{2}He
2	^{2}Li	^{4}Be											^{5}B	^{6}C	^{7}N	^{8}O	^{9}F	^{10}Ne
3	^{11}Na√	^{12}Mg√											^{13}Al√	^{14}Si√	^{15}P	^{16}S	^{17}Cl	^{18}Ar
4	^{19}K√	^{20}Ca√	^{21}Sc√	^{22}Ti√	^{23}V√	^{24}Cr√	^{25}Mn√	^{26}Fe√	^{27}Co√	^{28}Ni√	^{29}Cu√	^{30}Zn√	^{31}Ga	^{32}Ge	^{33}As√	^{34}Se√	^{35}Br√	^{36}Kr
5	^{37}Rb√	^{38}Sr√	^{39}Y√	^{40}Zr√	^{41}Nb	^{42}Mo√	^{43}Tc	^{44}Ru	^{45}Rh	^{46}Pd	^{47}Ag√	^{48}Cd√	^{49}In√	^{50}Sn√	^{51}Sb√	^{52}Te√	^{53}I	^{54}Xe
6	^{55}Cs√	^{56}Ba√	^{57}La√	^{72}Hf√	^{73}Ta√	^{74}W√	^{75}Re√	^{76}Os√	^{77}Ir√	^{78}Pt	^{79}Au√	^{80}Hg√	^{81}Tl√	^{82}Pb√	^{83}Bi√	^{84}Po	^{85}At	^{86}Rn
7	^{87}Fr	^{88}Ra	^{89}Ac															

Lanthanides	^{58}Ce√	^{59}Pr	^{60}Nd√	^{61}Pm	^{62}Sm√	^{63}Eu√	^{64}Gd√	^{65}Tb√	^{66}Dy√	^{67}Ho	^{68}Er	^{69}Tm	^{70}Yb√	^{71}Lu√
Actinides	^{90}Th√	^{91}Pa	^{92}U√	^{93}Np	^{94}Pu	^{95}Am	^{96}Cm	^{97}Bk	^{98}Cf	^{99}Es	^{100}Fm	^{101}Md	^{102}No	^{103}Lr

Total number of metals and metalloids 86 (include actinides and lanthanides).
Number determinable by neutron activation analysis [52] (indicated by tick)

Source: Own files

Table 1.15(a) Metallic and metalloid elements in Periodic Table determined by photon activation analysis in soil

Group / Period	I	II											III	IV	V	VI	VII	0
1	¹H																¹H	²He
2	²Li	⁴Be											⁵B	⁶C	⁷N	⁸O	⁹F	¹⁰Ne
3	¹¹Na	¹²Mg√											¹³Al√	¹⁴Si	¹⁵P	¹⁶S	¹⁷Cl	¹⁸Ar
4	¹⁹K√	²⁰Ca√	²¹Sc√	²²Ti√	²³V	²⁴Cr√	²⁵Mn	²⁶Fe√	²⁷Co√	²⁸Ni√	²⁹Cu	³⁰Zn√	³¹Ga	³²Ge	³³As√	³⁴Se√	³⁵Br	³⁶Kr
5	³⁷Rb	³⁸Sr√	³⁹Y	⁴⁰Zr√	⁴¹Nb	⁴²Mo	⁴³Tc	⁴⁴Ru	⁴⁵Rh	⁴⁶Pd	⁴⁷Ag	⁴⁸Cd	⁴⁹In	⁵⁰Sn	⁵¹Sb√	⁵²Te√	⁵³I	⁵⁴Xe
6	⁵⁵Cs	⁵⁶Ba	⁵⁷La	⁷²Hf	⁷³Ta	⁷⁴W	⁷⁵Re	⁷⁶Os	⁷⁷Ir	⁷⁸Pt	⁷⁹Au	⁸⁰Hg	⁸¹Tl	⁸²Pb√	⁸³Bi√	⁸⁴Po	⁸⁵At	⁸⁶Rn
7	⁸⁷Fr	⁸⁸Ra	⁸⁹Ac															

Lanthanides	⁵⁸Ce	⁵⁹Pr	⁶⁰Nd	⁶¹Pm	⁶²Sm	⁶³Eu	⁶⁴Gd	⁶⁵Tb	⁶⁶Dy	⁶⁷Ho	⁶⁸Er	⁶⁹Tm	⁷⁰Yb	⁷¹Lu
Actinides	⁹⁰Th	⁹¹Pa	⁹²U√	⁹³Np	⁹⁴Pu	⁹⁵Am	⁹⁶Cm	⁹⁷Bk	⁹⁸Cf	⁹⁹Es	¹⁰⁰Fm	¹⁰¹Md	¹⁰²No	¹⁰³Lr

Total number of metals and metalloids 86 (include actinides and lanthanides).
Number determinable by neutron activation analysis [16] (indicated by tick)

Source: Own files

Table 1.15(b) Metallic and metalloid elements in Periodic Table determined by photon activation analysis (PAA) in marine sediments

Group / Period	I	II										III	IV	V	VI	VIII	0
1	$_{1}$H															$_{1}$H	$_{2}$He
2	$_{3}$Li	$_{4}$Be										$_{5}$B	$_{6}$C	$_{7}$N	$_{8}$O	$_{9}$F	$_{10}$Ne
3	$_{11}$Na√	$_{12}$Mg√										$_{13}$Al	$_{14}$Si	$_{15}$P	$_{16}$S	$_{17}$Cl	$_{18}$Ar
4	$_{19}$K√	$_{20}$Ca√ $_{21}$Sc	$_{22}$Ti√	$_{23}$V	$_{24}$Cr√	$_{25}$Mn	$_{26}$Fe√	$_{27}$Co√	$_{28}$Ni√	$_{29}$Cu	$_{30}$Zn√	$_{31}$Ga√	$_{32}$Ge	$_{33}$As√	$_{34}$Se	$_{35}$Br	$_{36}$Kr
5	$_{37}$Rb	$_{38}$Sr√ $_{39}$Y	$_{40}$Zr√	$_{41}$Nb	$_{42}$Mo	$_{43}$Tc	$_{44}$Ru	$_{45}$Rh	$_{46}$Pd	$_{47}$Ag	$_{48}$Cd	$_{49}$In	$_{50}$Sn	$_{51}$Sb√	$_{52}$Te	$_{53}$I	$_{54}$Xe
6	$_{55}$Cs	$_{56}$Ba√ $_{57}$La	$_{72}$Hf	$_{73}$Ta	$_{74}$W	$_{75}$Re	$_{76}$Os	$_{77}$Ir	$_{78}$Pt	$_{79}$Au	$_{80}$Hg	$_{81}$Tl	$_{82}$Pb√	$_{83}$Bi	$_{84}$Po	$_{85}$At	$_{86}$Rn
7	$_{87}$Fr	$_{88}$Ra $_{89}$Ac															

Lanthanides	$_{58}$Ce	$_{59}$Pr	$_{60}$Nd	$_{61}$Pm	$_{62}$Sm	$_{63}$Eu	$_{64}$Gd	$_{65}$Tb	$_{66}$Dy	$_{67}$Ho	$_{68}$Er	$_{69}$Tm	$_{70}$Yb	$_{71}$Lu
Actinides	$_{90}$Th	$_{91}$Pa	$_{92}$U√	$_{93}$Np	$_{94}$Pu	$_{95}$Am	$_{96}$Cm	$_{97}$Bk	$_{98}$Cf	$_{99}$Es	$_{100}$Fm	$_{101}$Md	$_{102}$No	$_{103}$Lr

Total number of metals and metalloids 86 (include actinides and lanthanides).
Number determinable by PAA [17] (indicated by tick)

Source: Own files

Table 1.15(c) Metallic and metalloid elements in Periodic Table determined by photon activation analysis (PAA) in sludges

Group / Period	I	II	(transition metals)	III	IV	V	VI	VIII	0
1	^{1}H							^{1}H	^{2}He
2	^{2}Li	^{4}Be		^{5}B	^{6}C	^{7}N	^{8}O	^{9}F	^{10}Ne
3	^{11}Na√	^{12}Mg√		^{13}Al	^{14}Si√	^{15}P	^{16}S	^{17}Cl	^{18}Ar
4	^{19}K√	^{20}Ca√	^{21}Sc√ ^{22}Ti√ ^{23}V√ ^{24}Cr√ ^{25}Mn√ ^{26}Fe√ ^{27}Co√ ^{28}Ni√ ^{29}Cu ^{30}Zn√	^{31}Ga√	^{32}Ge	^{33}As√	^{34}Se	^{35}Br	^{36}Kr
5	^{37}Rb	^{38}Sr√	^{39}Y√ ^{40}Zr√ ^{41}Nb ^{42}Mo√ ^{43}Tc ^{44}Ru ^{45}Rh ^{46}Pd ^{47}Ag√ ^{48}Cd√	^{49}In√	^{50}Sn√	^{51}Sb√	^{52}Te√	^{53}I	^{54}Xe
6	^{55}Cs√	^{56}Ba√	^{57}La ^{72}Hf ^{73}Ta ^{74}W ^{75}Re ^{76}Os ^{77}Ir ^{78}Pt ^{79}Au ^{80}Hg√	^{81}Tl√	^{82}Pb√	^{83}Bi√	^{84}Po	^{85}At	^{86}Rn
7	^{87}Fr	^{88}Ra	^{89}Ac						

Lanthanides	^{58}Ce√	^{59}Pr	^{60}Nd	^{61}Pm	^{62}Sm	^{63}Eu	^{64}Gd	^{65}Tb	^{66}Dy	^{67}Ho	^{68}Er	^{69}Tm	^{70}Yb	^{71}Lu
Actinides	^{90}Th	^{91}Pa	^{92}U	^{93}Np	^{94}Pu	^{95}Am	^{96}Cm	^{97}Bk	^{98}Cf	^{99}Es	^{100}Fm	^{101}Md	^{102}No	^{103}Lr

Total number of metals and metalloids 86 (include actinides and lanthanides).
Number determinable by PAA [34] (indicated by tick)

Source: Own files

Qualitative analysis is greatly simplified by the presence of few peaks which occur in predictable positions and by the use of tabulated element/line markers which are routinely available from the computer-based analyser.

To date, the most successful method of combined background correction and peak deconvolution is to use the method of digital filtering and least squares (FLS) fitting of reference peaks to the unknown spectrum [26]. This method is robust, simple to automate and is applicable to any sample type.

The combination of the digital filtering and least squares peak deconvolution method and empirical correction procedures has application throughout elemental analysis. This approach is suitable for specimens of all physical types and is used in a wide selection of industrial applications.

1.1.6.1(b) Total reflection X-ray fluorescence spectrometry

The major disadvantage of conventional energy dispersive X-ray fluorescence spectrometry has been poor elemental sensitivity, a consequence of high background noise levels resulting mainly from instrumental geometrics and sample matrix effects. Total reflection X-ray fluorescence (TXRF) is a relatively new multi-element technique with the potential to be an impressive analytical tool for trace-elemental determinations for a variety of sample types. The fundamental advantage of TXRF is its capability to detect elements in the picogram range in comparison to the nanogram levels typically achieved by traditional energy-dispersive X-ray fluorescence spectrometry.

The problem in detecting atoms in the ng L^{-1} or sub-µg L^{-1} level is basically one of being able to obtain a signal which can be clearly distinguished from the background. The detection limit being given typically as the signal which is equivalent to three times the standard deviation of the background counts for a given unit of time. In energy-dispersive X-ray fluorescence spectrometry the background is essentially caused by interactions of radiation with matter resulting from an intense flux of elastic and Compton scattered photons. The background especially in the low-energy region (0–20 keV) is due in the main to Compton scattering of high-energy Bremsstrahlung photons from the detector crystal itself. In addition, impurities on the specimen support will contribute to the background. The Auger effect does not contribute to an increased background, as the emitted electrons, of different but low energy, are absorbed either in the beryllium foil of the detector entrance windows or in the air path of the spectrometer.

A reduction in the spectral background can be effectively achieved by X-ray total reflection at the surface of a smooth reflector material such as quartz. X-ray total reflection occurs when an X-ray beam impinges on a surface at less than the critical angle of total reflection. If a collimated X-ray beam impinges onto the surface of a plane smooth and polished reflector at an angle less than the critical angle, then total reflection occurs. In this case the angle of incidence is equal to the angle of reflection and the intensities of the incident and totally reflected beams should be equal.

The principles of TXRF were first reported by Yaneda and Horiuchi [27] and further developed by Aiginger and Wodbrauschek [28]. In TXRF the exciting primary X-ray beam impinges upon the specimen prepared as a thin film on an optically flat support at angles of incidence in the region of 2 to 5 minutes of arc below the critical angle. In practice the primary radiation does not (effectively) enter the surface of the support but skims the surface, irradiating any sample placed on the support surface. The scattered radiation from the sample support is virtually eliminated, thereby drastically reducing the background noise. A further advantage of the TXRF system, resulting from the new geometry used, is that the solid-state energy-dispersive detector can be accommodated very close to the sample (0.3 mm), which allows a large solid angle of fluorescent X-ray collection, thus enhancing a signal sensitivity and enabling the analysis to be carried out in air at atmospheric pressure.

The sample support or reflector is a 3 cm diameter wafer made of synthetic quartz or perspex. The water sample can be placed directly onto the surface. The simplest way to prepare liquid samples is to pipette volumes between 1 and 50 µL directly onto a quartz reflector and allow them to dry. For aqueous solutions the reflector can be made hydrophobic (e.g. by silicon treatment) in order to hold the sample in the centre of the plate. Suitable elements for calibration can be achieved by a simple standard addition technique.

Since Yaneda and Horiuchi [27] first reported the use of TXRF various versions have been developed [29–32]. Recently an X-ray generator with a fine focus tube and multiple reflection optics has been developed by Seifert & Co and coupled with an energy-dispersive spectrometer fitted with an Si(Li) detector and multi-channel analyser supplied by Link Analytical. The new system, which will be described later, known as the EXTRA II, represents the first commercially available TXRF instrument.

An example of the detection limits achieved by the Link Analytical EXTRA III (3 σ above background, counting timer 1000 s), are shown in Fig. 1.9 for the molybdenum anode X-ray tube and for excitation with the filtered Bremsstrahlung spectrum from a tungsten X-ray tube.

The data shown was obtained from diluted aqueous solutions which can be considered to be virtually free from any matrix effects. A detection limit of 10 pg for the 10 µL sample corresponds to a concentration of 1 µg L^{-1}. A linear dynamic range of four orders of magnitude is obtained for most elements: for example, lead at concentrations 2–20,000 µg L^{-1} using cobalt as an internal standard at 2000 µg L^{-1}.

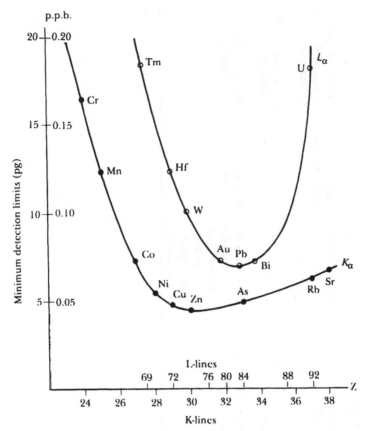

Fig. 1.9 Minimum detection limits of the TXRF spectrometer (Mo-tube 13 mA/60 kV; counting time 1000 s).
Source: Own files

The attractive features of TXRF can be summarised as follows:

- an inherent universal calibration curve is obtained as a smooth function of atomic number
- the use of internal single-element standardisation eliminates the need for matrix-matched external standards
- only small sample volumes are required (5–50 µL).
- the technique requires only a simple sample preparation methodology.

The attractive features of the TXRF technique outlined above suggest that TXRF has the potential to become a very powerful analytical tool for trace-elemental determinations applicable to a wide range of matrix types and may, indeed, compete with the inductively coupled plasma mass spectrometry.

Table 1.16 Energy dispersive and total reflection X-ray fluorescence spectrometers

Supplier	Model types	Type	Sample overlaps	Computer	Handling of peak	Element range	Detector
Link Analytical	XR 200/300	Energy dispersive	Solid and liquid	Yes	Filtered least squares technique	Atomic numbers 15–55 (Mo–Ba)	10 mm² 155 eV resolution Si (Li) detector
Pye Instruments (Philips)	PW 1404	Energy dispersive sequential	Solids and liquids	Yes	Various techniques	B–U	Argon flow scintillation
Seifert	Extra III	Energy dispersive and multiple	Solids and liquids	Yes	No correction for matrix effects required except for those in the range sodium to phosphorus	All elements beyond sodium	80-mm² 165 eV Si(Li) detector

Source: Own files

Instrumentation in energy and total reflection dispersive X–ray fluorescence spectrometry

Various suppliers of instruments are listed in Table 1.16.

Philips PW 1404 energy-dispersive X–ray fluorescence spectrometers
The Philips PW 1404 is a powerful versatile sequential X-ray spectrometer system developed from the PW 1400 series and incorporating many additional hardware and software features that further extend its performance.

All system functions are controlled by powerful microprocessor electronics, which make routine analysis a simple push-button exercise and provide extensive safeguards against operator error. The microprocessor also contains sufficient analytical software to permit stand-alone emergency operation, plus a range of self-diagnostic service-testing routines.

The main characteristics of this instrument are as follows:

- Identifies all elements from boron to uranium
- Choice of side window X-ray tubes allows optimum excitation for all applications
- New detectors and crystals bring improved light element performance
- 100 kv programmable excitation enhances heavy element detection
- Special calibration features give more accurate results:
    - auxiliary collimator provides high resolution
    - programmable channel mask reduces background
    - fast digital scanning speeds data collection
- Powerful software includes automatic peak labelling
- Compact one-cabinet system
- Distributed intelligence via five microprocessors
- High-frequency generator cuts running costs and improves stability
- New high-speed electronics allows operation at one million counts/s
- System self-selects analytical programs for unknowns
- Surface down sample presentation aids accurate analysis of liquids
- Small airlock speeds sample throughput, cuts helium costs
- Designed for laboratory automation
- Extended microprocessor control ensures simple, error-free operation
- Front panel continuously displays system status
- New generation software for DEC computers
- Computer dialogue in English, French, German, Spanish
- Colour graphics simplify results interpretation
- Extensive programming, reporting, editing facilities available.

The layout of the Philips PW 1404 instrument is shown in Fig. 1.10.

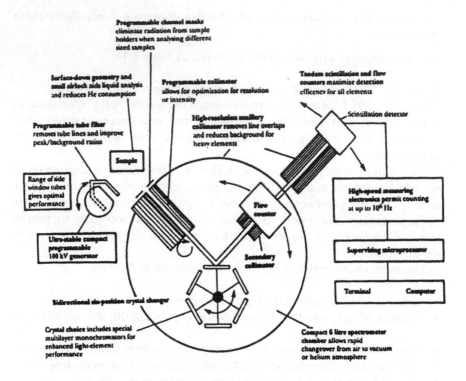

Fig. 1.10 Layout of Philips PW 1404 energy-dispersive X-ray fluorescence spectrometer
Source: Own files

Unique among XRF instruments the Extra II TXRF spectrometer yields lower limits of detection in the region of 10 pg (1 pg = 10^{-12}g) for more than sixty elements (Table 1.17). All elements upwards from sodium (Z = 11) in the periodic table may be determined. The inclusion of twin excitation sources, which may be switched electronically within a few seconds, assures optimum sensitivity for all detectable elements. The applicable concentration range is from per cent to below 1 µg L^{-1}. As little as 1 µg of sample is sufficient to determine elements at the milligram per litre level; calibration is necessary only once and is carried out during installation. The calibration will remain unchanged for a period of at least 12 months. Quantitative analysis is simple and uses the method of internal standardisation. No external standards are necessary. The method requires no correction of matrix effects for all elements except those in the range sodium to phosphorus. Empirical absorption-enhancement correction models may be applied to these light elements. Sample preparation for solutions and dispersions is very simple, requiring only a micropipette. Complete digestion of materials is not mandatory. Finely divided powders may be analysed providing they are homogeneous.

Table 1.17 Detection limits Seifert Extra II X-ray spectrometer

	Detection limit (pg)
Atomic numbers 18–38 (argon to strontium) and 53–57 (iodine–lanthanum) and 78–83 (osmium to bismuth)	<5
Chlorine atomic numbers 39–43 (yttrium to technetium), 47–52 (silver to tellurium), 65–71 (terbium to lutecium) and 90–92 (thorium to uranium)	5–10
Phosphorus (15) Sulphur (16) Ruthenium (44) Rhodium (45) Palladium (46) Neptunium (93) Plutonium (94)	10–30
Aluminium (13) Silicon (14)	30–100
Sodium (11) Magnesium (12	>100

Source: Own files

The Seifert EXTRA II total reflection X-ray fluorescence spectrometer has been applied to the determination of a range of elements in sea-water sediments and particulate matter [33]. Detection limits between 5 and 20 ng L^{-1} were achieved for most elements.

Berwick *et al.* [53] have reported a rapid on-site method for screening hazardous metallic wastes based on the use of a field-portable X-ray fluorescence analyser.

1.1.6.2 X-ray secondary emission spectrometry

This technique has been found to have limited applications in the analysis of soils (nickel and palladium), non-saline, marine and estuarine sediments. Thus heavy metals (cadmium, chromium, cobalt, lead, iron, manganese, nickel, zinc and copper) also arsenic, rubidium and strontium have been determined in non-saline sediments, six heavy metals (chromium, copper, iron, manganese, nickel and zinc) also calcium, aluminium, silicon, potassium and titanium in marine sediments and the same heavy metals plus calcium, aluminium, potassium, titanium and silicon in estuarine sediments. Total reflection X-ray spectrometry has been applied to studies of the speciation of metals in soils [54].

1.1.6.3 Electron probe microanalysis

This technique has been employed to measure the distribution of copper and zinc on the surface of soil and marine sediment particles. Electron probe microanalysis is a technique for identifying the nature of metallic inclusions in solid particles. The sample is bombarded with a very narrow beam of X-rays of known frequency and the back scattered electron radiation is examined. An image is produced of the distribution of elements of any particular atomic number. The emitted X-rays are detected and counted by a proportional counter to give a pulse height distribution curve.

Even though X-ray fluorescence is now widely used to analyse a large variety of samples, it does have some drawbacks. For instance the X-ray beam used is wide and this is of no great use for analysing tiny inclusions present in samples, and also does not allow point-by-point analysis on surfaces ('scanning analysis'). The first electron probe X-ray primary emission spectrometer was built in 1949. No doubt this encouraged the use of surface analyses, by allowing samples of very small dimensions to be studied. This was possible because the electron beam had a diameter of only about 1 μm. Although the small size of this beam permitted the analysis of some micro-inclusions in samples, and also multiple analyses by scanning, the main problem, which still remains unsolved, is that of the microhomogeneity and microtopography of the samples. Thus, whereas polishing the solid samples with a 30–100 μm grade abrasive is usually satisfactory for X-ray fluorescence spectrometry, a 0.25 μm grade abrasive or finer may be required for electron-probe microanalysis.

In principle, the difference between X-ray fluorescence spectrometry and electron-probe microanalysis lies in the fact that the analytical information is provided, in the first case, by secondary fluorescence X-rays, and in the second by primary X-rays, emitted as a result of the impact of the electron beam on the sample's electrons.

Owing to the small size of the electron beam on the one hand, and to the high sensitivity of the method on the other (a sensitivity which can go down to detection of 10^{-16} g), electron-probe microanalysis has found applications in many fields.

Some of the disadvantages of the electron-probe method may be overcome, as in other methods, by the use of complementary techniques. Such techniques can complete the results obtained by electron microprobe. For instance, the introduction of a proton microprobe [34] has led to a technique which is much more sensitive (by two orders of magnitude) than the electron microprobe, and may be used with very good results in geochemical and cosmo-chemical studies.

1.1.7 Stable isotope dilution methods

This technique has very limited applications *viz* the determination in soils of barium, caesium, calcium, potassium, rubidium and strontium. Finnigan MAT manufacture a range of mass spectrometric equipment for isotope ratio analysis.

The Thermionic Quadrupole (THQ™) mass spectrometer provides isotope ratio determination and trace-element analysis by stable isotope dilution. It allows the analysis of small samples with high sensitivity and precision, yet at a very low price.

1.1.8 Chromatographic procedures

1.1.8.1 Gas chromatography

Again, this technique has found limited applications in the analyses of soils (plutonium) and of non-saline sediments for elements which form volatile hydrides (*viz* antimony, arsenic, selenium also mercury), and for determining arsenic and selenium in marine sediments.

The basic requirements of a high–performance gas chromatograph are as follows:

- Sample is introduced to the column in an ideal state, i.e. uncontaminated by septum bleed or previous sample components, without modification due to distillation effects in the needle and quantitatively, i.e. without hold-up or adsorption prior to the column;
- The instrument parameters that influence the chromatographic separation are precisely controlled;
- Sample components do not escape detection, i.e. highly sensitive, reproducible detection and subsequent data processing are essential.

There are two types of separation column used in gas chromatography – capillary columns and packed columns.

Packed columns are still used extensively, especially in routine analysis. They are essential when sample components have high partition coefficients and/or high concentrations. Capillary columns provide a high number of theoretical plates, hence a very high resolution, but they cannot be used in all applications because there are not many types of chemically bonded capillary columns. Combined use of packed columns of different polarities often provides better separation than with a capillary column. It sometimes happens that a capillary column is used as a supplement in the packed-column gas chromatograph. It is best, therefore, to house the capillary and packed columns in the same column oven and use them routinely and the capillary column is used when more detailed information is required.

Conventionally, it is necessary to use a dual column flow line in packed-column gas chromatography to provide sample and reference gas flows. The electronic base-line drift compensation system allows a simple column flow line to be used reliably.

Advances in capillary column technology presume stringent performance levels for the other components of a gas chromatograph as column performance is only as good as that of the rest of the system. One of the most important factors in capillary column chromatography is that a high repeatability of retention times must be ensured even under adverse ambient conditions. These features combine to provide ±0.01 min repeatability for peaks having retention times as long as 2 h (other factors being equal).

Another important factor for reliable capillary column gas chromatography is the sample injection method. Various types of sample injection ports are available. The split/splitless sample injection port unit series is designed so that the glass insert is easily replaced and the septum is continuously purged during operation. This type of sample injection unit is quite effective for the analysis of samples having high boiling point compounds as the major components.

In capillary column gas chromatography, it is often required to raise and lower the column temperature very rapidly and to raise the sample injection port temperature. In one design of gas chromatograph, the Shimadzu GC14–A, the computer-controlled flap operates to bring in the external air to cool the column oven rapidly – only 6 min from 500 °C to 100 °C. This computer-controlled flap also ensures highly stable column temperature when it is set to a near-ambient point. The lowest controllable column temperature is about 26 °C when the ambient temperature is 20 °C.

Some suppliers of gas chromatography are listed in Table 1.18.

Instrumentation

Shimadzu gas chromatographs
This is a typical high-performance gas chromatograph version (see Table 1.18 for further details).

The inner chamber of the oven has curved walls for smooth circulation of air, the radiant heat from the sample injection port units and the detector oven is completely isolated. These factors combine to provide demonstrably uniform temperature distribution. (The temperature variance in a column coiled in a diameter of 20 cm is less than ±0.75 °K at a column temperature of 250 °).

When the column temperature is set to a near ambient temperature, external air is brought into the oven via a computer-controlled flap, providing rigid temperature control stability. (The lowest controllable column temperature is 24 °C when the ambient temperature is 18 °C and the injection port temperature is 250 °C. The temperature fluctuation is less than ±0.1 K even when the column temperature is set at 50 °C.

This instrument features five detectors (Table 1.18). In the flame ionisation detector, the high-speed electrometer, which ensures a very low noise level, is best suited to trace analysis and fast analysis using a capillary column.

Samples are never decomposed in the jet, which is made of quartz.

Carrier gas, hydrogen, air and make-up gas are separately flow-controlled. Flow rates are read from the pressure flow-rate curves.

In the satellite system, one or more satellite gas chromatographs (GC–14 series) are controlled by a core gas chromatograph (e.g. GC–16A series). Since the control is made externally, the satellite gas chromatographs are not required to have control functions (the keyboard unit is not necessary).

When a GC–16A series gas chromatograph is used as the core, various laboratory-automation-orientated attachments such as bar-code reader and a magnetic-card reader become compatible: a labour-saving system can be built, in which the best operational parameters are automatically set. Each satellite gas chromatograph (GC–14A series) operates as an independent instrument when a keyboard unit is connected.

The IC card operated gas chromatography system consists of a GC–14A series gas chromatograph and a C–R5A Chromatopac data processor. All of the chromatographic and data processing parameters are automatically set simply by inserting the particular IC card. This system is very convenient when one GC system is used for the routine analysis of several different types of samples.

One of the popular trends in laboratory automation is to arrange for a personal computer to control the gas chromatograph and to receive data from the GC to be processed as desired. Bilateral communication is made via the RS–232C interface built in a GC–14A series gas chromatograph. A system can be built to meet requirements.

A multi-directional gas chromatography system (multi-stage column system) is effective for analysis of difficult samples and can be built up by connecting several column ovens, i.e. tandem GC systems, each of which has independent control functions such as for temperature programming.

The Shimadzu GC–15A and GC–16A systems are designed not only as independent high-performance gas chromatographs but also as core instruments (see above) for multi-gas-chromatography systems (i.e. several gas chromatographs in the laboratory linked to a central management system) or computerised laboratory automation systems. The GC–16A has a keyboard, the GC–15A does not. Other details of these instruments are given in Table 1.18. The Shimadzu GC–8A range of instruments do not have a range of built-in detectors but are ordered either as temperature programmed instruments with TCD, FID or FPD detectors or as isothermal instruments with TCD, FID or ECD detectors.

Table 1.18 Commercial gas chromatographs

Manufacturer	Model	Packed column	Capillary column	Detectors	Sample injection point system	Keyboard control	Link to computer	Visual display	Printer	Core instrument amenable to tap automation	Temperature programming/ isothermal	Cryogenic unit (sub-ambient chromatography)
Shimadzu	GC–14A	Yes	Yes	FID ECD FTD FPD TCD (all supplied)	1. Split-splitters 2. Glass insert for single column 3. Glass insert to dual column 4. Cool on column system unit 5. Moving needle system 6. Rapidly ascending temperature vaporiser	Yes	Yes	No	No	No	Yes/Yes	No
Shimadzu	GC–15A	Yes	Yes	FID ECD FTD FPD TCD	1. Split-splitters 2. Direct sample injection (capillary) column 3. Standard sample injector (packed column) 4. Moving precolumn system (capillary columns) 5. On column (capillary columns)	No	Yes	Yes	Yes	No	Yes/Yes	No

Instrument		Detectors	Injection						
Shimadzu–GC 16A	Yes	FID FCD FTD FPD TCD (all supplied)	Split–splitters	Yes	Yes	Yes	No	Yes/Yes	No
Shimadzu–GC 8A	Yes	FID FCD FPD TCD Single detector instruments detector chosen on purchase	1. Point for packed columns 2. Point for capillary columns 3. Split–splitters	Yes / Not built in	Optional No	Optional No	Optional No	Yes temp prgmg GC 8APT (TCD detector) GC 8APF (FID detector) GV 8APFD (FID detector) isothermal: GC 8AIT (TCD detector) GC 8AIF (FID detector) GC 8AIE (ECD detector)	No
Perkin–Elmer 8410	Yes	Single detector (chosen on purchase) FID ECD FTD FPD TCD	1. Flash vaporisation 2. Split-splitless injection 3. Manual or automatic gas sampling valves 4. Manual or automatic liquid sampling valves	No	No	No	No	Yes/Yes	Yes, down to −80°C

Table 1.18 continued

Manu-facturer	Model	Packed column	Capillary column	Detectors	Sample injection point system	Keyboard control	Link to computer	Visual display	Printer	Core instrument amenable to tap automation	Temperature programming/isothermal	Cryogenic unit (sub-ambient chromatography)
Perkin-Elmer	8420	No	Yes	Single detector (chosen on purchase) FID ECD FTD FPD TCD	1. Programmable temperature vaporiser 2. Split–splitless injector 3. Direct on column injector	No	No	No	No	No	Yes/Yes	Yes, down to –80°C
Perkin-Elmer	8400 and 8500	Yes	Yes	Dual detector instrument (detectors chosen from following) FID ECD FTD FPD TCD	Can be fitted with any combination of above injection systems	Yes	Yes	Yes	Yes GP 100 printer plotter)	Yes	Yes/Yes	Yes, down to –80°C
Perkin-Elmer	8700	Yes	Yes	FID ECD FPD TCD	1. Flash vaporisation 2. Split–splitless 3. Programmable temperature vaporiser	Yes	Yes	Yes	Yes	–	Yes/Yes	Yes

| Nordion | Micromat HRGC 412 | No | Yes | Dual simultaneous detector combinations from the following: FID ECD FTD Photoionisation Hall E.C. | 1. Split–splitless 2. On-column injector | Hall E.C. photo-ionisation dual detector instrument (Detectors chosen from above list) 4. Gas sampling valve 5. Liquid sampling valve | Yes | Yes | Yes | — | Yes/Yes | No |
| Siemens | SiChromat 1–4 (single oven) SiChromat 2–8 (dual oven for multi-dimensional GC) | Yes | Yes | FID ECD FTD FPD TCD Helium detector | 1. Liquid–liquid packed columns 2. Split–splitless 3. Temperature programmable 4. On-column 5. Liquid injector valve on-line 6. Gas injection valve 7. Rotary as injection valve | | No | Yes | Yes | — | Yes/Yes | No |

Source: Own files

Perkin–Elmer supply a range of instruments including the basic models 8410 for packed and capillary work and the 8420 for dedicated capillary work, both supplied on purchase with one of the six different types of detection (Table 1.18). The models 8400 and 8500 are more sophisticated capillary column instruments capable of dual detection operation with the additional features of keyboard operation, screen graphics method storage, host computer links, data handling and compatibility with laboratory automation systems. Perkin–Elmer supply a range of accessories for these instruments including an autosampler (AS–8300), an infrared spectrometer interface, an automatic headspace accessory (HS–101 and H5–6), an autoinjector device (AI–I), also a catalytic reactor and a pyroprobe (CDS–190) and automatic thermal desorption system (ATD–550) (both useful for examination of sediments).

The Perkin–Elmer 8700, in addition to the features of the models 8400 and 8500, has the ability to perform multi-dimensional gas chromatography.

The optimum conditions for capillary chromatography of material heart cut from a packed column demand a highly sophisticated programming system. The software provided with the model 8700 provides this, allowing methods to be linked so that pre-column and analytical column separations are performed under optimum conditions. Following the first run in which components are transferred from the pre-column to the on-line cold trap, the system will reset to a second method and, on becoming ready, the cold trap is desorbed and the analytical run automatically started.

Other applications of the model 8700 system include fore-flushing and back-flushing of the pre-column, either separately or in combination with heart cutting, all carried out with complete automation by the standard instrument software.

There are many other suppliers of gas chromatography equipment, some of which are discussed further in Table 1.18.

1.1.8.2 Ion exchange chromatography

This technique has been used to determine arsenic in marine and estuarial sediments. Ion exchange chromatography is based upon the different affinity of ions for the stationary phase. The rate of migration of the ion through the column is directly dependent upon the type and concentration of ions that comprise the eluent. Ions with low or moderate affinities for the packing generally prove to be the best eluents. Examples are hydroxide and carbonate eluents for anion separations.

1.1.8.3 Thin layer chromatography

This technique has been applied to the determination of lead in sludges.

1.1.9 Ion selective electrodes

Application of ion selective electrodes is limited to the determination of ammonium ions in soil.

Ion selective electrode technology is based on the simple measuring principle consisting of a reference electrode and a suitable sensing or indicator electrode, dipped in the sample solution and connected by a sensitive voltameter. The sensing electrode responds to a difference between the composition of the solution inside and outside the electrode and requires a reference electrode to complete the circuit.

The Nernst equation, $E = E_0 + S \log C$, which gives the relationship between the activity or concentration (C) contains two terms which are constant for a particular electrode. These are E_0 (a term based on the potentials which remain constant for a particular sensing/reference electrode pair) and the slope S (which is a function of the sign and valency of the ion being sensed and the temperature). In direct potentiometry, it has to be assumed that the electrode response follows the Nernst equation in the sample matrix and in the range of measurement. E_0 and slope are determined by measuring the electrode potential in two standard solutions of known composition and the activity of the ion in the unknown sample is then calculated from the electrode potential measured in the sample.

Reference electrodes are of two types – single function and double function. Indicating or sensing electrodes are of four types:

- solid state
  Determination of Br^{1-}, Cd^{2+}, Cl^{1-}, Cu^{2+}, CN^{1-}, F^{1-}, I^{1-}, Pb^{2+}, Redox, silver/sulphide $^{1-}$, Na^+, CNS^{1-}.
- liquid membrane
  Ca^{2+}, divalent (hardness), fluoroborate, NO_3^{1-}, K^4, ClO_4^{1-}, HF, surfactants
- residual chlorine
- glass sodium

Variables which effect precise measurement by ion selective electrodes are the following:

- concentration range
- ionic strength – an ionic strength adjuster is added to the samples and standards to minimise differences in ionic strength
- temperature
- pH
- stirring
- interferences
- complexation

Traditionally electrodes have been used in two basic ways, direct potentiometry and potentiometric titration. Direct potentiometry is usually used for pH measurement and for measurement of ions like sodium, fluoride, nitrate and ammonium, for which good selective electrodes exist.

Direct potentiometry is usually done by manually preparing ionic activity standards and recording electrode potential in millivolts, using a high–impedance millivoltmeter and plotting a calibration graph on semilogarithmic graph paper (or using a direct reading pH/ion meter which plots the calibration graph internally).

Direct potentiometry is an accurate technique but the precision and repeatability are limited because there is only one data point. Electrodes drift and potential can rarely be reproduced to better than ±0.5 mV so that the best possible repeatability in direct measurement is usually considered to be ±2%.

Orion, the leading manufacturers of ion-selective electrodes, supply equipment for both direct potentiometry (EA 940, EA 920, SA 720 and SA 270 meters) and potentiometric titration (Orion 90 autochemistry system) (Table 1.19).

1.1.10 α, β and γ spectrometry

α spectrometry has been employed to determine the actinides, americium, californium, curium and plutonium in soils.

β spectrometry has been used to determine strontium 90 in soil.

γ spectrometry has been used to determine the lanthanides neodymium lanthanum, cerium, dysprosium, lutecium, samerium, tamerium, terbium and ytterbium, also the actinides uranium and thorium in soils and sediments.

The production of electricity in many parts of the world is based upon the growth of the nuclear power industry. This has been accompanied by increasing public concern, heightened by events such as Three Mile Island and Chernobyl and has led to a demand for improved standards of radiation protection. One method of improving protection is to establish radioactivity monitoring schemes in the water industry. Now that it is clear that Water Authorities should take radioactivity into account when carrying out their statutory duty to supply wholesome water, people in the industry are expressing a greater interest in radioactivity monitoring [35].

Monitoring schemes should be capable of detecting both man–made and natural radionuclides in soils, sediments and sludges. Natural radionuclides include radon, radium and uranium. The radiation dose attributable to presence of these radionuclides is frequently far more significant than that due to man-made radioactivity and may go undetected if incorrect monitoring equipment is used.

Table 1.19 Orion/pH/ISE meter features chart

Feature	Orion pH/SE meters			
	EA 940	EA 920	SA 720	SA 270
pH	√	√	√	
Direct concentration readout in any unit	√	√	√	√
mV	√	√	√	√
Rel mV	√	√	√	
Temperature	√	√	√	√
Oxygen	√	√		√
Redox	√	√	√	√
Dual electrode inputs	√	√		
Expandable/upgradable	√	√		
Automatic anion/cation electrode recognition	√	√	√	√
Multiple point calibration	√			
Incremental analytical techniques	√	√¹		
Multiple electrode memory	√	√		
Prompting	√	√	√	√
Ready indicator	√	√	√	
Resolution and significant digit selection	√	√	√	√
pH autocal	√	√	√	
Blank correction	√	√		
Multiple print option	√	√		
Recorder output	√	√	√	
RS–232C output	√	√	√	
Adjustable ISO	√	√	√	√
Automatic temperature compensation-line and battery operation	√	√	√	√

¹ With PROM upgrade.
Source: Own files

Erring on the side of extreme safety the WHO have provided values in becquerels (Bq) for gross alpha and gross beta activity of 0.1 and 1 Bq 1 $^{-1}$ respectively. These levels were derived by considering the most radiologically hazardous alpha and beta emitters likely to be encountered in water (radium 226 and strontium 90 respectively). The values are therefore very conservative below which water can be considered potable without any further radiological examination and correspond to a maximum dose commitment of about 0.05 mSv year $^{-1}$ [36].

In the aftermath of the Chernobyl accident it was recognised that there was a need for unified approach in the EEC to controlling the levels of radioactivity in water, soils and sediments [37]. Therefore, new levels have been derived, based on an annual dose limit of 5 mSv.

After the Chernobyl accident, many measurements were made of the levels of radioactivity in water resources in western Europe and the USSR. In western Europe the radioactive burden of surface water remained low enough to leave drinking water supplies unaffected. The levels of radioactivity in surface waters remained low and the effect of water treatment was to remove a proportion of the small amount of radioactivity present in the raw water and to concentrate it in the waste sludge. The resulting waste sludges were comparatively radioactive. A similar effect was observed in sewage sludges, although the levels of activity in the sludge were at least an order of magnitude lower. In the USSR and closer to the site of the reactor accident, water supplies were affected and it was necessary to carry out a series of major remedial measures to limit the contamination.

Natural radioactivity

To put man-made radiation into perspective it is important to realise that in normal circumstances, natural radioactivity contributes less than 2% of the average person's radiation dose [35].

The natural radionuclides of most concern to drinking-water supplies are radium, uranium and radon. Many naturally occurring radionuclides such as uranium 238, uranium 235, thorium 232 and potassium 40 are very slow to decay and represent the remains of the radioactivity produced before or when the earth was formed. Most of the other natural radioisotopes consist of three series of radioisotopes supported by the decay of uranium 238, uranium 235 and thorium 232. More than one-half the radionuclides in these three series are alpha emitters with decay energies of 4–8.9 MeV.

The concentration of a particular radionuclide in the water supply depends upon the concentration of the parent radionuclide in the surrounding rock, the isotope's half–life and its geological setting. The geological setting affects the solubility of the radionuclide, the rate at which the radionuclide escapes from the source rock and the adsorption of the radionuclide by surrounding materials. Ground waters are likely to contain the highest concentrations of natural radionuclides.

Uranium

An excess of uranium 234 occurs in natural waters (higher than that produced by the decay of uranium 238) due to the alpha recoil propelling the uranium 234 into solution. Analytical reports often quote the ratio in water of uranium 234/uranium 238 and this ranges from about 1 to about 30. In areas of granite, around Devon, water supplies have been found to contain a little under 2 mg $^{-1}$ of uranium.

The decay of uranium 238 gives rise to series of radioisotopes and these include the isotopes of radium 226 and radon 222. Other isotopes of radon and radium are formed in two other decay series but usually the predominant radioisotope in water is radon 222. Typically, the levels of radon 222 in water are three orders of magnitude higher than those of radium 226.

Radon 222

Radon 222 is formed in rock strata by the alpha decay of radium 226. The radon gas diffuses through the imperfections in the rock's structure and into the surrounding air or water. The physical condition of the rock matrix appears to play a greater role than the concentration of the parent radionuclide. It has been noted that weathered igneous rocks are most likely to give rise to high radon concentrations. Radon is very soluble in water and concentrations as high as 10,000 Bq 1 $^{-1}$ have been recorded in some ground waters in the USA.

If radon is present in the water together with radium, then the radon is said to be supported. This is because the decay of the long–lived radium produces a continuous supply of new radon. For the first few hours after sealing a sample of water containing radon 222 there is an increase in alpha activity. This results from the in-growth of the short-lived daughter isotopes such as polonium 218 (half life 3.05 min) from the decay of radon (half life 3.82 days).

Radon in water presents the dual pathway of exposure of individuals by ingestion from the direct consumption of water and by inhalation exposure when the radon gas emanates from water.

The primary cause of elevated concentrations of airborne radon in most homes is the entry of soil gas through the foundations and walls, although in some areas groundwater can be a major source. The effective dose from drinking water has been estimated to be between 1 and 12% of the total annual effective dose from all indoor radon [38].

Instrumentation for alpha, beta and gamma measurements

A wide range of instrumentation is available (Appendix 1) for measuring alpha, beta and gamma radiation by the gross or more detailed methods discussed above. Individual spectrometers for each type of radiation or combined alpha-beta-gamma counting systems are available. Discussion here will be limited to the Canberra models 2401 and 2401F (manual) and models 2400 and 2400F (automatic) low level alpha-beta-gamma counting system, as this has been particularly recommended for use in the water industry. This system provides a highly efficient sealed proportional sample counter along with a sealed anti-coincidence guard detector to provide extremely low background sample counting. Combined with a microprocessor-based system controller, the 2400 provides automatic sample changing for up to 100 samples, automatic background, crosstalk, concentration, chi-square, data average and user-entered calculations, all in addition to the operator-definable counting sequences.

This system is ideally suited for measuring nuclides such as ^{14}C, ^{228}Ra, ^{90}Sr, ^{99}Tc, ^{226}Ra, ^{210}Po, ^{230}Th, etc. For special counting requirements, ultra-thin window-flow sample detectors are available, providing the user with very efficient detection of environmental alpha emitters. The system performs simultaneous alpha, beta and gamma counting.

The exclusive features and operating characteristics of the automatic model 2400 alpha-beta-gamma counting system are also available in the efficient single-sample manual model 2401. The system employs two main counting channels for alpha and beta events. Energy discrimination is accomplished through a programmable single-channel analyser contained in the controller. Cosmic and other background events are excluded from either sample channel through anti-coincidence circuitry connected to the guard detector. Live-time correction is provided during all guard channel events. An integral detector assembly is included in each system.

Gross radioactivity measurements are a simple and fairly inexpensive method of assessing the level of radioactivity in water. If the results are below the World Health Organisation (WHO) levels of gross alpha-gross beta activity discussed below, generally the water sample can be classified as fit for consumption without further analysis.

It is important to remember that radioactive isotopes such as tritium and radon will not be detected by these methods, and should the level of radioactivity detected exceed the WHO limits then the samples will have to be analysed for these.

Gross measurements

Beta ray spectrometers

The gross beta activity of samples can be measured by the use of a Geiger counter or proportional counter and scaler [39]. The sample is evaporated to low volume and then to dryness on a tray or planchette which is presented to the counter. Results have to be corrected for background and efficiency of detection.

The gross alpha activity of a sample can be measured in a similar manner using a zinc sulphide scintillation counter or proportional counter, but the problems of the self-absorption of alpha particles are considerably greater than those encountered with beta particles and this must be allowed for in the calculations of the analytical figure.

Gamma ray spectrometers

Gamma ray spectrometers are used to identify and quantify the level of individual gamma-emitting radioisotopes in a sample. Basically there are two types in common use which differ in the nature of the detector. The type based on the sodium iodide type of detector is the least expensive but suffers from the considerable drawback that the spectrum produced is very poorly resolved and this limits the ability of the technique to identify gamma emitters. The type based on the hyper-pure germanium detector has excellent resolution, but has to be kept at liquid nitrogen temperatures and this type of detector is best suited to the type of analysis required in the water industry.

Evidence suggests that traditional water-treatment processes are effective in removing a significant proportion of radionuclides from raw water supplies although the type of treatment, chemical nature of the water and the nature of the radionuclides all affect removal efficiency.

Estimates of the removal efficiency of radionuclides by water treatment processes are shown in Table 1.20. Generally speaking radionuclides such as those of ruthenium and caesium, which are quickly absorbed onto fine particulate matter in the water, seem to be readily removed by filtration, coagulation and settling processes while the more soluble radionuclides such as those of strontium and iodine pose a greater problem, and ion exchange resins or reverse osmosis may be required.

The Chernobyl accident provided information about the removal efficiencies under 'real life' conditions. At Arnfield treatment plant, east of Manchester in the UK, removal efficiencies of 80% gross beta activity were recorded, and in the Netherlands removal efficiencies measured at six surface water treatment plants were found to vary from 50 to 84%.

Gamma spectrometry is undoubtedly the technique to use when it is required to identify and determine individual radioactive isotopes in water samples, sludges or sediments. The multi-channel analyser trace in Fig. 1.11 shows the presence of ^{137}Cs, ^{134}Cs and naturally occurring ^{40}K.

1.2 Summary of methodologies for determining anions

1.2.1 Titration procedures

Titration procedures have been employed for the determination of nitrate and sulphate in soils. The titration process has been automated so that batches of samples can be titrated non–manually and the data processed and reported via printouts and screens. One such instrument is the Metrohm 670 titroprocessor. This incorporates a built-in control unit and the sample changer so that up to nine samples can be automatically titrated. The 670 titroprocessor offers incremental titrations with variable or constant-volume steps (dynamic or monotonic titration). The measured value transfer in these titrations is either drift controlled (equilibrium titration) or effected after a fixed waiting time; pK determinations and fixed end points (e.g. for specified standard procedures) are naturally included. End-point titrations can also be carried out.

Sixteen freely programmable computational formulae with assignment of the calculation parameters and units, mean-value calculations and arithmetic of one titration to another (via common variables) are available. Results can be calculated without any limitations.

The 670 titroprocessor can also be used to solve complex analytical tasks. In addition to various auxiliary functions which can be freely programmed, up to four different titrations can be performed on a single sample.

Table 1.20 Removal efficiencies of radionuclides by water-treatment processes

Water treatment	^{51}Cr	^{32}P	^{90}Sr	^{91}Y	^{106}Ru	^{131}I	^{137}Cs	^{114}Ce	^{222}Rn	^{226}Ra	^{239}Pu	Particulate-associated radioactive compounds
Aeration									90–100			
Chemical coagulation, settling and/or filtration	10–98	68–99	0–70	80–91	77–96	0–44	0–6	90–84		90–100	90	85–90
Slow and filtration		80–99	0–5			50–99	50	99				
Softening with lime soda			20–80	90	50	0–10	0–20		85			
Ion exchange	99		98.5	75–99		0	99		95			
Reverse osmosis			99						92–95			

Values are only estimates based on information available.

Source: Own files

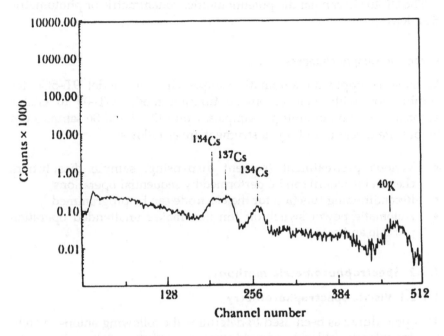

Fig. 1.11 Gamma spectrography indicating fall-out products in environmental sample
Source: Own files

In addition to the fully automated 670 system, Metrohm also supply simpler units with more limited facilities which nevertheless are suitable for more simple titrations. Thus the model 682 titroprocessor is recommended for routine titrations with automatic equivalence pointer cognition or to preset end points. The 686 titroprocessor is a lower–cost version of the above instrument again with automatic equivalence point recognition and titration to preset end points.

Mettler produce two automatic titrimeters suitable for use in the water laboratory, the DL 40 GP memotitrator and the lower-cost DL 20 compact titrator. Features available on the LD 40 GP include absolute and relative end-point titrations, equivalence point titration, back-titration techniques, multi–method applications, dual titration, pH stating, automatic learn titrations, automatic determination of standard deviation and means, series titrations, correction to printer, acid balance analogue output for recorder and correction to the laboratory information system. Up to 40 freely definable methods can be handled and up to 20 reagents held on store. Six control principles can be invoked. The DL 20 can carry out absolute (not relative) end-point titrations and equivalence point titrations, back-titration, series titrations and correction to printer and balance and the laboratory information system. Only one freely definable method is available. Four control principles can be invoked.

The DL 40 GP can handle potentiometric, voltammetric or photometric titrations.

Automatic sample changers

Mitsubishi supply an automatic sample changer Model 9T–5Gc for combination with their automatic titration model GT–05 to enable automatic titration of multiple samples. Up to 12, 24 and 36 samples can be prepared. Features of this instrument are as follows:

- Various pretreatment (solvent dispensing, sample dissolution, chemical reaction) can be performed by sequential operations
- Reconditioning function for the electrode can be programmed
- Automatic power switch-off function makes unattended operation possible and safe.

1.2.2 Spectrophotometric methods

1.2.2.1 Visible spectrophotometry

This procedure has been used to determine the following anions in soils, borate, chlorate, chloride, cyanide, iodide, nitrate, nitrite, phosphate, sulphate, vanadate and sulphide. Nitrate has also been determined in sediments and borate, cyanide, fluoride, nitrate, nitrite and phosphate in sludges.

Equipment for this technique is discussed in section 1.1.2.1.

1.2.2.2 Spectrofluorimetry

This technique has been used to determine borate in soils. Equipment is discussed in section 1.1.2.2.

1.2.2.3 Flow injection analysis

This technique has been used to determine bromide, nitrate and sulphate in soils. Instrumentation is discussed in section 1.1.2.3.

1.2.4 Atomic absorption spectrometry

This technique has been used as an indirect method for the determination of sulphate and sulphide in soil and cyanide in sludges. Equipment is discussed in section 1.1.3.1.

1.2.5 Neutron activation analysis

This technique has been applied to the determination of bromide in soils. Equipment is discussed in section 1.1.4.

1.2.6 Chromatographic procedures

1.2.6.1 Gas chromatography

This technique has been used in soil analysis (bromide, nitric and nitrous oxide) and sediment analysis (sulphide). Equipment is discussed in section 1.1.8.1.

1.2.6.2 High-performance liquid chromatography

This technique has been used to determine phosphate in sludges. Modern high-performance liquid chromatography has been developed to a very high level of performance by the introduction of selective stationary phases of small particle size, resulting in efficient columns with very large plate numbers per litre.

The most commonly used chromatographic mode used in high-performance liquid chromatography is reversed-phase chromatography. Most common reversed phase chromatography is performed using bonded silica-based columns, thus inherently limiting the operating pH range to 2.0–7.5. High sensitivity detection of non-chromophoric ions can be achieved by combining the power of suppressed conductivity detection with these columns and this is usually a superior approach to using refractive index or low ultra-violet wavelength detection. Often detectors that have been used include rapid diode array detectors and electrochemical detectors. Various companies supply equipment for high-performance liquid chromatography including Dionex, Perkin Elmer, Kontrol, Shimadzu, LKB, Cecil Instruments, Vorian, Isio, Hewlett–Packard, Applied Chromatography Systems, Roth Scientific, PSA Inc (see Appendix 1 for further details).

1.2.6.3 Ion chromatography

This technique is extremely useful for the sensitive determination of a range of anions (chloride, fluoride, nitrite, nitrate, phosphate, selenite, sulphate, tungstate and molybdate) in soil, sulphate in sediments and nitrate and tungstate in sludges.

This technique developed by Small *et al.* [40] in 1975 is usually employed for the separation and determination of mixtures of anions. The technique uses specialised ion–exchange columns and chemically suppressed conductivity detection. Advances in column and detection technologies have expanded this capability to include wider range of anions as well as organic ions. These recent developments, discussed below, provide the chemist with a means of solving many problems that are difficult, if not impossible, using other instrumental methods. Ion chromatography can analyse a wide variety of organic and inorganic anions more easily than either atomic absorption spectrometry or inductively coupled plasma techniques.

At the heart of the ion chromatography system is an analytical column containing an ion-exchange resin on which various anions (and/or cations) are separated before being detected and quantified by various detection techniques such as spectrophotometry, atomic absorption spectrometry (metals) or conductivity (anions).

Ion chromatography is not restricted to the separate analysis of only anions or cations, and, with the proper selection of the eluent and separator columns, the technique can be used for the simultaneous analysis of both anions and cations.

The original method for the analysis of mixtures of anions used two-columns attached in series packed with ion-exchange resins to separate the ions of interest and suppresses the conductance of the eluent, leaving only the species of interest as the major conducting species in the solution. Once the ions were separated and the eluent suppressed, the solution entered a conductivity cell, where the species of interest were detected.

The analytical column is used in conjunction with two other columns, a guard column which protects the analytical column from troublesome contaminants, and a pre-concentration column.

The intended function of the pre-concentration column is twofold. First, it concentrates the ions present in the sample, enabling very low levels of contaminants to be detected. Second, it retains non-complexed ions on the resin, while allowing complexed species to pass through.

Dionex Series 40000I Ion Chromatographs
Some of the features of this instrument are:

- chromatography module;
- up to six automated values made of chemically inert, metal-free material eliminate corrosion and metal contamination;
- liquid flow path is completely compatible with all HPLC solvents;
- electronic valve switching, multi-dimensional, coupled chromatography or multi-mode operation;
- automated sample clean up or pre-concentration;
- environmentally isolates up to four separator columns and two suppressors for optimal results;
- manual or remote control with Dionex Autoion 300 or Autoion 100 automation accessors;
- individual column temperature control from ambient to 100 °C optional.

Dionex Ion-Pac Columns
Features are:

- polymer ion exchange columns are packed with new pellicular resins for anion or cation exchange applications;
- 4u polymer ion exchange columns have maximum efficiency and minimum operating pressure for high-performance ion and liquid chromatography applications;
- new ion exclusion columns with bifunctional cation exchange sites offer more selectivity for organic acid separations;
- neutral polymer resins have high surface area for reversed phase ion-pair and ion-suppression applications without pH restriction;
- 5 and 10u silica columns are optimised for ion-pair, ion suppression and reversed phase applications.

Micromembrane suppressor
The micromembrane suppressor makes possible detection of non-UV absorbing compounds such as inorganic anions and cations, surfactants, antibiotics, fatty acids and amines in ion-exchange and ion-pair chromatography.

Two variants of this exist: the anionic (AMMS) and the cationic (CMMS) suppressor. The micromembrane suppressor consists of a low dead volume eluent flow path through altering layers of high-capacity ion-exchange screens and ultra-thin ion exchange membranes. Ion-exchange sites in each screen provide a site-to-site pathway for eluent ions to transfer to the membrane for maximum chemical suppression.

Dionex anion and cation micromembrane suppressors transform eluent ions into less conducting species without affecting sample ions under analysis. This improves conductivity detection, sensitivity, specificity and baseline stability. It also dramatically increases the dynamic range of the system for inorganic and organic ion chromatography. The high ion-exchange capacity of the MMS permits changes in eluent composition by orders of magnitude, making gradient ion chromatography possible.

In addition, because of the increased detection specificity provided by the MMS sample, preparation is dramatically reduced, making it possible to analyse most samples after filtering and dilution.

Conductivity detector
Features include:

- high sensitivity detection of inorganic anions, amines, surfactants, organic acids, Group I and II metals, oxy-metal ions and metal cyanide complexes (used in conjunction with MMS);
- bipolar-pulsed excitation eliminates the non-linear response with concentration found in analogue detectors;

- microcomputer-controlled temperature compensation minimises the baseline drift with changes in room temperature.

UVVis detector
Important factors are:

- high sensitivity detection of amino acids, metals, silica, cheating agents, and other UV absorbing compounds using either post-column reagent addition or direct detection;
- non-metallic cell design eliminates corrosion problems;
- filter-based detection with selectable filters from 214 to 800 nm;
- proprietary dual wavelength detection for ninhydrin-detectable amino acids and PAR-detectable transition metals.

Optional detectors
In addition to the detectors show, Dionex also offer visible fluorescence and pulsed amperometric detectors for use with the series 4000i.

Dionex also supply a wide range of alternative instruments, e.g. single channel (2010i) and dual channel (2020i). The latter can be upgraded to an automated system by adding the Autoion 100 or Autoion 300 controller to control two independent ion chromatograph systems. They also supply a 2000i series equipped with conductivity pulsed amperometric, UV-visible and a typical separation of anions achieved by this technique is shown in Fig. 1.12.

The Dionex AS4A is an analytical anion-exchange separator column designed for optimum performance in separations of common inorganic ions. Seven common anions (F, Cl, NO_2, Br, NO_3, PO_4^{3-} and SO_4^{2-}) can be determined in six minutes (Fig. 1.12(a)).

1.2.7 Ion selective electrodes

Ion selective electrodes have been used to determine nitrate in soil and nitrate chloride and sulphide in sludges. Instrumentation is discussed in section 1.1.9.

1.3 High purity laboratory water

In the determination of metals and anions in environmental solid samples it is essential to use water of suitable purity in the analyses. When extremely low concentrations are being determined water with a very low blank is required in order to achieve the desired sensitivity. Laboratory water purification has undergone dramatic changes in the last decade. Chemists, life scientists and medical technologists are now routinely concerned with impurity levels impossible to measure 10 years ago.

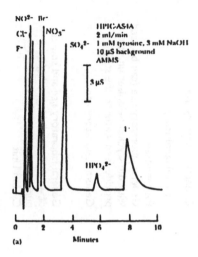

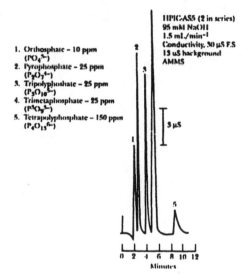

Fig.1.12 Ion chromatograms obtained with Dionex instrument using (anodic) AMMS and CMMS micromembrane suppression: (a) anions with micromembrane suppressor; (b) separation of polyphosphates using micromembrane suppressor
Source: Own Files

As a result, distilled water purity is marginal for much of today's analytical work (Table 1.21). In addition, soaring energy and maintenance costs are making distillation a poor choice. Today, reverse osmosis, deionisation, carbon absorption and membrane microfiltration are all, in some respects, superior to distillation.

Table 1.21 Water purification units

Supplier	Model	Sensitivity ($M\Omega$ cm)	Conductivity ($\mu s\ cm^{-1}$)	Applications
A. Distilled water				
Manestry	L4 still	—	—	Pyrogen-free General laboratory use
Hamilton	stills	—	—	General laboratory use
S. Bibby	Aquatron	—	0.1–0.2	General laboratory use
Fistreem	Cyclon	—	≤1	Pyrogen-free General laboratory use (can be used with the Cyclon deioniser unit to provide feed water to the still)
Jencons	Autostill range	—	0.5–2.0	General laboratory use
B. Reverse osmosis water				
Fistreem	RO60	—	—	General laboratory use and source of water for polishing to high-purity water using the Fistreem Cyclon Unit
Elga	Prima	—	—	General laboratory use and source water for polishing to high-purity water using the Elga stat UHP or UHW
Millipore	Milliro system	18 at 25 °C	—	General laboratory use and source of water for polishing to high-purity water using the Millipore Milli-Q-system

C. *Reverse osmosis – deionisation water*

Elga	Elgastat Spectrum	Up to 18	0.05	HPLC atomic absorption spectrometry; tissue culture, spectrometry, flame photometry Reference and buffer solutions: general laboratory use
Millipore	Milli R/Q	2–5	–	Spectrophotometry liquid–liquid extraction potentiometers, volumetric analysis

D. *Reverse osmosis – multi-column water*

Gelman	Water 1 system	18	0.055	Atomic absorption spectrometry; emission spectrometry; HPLC
Fistreem	Nonpure Ultrapure Water system	Up to 18	0.055	Atomic absorption spectrometry; HPLC; spectrophotometers: flame photometry; reference and buffer solutions: tissue culture; enzymology, haematology
Elga	Elgastat UHP and Elgastat UHQ	Precede by reverse osmosis: 5 columns for removal of metals, anions, cations, organics, particulates, pyrogens, bacteria and colloids	18 at 25 °C	HPLC, ion chromatography, fluorescence analysis, total organic carbon, microbiology
Millipore	Milli-Q system	Precede by reverse osmosis: 5 columns for removal of metals, anions, cations, organics, particulates, pyrogens, bacteria and colloids	18 at 25 °C	Atomic absorption spectrometry, HPLC, total organic carbon, enzymology, tissue culture

Source: Own files

To enable users of high-purity water to define their needs more precisely, several professional organisations have established standards for certain classes of use.

With this system it is possible to group laboratory water purification systems according to the classes of use.

Laboratory grade water: types III, IV

Water of this quality has been prepared traditionally by single stage distillation (Table 1.21). It is adequate for most general laboratory use including qualitative analysis, non-critical media and reagents and glassware washing. Another use for type III/IV water is the pre-treatment of water prior to reagent grade polishing. Reverse osmosis is now a more practical choice than distillation for general laboratory water (Table 1.22). It is far less expensive, more dependable and almost maintenance-free because contaminants are continuously flushed away with the reject stream.

Analytical grade water: type II

This grade of water has a resistivity of at least $1\,M\Omega\,cm$ at 25 °C and is suitable for all but the most critical procedures including spectrophotometry, liquid–liquid extraction potentiometry and volumetric analysis. Traditionally, water of this quality has been produced by single and double distillation. Increasingly, nowadays, this quality is achieved by equipment based on a combination of reverse osmosis and ion-exchange and final microfiltration (Table 1.21).

Water produced by this method is not of the highest quality and cannot be classified as ultra-high quality. Whereas, for example, it might suffice for use in the atomic absorption determination of metals in solid samples at the milligram per litre level, it would be insufficiently pure when carrying out analyses at the microgram per litre level or lower.

Reagent grade water: type I

Type I water has a resistivity of $18\,M\Omega$ at 25 °C. This type of water is particularly recommended for procedures that require freedom from trace impurities at the very limits of detection, such as atomic absorption analysis at the microgram or nanogram level, UV and IR spectroscopy, voltammetry, specific ion electrodes, HPLC and total organic carbon measurements in the range $10\text{–}100\,\mu gL^{-1}$. Such water quality is a fundamental element in the biological work.

Producing water of this quality always requires several stages of treatment because no single process is capable of removing all contaminants. In order to take a heavy load off the columns used in such equipment and to enhance their life and to ensure the highest quality of the final product, it is recommended that the input water is first treated by reverse osmosis to first remove gross impurities. Several such systems are summarised in Table 1.22.

Table 1.22 Ultra high-purity water multi-cartridge systems

Column	Elgastat UHP	Elgastat UHQ	Fistream Nanopure II	Millipore Milli-CI system
1	Organics removal	Reverse osmosis (built-in)	Organics removal	Particle/bacteria removal
2	Inorganic ion removal	Organics removal	Colloids removal	Organics removal
3	Ultramicro filtration	Inorganics removal	Inorganics removal	Inorganics removal
4	Photo-oxidation (organics removal)	Microfiltration particle/ bacteria removal	Ultra-filtration (particle/ bacteria removal	Final trace organics removal
5	Ultrafiltration (particle/ bacteria removal)	Photo-oxidation (organics removal)	–	Ultrafiltration (pyrogen removal)
6	–	–	–	Membrane filtration (micro-organics removal)
	Removal of organics		Activated carbon ultraviolet radiation	
	Removal of inorganics		Ion-exchange resin	
	Removal of colloids, micro-organisms, particles		Ultra-microfiltration, macroreticular resins	
	Removal of particles and bacteria		Ultrafiltration on membrane filters	

Source: Own files

1.4 Rationale

If the reader finds that a method is not listed for determining a particular species in the particular type of solid sample, then by examination of the table he may find a method that is listed under another type of solid sample that is applicable to the type of sample in which he is interested. Thus, if a method is not available for determining a particular species in say soil, he may find that one is listed under sediments. Obviously this approach will not always be applicable.

1.4.1 Analysis of soils

1.4.1.1 Determination of cations

Table 1.23 reviews methods used for the determination of cations in soils. In these tables analytical techniques are cross-referenced with the species determined and the section number in the book.

By far the most extensively used techniques are inductively coupled plasma atomic emission spectrometry and atomic absorption spectrometry (respectively 24 and 21 determinands), neutron activation analysis (19 determinands), emission spectrometry (12 determinands), differential pulse anodic scanning voltammetry (11 determinands) and X-ray fluorescence spectroscopy (10 determinands). Spectrophotometric methods (16 determinands) are fairly commonplace but do not have the sensitivity needed for low level analysis.

Strangely, inductively coupled plasma mass spectrometry has not yet made the inroads into metals analysis in soils that it has made into sediment analysis (see Tables 1.23, 1.25 and 1.26), but no doubt this situation will be corrected as soil analysts learn to appreciate the attractiveness and advantages of this technique. Other, lesser used techniques that have been applied to soil analysis include spark source mass spectrometry, α spectrometry, stable isotope dilution methods (5–6 determinands), molecular absorption spectrometry (4 determinants), spectrofluorimetric methods, flow injection analysis and X-ray spectrometry (2 determinands) and, finally, techniques for which only one application has been described in the literature (plasma emission spectrometry, polarography, ion-selective electrodes, gas chromatography, flame photometry and ß spectrometry).

Of the 58 metallic elements in the periodic table (including silicon and boron and excluding the 14 lanthanides and 14 actinides), methods have been described for the determination of 38 metals. Of the 14 lanthanides only tamerium and terbium are mentioned and of the fourteen actinides only uranium, plutonium and thorium are mentioned. Metals for which methods have not yet been described are spread fairly evenly across the seven periods of the periodic table as illustrated below.

Group	Metals for which methods have not been discussed in literature
2	lithium
3	sodium, magnesium
4	scandium, gallium, germanium
5	strontium, niobium, ruthenium, rhodium, palladium
6	lanthanum, hafnium, tantalum, rhenium, iridium, gold, polonium
7	francium, radium, actinium
Lanthanides	cerium, praseodymium, promethium, neodymium, samarium, luropium, gadolinium, dysprosium, holmium, erbium, ytterbium, lutecium
Actinides	protoactinium, neptunium, americium, curium, berkelium, californium, einsteinium, fermium, mendelevium, nobelium, laurencium

1.4.1.2 Determination of anions

Methods are available for the determination of most of the common anions in soil (Table 1.24). Spectrophotometric methods feature largely while the increasing use of ion chromatography for the analysis of mixtures of anions is very apparent.

1.4.2 Analysis of non-saline sediments

1.4.2.1 Determination of cations

Table 1.25 reviews methods for the determination of cations in non–saline sediments. The largest number (48) of metals have been determined by neutron activation analysis. This technique, however, cannot usually be practiced in the in-house laboratory. Inductively coupled plasma atomic emission spectrometry, inductively coupled plasma mass mass spectrometry and atomic absorption spectrometry (respectively 26, 21 and 21 determinands) are the most commonly used techniques. γ-ray spectrometry and X-ray spectrometry are fairly commonly used (both 13 determinands). Spectrophotometry (6 determinands) has limited applications, as have X-ray fluorescence spectroscopy, gas chromatography and flow injection analysis (all 4 determinands). Other techniques with limited applications include spectrofluorimetric methods and high-performance liquid chromatography (both 3 determinands), potentiometry and anodic scanning voltammetry (both 2 determinands) and mass spectrometry, cathodic scanning voltammetry and differential pulse polarography (all one determinand).

Of the 58 metallic elements in the period tables (including boron and silicon but excluding the fourteen actinides and 14 lanthanides) methods have been described for the determination of 48 metals. Of the 14 lanthanides methods have been described for the determination of all with the exception of promethium, praseoactinium, holmium and erbium. Of the 14 actinides methods have been described for the determination of all metals up to and including americium.

Table 1.23 Methods for the determination of cations in soil

Soil	Spectro-photo	Spectro-fluoro	FIA	AAS	ICPAES	ICPMS	Plasma emission	Spark source	Emission spec	PAA
Al	2.1.1		2.1.2	2.1.3 2.60.2.2	2.1.4 2.60.2.2				2.1.5 2.60.5.1	2.1.6 2.60.4.1
NH₄	2.3.1		2.3.3							
Sb				2.4.1 2.60.1.6	2.4.2 2.60.2.4			2.4.4 2.60.6.1		2.4.3 2.60.4.1
As	2.5.1			2.5.2 2.60.1.4 2.60.1.6	2.5.3 20.60.2.4	2.5.4 2.60.2.2				2.5.5 2.60.4.1
Ba					2.6.1 2.60.2.2				2.6.2 2.60.5.1	
Be							2.7.1		2.7.1 2.60.5.1	
Bi				2.8.1 2.60.1.6	2.8.2 2.60.2.4					
B	2.9.1				2.9.2 2.60.2.3					
Cd				2.10.1 2.60.1.1 2.60.1.3 2.60.1.4 2.60.1.5	2.10.2 2.60.2.1 2.60.2.2	2.10.2 2.60.2.2			2.10.5 2.60.5.1	
Cs								2.11.1 2.60.6.1		
Ca					2.12.2 2.60.2.2					2.12.4 2.60.4.1

Table 1.23 continued

Soil	Spectro-photo	Spectro-fluoro	FIA	AAS	ICPAES	ICPMS	Plasma emission	Spark source	Emission spec	PAA
Cr	2.15.1			2.15.2 2.60.1.1 2.60.1.3 2.60.1.4	2.15.3 2.60.2.1 2.60.2.2			2.15.6 2.60.6.1	2.15.8 2.60.5.1	2.15.5 2.60.4.1
Co	2.16.1			2.16.2 2.60.1.1 2.60.1.5	2.16.3 2.60.2.1 2.60.2.3				2.16.7 2.60.5.1	2.16.4 2.60.4.1
Cu				2.17.1 2.60.1.1 to 2.60.1.5	2.17.2 2.60.2.1				2.17.7 2.60.5.1	
In				2.2.1.1 2.60.1.5						
Fe	2.23.1			2.2.3.2 2.60.1.1 2.60.1.2	2.2.3.3 2.60.2.1 2.60.2.2				2.23.7 2.60.5.1	2.2.3.4 2.60.4.1
Pb	2.25.1	2.25.3		2.25.2 2.60.1.1 2.60.1.3 2.60.1.4 2.60.1.5	2.25.4 2.60.2.1	2.25.4 2.60.2.2			2.25.9 2.60.5.1	2.25.8 2.60.4.1
Mg				2.26.1 2.60.2.2	2.26.2 2.60.2.2					2.26.3 2.60.4.1
Mn	2.27.1			2.27.2 2.60.1.1 2.60.1.2	2.27.3 2.60.2.1 2.60.2.2				2.27.6 2.60.5.1	

Table 1.23 continued

Soil	Spectro-photo	Spectro-fluoro	FIA	AAS	ICPAES	ICPMS	Plasma emission	Spark source	Emission spec	PAA
Hg	2.28.1			2.28.2 2.60.1.4 2.60.1.5						
Mo	2.29.1			2.29.2 2.60.1.3 2.60.1.4 2.60.1.5	2.29.3 20.6.2.3				2.29.5 2.60.5.1	
Ni	2.31.1			2.31.2 2.60.1.1 2.60.1.3 2.60.1.4 2.60.1.5	2.31.3 2.60.2.1			2.31.4 2.60.1	2.31.8 2.60.5.1	2.31.7 2.60.4.1
Pt					2.33.1					
Sr					2.4.4.1 2.60.2.2				2.44.2 2.60.5.1	2.44.5 2.60.4.1
Tm									2.45.1 2.60.5.1	
Tc										
Te						2.47.1			2.4.8.1 2.60.5.1	
Tb									2.49.1 2.60.5.1	
Tl				2.50.1					2.50.3 2.60.5.1	
Th						2.51.1			2.51.3 2.60.5.1	

Table 1.23 continued

Soil	Spectro-photo	Spectro-fluoro	FIA	AAS	ICPAES	ICPMS	Plasma emission	Spark source	Emission spec	PAA
Sn								2.52.1 2.60.6.1		
Pu					2.34.1 2.36.2 2.60.2.2					
K										
Rb								2.38.1 2.60.6.1		
Se		2.40.1		2.40.2 2.60.1.4 2.60.1.5 2.60.1.6	2.40.4 2.60.2.4					
Si				2.41.1	2.41.2 2.60.2.2					
Ag				2.42.2 2.60.1.5 2.43.1						
Na										
Ti	2.53.1				2.53.2 2.60.2.2				2.53.3 2.60.5.1	2.53.4 2.60.4.1
W	2.54.1								2.54.2 2.60.5.1	
U	2.55.1					2.55.2			2.55.4 2.60.5.1	2.55.3 2.60.4.1
V	2.56.1				2.56.2 2.60.2.2				2.56.3 2.60.5.1	
Y									2.57.1 2.60.5.1	

Table 1.23 continued

Soil	Spectro-photo	Spectro-fluoro	FIA	AAS	ICPAES	ICPMS	Plasma emission	Spark source	Emission spec	PAA
Zn				2.58.1 2.60.1.1 2.60.1.3 2.60.1.4 2.60.1.5	2.58.2 2.60.2.1 2.60.2.2	2.58.2 2.60.2.2			2.58.7 2.60.5.1	2.58.6 2.60.4.1
Zr									2.59.1 2.60.5.1	2.59.2 2.60.4.1

Spectro photo = spectrophotometry; Spectro fluoro = spectrofluorimetry; FIA = Flow injection analysis; AAS = Atomic absorption spectrometry; ICPAES = Inductively coupled plasms atomic emission spectrometry; ICPAES = Inductively coupled plasma mass spectrometry; Plasma emission = plasma emission spectrometry; Spark source = spark source mass spectrometry; Emission spec = emission spectrometry; PAA = photon activation analysis

Table 1.23 continued

Soil	Polar	α spec	Select elect	NAA	Stable isotope dil.	Molec. absorp.	Diff. pulse	XRF	X-ray analysis	GLC	Flame photo	β Spec	Misc
Al													2.1.7
Am		2.2.1 2.60.11.1											2.2.2
NH₄		2.3.4											
Sb				2.4.3 2.60.3.1									2.4.5
As				2.5.5 2.60.3.1 2.60.3.2									2.5.6
Ba					2.6.3 2.60.9.1								
Bi													2.8.3
B						2.9.3							
Cd							2.10.3 2.60.7.1	2.10.4 2.60.8.1					2.10.6
Cs					2.11.2 2.60.9.1								
Ca					2.12.3 2.60.9.1								
Cf		2.13.1 2.60.11.1											
Ce				2.14.1 2.60.3.3									
Cr				2.15.4 2.60.3.1			2.15.6	2.15.7 2.60.8.1					2.15.9

Table 1.23 continued

Soil	Polar	α spec	Select. elect	NAA	Stable isotope dil.	Molec. absorp.	Diff. pulse	XRF	X-ray analyser	GLC	Flame photo	β Spec	Misc
Co				2.16.4 2.60.3.1			2.16.5 2.60.7.1	2.16.6 2.60.8.1					
Cu				2.17.6			2.17.3 2.60.7.1	2.17.4 2.60.8.1					2.17.8
Cm		2.18.1 2.60.11.1											
Eu				2.19.1 2.60.3.3									
Hf				2.20.1 2.60.3.1									
Ir				2.22.1									
Fe				2.23.4 2.60.3.1			2.23.5 2.60.7.1	2.23.6 2.60.8.1					
La				2.24.1 2.60.3.3									
Pb	2.25.6						2.25.5 2.60.7.1	2.25.7 2.60.8.1					2.25.10
Mn							2.27.4 2.60.7.1	2.27.5 2.60.8.1					
Hg				2.28.4			2.28.3						2.28.5
Mo				2.29.4 2.60.3.1									2.29.6
Np													2.30.1
Ni							2.31.5 2.60.7.1	2.31.6 2.60.8.1	2.31.6				2.44.6

Table 1.23 continued

Soil	Polar	α spec	Select. elect.	NAA	Stable isotope dil.	Molec. absorp.	Diff. pulse	XRF	X-ray analyser	GLC	Flame photo	B Spec	Misc
Pd													
Pt		2.34.3		2.33.2					2.32.1				
Pu										2.34.2			2.34.4
Po													2.35.1
K					2.36.3 2.60.9.1						2.36.1		
Ra					2.38.2								2.37.1
Rb					2.60.9.1								
Sc				2.39.1 2.60.3.1									
Se				2.40.5 2.60.3.1 2.60.3.2				2.40.3					2.40.6
Sr					2.44.3 2.60.9.1							2.44.4	2.44.6
Tm				2.45.2 2.60.3.3									
Ta				2.46.1 2.60.31									
Tc													2.47.2

Table 1.23 continued

Soil	Polar	α spec	Select elect	NAA	Stable isotope dil.	Molec. absorp.	Diff. pulse	XRF	X-ray analyser	GLC	Flame photo	β Spec	Misc
Tb				2.4.9.2 2.60.3.3									
Tl				2.51.2 2.60.3.1									2.50.4
Th							2.50.2						2.51.4
Sn				2.55.3 2.60.3.1									2.52.2
U													2.55.5
Zn							2.58.3 2.60.7.1	2.58.4 2.60.8.1					2.58.8

Polar = polarography; α spec. = α spectrometry; Select. elect. = ion selective electrode; NAA = neutron activation analysis; Stable isotope dil. = stable isotope dilution; Molec. absorp. = Molecular absorption spectrometry; Diff. pulse = differential pulse anodic stripping voltammetry; XRF = X-ray fluorescence; X-ray analyser = X-ray spectroscopy; GLC = gas chromatography; Flame phot = flame photometry; β Spectrom = β spectrometry; Misc. = miscellaneous

Source: Own files

Table 1.24 Methods for the determination of anions in soil

Soil	Titr.	Spectro-photo	Spectro-fluro	FIA	AAS	Molec. emisn.	NAA	Select elect	Potent	GLC	IC	Micro-diff.	Chemi.	Gaso-metric	Misc
BO_3		7.1.1	7.1.2												
Br				7.2.1			7.2.2			7.2.3					
CO_3														7.3.1	
ClO_3		2.4.1													
Cl		7.5.1							7.5.2		7.5.3				
CN		7.6.1													
F											7.7.1				
I		7.8.1													
M_oO_4											7.9.1				
NO_3	7.10.1	7.10.2		7.10.3				7.10.5			7.10.6	7.10.4			
NO/NO_2										7.11.1					
NO_2		7.12.1									7.12.2				
PO_4		7.13.1									7.13.2				7.13.3
SeO_3											7.14.1				
SO_4	7.15.1	7.15.2		7.15.3	7.15.4	7.15.5									7.15.7
S		7.16.1			7.16.2										7.16.3
WO_4											7.17.1				
VO_4		7.18.1													

Titr. = titration; Spectrophoto. = spectrophotometry; Spectrofluro = spectrofluorimetry; FIA = Flow injection analysis; AAS = atomic absorption spectrometry; Molec. emisn. = Molecular emission cavity analysis; NAA = neutron activation analysis; Select. elect. = ion selective electrode; Potent. = potentiometric; GLC = gas chromatography; IC = ion chromatography; Microdiff. = microdiffusion; Chemi. = chemiluminescence; Gasometric = gasometric

Source: Own files

Table 1.25 Methods for the determination of cations in non-saline sediments

Non-saline sediment	Spectro-photo	Spectro-fluoro	FIA	AAS	ICPAES	ICPMS	Plasma emission	Diff. pulse polar	Potent-iometry	ASV	CSV
Al				3.1.1 3.67.2.3 3.67.2.4	3.1.2 3.67.3.1		3.1.4 3.67.6.1				
Sb	3.3.1				3.3.2 3.67.3.2	3.3.2 3.67.4.1					
As	3.4.1			3.4.2 3.67.2.6	3.4.3 3.67.3.2	3.4.4 3.67.4.1					
Ba					3.5.1 3.67.3.1						
Be				3.6.1 3.67.2.4							
Bi				3.7.1	3.7.2 3.67.3.2						
Cd			3.8.1 3.67.1.1	3.8.2 3.67.2.1 3.67.2.2 3.67.2.4 3.67.2.5	3.8.3 3.67.3.1	3.8.4 3.67.4.1					
Ca			3.9.1	3.9.1 3.67.2.4	3.9.2 3.67.3.1						
Ce						3.11.2					
Cr				3.12.1 3.67.2.1 3.67.2.4	3.12.2 3.67.3.1		3.12.4 3.67.6.1				

Table 1.25 continued

Non-saline sediment	Spectro-photo	Spectro-fluoro	FIA	AAS	ICPAES	ICPMS	Plasma emission	Diff. pulse polar	Potent-iometry	ASV	CSV
Co				3.13.1 3.67.2.1 3.67.2.4							
Cu			3.14.1 3.67.1.1	3.14.2 3.67.2.1 3.67.2.3 3.67.2.4			3.14.3 3.67.6.1		3.14.4 3.67.7.1		
Dy						3.15.2 3.67.4.2					
Europium						3.16.2 3.67.4.2					
Gd						3.17.2 3.67.4.2					
Ga				3.18.1							
In					3.21.1 3.67.3.3						
Fe				3.23.1 3.67.2.1 3.67.2.3 3.67.2.4	3.23.2 3.67.3.1		3.23.4 3.67.6.1				
La						3.24.2 3.67.4.2					

Table 1.25 continued

Non-saline sediment	Spectro-photo	Spectro-fluoro	FIA	AAS	ICPAES	ICPMS	Plasma emission	Diff. pulse polar	Potent-iometry	ASV	CSV
Pb	3.24.5.1		3.25.2 3.67.1.1	3.25.3 3.67.2.1 3.67.2.4 3.67.2.5		3.25.4 3.67.4.1	3.25.8 3.67.6.1		3.67.7	3.25.9 3.67.7.1	3.25.6 3.67.7.1
Li				3.26.1 3.62.2.4							
Lu						3.27.2 3.67.4.2					
Mg				3.28.1 3.67.2.4	3.28.2 3.67.3.1						
Mn				3.29.1 3.67.2.1 3.67.2.3 3.67.2.4	3.29.2 3.67.3.1		3.29.4 3.67.6.1				
Hg	3.30.2			3.30.3	3.30.4	3.30.5 3.32.2 3.67.4.2				3.30.7	
Nd											
Ni				3.34.1 3.67.2.1 3.67.2.4			3.34.4 3.67.6.1				
Pt					3.37.1 3.67.3.3						

Table 1.25 continued

Non-saline sediment	Spectro-photo	Spectro-fluoro	FIA	AAS	ICPAES	ICPMS	Plasma emission	Diff. pulse polar	Potent-iometry	ASV	CSV
Pu						3.38.1					
K					3.40.1 3.67.3.1						
Re					3.42.1 3.67.3.3						
Sm						3.67.4.2 3.45.2					
Se		3.47.1		3.47.2 3.67.2.6	3.47.3 3.67.3.2	3.47.4					3.47.6
Si					3.48.1 3.67.3.1						
Ag				3.49.1 3.67.2.2							
Na					3.50.1 3.67.3.1						
Sr					3.51.1 3.67.3.1						
Tm						3.52.2					
Tb						3.55.2 3.67.4.2					

Table 1.25 continued

Non-saline sediment	Spectro-photo	Spectro-fluoro	FIA	AAS	ICPAES	ICPMS	Plasma emission	Diff. pulse polar	Potent-iometry	ASV	CSV
Tl				3.56.1 3.67.2.5		3.56.2 3.67.4.1					
Th		3.57.1				3.57.2					
Sn				3.58.1	3.58.2						
Ti					3.59.1 3.67.3.1						
W	3.60.1										
U		3.61.1				3.61.2					
V	3.62.1				3.62.2 3.67.3.1			3.62.4			
Yb						3.63.2 3.67.4.2					
Zn				3.64.1 3.67.2.1 3.67.2.3 3.67.2.4	3.64.2 3.67.3.1	3.64.3 3.67.4.1	3.64.5 3.67.6.1				

Spectrophoto = spectrophotometry; Spectrofluoro = spectrofluorimetry; FIA = flow injection analysis; AAS = atomic absorption spectrometry; ICPAES = Inductively coupled plasma atomic emission spectrometry; ICPMS = inductively coupled plasma mass spectrometry; plasma emission = plasma emission spectrometry; Diff. pulse polar = differential pulse polarography; Potentiometry; potentiometry; ASV = Anodic stripping voltammetry; SCV = cathodic stripping voltammetry

Source: Own files

Table 1.25 continued

Non-saline sediments	NAA	XRF	γ ray	m/s	GLC	HPLC	X-ray	Misc
Al	3.1.3 3.67.5.1							3.2.1
Am								
Sb	3.3.3 3.67.5.1				3.3.4 3.67.11.1			3.3.5
As	3.4.5 3.67.5.1	3.4.6 3.67.8.1			3.4.8 3.67.11.1		3.4.7 3.67.9.1	3.4.9
Ba	3.5.2 3.67.5.1							
Cd		3.8.5					3.8.6 3.67.9.1	
Ca	3.9.3 3.67.5.1							
Cs	3.10.1 3.67.5.1							
Ce	3.11.1 3.67.5.1		3.11.3 3.67.10.1					
Cr	3.12.3 3.67.5.1						3.12.5 3.67.9.1	3.12.6
Co	3.13.2 3.67.5.1						3.13.3 3.67.9.1	
Cu							3.14.5 3.67.9.1	3.14.6

Table 1.25 continued

Non-saline sediments	NAA	XRF	γ ray	m/s	GLC	HPLC	X-ray	Misc
Dy	3.15.1 3.67.5.1		3.15.3 3.67.10.1					
Eu	3.16.1 3.67.5.1		3.16.3 3.67.10.1					
Gd	3.17.1 3.67.5.1		3.17.3 3.67.10.1					
Au	3.19.1 3.67.5.1							3.19.2
Hf	3.20.1 3.67.5.1							
In	3.21.1 3.67.5.1							
Ir	3.22.1 3.67.5.1							
Fe	3.23.3 3.67.5.1					3.23.5	3.23.5 3.67.9.1	
La	3.24.1 3.67.5.1		3.24.3 3.67.10.1					
Pb	3.25.5 3.67.5.1	3.25.7					3.25.10 3.67.9.1	3.25.11
Lu	3.27.1 3.67.5.1		3.27.3 3.67.10.1					
Mg	3.28.3 3.67.5.1							
Mn	3.29.3 3.67.5.1						3.29.5 3.67.9.1	

Table 1.25 continued

Non-saline sediments	NAA	XRF	γ ray	m/s	GLC	HPLC	X-ray	Misc
Hg	3.30.6 3.67.5.1				3.30.10	3.30.9	3.30.8 3.67.9.1	30.30.11
Mo	3.31.1 3.67.5.1							
Nd	3.32.1 3.67.5.1		3.32.3 3.67.10.1					
Np								3.33.1
Ni	3.34.2 3.67.5.1						3.34.3 3.67.9.1	
Os	3.35.1 3.67.5.1							
Pd	3.36.1 3.67.5.1							
Pt	3.37.2 3.67.5.1							
Po								3.39.1
Pu								3.38.2
K	3.67.5.1							
Ra				3.64.1				
Rb							3.43.1 3.67.9.1	3.4.1.2
Ru	3.44.1 3.67.5.1		3.45.3 3.57.10.1					
Sm	3.45.1 3.67.5.1							

Table 1.25 continued

Non-saline sediments	NAA	XRF	γ ray	m/s	GLC	HPLC	X-ray	Misc
Sc	3.46.1 3.67.5.1							
Se	3.47.5 3.67.5.1				3.47.8	3.47.7		3.47.9
Ag	3.49.2 3.67.5.1							
Na	3.50.2 3.67.5.1							
Sr	3.51.2 3.67.5.1						3.51.3 3.67.9.1	3.51.4
Tm	3.52.1 3.67.5.1		3.52.3 3.67.10.1					
Ta	3.53.1 3.67.5.1							3.54.1
Tc								
Tb	3.55.1 3.67.5.1		3.55.3 3.67.10.1					
Tl	3.57.3 3.67.5.1		3.57.4 3.67.10.1					3.56.3
Th								3.57.5
Ti	3.59.2 3.67.5.1							
W	30.6.2 3.67.5.1							

Table 1.25 continued

Non-saline sediments	NAA	XRF	γ ray	m/s	GLC	HPLC	X-ray	Misc
U	3.61.3 3.67.5.1	3.61.4	3.61.5 3.67.10.1					3.61.6
V	3.62.3 3.67.5.1							
Yb	3.63.1 3.57.5.1		3.63.3 3.67.10.1					
Zn	3.64.4 3.67.5.1						3.64.6 3.67.9.1	3.64.7
Zr	3.65.1 3.67.5.1							

NAA = neutron activation analysis; XRF = X-ray fluorescence spectroscopy; γ ray spec = γ-ray spectroscopy; m/s = mass spectrometry; GLC = gas chromatography; X-ray spec = X-ray spectrometry; Misc. = miscellaneous

Source: Own files

Metals for which methods have not been described are as follows:

Group	Metals for which methods have not been described in literature
2	boron
4	germanium
5	yttrium, niobium, technetium, rhodium, tellurium
6	hafnium, francium, actinium
Lanthanides	promethium, prasoactinium, holmium, erbium.
Actinides	curium, berkelium, californium, einsteinium, fermium, mendelevium, nobelium, laurencium

1.4.3 Analysis of saline sediments

1.4.3.1 Determination of cations

Table 1.26 reviews methods for the determination of cations in saline sediments. Again, the most popular methods of analysis are inductively coupled plasma atomic emission spectrometry (29 determinands) and atomic absorption spectrometry (22 determinands) followed by neutron activation analysis (8 determinants) and photon activation analysis (17 determinands). Secondary emission X-ray spectrometry has been applied to the determinands of 11 determinands. Spectrophotometric methods are used when sensitivity is not a limitation (7 determinands). Lesser used methods include anodic scanning voltammetry (4 determinands). Spectrofluorimetry, column chromatography and electron probe microanalysis are techniques that have found very limited applications in saline sediment analyses.

Of the 58 metallic elements in the periodic table (including boron and silicon and excluding the 14 lanthanides and 14 actinides) methods have been described for the determination of 39 metals. Of the 14 lanthanides methods have been described for the determination of cerium, ytterbium and dysprosium. Of the 14 actinides methods have been described only for the determination of uranium.

Methods have not been described for the following metals:

Group	Metals for which methods have not been described in the literature
2	lithium, boron
4	germanium
5	niobium, technecium, ruthenium, palladium, indium, tellurium
6	hafnium, tantalum, tungsten, rhenium, osmium, gold, polonium
7	francium, radium, actinium
Lanthanides	praseodynium, neodynium, promethium, samarium, europium, gadolinium, terbium, holmium, erbium, tamerium, lutecium
Actinides	thorium, praseoactinium, neptunium, plutonium, americium, curium, berkelium, californium, einsteinium, fermium, mendelevium, nobelium, laurencium

1.4.4 Analysis of estuarine sediments

1.4.4.1 Determination of cations

Table 1.27 reviews methods of analysis of estuarine sediments. Inductively coupled plasma atomic emission spectrometry (17 determinands), atomic absorption spectrometry (12 determinands) and X-ray secondary emission spectrometry (11 determinands) are the methods of choice for the analysis of estuarine sediments, a situation similar to that occurring in the case of non–saline and saline sediments.

Of the 58 metallic elements (including silicon and boron and excluding the 14 lanthanides and 14 actinides), methods have been described for the determination of 22 metals. Methods have been described for only three of the 14 lanthanides viz cerium, dysprosium and ytterbium. No work seems to have been published on the determination of actinides in estuarine sediments. No doubt, many of the methods described for the analysis of saline sediments would be applicable to the analysis of estuarine sediments in which case up to 41 metals (excluding lanthanides and actinides) could be determined.

1.4.4.2 Determination of anions

Work on the determination of anions to saline and non-saline sediments is limited to the determination of nitrate by spectrophotometric and chemiluminescence techniques, sulphide by gas chromatography and sulphate by ion chromatography (Table 1.28).

1.4.5 Analysis of sludges

1.4.5.1 Determination of cations

Table 1.29 reviews methods of analysis of sludges.

Atomic absorption spectrometry (24 determinands), inductively coupled plasma atomic emission spectrometry (7 determinands) and differential pulse anodic scanning voltammetry (7 determinands) are the most commonly used methods of analysis. Those who have access to neutron activation analysis (42 determinands) and photon activation analysis (34 determinands) have used these techniques. Other techniques with few applications include spectrophotometric and spectrofluorimetric methods, thin-layer chromatography and potentiostripping methods.

Of the 58 metallic elements (including boron and silicon and excluding 14 lanthanides and 14 actinides) methods have been described for the determination of 41 metals. Methods have been described for the determination of the following lanthanides: cerium, dysprosium, europium, gadolinium, lutecium, neodymium, samarium, terbium and ytterbium. Of the 14 actinides only methods for the determination of uranium and thorium have been discussed.

Table 1.26 Methods for the determination of metals in saline sediments

	Spectro-photo	Spectro-fluoro	AAS	ICPAES	ICPMS	ASV	NAA	PAA	GLC	Column chromat.	X-ray s.e.	Electron probe	Misc
Al				4.1 4.44.3.1 4.44.3.2							4.1.2 4.44.8.1		
Sb				4.2.1 4.44.3.3	4.2.1 4.44.4.1		4.2.2 4.44.5.1	4.2.3 4.44.7.1					
As	4.3.1		4.3.2 4.44.2.3 4.4.1	4.3.3 4.44.3.3				4.3.4 4.44.7.1	4.3.5	4.3.6			4.3.7
Ba								4.4.2 4.44.7.1					
Be			4.5.1 4.44.2.2	4.5.2 4.44.3.1									
Bi			4.6.1	4.6.2 4.44.3.3									
Cd			4.7.1 4.44.2.1 4.44.2.2	4.7.2	4.7.2 4.44.4.1								
Cs							4.8.1 4.44.5.1						
Ca			4.9.1 4.44.2.1	4.9.2 4.44.3.1				4.9.3 4.44.7.1			4.9.4 4.44.8.1		
Ce				4.10.1 4.44.3.2									
Cr			4.11.1 4.44.2.1 4.44.2.2	4.11.2 4.44.3.2	4.11.2 4.44.4.1		4.11.3 4.44.5.1	4.11.4 4.44.7.1			4.11.5 4.44.8.1		

Table 1.26 continued

	Spectro-photo	Spectro-fluoro	AAS	ICPAES	ICPMS	ASV	NAA	PAA	GLC	Column chromat.	X-ray s.e.	Electron probe	Misc
Co	4.12.1 4.44.1.1		4.12.2 4.44.2.1 4.44.2.2	4.12.3 4.44.3.1 4.44.3.2			4.12.4 4.44.5.1	4.12.5 4.44.7.1					
Cu	4.13.1 4.44.1.1		4.13.2 4.44.2.1 4.44.2.2	4.13.3 4.44.3.1 4.44.3.2		4.13.4 4.44.6.1					4.13.5 4.44.8.1	4.13.6 4.44.9.1	4.13.7
Dy				4.14.1 4.44.3.2									
Ga				4.15.1 4.44.3.2									
Ir			4.16.1										
Fe			4.17.1 4.44.2.1	4.17.2 4.44.3.1 4.44.3.2			4.17.3 4.44.5.1	4.17.4 4.44.7.1			4.17.5 4.44.8.1		4.17.6
Pb			4.18.1 4.44.2.1 4.44.2.2	4.18.2 4.44.3.1	4.18.2 4.44.4.1	4.18.3 4.44.6.1		4.18.4 4.44.7.1					
La				4.19.1 4.44.3.2									
Mg				4.20.1 4.44.3.1				4.20.2 4.44.7.1					
Mn			4.21.1 4.44.2.1	4.21.2 4.44.3.1 4.44.3.2							4.21.3 4.44.8.1		
Hg		4.22.1	4.22.2		4.22.3								4.22.4

Table 1.26 continued

	Spectro-photo	Spectro-fluoro	AAS	ICPAES	ICPMS	ASV	NAA	PAA	GLC	Column chromat.	X-ray s.e.	Electron probe	Misc
Mo	4.23.1				4.23.2 4.44.4.1								4.24.5
Ni			4.24.1 4.44.2.1 4.44.2.2	4.24.2 4.44.3.1 4.44.3.2	4.24.2 4.44.4.1			4.24.3 4.44.7.1			4.24.4 4.44.8.1		
Pt			4.25.1										
K								4.26.1 4.44.7.1			4.26.2 4.44.8.1		
Re			4.27.1										
Rb							4.28.1 4.44.5.1						
Sc	4.29.1			4.29.2			4.29.2						
Se	4.30.1		4.30.2 4.44.2.3	4.30.3 4.44.3.3					4.30.4				
Si				4.31.1 4.44.3.2							4.31.2 4.44.8.1		
Ag			4.32.1										
Na				4.33.1 4.44.3.1				4.33.2 4.44.7.1					
Sr					4.34.3 4.44.4.1		4.34.1 4.44.5.1	4.34.2 4.44.7.1					
Tl			4.35.1		4.35.2 4.44.4.1	4.35.1							

Table 1.26 continued

	Spectro-photo	Spectro-fluoro	AAS	ICPAES	ICPMS	ASV	NAA	PAA	GLC	Column chromat	X-ray s.e.	Electron probe	Misc
Sn		4.36.1			4.36.2 4.44.4.1								
Ti			4.44.3.2	4.37.1 4.44.3.2				4.37.2 4.44.7.1			4.37.3 4.44.8.1		
U					4.44.4.1 4.38.2			4.44.7.1 4.38.1					
V				4.39.1 4.44.3.1 4.44.3.2									
Yb				4.40.1 4.44.3.2									
Y				4.41.1									
Zn			4.42.1 4.44.2.1	4.42.2 4.44.3.1 4.44.3.2	4.42.2 4.44.4.1	4.44.6.1		4.42.3 4.44.7.1			4.42.4 4.44.8.1	4.42.5 4.44.9.1	
Zr	4.43.1			4.43.2 4.44.3.2				4.43.3 4.44.7.1					

Spectrophoto = spectrophotometry; Spectrofluoro = spectrofluorimetry; AaS = atomic absorption spectrometry; ICPAES = inductively coupled plasma atomic emission spectrometry; ICPMS = inductively coupled plasma mass spectrometry; ASV = anodic stripping voltammetry; NAA = neutron activation analysis; PAA = photon activation analysis; GLC = gas chromatography; column chromat. = column chromatography; X-ray s.e. = X-ray second emission spectrometry; electron probe = electron probe microanalysis; misc = miscellaneous

Source: Own files

Table 1.27 Methods for the determination of metals in estaurine sediments

	AAS	ICPAES	Column chromat.	X-ray second. emission spectro.	Misc.
Al		5.1.1 5.26.2.1		5.1.2 5.26.4.1	
As	5.2.1		5.2.1		
Be	5.3.1 2.26.1.2				
Cd	5.4.1 5.26.1.1 5.26.1.2				
Ca				5.5.1 5.26.4.1	
Ce		5.6.1 5.26.2.1 5.26.1.2			
Cr	5.7.1 5.26.1.1 5.26.1.2	5.7.2 5.26.2.1		5.7.3 5.26.4.1	
Co	5.8.1 5.26.1.1 5.26.1.2	5.8.2 5.26.2.1 5.26.1.2			
Cu	5.9.1 5.26.1.1 5.26.1.2	5.9.2 5.26.2.1		5.9.3 5.26.4.1	
Dy		5.10.1 5.26.2.1			
Fe	5.11.1 5.26.1.1	5.11.2 5.26.2.1		5.11.3 5.26.4.1	5.11.4
La		5.12.1 5.26.2.1			
Pb	5.13.1 5.26.1.1 5.26.1.2				
Mn	5.14.1 5.26.1.1	5.14.2 5.26.2.1		5.14.3 5.26.4.1	
Hg					5.15.1
Ni	5.16.1 5.26.1.1 5.26.1.2	15.16.2 5.26.2.1		5.16.3 5.26.4.1	
K				5.17.1 5.26.4.1	
Se	5.18.1				
Si		5.19.1 5.26.2.1		5.19.2 5.26.4.1	
Ti		5.20.1 5.26.2.1		5.20.2 5.26.4.1	
V		5.21.1 5.25.2.1			

Table 1.27 continued

	AAS	ICPAES	Column chromat.	X-ray second. emission spectro.	Misc.
Y		5.22.1			
		5.26.2.1			
Y		5.23.1			
		5.26.2.1			
Zn	5.24.1	5.24.2		4.24.3	
	5.26.1.1	5.26.2.1		5.26.4.1	
Zr		5.25.1			
		5.26.2.1			

AAS = atomic absorption spectrometry; ICPAES = inductively coupled plasma atomic emission spectrometry; X-ray second. emission = X-ray second emission spectrometry; Misc. = miscellaneous
N.B. The following additional metals which are determinable in saline sediments (Table 1.26) could in many cases be determined in estaurine sediments: antimony, barium, bismuth, caesium, gallium, iridium, magnesium, molybdenum, platinum, rhenium, rubidium, silicon, silver, sodium, strontium, thallium, tin, uranium and vanadium.

Source: Own files

Table 1.28 Methods for the determination of anions in sediments

	Spectro-photometry	Chemilum-inescence	Gas chromat-ography	Ion chromat-ography	Misc.
Nitrate	8.1.1	8.1.2			
Sulphate				8.2.1	
Sulphide			8.3.1		8.3.2

Source: Own files

Methods have not been described for the following metals:

Group	Metals for which methods have not been described in the literature
2	lithium, boron
4	gallium, germanium
5	niobium, technecium, ruthenium, rhodium, palladium
6	rhenium, osmium, platinum, polonium
7	francium, radium, actinium
Lanthanides	praseoactinium, promethium, holmium, erbium, tamerium
Actinides	protactinium, neptunium, plutonium, americium, curium, berkelium, californium, einsteinium, fermium, mendelevium, nobelium, laurencium

1.4.5.2 Determination of anions

Spectrophotometric methods have been described for the determination of borate, cyanide, fluoride, nitrate, nitrite and phosphate, Table 1.30.

Table 1.29 Methods for the determination of metals in sludges

Sludge	Spectro-photo.	Spectro-fluoro	AAS	ICPAES	Diff. pulse polar.	Potentio-metric str.	NAA	PAA	TLC	Misc
Al			6.1.1 6.53.1.3 6.53.1.5				6.1.2 6.53.3.1			
Sb			6.2.1 6.53.1.6				6.2.2 6.53.3.1	6.2.3 6.53.4.1		
As			6.3.1 6.53.1.3 6.53.1.6	6.3.2			6.3.3 6.53.3.1	6.3.4 6.53.4.1		
Ba							6.4.1 6.53.3.1	6.4.2 6.53.4.1		
Bi			6.5.1 6.53.1.6				6.5.2 6.53.3.1	6.5.3 6.53.4.1		
Cd		6.6.1	6.6.2 6.53.1.1– 6.53.1.4	6.6.3 6.53.2.1	6.6.6 6.53.6.1		6.6.4 6.53.3.1	6.6.5 6.53.4.1		
Cs							6.7.1 6.53.3.1	6.7.2 6.53.4.1		
Ca			6.8.1 6.53.1.2 6.53.1.5				6.8.2 6.53.3.1	6.8.3 6.53.4.1		
Ce							6.9.1 6.53.3.1	6.9.2 6.53.4.1		
Cr			6.10.1 6.53.1.1 6.53.1.2	6.10.2 6.53.2.1	6.10.5 6.53.6.1	6.10.3 6.53.3.1	6.10.4 6.53.4.1	6.10.6		

Table 1.29 continued

Sludge	Spectro-photo.	Spectro-fluoro	AAS	ICPAES	Diff. pulse polar.	Potentio-metric str.	NAA	PAA	TLC	Misc
Co			6.11.1 6.53.1.1 6.53.1.2 6.53.1.7				6.11.2 6.53.3.1	6.11.3 6.53.4.1		
Cu			6.12.1 6.53.1.1 6.53.1.2	6.12.2 6.53.2.1	6.12.4 6.53.6.1	6.12.5 6.53.7.1	6.12.3 6.53.3.1			6.12.6
Dy							6.13.1 6.53.3.1			
Eu							6.14.1 6.53.3.1			
Gd							6.15.1 6.53.3.1			
Au							6.16.1 6.53.3.1			
Hf							6.17.1 6.53.3.1			
In							6.18.1 6.53.3.1	6.18.2 6.53.4.1		
Ir							6.19.1 6.53.3.1			
Fe			6.20.1 6.53.1.1 6.53.1.2 6.53.1.5		6.20.4 6.53.6.1		6.20.2 6.53.3.1	6.20.3 6.53.4.1		

Table 1.29 continued

Sludge	Spectro-photo.	Spectro-fluoro	AAS	ICPAES	Diff. pulse polar.	Potentio-metric str.	NAA	PAA	TLC	Misc
La							6.21.1 6.53.3.1			
Pb			6.22.1 6.53.1.1– 6.53.1.4	6.22.2 6.53.2.1	6.22.5 6.53.6.1	6.22.6 6.53.7.1	6.22.3 6.53.3.1	6.22.4 6.53.4.1	6.22.7	
Lu							6.23.1 6.53.3.1			
Mg			6.24.1 6.53.1.2 6.53.1.5				6.24.2 6.53.3.1	2.24.3 6.53.4.1		
Mn			6.25.1 6.53.1.1 6.53.1.2 6.53.1.7		6.25.4 6.53.6.1		6.25.2 6.53.3.1	6.25.3 6.53.4.1		
Hg			6.26.1 6.53.1.3 6.53.1.4				6.26.2 6.53.3.1	6.26.3 6.53.4.1		6.26.4
Mo	6.27.1		2.67.4 6.53.1.7				6.27.2 6.53.3.1	6.27.3 6.53.4.1		
Nd							6.28.1 6.53.3.1			
Ni			6.29.1 6.53.1.1 6.53.1.2	6.29.2 6.53.2.1			6.29.3 6.53.3.1	6.29.4 6.53.4.1		
K			6.30.1 6.53.1.2				6.30.2 6.53.3.1	6.30.3 6.53.4.1		

Table 1.29 continued

Sludge	Spectro-photo.	Spectro-fluoro	AAS	ICPAES	Diff. pulse polar.	Potentio-metric str.	NAA	PAA	TLC	Misc
Rb							6.31.1 6.53.3.1	6.31.2 6.53.4.1		
Sm							6.32.1 6.53.3.1			
Sc							6.33.1 6.53.3.1	6.33.2 6.53.4.1		
Se			6.34.1				6.34.2 6.53.3.1	6.34.3 6.53.4.1		
Si	6.35.1						6.35.2 6.53.3.1	6.35.3 6.53.4.1		
Ag			6.36.1 6.53.1.3 6.53.1.7				6.36.2 6.53.3.1	6.36.3 6.53.4.1		
Na			6.37.1 6.53.1.2				6.37.2 6.53.3.1	6.37.3 6.53.4.1		
Sr							6.38.1 6.53.3.1	3.38.2 6.53.4.1		
Ta							6.39.1 6.53.3.1			
Tb							6.40.1 6.53.3.1			
Te			6.41.1 6.53.1.6				6.41.2 6.53.3.1	6.41.3 6.53.4.1		
Tl			6.42.1 6.53.1.6				6.42.2 6.53.3.1	6.42.3 6.53.4.1		

Table 1.29 continued

Sludge	Spectro-photo.	Spectro-fluoro	AAS	ICPAES	Diff. pulse polar.	Potentio-metric str.	NAA	PAA	TLC	Misc
Th							6.43.1 6.53.3.1			
Sn			6.44.1 6.53.1.7				6.44.2 6.53.3.1	6.44.3 6.53.4.1		
Ti							6.45.1 6.53.3.1	6.45.2 6.53.4.1		
W							6.46.1 6.53.3.1			
V			6.47.1 6.53.1.3 6.53.1.6				6.47.2 6.53.3.1	6.47.3 6.53.4.1		
U							6.48.1 6.53.3.1			
Yb							6.49.1 6.53.3.1			
Y							6.50.1 6.53.3.1	6.50.2 6.53.4.1		
Zn	6.51.1		6.51.2 6.53.1.1 6.53.1.2 6.53.1.3	6.51.1.3 6.53.2.1	6.51.6 6.53.6.1		6.51.4 6.53.3.1	6.51.5 6.53.4.1		
Zr							6.52.1 6.53.3.1	6.52.2 6.53.4.1		

Spectrophoto. = spectrophotometry; Spectrofluoro = Spectrofluorimetry; AAS = atomic absorption spectrometry; ICPAES = industively coupled plasma atomic emission spectrometry; Potentiometric. str. = Diff. pulse polar. = Differential pulse polarography; Potentiometric stripping; NAA = neutron activation analysis; PAA = photon activation analysis; TLC = Thin-layer chromatography; Misc. = miscellaneous

Source: Own files

Table 1.30 Methods for determination of anions in sludges

Sludge	Spectro-photometry	Ion selective electrode	High-performance liquid chromatography	Ion chromatography	Drager tube	Atomic absorption spectrometry
BO₃	9.1.1					
Cl		9.2.1				
CN	9.3.1					9.3.2
F	9.4.1					
NO₃	9.5.1	9.5.2				
NO₂	9.6.1					
PO₄	9.7.1		9.7.2			
S		9.8.1			9.8.2	
WO₄				9.9.1		

Source: Own files

References

1 Ruzicka, J. and Hansen, E.A. *Analytical Chemistry*, **78**, 145 (1975).
2 Goulden, P.D. and Brooksbank, P. *Analytical Chemistry*, **46** ,1431 (1974).
3 Stockwell, P.P. In *Topics in Automatic Chemical Analysis*, Horwood, Chichester (1979).
4 Dennis, A.L. and Porter, D.G. *Journal of Automatic Chemistry*, **2**,134 (1981).
5 Thompson, M., Pahlavanpour, B. and Thorne, L. *Analyst* (London), **106**, 467 (1981).
6 Greenfield, S. Jones, I.L. and Berry, C.T. *Analyst* (London), **89**, 713 (1964).
7 Wendt, R.H. and Fassel, V.A. *Analytical Chemistry*, **37**, 920 (1965).
8 Scott, R.H. *Analytical Chemistry*, **46**, 75 (1974).
9 Thompson, M. and Walsh, J.N. In *A Handbook of Inductively Coupled Plasma Spectrometry*, Blackie, London & Glasgow, p55 (1983).
10 Suddendorf, R.F. and Boyer, K.W. *Analytical Chemistry*, **50** ,1769 (1978).
11 Sharp, B.L. *The conespray nebulizer*, British Technology Group, Patent Assignment No 8,432,338.
12 Gunn, A.M., Milland, D.L. and Kirkwright, G.F. *Analyst* (London), **103**, 1066 (1978).
13 Matusierucz, H. and Barnes, R.M. *Applied Spectroscopy*, **38**, 745 (1984).
14 Tikkanen, M.W. and Niemczyk, T.M. *Analytical Chemistry*, **56**, 1997 (1984).
15 Salin, E.D. and Horlick, G. *Analytical Chemistry*, **51**, 2284 (1979).
16 Salin, E.D. and Szung, R.L.A. *Analytical Chemistry*, **56**, 2596 (1984).
17 Gray, A.L. *Analyst* (London), **100**, 289 (1975).
18 Honk, R.S. *Analytical Chemistry*, **52**, 2283 (1980).
19 Date, A.R. and Gray, A.L. *Analyst* (London), **106**, 1255 (1981).
20 Date, A.R. and Gray, A.L. *Spectrochimica Acta*, **38 B**, 29 (1983).
21 Gray, A.L. and Date, A.R. *Analyst* (London), **108**, 1033 (1983).
22 Gray, A.L. *Spectrochimica Acta*, **41 B**, 151 (1986).
23 Douglas, D.J., Quan, E.S.K. and Smith, R.G. *Spectrochimica Acta*, **38 B**, 39 (1983).
24 Harlick, G. Winter Conference on Plasma Spectrometry, Hawaii January, Paper 2 (1986).
25 Pickford, C.J. and Brown, R.M. *Spectrochimica Acta*, **41 B**, 185 (1986).
26 Stathan, P.J. *Analytical Chemistry*, **49**, 2149 (1977).
27 Yaneda, Y. and Horiuchi, T. *Development Scientific Instruments*, **42**, 1069 (1971).
28 Aiginger, H. and Wodbrauschek, P. *Nuclear Instruments and Methods*, **114**, 157 (1974).
29 Knoth, J, and Schwenke, H. *Fresenius Zeitschrift Für Analytisch Chemie*, **291**, 200 (1978).
30 Knoth, J. and Schwenke, H. *Fresenius Zeitschrift Für Analytisch Chemie*, **201**, 7 (1980).
31 Schwenke, H. and Knoth, J. *Nuclear Instruments and Methods*, **193**, 239 (1982).
32 Pella, P.A. and Dobbyn, R.C. *Analytical Chemistry*, **60**, 684 (1988).
33 Prange, A., Knoth, J., Stobel, R.B., Bodekker, H. and Kramer, K. *Analytica Chimica Acta*, **195**, 275 (1987).
34 Bosch, F., El Goresy, A., Martin, B., Pouh, B. *et al*. *Science*, **199**, 765 (1978).
35 Castle, R.G. *Journal of Institute of Environmental Management*, **275**, 2nd June (1988).
36 Newstead, S. *Radioactivity in Drinking Water*. Paper submitted to the Joint Working Group for Radioactivity Releases affecting the water industry, HM Inspectorate of Pollution (1988).
37 Commission for European Communities. Proposal for a Communal Regulations (Euratom) Commission COM (87) 281 Final (1987).

38 Cothern, C.R. *Health Physics*, **50**, 33 (1986).
39 Nadkarni, R.A. *American Laboratory*, **13** ,22 (1981).
40 Small, H., Stevens, T.S. and Bauman, W.C. *Analytical Chemistry*, **47**, 1801 (1975).
41 Kuchn, D.G., Brandvig, R.L., Lundeen, D.C. and Jefferson, R.H. *International Laboratory*, **82**, September (1986) .
42 Haynes, W. *Perkin–Elmer Atomic Absorption Newsletter*, **17**, 49 (1978).
43 Dalton, E.F. and Melanoski, A.J. *Journal of Official Analytical Chemists*, **52**, 1035 (1969).
44 Adrian, W.A. *Perkin–Elmer Atomic Absorption Newsletter*, **10**, 96 (1971).
45 Park Manual, 207 M Park Instruments Co, 211 Fifty-third Street, Moline, IL 62165.
46 Rence, B.W. *Automatic Chemistry*, **105**, 1137 (1980).
47 Revenz, K. and Hasty, E. Recovery Study using an elevated pressure temperature microwave dissolution technique, presented at the 1987 Pittsberg Conference and Exposition on Analytical Chemistry and Applied Spectroscopy March (1987).
48 Nadkarni, R.A. *Analytical Chemistry*, **56**, 2233 (1984).
49 Zamzow, D.S., Baldwin, D.P., Weeks, J., Bajie, S.J. and D'Silva, A.P. *Environmental Science and Technology*, **28**, 352 (1994).
50 Taylor, H.E., Garbarino, J.R., Murphy, D.M. and Beckett, R. *Analytical Chemistry*, **64**, 2036 (1992).
51 Duckworth, D.C., Barshick, C.M. and Smith, D.H. *Journal Analytical Atomic Spectroscopy*, **8**, 875 (1993).
52 Yamamoto, K.Y., Cremers, D.A., Ferris, M.A. and Foster, L.E. *Applied Spectroscopy*, **50**, 222 (1996).
53 Berwick, M., Berry, P.F., Voots, G.R. *et al. Advances in X-ray Analysis*, **30B**, 1047 (1992).
54 Battista, G.A., Gerhasi, R., Degetto, S. and Sbrignadello, G. *Spectrochimica Acta Part B*, **48B**, 217 (1993).
55 Schramel, P. *Mikrochimica Acta*, **3**, 355 (1989).
56 Dolan, R., Van Loon, J., Templeton, D. and Pandyn, A. *Fresenius Zeitschrift für Analytische Chemie*, **336**, 49 (1990).
57 Cresser, M. Analytical *Spectroscopy Libe Atomic Absorption Spectrometry*, **5**, 515 (1991).
58 Raab, G.A., Enwall, R.E., Cole, W.G., Kuharic, C.A. and Duggan, J.S. *Environmental Science Research*, **42** (Chem Prot Environ), 155 (1991).

Chapter 2

Determination of Metals in Soils

2.1 Aluminium

2.1.1 Spectrophotometric methods

An early spectrophotometric method [1] for aluminium in soil involves the use of a Technicon sample changer, proportioning pump and automatic colorimeter. The method is based on the measurement of the rate of colour development in the reaction between aluminium and xylenol orange in ethanolicmedia. The calculation graph is rectilinear up to 2.7 mg L^{-1} aluminium and the coefficient of variation 4.5%.

2.1.2 Flow injection analysis

Flow injection analysis has been used to determine aluminium in soil. Reis *et al.* [2] studied the spectrophotometric determination of aluminium in soil using merging zones and sequential addition of pulsed reagents.

Tecator [3] have described a flow injection method for the determination of 0.5–100 mg L^{-1} aluminium in 0.1M potassium chloride extracts of soils in which the acidified soil extract is injected into a carrier stream which has the same composition as the sample matrix (ie 0.1M KCl) and merged with a masking solution for iron (R1 in Fig. 2.1, hydroxylamine and 1.10 phenanthroline monohydrate or o-phenanthroline hydrochloride) and subsequently with the colour reagent for aluminium (R2, pyrocatechol violet) and a buffer (R3, aqueous hexamethylene tetramine). The coloured complex formed between aluminium and pyrocatethol violet is measured at 585 nm. Repeatability is 1% RSD.

In addition to the above method, based on the use of pyrocatechol violet, Tecator also describe a flow injection analysis for determining 0.5–0.5 mg L^{-1} aluminium in soil extracts based on the measurement of the chromazurol-aluminium complex at 570nm[4, 5].

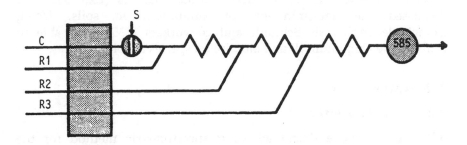

Fig. 2.1 Determination of aluminium in soil by flow injection analysis Range 0.1–2.0 mg L^{-1} extract solution. 0.5–100 mg L^{-1} Al in the soils sample

Source: Tecator Ltd. [3]

2.1.3 Atomic absorption spectrometry

Ross *et al.* [35] analysed samples of soil leachates from laboratory columns and of soil pore water from field porous cup lysimeters for aluminium by atomic absorption spectrometry under two sets of instrumental conditions. Method 1 employed uncoated graphite tubes and wall atomisation; method 2 employed a graphite furnace pyrolitically-coated platform and tubes. Aluminium standards were prepared and calibration curves used for the colorimetric quantification of aluminium. Method 1 gave results which compared favourably with method 2 in both sensitivity and interference reduction for samples containing 1–15 uM aluminium.

2.1.4 Inductively coupled plasma atomic emission spectrometry

The determination of aluminium is discussed under multi-metal analysis in section 2.60.2.2.

2.1.5 Emission spectrometry

The determination of aluminium is discussed under multi-metal analysis in section 2.60.5.1.

2.1.6 Photon activation analysis

The determination of aluminium is discussed under multi-metal analysis in section 2.60.4.1.

2.1.7 Miscellaneous

Mitrovic *et al.* [6] and Kozuh *et al.* [7] have carried out aluminium speciation studies on soil extracts. Various markers [232–234] have discussed the determination of aluminium in soils. Using isotachoelectrophoresis Schmidt and co-workers [235] were able to differentiate aluminium III and aluminium species in soil leachates.

2.2 Americium

2.2.1 α spectrometry

Sill *et al.* [8] have discussed an α spectrometric method for the determination of americium and other alpha-emitting nucleids including curium and californium in potassium fluoride-pyrosulphate extracts of soils. This method is discussed further in section 2.60.11.1. Sekine [126] used α spectrometry to determine americium in soil with a chemical recovery of 60–70%.

2.2.2 Miscellaneous

Joshi [236] and Livens *et al.* [237] have discussed methods for the determination of americium-241 in soils.

2.3 Ammonium

2.3.1 Spectrophotometric method

Keay and Menage [9] have described an automated method for the determination of ammonium and nitrate in 2M potassium chloride extracts of soil. In this method a sample of soil (2g) is shaken for 1h with 2N-potassium chloride (20 ml) and the filtrate is distilled, in the Auto-Analyser, with a 0.25% suspension of magnesium oxide; the ammonia evolved is absorbed in 0.1N-hydrochloric acid and determined spectrophotometrically at 625nm by the indophenol method. The sum of ammonium plus nitrate is determined similarly, but with addition of 4.5% titanous chloride solution before distillation; this reduces nitrate but not nitrite.

Waughman [14] has described a microdiffusion method for the determination of ammonium and nitrate in soils.

Nitrate in the sample solution is reduced to ammonia by titanous sulphate and the ammonia is then released from the solution and diffused and absorbed onto a nylon square impregnated with dilute sulphuric acid. The nylon is then put into a solution which colours quantitatively when ammonia is present and a spectrophotometer is used to measure the colour.

2.3.2 Gasometric method

Alder *et al.* [10] describe a method for determining low levels of ammonium ions in solution in which the ammonium ion is oxidised with sodium hypobromite in alkaline medium, the evolved nitrogen is passed into an argon plasma,

$$2NH_3 + 3Na\ Br = 3Na\ Br + 3H_2O + N_2.$$

The nitrogen–hydrogen emission intensity produced in the plasma at 336nm is monitored. A practical detection limit of 0.1 μg nitrogen per ml for 5 ml aqueous sample solutions was obtained. The method has been applied to the determination of the exchangeable ammonium content of soil samples.

The instrumental system employed utilised a 2kW crystal-controlled radiofrequency generator operating at 27 MHz (International Plasma Corporation, model 120–27) and a 1m plane grating scanning monochromator (Monospek 1000, Rank Hilger Ltd, Margate, Kent). Details of the instrumental system are given in Table 2.1. A demountable plasma torch with tangential argon inlets and sample introduction from a central injector tube was used. The outer quartz tubing was extended to a height of 40 mm above the work coil to prevent entrainment of atmospheric nitrogen into the discharge.

The nitrogen generation apparatus is shown in Fig. 2.2. A three-way tap allowed the system to be flushed free of air before the injector gas was introduced to the plasma. Addition of hypobromite reagent solution to the sample was achieved by rotation of the glass bulb in the sidearm through 180°.

Procedure

A 5 ml sample containing ammonium ion was placed in the nitrogen generation flask. The 9 ml glass bulb was filled with the argon degassed hypobromite solution (100 ml 2% available chlorine hypochlorite plus 1 ml 10% potassium bromide) and inserted in the sidearm of the apparatus. After flushing the system free of air, argon was passed through the flask to the plasma injector inlet at 230 cm^3 min^{-1}. The hypobromite solution was then added to the sample by inverting the glass bulb; the NH 336.0nm emission intensity was measured in the plasma at a height of 30 mm above the work coil. The plasma operating parameters employed in the study are shown at Table 2.2. These were chosen to optimise the signal-to-background intensity ratio at the NH band head at 336.0nm. Typical analytical signals obtained for ammonium ion solutions containing 2, 4 and 10 μg N m L^{-1} are shown in Fig. 2.3. The total signal duration under the conditions employed was approximately 30 s.

Table 2.1 Instrumentation

Plasma power supply	IPC model 120–27. Operating frequency 27.12 MHz; power output 0–2kW continuously variable. Work coil 1½ turns 6 mm o.d. copper tubing.
Spectrometer	Hilger Monospek 1000. Czerny–Turner scanning monochromator with grating (1200 lines mm⁻¹) blazed at 300nm; reciprocal linear dispersion 01.8nm mm⁻¹.
Optics	Plasma imaged in 1:1 ratio onto entrance slit with two 7.5 cm focal length × 5 cm diameter fused silica lenses.
Readout	Signal from EMI 6256B photomultiplier tube displayed on Servoscribe chart recorder.
Plasma torch	Demountable fused silica torch with brass base. Coolant gas tubing, 21 mm o.d. Plasma gas tubing, 17 mm o.d. Injector tube, 6 mm o.d.

Source: Elsevier Science Publishers, Netherlands [10]

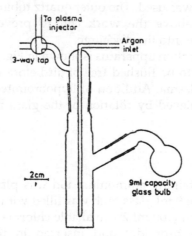

Fig. 2.2 Apparatus employed for generation of nitrogen from aqueous samples.
Source: Elsevier Science Publishers, Netherlands [10]

Table 2.2 Plasma operating conditions

Net forward r.f. power	1000W
Spectrometer slits	20 µ entrance and exit slits
Argon coolant gas flow rate	127000 cm³ min⁻¹
Argon plasma gas flow rate	1000 cm³ min⁻¹
Argon sample transport flow rate	2300 cm³ min⁻¹
Viewing height	30 mm above work coild

Source: Elsevier Science Publishers, Netherlands [10]

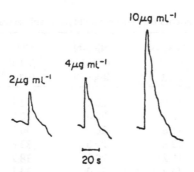

Fig. 2.3 Typical analytical signals at 336.0nm at different concentrations of ammonium ion (expressed as µg N mL $^{-1}$)
Source: Elsevier Science Publishers, BV, Netherlands [10]

A calibration graph, obtained by plotting the peak emission intensity at 336.0nm vs concentration, was found to be linear over the range 1–2000µg N m L $^{-1}$ with a slope of 0.75 when plotted on logarithmic axes. At concentrations of 3000 µg mL $^{-1}$ or greater, the evolved nitrogen extinguished the plasma, and at concentrations below 1 µg mL $^{-1}$ the use of the peak emission intensity as the analytical signal became unreliable. The integrated emission intensity for the whole signal at 336.0nm gave a linear response with concentration over the range 0.1–2 µg N mL $^{-1}$ and so this method of calibration was employed at these levels.

The precision of the method was estimated by repetitive determination of the nitrogen liberated from sample solutions containing 100 µg N mL $^{-1}$. A relative standard deviation of 0.02 was obtained for the determination of nitrogen in pure aqueous solutions.

The limit of detection of the method depended not on the background noise level but on the reproducibility of the blank determination. The concentration required to produce a signal equal to three times the standard deviation on the blank above the mean blank was found to be 0.1 µg N mL $^{-1}$ for a 5 ml sample.

As an application of the method the exchangeable ammonium-nitrogen content of six soil samples was determined. The soil samples, supplied by the Macaulay Institute for Soil Research (Aberdeen), were first air dried at 25°C and sieved to 2 mm.

The ammonium ion was extracted by the method described by Bremner [11]. A 10 g sample of each of the soils was shaken with 40 ml of neutral 2M potassium chloride solution for 1h. After the extracts had settled, 5 ml aliquots were removed from the clear supernatant liquid and analysed for ammonium nitrogen.

The results of the analyses are shown in the first column of Table 2.3. The values obtained are all within the expected range for exchangeable ammonium in soils. No interference was observed in the case of Na⁺, K⁺,

Table 2.3 Recovery experiments on potassium chloride soil extracts

Soil sample	NH_4–N in soil ($\mu g\ g^{-1}$)	NH_4–N added ($\mu g\ g^{-1}$)	NH_4–N found ($\mu g\ g^{-1}$)	Recovery (%) ($\mu g\ g^{-1}$)
Shaggart	10.0	20.0	32.1	107
Drumforber	10.1	20.0	30.2	100
Ardconnon	11.6	20.0	30.7	97
Smiddyhowe Wartie	13.6	20.0	33.8	101
Waulkmill Strachan	15.2	20.0	38.2	108
Logie Newton	16.2	20.0	34.1	94

Source: Elsevier Science Publishers, Netherlands [10]

Ca^{2+}, Mg^{2+}, Cl^-, NO_3^- and SO_4^{2-}. With hypobromite reagent, however, where a precipitate was formed on addition of the hypobromite reagent, the rate of evolution of nitrogen was found to be reduced, resulting in a reduction in the emission peak height. Such samples could still be analysed if the peak area was used as the analytical signal since this was unaffected by precipitate formation.

2.3.3 Flow injection analysis

Tecator Ltd [12] have described a flow injection analysis method for the determination of 0.2–1.4 mg L^{-1} (as NH_3N) of ammonia nitrogen in soil samples extractable by 2M potassium chloride. The soil suspension in 2M potassium chloride is centrifuged and filtered and introduced into the flow injection system for analysis of ammonia (and nitrate) one parameter at a time. Ammonia is determined by the gas diffusion principle in which a PTFE membrane is mounted in the gas diffusion cell.

2.3.4 Ammonia selective electrode

HMSO (UK) [13] have published a method for the determination of ammonia, nitrate and nitrite in potassium chloride extracts of soil extracts. An aliquot of the extract is made alkaline and the released ammonia, originating from ammonium ion, is determined either with an ammonia selective probe or, after removal by distillation, by titration.

2.4 Antimony

2.4.1 Atomic absorption spectrometry

The determination of antimony is discussed in section 2.5.2 [28], also under multi-metal analysis in section 2.60.1.6.

2.4.2 Inductively coupled plasma atomic emission spectrometry

The determination of antimony is discussed under multi-metal analysis in section 2.60.2.4.

2.4.3 Neutron activation analysis and photon activation analysis

The determination of antimony is discussed under multi-metal analysis in sections 2.60.3.1 (neutron activation analysis) and 2.60.4.1 (photon activation analysis).

2.4.4 Spark source mass spectrometry

The determination of antimony is discussed under multi-metal analysis in Section 2.60.6.1.

2.4.5 Miscellaneous

Chikhaikar *et al.* [15, 238] have discussed the speciation of antimony in soil extracts and soils. Asami *et al.* [239] have reviewed methods for the determination of antimony in soils.

2.5 Arsenic

Arsenic occurs naturally in the earth's crust, but a considerable amount of arsenic is added to the human environment through its uses in wood preservatives, sheep dips, fly paper, arsenical soaps, rat poison, glass additives, dye pigment for calico prints, wallpaper, lead shot and pesticides. During 1971, the estimated production of organoarsenical herbicides such as monosodium methanearsenate, disodium methanearsenate and hydroxydimethylarsine oxide (cacodylic acid) in the USA was 10.7×10^8 kg [9, 10, 16, 17]. Generally, soils contain about 5.0 ppm of arsenic, but soils with a known history of arsenic application average about 165 ppm [18]. In some places such as Buns, Switzerland and Wiatapu Valley, New Zealand, the arsenic level in the soil may reach 10^4 ppm [19]; a substantial portion of arsenic in soil and soil-like material (sediment, clay, sand, etc.) is expected to be found in soluble form and probably can be dislodged easily by the action of water moving through the soil. Soluble forms of arsenic are relatively more mobile in the environment and pose a greater potential for contaminating both ground water and surface water. Soluble forms of arsenic from soil and soil-like material are likely to enter a bioconversion chain through their initial uptake by vegetation.

2.5.1 Spectrophotometric methods

The limitations of the Gutzeit method for determining arsenic are well known. The spectrophotometric molybdenum blue or silver diethyl-dithiocarbamate procedures tend to suffer from poor precision and accuracy as shown in collaborative studies [20, 21]. Sandhu [22] has described a spectrophotometric method for the direct determination of hydrochloric acid-releasable inorganic arsenic in soils and sediments. The method provides reliable data on the quantitative recovery of 2.0 µg of arsenic (V) added to 5.0 g (0.4 mg kg^{-1}) of soil, clay, sand and sediment samples. The method is simple, reliable and relatively rapid; 24 samples can be analysed in about 1h. It does not require elaborate equipment and can be routinely used for the quantitative determination of arsenic in soil and soil-like material. The detection limit has been established as 0.5 µg of arsenic. The extent of ionic interference in the use of this method for arsenic determination in soil was also quantitatively evaluated.

The native forms of hydrochloric acid-releasable arsenic in soil, clay, sand and sediment were determined in the untreated samples. Ten grams of each of these materials, followed by 50.0 ml of de-ionised water, 7.5 ml of concentrated (12M) hydrochloric acid, 2.0 ml of potassium iodide solution (150 g L^{-1}) and 0.7 ml of tin (II) chloride solution (400 g L^{-1} of SnCl$_2$2H$_2$O in concentrated hydrochloric acid) were transferred into an arsine generator and allowed to stand for about 30 min, with occasional manual shaking, in order to reduce the arsenic to the trivalent state (the hydrochloric acid concentration of the mixture in the generator is about 1.50M). Three grams of 20–30 mesh zinc were added to the arsine generator, and the generator was immediately connected to the absorber assembly, which was equipped with a lead acetate-impregnated glass-wool scrubber, and contained 4.0 ml of silver diethyldithiocarbamate (SDDC) reagent (1g of AgSCSN(C$_2$H$_5$)$_2$ in 200 ml of pyridine). The arsine reacts with the SDDC and produces a red complex, which is measured at 535 nm against a reagent blank (SDDC solution treated in the absorber tube according to the experimental procedure, but without soil or arsenic). An absorbance calibration graph using 0.0, 1.0, 2.0, 4.0 and 5.0 µg of arsenic soil was prepared. The recovery of arsenic added to various soils was negligible when the samples were analysed, after treatment for 24h with cacodylic acid. However, the recovery of arsenic from the same set of samples improved significantly when they were analysed one week later. The recovery of arsenic from the soil samples is highly variable (Table 2.4) and appears to depend on the soil characteristics. If methylarsines do not react with SDDC, as seems apparent from the behaviour of cacodylic acid in de-ionised water, then the organoarsenical that was added to the soil must be mineralised for the generation of SDDC-intractable arsine. The recovery of organoarsenical added to various soils varied from 98.0% (in Dothan$_{AP}$) to 11.5% (Dothan$_{B22}$) and

Table 2.4 Recovery of cacodylic acid arsenic from water and soil samples

Sample	Arsenic added	Arsenic recovered*					
		24 h			I week		
		µg	%	Standard deviation %	µg	%	Standard deviation %
Water	10 µg per 50 ml	0.26	2.60	2.7	0.28	2.8	2.4
Dothan_AP	2 µg per 5 g	0.03	1.50	3.0	1.96	98.0	2.8
Dothan _A2	2 µg per 5 g	0.08	4.00	2.9	0.64	32.0	2.6
Goldsboro_AP	2 µg per 5 g	0.10	5.00	2.6	0.59	29.5	2.8
Goldsboro_B21	2 µg per 5 g	0.13	6.50	2.9	0.52	26.0	3.1

* Recovered = Total − Native

Source: Royal Society of Chemistry [22]

suggests that some soils are more effective than others in their potential to mineralise cacodylic acid. It must be mentioned that some of the inorganic arsenicals (arsenic sulphide, aluminium arsenate, etc) are either insoluble or only sparingly soluble in 1.50M hydrochloric acid. Consequently, the technique determines various forms of inorganic arsenic released by 1.50M hydrochloric acid in soil-like materials.

Merry and Zarcinas [29] have described a silver diethyldithiocarbamate method for the determination of arsenic and antimony in soil. The method involves the addition of sodium tetrahydroborate to an acid digested sample which has been treated with hydroxylammonium chloride to prevent formation of insoluble antimony compounds. The generated arsine and stibine react with a solution of silver diethyldithiocarbamate in pyridine in a gas washtube (Fig. 2.4).

Absorbance is measured twice at wavelengths of 600 and 504nm.

At 600nm the concentration of arsenic can be determined because the Sb–Ag DDTC complex does not absorb light of this wavelength. At 504nm the molar absorptivity of the antimony complex with Ag DDTC reaches its maximum value but there is also appreciable light absorbance from the As–Ag DDTC complex at this wavelength. The antimony concentration can be calculated from the total extinction value measured at 504nm by substraction of the extinction value (at 504nm) that corresponds to the already determined arsenic concentration. It is clear that calibration curves of arsenic at 504 and 600nm and antimony at 504nm are necessary to perform the calculation.

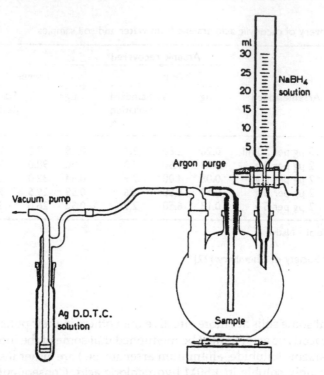

Fig. 2.4 Glass equipment for the spectrophotometric determination of arsenic and antimony by the silver diethyldithiocarbamate method
Source: Royal Society of Chemistry. [29]

2.5.2 Atomic absorption spectrometry

Atomic absorption spectrometry has applied extensively to the determination of arsenic in soils [23–31, 35]. Forehand *et al.* described a method, discussed below, for the determination of down to 0.8 mg L^{-1} arsenic in sandy soils.

Acid digestion

Weigh accurately 100 g of soil sample (adequately prepared by mixing, grinding and sifting through a quarter-inch mesh screen) into a 250 ml Erlenmeyer flask. Add 100 ml of 9.6 N HCl. (For very basic soils, add additional acid.) Swirl the flask to ensure thorough mixing, and allow the mixture to stand at room temperature for at least 12h.

Reduction of As(V) to As(III)

Decant 50 to 75 ml of the clear supernatant, depending upon the amount of arsenic present, into a 500 ml Erlenmeyer flask with a ground glass stopper. Add water, usually about 30 ml, to adjust the acid strength to about 6N. (It was found that the subsequent reduction was the most efficient at acid concentrations between 5 and 7N.) Then add 4 ml of 50% stannous chloride solution and 5 ml of the potassium iodide solution. Mix well by swirling, and allow the mixture to stand at room temperature for 15 min.

Solvent extraction

To the mixture add concentrated hydrochloric acid to adjust the acid strength to 9 N. The hydrochloric acid concentration was critical to the extractability of As(III). Add 50 ml of benzene, stopper, and shake the mixture for 1h on a mechanical shaker. After shaking, decant the mixture into a 250 ml separatory funnel. Allow the layers to separate, and drain and discard the bottom (acid) layer.

To the benzene solution in the same 250 ml separatory funnel, add accurately 15.0 ml of water and shake vigorously for about one minute. Allow the layers to separate, drain the bottom (water) layer into a 50 ml Erlenmeyer flask and stopper. This water solution, containing As(III), is ready for the atomic absorption measurements.

A Perkin–Elmer Model 303 Atomic Absorption Spectrophotometer was used for measuring the concentration of arsenic. A procedure was followed employing argon-hydrogen-entrained air flame and a premix burner with a three-slot head. Absorbance measurements were made at the arsenic line of 193.7nm. The instrument settings are: Damping 2; slit width 0.7nm; argon pressure 40 lb/in ^{2}, flowmeter reading 9 (about 15 L/min); hydrogen pressure 20 lb/in ^{2}, flowmeter reading 4.5 (about 6 L/min).

Arsenic recoveries between 85 and 90% were obtained for naturally occurring arsenic in sandy soils containing 91–98% sand, recoveries reducing as the sand content decreases from 90% to 23% (Table 2.5).

Recoveries of arsenic from sandy soils (90% sand) spiked with sodium meta-arsenite, sodium meta-arsenate, arsenic trioxide and arsenic pentoxide ranged between 86 and 92%.

The determination of arsenic by atomic absorption spectrometry with thermal atomisation and with hydride generation using sodium borohydride has been described by Thompson and Thomerson [24] and it was evident that this method could be modified for the analysis of soil. Thompson and Thoresby [25] have described a method for the determination of arsenic in soil by hydride generation and atomic absorption spectrophotometry using electrothermal atomisation. Soils are decomposed by leaching with a mixture of nitric and sulphuric acids or fusion with pyrosulphate. The resultant acidic sample solution is made to react with sodium borohydride and the liberated arsenic hydride is swept into an electrically heated tube mounted on the optical axis of a simple, laboratory constructed atomic absorption apparatus.

Table 2.5 Recovery of arsenic from soils (<90% sand)

Soil sample No.	Composition				μg As		Av rel error, %	Av recovery %
	Sand	Silt	Clay	Found	Added	Found after spiking		
IV	89.6	3.2	6.4	113	100	173.2; 182.9;	6.9	70.2
					300	320.8; 310.3;	1.3	67.3
					500	416.3; 422.8;	2.0	62.8
					1000	796.6; 811.2;	1.5	70.2
					2000	1467.2; 1525.0;	3.4	71.8
					5000	3230.5; 3432.0;	2.0	66.4
V	59.0	17.0	24.0	120	100	170.4; 187.7;	6.0	58.7
					300	312.8; 273.0;	6.5	60.7
					500	358.3; 371.3;	1.5	50.0
					1000	767.2; 772.4;	0.5	65.3
					2000	1341.8; 1384.2;	2.2	60.5
					5000	3148.0; 3180.5;	0.9	60.2
VI	22.8	3.2	74.0	203	100	239.1; 245.6;	4.7	43.6
					300	368.0; 341.0;	3.1	50.7
					500	449.8; 443.3;	3.1	51.1
					1000	668.5; 7808.5; 76.08	4.5	53.4
					2000	1350.0; 1293.8;	1.5	56.8
					5000	2980.5; 2996.5;	1.0	56.5

Source: American Chemical Society [23]

The advantages of high sensitivity, rapid analysis and simplicity of equipment are discussed, and the results for both types of sample material are compared with values obtained by use of the molybdenum-blue method.

Fusion procedure

Weigh 0.100 g of a dried soil sample, sieved to pass a 17 μm aperture, into a borosilicate glass test tube, to it add 0.5 g of powdered potassium pyrosulphate, then mix and fuse the solids until a quiescent melt is obtained. Leach the cooled melt by warming it with 10 ml of 15N sulphuric acid and transfer the solution into a 50 ml calibrated flask, washing the siliceous residue thoroughly with water. Oxidise the arsenic to the pentavalent state by addition of a few drops of 0.1M potassium permanganate (just enough to colour the solution pink), cool the solution and dilute it to volume with water. Dilute an aliquot of this sample stock solution appropriately with 3N sulphuric acid to obtain an arsenic concentration of not greater than 0.1 μg mL^{-1}.

Acid–leach procedure

Weigh 0.100 g of dried soil, sieved to pass a 177 μm aperture, into a wide-necked 250 ml conical flask and to it add 20 ml of concentrated nitric acid and 10 ml of 15 N sulphuric acid. Evaporate the solution on a hotplate to low volume, with further dropwise additions of concentrated nitric acid if necessary in order to prevent charring of organic matter. When the oxidation of organic matter is complete, increase the temperature until fumes of sulphur trioxide are evolved for a few minutes, but avoid an excessive loss of the acid. Cautiously dilute the cooled solution to about 25 ml with water and transfer into a 50 ml calibrated flask, washing the siliceous residue thoroughly with water; when the solution is at room temperature, dilute it to volume with water. Dilute an aliquot of this sample solution with 3N sulphuric acid to obtain an arsenic concentration of not greater than 0.1 μg mL^{-1}.

A schematic diagram of the apparatus is given in Fig. 2.5. The hydride-generator cell, with a volume of approximately 60 ml, was constructed from a sealed Quickfit B24/29 socket fitted with an angled side-arm near to its base. A Drechsel head with matching joint was used to admit nitrogen carrier gas and vent the arsenic hydride to the furnace tube. The sample was introduced by means of an Eppendorf pipette, which formed a gas-tight seal with a short length of plastic tubing attached to the side-arm.

The silica furnace tube (constructed to dimensions described by Thompson and Thomerson [24]) was wound with a heating element of flat cross-section of the type used in domestic irons. Approximately half

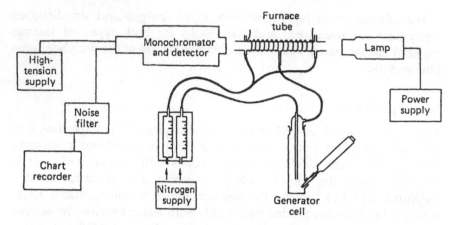

Fig. 2.5 Schematic diagram of apparatus for determining arsenic in soil
Source Royal Society of Chemistry [25]

of the length of a 750 W element, operated at 120 V from a variable output transformer, was found to be adequate to give an optimum working temperature of 850–900°C, Asbestos string was used in order to secure and insulate the wire as ceramic cement tended to shrink and distort the tube.

The arsenic hollow cathode lamp, used as the source of the 193.7nm resonance radiation, was of the rectangular slotted cathode type. The most reproducible results were obtained by using a wide monochromator slit width of 0.5 mm, thus observing a large area of the furnace type and minimising errors caused by inhomogeneity of the absorbing atom population. However, this slit width degraded the resolution (simulating the performance of a cheap monochromator) and dictated the use of an arsenic lamp with a high line to background ratio. Fig. 2.6 shows the lamp spectrum at this slit width and the proximity of an unidentified line, assumed to be that of neon, to the 193.7nm arsenic resonance line.

The absorption signal for arsenic is dependent on the oxidation state of the element prior to the hydride-generation stage. The ratio of signals for the two oxidation states varies with the major anion present. This behaviour could be indicative of anionic complexation of the arsenic, or simply a result of differences in the reaction kinetics for the reduction mechanism. Solutions in sulphuric acid were found to give more consistent results. The chemical procedure recommended here resulted in a solution of arsenic(V) in 3N sulphuric acid, and the absorption signal was 54% of that for arsenic(III) in the same solution. No interference was detected for cadmium, copper, lead, zinc, magnesium, calcium, iron and aluminium and the average arsenic recovery over the calibration range was 99.85%.

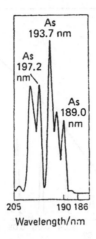

Fig. 2.6 Arsenic hollow-cathode lamp spectrum
Source: Royal Society of Chemistry [25]

Before beginning the analysis of a batch of samples, the apparatus was switched on and allowed to stabilise for 10–20 min. During this time, the alignments of the hollow cathode lamp, furnace tube and the monochromator were checked and the gas flow rates adjusted as has been previously described. The reference emission signal from the lamp (ie the 193.7nm resonance line) was then selected at the monochromator and displayed on the chart recorder.

The cell was charged with 1 ml of freshly prepared sodium borohydride reagent and the Drechsel head replaced. An Eppendorf pipette containing 1 ml of sample solution, or calibration standard, was then inserted into the side-arm seal. After allowing a few seconds for the ingressed air to be swept from the system (as indicated by the recorder pen returning to the maximum emission signal datum), the sample was injected into the cell. The absorption signal was instantaneously observed as a pulse at the recorder. The cell was then carefully rinsed with water and recharged with borohydride ready for the next sample.

Calibration was carried out by injecting solutions in 3N sulphuric acid containing known amounts of arsenic within the range 0.005–0.10 µg ml $^{-1}$. It was normal practice to precede and follow each batch of samples with a complete series of calibration standards and reagent blanks. Further calibration solution injections were interspersed with the samples and the averaged measurements of each standard used to prepare a calibration graph. Fig. 2.7 shows typical calibration standard traces and an averaged calibration graph corrected for a reagent blank.

The absorption signals were measured as peak heights from the recorder traces and calculated as percentages of the reference signal. The percentage absorptions, after blank correction, were converted into

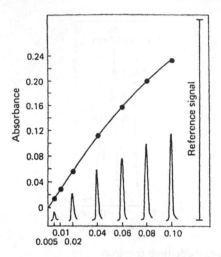

Fig. 2.7 Typical calibration standard traces and averaged graph
Source: Royal Society of Chemistry [25]

Table 2.6 Comparison of results for soils by atomic absorption spectrometry and molybdenum blue methods

Sample	Arsenic found µg g^{-1}	
	atomic absorption	molybdenum blue
I	299	238
2	375	305
3	293	259
4	189	155
5	103	71
6	467	449
7	207	172

Source: Royal Society of Chemistry [25]

absorbances and the calibration graph drawn. Blank corrections were also made to the sample determinations.

This method is capable of determining down to 0.001 mg L^{-1} of arsenic with a precision of 4.8 to 10.7% RSD.

In Table 2.6 are compared results obtained by this method with those obtained by a molybdenum blue spectrophotometric method [115, 116]. Values obtained by atomic absorption spectrometry are higher than those obtained by the molybdenum blue method and this is believed to reflect the greater inherent accuracy of the former method.

Table 2.7 Instrumental parameters for AA measurement for As and Sb with the hydride generating technique

Parameters		As	Sb
Wavelength		193.7	217.6
Slit width (nm)		0.7	0.2
Random noise suppression (s)		0.4	0.4
Quartz cell temperature	MHS–1 (C)	1000	1000
Programme	MHS–1	1	1
Pump vollume (NaBH₄)	MHS–1 (ml)	2.5	2.5
Sample volume	MHS–1 (ml)	25.0	25.0

Source: Springer Verlag Chemie GmbH, Germany [28]

Table 2.8 Instrumental parameters for graphite furnace AA determination of As and Sb (in presence of added nickel)

Parameters	As	Sb
Wavelength (mn)	193.7	217.6
Slith width (nm)	0.7	0.2
Ashing temperature (°C)	1300	1000
Ashing time (s)	30	30
Miniflow during atomisation (ml/min)	50	50
Deuterium background correction	Yes	Yes
Atomisation temperature (°C)	2600	2500 (maximum power)
Atomisation time (s)	4	4

Source: Spring Verlag Chemic GmbH, Germany [28]

To avoid problems previously encountered with flame atomic absorption spectrometry of arsenic and also with flameless methods such as that in which the element is converted to arsine, Ohta and Suzuki [27] proposed an alternative method based on electrothermal ionisation with a metal micro-tube atomiser. Effective atomisation can be achieved by the addition of thiourea to the arsenic solution or by preliminary extraction of the arsenic-thionalide complex. The second method is recommended for soil samples so as to avoid interference due to the presence of trace elements.

Haring et al. [28] determined arsenic and antimony by a combination of hydride generation and atomic absorption spectrometry. These workers found that compared to the spectrophotometric technique, the atomic absorption spectrophotometric technique with a heated quartz cell suffered from interferences by other hydride forming elements.

The recommended procedure for the determination of arsenic and antimony involves the addition to the sample of 20 ml of concentrated hydrochloric acid, 1 g of potassium iodide and 1 g of ascorbic acid. This solution should be kept at room temperature for at least five hours before initiation of the programmed MH 5–1 hydride generation system, ie before addition of ice cold 10% sodium borohydride, 5% sodium hydroxide. The instrument parameters are given in Tables 2.7 and 2.8 respectively for the hydride generation, atomic absorption technique and the graphite furnace atomic absorption spectrometric technique. In the hydride generation technique the evolved metal hydrides are decomposed in a heated quartz cell prior to determination by atomic absorption spectrometry. The hydride method offers improved sensitivity and lower detection limits compared to graphite furnace atomic absorption spectrometry. However, the most important advantage of hydride generating techniques is prevention of matrix interference which usually is very important in the 200nm area.

Stibine and arsine generation by sodium borohydride is influenced by the valence state of arsenic and antimony.

In fact the reduction of As(V) and Sb(V) to their hydrides is not completed during the time that the borohydride reacts in the acidic solution to which it is added. Transformation of the trivalent arsenic and antimony species to their hydrides is much faster and complete under the same conditions. A pre-reduction of As(V) and Sb(V) to As(III) and Sb(III) with iodide and ascorbic acid in hydrochloric acid medium was chosen as a suitable pre-treatment of the samples.

In this case the ascorbic acid served to prevent the oxidation of I^- to I_3^- by air or Fe(III) because the presence of I_3^- in the samples can lead to erroneous results by absorption of light at the same wavelength. Furthermore, potassium iodide has a masking effect on interferences by a large number of other metal ions. In this study the pre-reduction of As(V) and Sb(V) was investigated at different concentrations of the hydrochloric acid medium.

Additions of As(III), As(V), Sb(III) and Sb(V) to surface water, drinking water and sea water were almost completely recovered when the determinations were carried out in 6% hydrochloric acid. In 6% hydrochloric acid the pre-reduction takes less time and no difference in recovery between the trivalent and pentavalent species could be detected. The temperature of the quartz cell is an important factor in the sensitivity of the determination of the hydride forming elements. A temperature of 1000°C was adequate to ensure reproducible results.

In this study interferences from copper and lead in the concentration range up to 1 mg L $^{-1}$ and from calcium, magnesium, potassium, sodium and manganese in the concentration range up to 0.5 gL $^{-1}$ were examined. None of these metals in the concentration ranges mentioned were found to have any effect on the arsenic and antimony determination.

However, a suppression of the arsine and/or stibine formation was noticed if other hydride forming elements like selenium, bismuth and tin were present in the sample. Also a mutual interference between

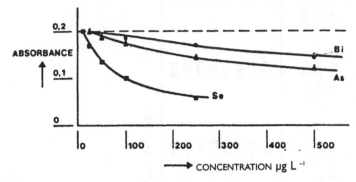

Fig. 2.8 Effect of As, Se and Bi on the absorbance signal of 2.5 µg L⁻¹ Sb. •Bi; ▲ As(III);
■ Se(IV)
Source: Springer Verlag Chemie GmbH, Germany [28]

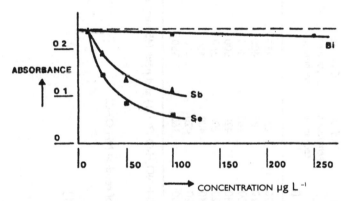

Fig. 2.9 Effect of Se, Sb and Bi on the absorbance signal of 2.5 µg L⁻¹ As. •Bi; ▲ Sb(III);
■ Se(IV)
Source: Springer Verlag Chemie GmbH, Germany [28]

the arsenic and antimony determination was found. In Fig. 2.8 the influence of various concentrations of bismuth, selenium and arsenic on the absorbance signal of 2.5 µgL⁻¹ antimony is shown. A similar graphical presentation is given in Fig. 2.9 for interferences on the determination of arsenic.

The antimony absorbance signal (2.5 µg L⁻¹ Sb) was found to decrease by more than 10% if the amount of selenium present in the sample is eight times higher than the antimony concentration. Arsenic and bismuth interfere with the antimony determination only in case of 30-resp 70-fold concentration of antimony. Formation of arsine from a solution of 2.5 µg L⁻¹ arsenic is suppressed at a five-fold concentration of both selenium and antimony. An explanation of the observed interferences in

Table 2.9 Result of intercomparison experiments of As determination in soil* samples

Arsenic all concentrations in µg L^{-1}	Not neutralised samples*				Neutralised samples*			
	Colorimetric technique		Hydride generation technique	Graphite furnace technique	Colorimetric technique		Hydride generation technique	Graphite furnace technique
	α	β			α	β		
Sample A	110	220	220	250	130	280	200	250
Sample A + 200 µg L^{-1} As	210	420	380	400	210	420	380	400
Sample B	170	300	260	300	170	320	260	300
Sample B + 200 µg L^{-1} As	240	540	420	500	460	560	420	500
Sample C	110	220	200	250	110	200	200	250
Sample C + 200 µg L^{-1} As	180	440	400	450	180	440	400	450
Blank	<10	<20	<20	<50	<10	<20	<20	<50
Standard 400 µg L^{-1} As	310	400	360	350	280	460	360	350

* 25.0 g of soil was refluxed (1h) with 100 ml of HCl-HNO$_3$H$_2$O (3:1:1) and next (45 min) with 20 ml of 50% NH$_3$OHCl

α without KI addition

β with KI addition: 3% KI

Not neutralised samples – unknown concentration of residual NH$_3$OHCl

Neutralised samples – 2% NH$_3$OHCl

Source: Springer-Verlag Chemie GmbH, Germany [28]

Table 2.10 Result of intercomparison experiments of Sb determination in soil* samples

Antimony all concentrations in µg L^{-1}	Not neutralised samples*				Neutralised samples*			
	Colorimetric technique		Hydride generation technique	Graphite furnace technique	Colorimetric technique		Hydride generation technique	Graphite furnace technique
	α	β			α	β		
Sample A	<10	80	80	100	<10	60	80	100
Sample A + 200 µg L^{-1} Sb	10	260	320	300	50	240	300	300
Sample B	120	800	800	750	160	800	800	700
Sample B + 200 µg L^{-1} Sb	260	1020	1000	1000	280	1000	1050	1000
Sample C	70	340	420	400	90	280	420	400
Sample C + 200 µg L^{-1} Sb	160	520	620	600	170	480	600	600
Blank	<10	<20	<20	<50	<10	<20	<20	<50
Standard 400 µg L^{-1} Sb	120	380	400	350	110	360	400	350

* 25.0 g of soil was refluxed (1h) with 100 ml of HCl-HNO$_3$–H$_2$O (3:1:1) and next (45 min) with 20 ml of 50% NH$_3$OHCl
α without KI addition
β with KI addition: 3% KI
Not neutralised samples – unknown concentration of residual NH$_3$OHCl
Neutralised samples – 2% NH$_3$OHCl

Source: Springer Verlag Chemic GmbH, Germany [28]

terms of a competition for the reaction with sodium borohydride does not seem to be satisfactory because of the excess concentration of the sodium borohydride reagent. It is more likely that the phenomenon is caused by the short lifetime of the hydrogen radical.

In Tables 2.9 and 2.10 are shown an interparison of results obtained in determinations of arsenic (Table 2.9) and antimony (Table 2.10) in soil by the three methods discussed above. Results are in reasonable agreement with each other. The need for potassium iodide treatment was clearly illustrated.

Jiminez de Blas et al. [31] have reported a method for the determination of total arsenic in soils based on hydride generation atomic absorption spectrometry and flow injection analysis. The method gave good recoveries and had a detection limit below 1 µg L^{-1} for an injection volume of 160 µL.

A UK standard method also discusses the determination of arsenic in soil by atomic absorption spectrometry [30]. The determination of arsenic in soils by atomic absorption spectrometry is also discussed under multi-metal analysis in sections 2.60.1.4 and 2.60.1.6.

2.5.3 Inductively coupled plasma atomic emission spectrometry

Hydride generation inductively coupled plasma atomic emission spectrometry has been used to determine arsenic in soils. This technique was found to greatly reduce sample preparation time [241]. The determination of arsenic is discussed under multi-metal analysis in section 2.60.2.4.

2.5.4 Inductively coupled plasma mass spectrometry

Lasztity et al. [240] have reported on inductively coupled plasma mass spectrometric methods for the determination of total arsenic in soils. The determination of arsenic in soils is also discussed under multi-metal analysis in section 2.60.2.2.

2.5.5 Neutron activation analysis and photon activation analysis

The determination of arsenic is discussed under multi-metal analysis in sections 2.60.3.1, 2.60.3.2 (neutron activation analysis) and section 2.60.4.1 (photon activation analysis).

2.5.6 Miscellaneous

Agemian and Bedak [32] have described a semi–automated method for the determination of total arsenic in soils. Chappell et al. [34] have described an inexpensive but effective method for the quantitative determination of arsenic species in contaminated soils. Chappell found that the extraction efficiency varied with the ratio of soil to acid and with the concentration of the acid. Rurikova and Beno [242] accomplished speciation of arsenic III and arsenic V in soils by cathodic stripping voltammetry. Wenclawiak and Krah [33] used reactive supercritical fluid extraction in speciation studies of inorganic and organic arsenic in soils. In this method, derivatisation with thioglycolic acid methyl ester was performed in supercritical carbon dioxide. Various other workers [243–245] have discussed the determination of arsenic in soils.

2.6 Barium

2.6.1 Inductively coupled plasma atomic emission spectrometry

The determination of barium is discussed under multi-metal analysis in section 2.60.2.2.

2.6.2 Emission spectrometry

The determination of barium is discussed under multi-metal analysis in section 2.60.5.1.

2.6.3 Stable isotope dilution

The determination of barium is discussed under multi-metal analysis in section 2.60.9.1.

2.7 Beryllium

2.7.1 Plasma emission spectrometry

The determination of beryllium is discussed under multi-metal analysis in section 2.60.5.1

2.8 Bismuth

2.8.1 Atomic absorption spectrometry

The determination of bismuth is discussed under multi-metal analysis in sections 2.60.1.5, 2.60.1.6 and 2.60.1.7.

2.8.2 Inductively coupled plasma atomic emission spectrometry

The determination of bismuth is discussed under multi-metal analysis in section 2.60.2.4.

2.8.3 Miscellaneous

Asami *et al.* [239] have reviewed methods for the determination of bismuth in soils.

2.9 Boron

2.9.1 Spectrophotometric methods

Spectrophotometric methods have been used to determine water soluble boron in soils. In one method [36] the soil is extracted with boiling water then converted to fluoroborate which is evaluated spectrophotometrically as the methylene blue complex.

Aznarez *et al.* [37] have described a spectrophotometric method using curcumin as chromopore for the determination of boron in soil. Boron is extracted from the soil into methyl isobutyl ketone with 2–methyl-pentane–2. In this method 0.2–1 g of finely ground soil is digested with 5 ml concentrated nitric–perchloric acid (3 + 1) in a PTFE lined pressure bowl for two hours at 150 °C. The filtrate is neutralised with 5M sodium hydroxide and diluted to 100 ml with hydrochloric acid 1 + 1. This solution is triple extracted with 10 ml of methyl isobutyl ketone to remove iron interference. This solution is then extracted with 10 ml 2–methyl-pentane–2,4 diol and this extract dried over anhydrous sodium sulphate. The development of the colour is carried out in the organic phase used for extraction by the addition of curcumin in glacial acetic acid and phosphoric acid as dehydrating agent. Spectrophotometric evaluation is carried out at 510nm.

The effect of those ions most frequently present in soils on the boron determinations is shown in Table 2.11. The interference of iron at concentrations higher than 7×10^{-5} M can be eliminated as the chloro complex by extraction with methyl isobutyl ketone. The total elimination of Fe(III) was not necessary as the phosphoric acid masked the residual Fe(III) in the boric acid–curcumin reaction.

The results of boron determinations in soils by the spectrophotometric method are shown in Table 2.12. Soil samples were provided by Aula Dei, Experimental Station of the Consejo Superior de Investigaciones Cientificas (CSIC) and correspond to alluvial ground or soils with a high limestone content (35–45% of calcium carbonate).

In a further spectrophotometric method [38, 39] for water soluble boron in soil, boron is extracted from soil with boiling water. Borate in the extract is converted to fluoroborate by the action of orthophosphoric acid and sodium fluoride. The concentration of fluoroborate is measured spectrophotometrically as the blue complex formed with methylene blue and which is extracted into 1, 2–dichloroethane. Nitrates and nitrites interfere; they are removed by reduction with zinc powder and orthophosphoric acid.

Table 2.11 Effect of foreign ions on boron determination

Determination of 58.3 µg of boron by the spectrophotometric method and 2.38 µg of boron by the fluorimetric method

Foreign ion	Maximum concentration tested without giving interference/M	
	Spectrophotometric method	Fluorimetric method
Cl⁻	7.2	7.2
SO_4^{2-}	2.5	2.5
NH_4^+	1.0	0.8
Na^+	0.2	0.2
K^+, Ca^{2+}, Al^{3+}	0.5	0.4
NO_3^-, NO_2^-, Cr^{3+}	0.2	0.2
Mg^{2+}, HCO_3^-, CO_3^{2-}	0.05	0.05
Mn^{2+}	0.05	0.01
F⁻	0.02	0.01
Cu^{2+}, Zn^{2+}, Ni^{2+}	4×10^{-3}	4×10^{-3}
Sr^{2+}, PO_4^{3-}, SIO_3^{2-}, Br⁻	3×10^{-3}	3×10^{-3}
Fe^{3+}	$7 \times 10^{-5*}$	$4 \times 10^{-5*}$
Fe^{3+} (by elimination with three 10 ml HBMK washes)	0.1	0.1

*Tolerance limit (M) as the concentration level at which the interferent causes anerror of not more than ± 2% (spectrophotometric method) or ±3% (fluorimetric method)

Source: Royal Society of Chemistry [37]

Table 2.12 Determination of boron in soils by the spectrophotometric method

Soil No.	Source	Mean boron content ± standard deviation*/µg g⁻¹	Boron added to spiked samples/µg	Mean recovery†%
1	El Bayo (H_1)	249 ± 3.4	250	97.2
2	El Bayo (H_2)	181 ± 1.3	200	104.0
3	El Bayo (H_3)	260 ± 2.4	250	101.6
4	El Bayo (H_4)	288 ± 1.0	300	102.7
5	El Espinal (H_1)	78 ± 0.6	80	98.8
6	Aula Dei (H_1)	165 ± 0.5	150	100.7
7	Villamayor (H_1)	79 ± 0.3	80	98.8
8	Villamayor (H_2)	94 ± 0.4	100	99.0

*Eight determinations
†Three determinations

Source: Royal Society of Chemistry [37]

2.9.2 Inductively coupled plasma atomic emission spectrometry

Inductively coupled plasma optical emission spectrometry has been applied to the determination of boron in soil extracts in amounts down to 0.05 mg L $^{-1}$ [40].

To extract boron from the sample a 50 g sample of air dried soil was boiled under reflux with 100 ml of water for 10 min. After partial cooling, the extract was filtered through an 18.5 cm Whatman No 3 filter paper into a conical beaker and a 40 ml aliquot of the filtrate transferred to a silica beaker. The solution was taken to dryness and the residue was oxidised twice with 10 ml of 6% hydrogen peroxide solution. The residue was then diluted to 50 ml.

Boron has a very simple ICPAES spectrum with the sensitive doublet at 249.7 and 249.8nm being the only useful analytical lines. Between 0.4 and 0.7 mg L $^{-1}$ boron was found in soil extracts. Zacinas and Cartwright [41] studied the acid dissolution of boron from soils prior to determination by inductively coupled plasma atomic emission spectrometry. The application of this technique in the multi-metal analysis of boron is also discussed in section 2.60.2.3.

2.9.3 Molecular absorption spectrometry

In addition to the spectrophotometric method discussed in section 2.9.1, Aznarez et al. [37] have described a method based on the molecular fluorescence of boron with dibenzoylmethane. The preliminary soil digestion and extraction procedures are identical to those described earlier. The relative fluorescence intensity of the boron complex is measured at 400nm with excitation at 390nm and quinine sulphate as reference. The calibration graph was linear in the range 0.5–5 μg of boron in aqueous solution (20–200 μg L $^{-1}$ of boron in the final solution to be measured). The detection limit and precision were 1 μg L $^{-1}$ of boron and 3% for 10 replicate determinations of 1.2 μg of boron, respectively. Interference by foreign ions is minimal (see Table 2.11).

2.10 Cadmium

Cadmium is readily taken up by most plants. The occurrence of cadmium in motor oils, car tyres, phosphorus fertilisers and zinc compounds explains its accumulation in soils; the cadmium contents of soils in non-polluted areas are below 1 ppm, but values as high as 50 ppm can be found [45].

2.10.1 Atomic absorption spectrometry

The determination of cadmium by graphite furnace atomic absorption spectrometry is especially difficult because cadmium is a volatile element, and matrix constituents cannot be removed by charring without a loss of cadmium. The use of selective volatilisation often makes it possible to obtain a cadmium peak before the background has risen to such a high value that it interferes with the cadmium measurement. Another unrecognised source of interference is char loss resulting from the salt matrix. Although uncoated graphite tubes can be used for the determination of cadmium because of its volatility, some workers have found that pyrolytically coated tubes give better results when cadmium is determined in the presence of high contents of alkali and alkaline–earth elements [113]. Many studies of the determination of cadmium in soil extracts have been reported, but a chelation–extraction step has always been used prior to determination by graphite furnace atomic absorption spectrometry in order to reduce matrix interferences and to improve detection limits [49, 104, 114–118].

Atomic absorption spectrometry with [19] or without [20–22] preliminary solvent extraction of metal has been applied extensively to the determination of cadmium in soils [42–44]. In an early standard official method [42] the sieved soil sample is digested with hot nitric perchloric acids then the filtered extract dissolved in hydrochloric acid. The extract is evaluated at 228.8nm using a cadmium hollow cathode lamp with a spectral width of 0.6nm.

Acetic acid extractable cadmium in soil [43] is determined on a 0.5M acetic extract. This is removed from an acid solution as its pyrrolidine dithiocarbamate complex by extraction into chloroform. The chloroform is removed by evaporation and the organic matter destroyed by wet oxidation. This residue is dissolved in hydrochloric acid and cadmium determined by atomic absorption spectrometry utilising the 228.8nm emission.

Berrow and Stein [75] have described a procedure based on digestion with aqua regia followed by atomic absorption spectrometry for the determination of cadmium (also iron and zinc) in soils. The soil sample was air dried at a maximum temperature of 30°C and sieved through a 2 mm sieve. The sieved soil was mixed, coned and quartered and about 30 g ground in an agate planetary mill for 30 min to <150 µ size. 3 g of soil was weighed into a 100 ml flask and 2–3 ml water added, then 7.5 ml concentrated by hydrochloric acid and 2.5 ml concentrated nitric acid and added per g of dry sample. The flask is covered and left to digest at 20°C for 16 h. A 30 cm reflux condenser is attached to the top of the flask which is then boiled gently for 2 h on a temperature controlled electrothermal extraction apparatus. After cooling the condenser is rinsed with 30 ml water and the solution filtered into 100 ml calibrated flask. The filter paper is rinsed five times with a few millilitres of warm (250°C) 2M nitric acid. After cooling the flask contents are made up to 100 ml with 2M nitric acid. The solution was then analysed for cadmium by atomic absorption spectrometry equipped with a single slot burner and an air acetylene

Table 2.13 Analysis of Canadian reference soils

Soil	Mean determined mg kg⁻¹	Rsd	Recommended value, mg kg⁻¹ oven dry basis	Solubility value % on
SO–1	0.12	33	0.15	80
SO–2	0.11	32	0.18	61
SO–3	0.07	4.2	0.14	50
SO–4	0.38	8.3	0.42	90

Source: Royal Society of Chemistry [75]

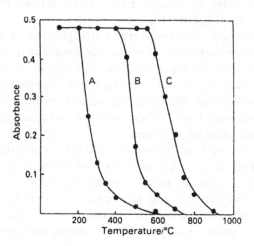

Fig. 2.10 Effect of analyte modification on the volatility of Cd in DTPA extract. A, DTPA soil extract; B, DTPA soil extract +1% HNO₃; and C, DTPA soil extract + 1% HNO₃–1000 μg mL⁻¹ Mo–1.5% H₂O₂

Source: Royal Society of Chemistry [75]

flame using a 228.8nm hollow cathode lamp. A relative standard deviation for cadmium of 3.4–5.2% was obtained. The cadmium contents obtained for Canadian Reference soil samples are compared with recommended values in Table 2.13.

Baucells [44] applied graphite furnace atomic absorption spectrometry to the determination of cadmium in soils with a precision of 0.4% at the 69 μL⁻¹ cadmium level. The decomposition curve for cadmium in a soil extract is shown in Fig. 2.10. The loss of cadmium during the charring cycle was high, preventing the use of any char in the atomisation process in order to remove the organic matrix or minimise interference effects. Baucells [44] used a procedure described by Henn [119] in which

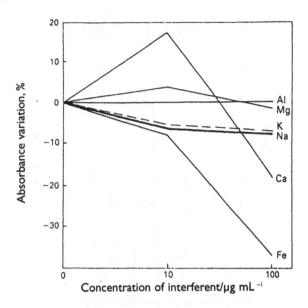

Fig. 2.11 Absorbance variation of Cd (2 μg L⁻¹) as a function of interferent metal concentrations
Source: Royal Society of Chemistry [75]

cadmium is converted to heteropolymolybdate anion with a central metal atom.

Fig. 2.10 shows the effect of the addition of 1% nitric acid – 100 μg mL⁻¹ molybdenum – 1.5% hydrogen peroxide on the volatility of cadmium when applied to a soil extract. The loss of cadmium when molybdenum is present begins at 550°C; this permits a correct decomposition treatment.

The absorbance remained constant when a temperature of 1000°C was exceeded, and 1200°C was chosen for good reproducibility. At this temperature, the background absorbance was zero; the matrix was atomised at higher temperatures (>1600°C).

The low atomisation temperature used and the presence of heavy metals in the samples required a subsequent clean-up step at a high temperature (3000°C) to avoid erroneous readings.

The minimum atomisation temperature that gives maximum absorbance (1000°C) is unexpectedly low relative to the temperatures cited by other workers, which range from 1600 to 2000°C. Baucells *et al.* [44] also obtained the same temperature (1000°C) for cadmium in 1% nitric acid.

An interference study was carried out with calcium, magnesium, sodium, potassium, iron and aluminium (as nitrates) at two concentration levels, 10 and 100 μg mL⁻¹, using a 2 μg mL⁻¹ cadmium solution. The

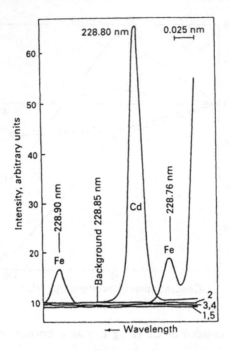

Fig. 2.12 Wavelength scans in the vicinity of the Cd I 228.802nm line. Cd I µg mL^{-1}, and Fe, 500 µg mL^{-1}. Other interferents: 1, 1000 µg mL^{-1} Al; 2, 100 µg mL^{-1} Ca; 3, 100 µg mL^{-1} Mg; 4, 1000 µg mL^{-1} Na; and 5, 1000 µg mL^{-1} K. Slit widths, 30 µm entrance and 40 µm exit.
Source: Royal Society of Chemistry [44]

results obtained are shown in Fig. 2.11. The most important interference effects observed are due to calcium and iron. The calcium concentration in the samples is almost constant (except when using very alkaline soils) and is due to the presence of calcium chloride in the extraction solution. The maximum iron concentration in the dilute solutions is 6.4 µg mL^{-1}.

The determination of cadmium by atomic absorption spectrometry is also discussed under multi-metal analysis in sections 2.60.1.1, 2.60.1.3, 2.60.1.4 and 2.60.1.5.

2.10.2 Inductively coupled plasma atomic emission spectrometry and inductively coupled plasma mass spectrometry

The application of inductively coupled plasma atomic emission spectrometry and graphite furnace atomic absorption spectrometry to the determination of cadmium (and molybdenum) in soils has been discussed by Baucells *et al.* [44]. Baucells *et al.* chose the 228.802nm cadmium line because it is well resolved from the 228.763nm iron line with the spectrometer used in this work (Fig. 2.12). Background measurement

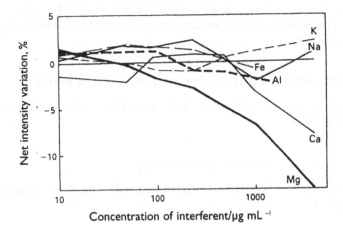

Fig. 2.13 Net line intensity variation (%) of Cd (1 μg mL −1_ as a function of interferent metal concentration
Source: Royal Society of Chemistry [44]

could only be carried out at +0.05nm. These workers obtained good agreement between cadmium values obtained by direct graphite furnace atomic absorption spectrometry and inductively coupled plasma atomic emission spectrometry. Chelation extraction procedures that require extensive sample handling are avoided.

Problems in the direct determination of cadmium in soil extracts by graphite furnace atomic absorption spectrometry are overcome by the use of a low atomisation temperature of 1200°C (mini-furnace or high heating rate of >2000°C s⁻¹), the addition of molybdenum, hydrogen peroxide and nitric acid, as a matrix modifier and accurate optimisation of the instrumental parameters.

The addition of ammonium dihydrogen phosphate and ammonium sulphate, normally used as matrix modifiers in the determination of cadmium, is not recommended with this type of sample because of the appearance of multiple peaks.

Inductively coupled plasma atomic emission spectrometry has proved to be an excellent technique for the direct analysis of soil extracts because it is precise, accurate and not time consuming, the level of matrix interference being very low. Of course, the graphite furnace technique yields better detection limits than the inductively coupled plasma procedure.

Interference effects on the determination of cadmium by the inductively coupled plasma technique are illustrated in Fig. 2.13.

Other applications of inductively coupled plasma atomic emission spectrometry has been discussed [107, 120, 121]. Further discussion of the determination of cadmium in multi-element analyses by this technique of soils is given in sections 2.60.2.1 (inductively coupled plasma atomic emission spectrometry) and section 2.60.2.2 (inductively coupled plasma mass spectrometry).

2.10.3 Differential pulse anodic stripping voltammetry

The determination of cadmium is discussed under multi-metal analysis in section 2.60.7.1.

2.10.4 X-ray fluorescence spectroscopy

The determination of cadmium is discussed under multi-metal analysis in section 2.60.8.1.

2.10.5 Emission spectrometry

The determination of cadmium is discussed under multi-metal analysis in section 20.60.5.1.

2.10.6 Miscellaneous

Lewin and Beckett [46] have shown that cadmium added to soils treated with sewage will quickly divide between a number of different forms of combination from some of which it can become available to plants during crop growth. These workers investigated reagents which can extract cadmium and make it available for analysis. Acidified fluoride and EDTA were effective extractants.

Christensen and Lun [47] developed a speciation procedure using a cation-exchange resin (Chelex 100) in a sequential batch/column/batch system for determining free divalent cadmium and cadmium complexes of various stabilities at the cadmium concentrations typically found in landfill leachates (less than 100 µg per litre). Results obtained on standardised solutions containing cadmium and on two actual leachates are included. The leachates had only a small percentage of free divalent cadmium and a large percentage of labile complexes.

Turner et al. [48] discussed the limitations in research on adsorption of trace metals on soils owing to inadequate control of composition and pH of the equilibrium solution. The use of chelating resins is suggested to establish and maintain constant pH and metal activity in a solution of constant ionic strength and composition. Details are given of the preparation of suitable resins and the experimental procedure used to investigate the adsorption of cadmium on iron gel and on organic matter over a range of cadmium:calcium ratios similar to those found under normal soil solution concentrations. The suggested method was more difficult and more time-consuming than conventional equilibration

Table 2.14 Recovery of cadmium and zinc added to calcium chloride extracts of soils

Soil	Added/ µg	Cadmium Recovered/ µg	Recovery, %	Zinc Recovered/ µg	Recovery, %
Egmont black loam	0	0.05	–	0.75	–
	1.00	1.15	110	1.85	110
	2.50	2.80	110	3.25	100
Makerau peaty silt loam	0	0.05	–	1.00	–
	1.00	1.10	105	1.95	95
	2.50	2.65	104	3.45	98
Tokomaru silt loam	0	0.01	–	1.15	–
	1.00	1.00	100	2.10	95
	2.50	2.60	104	3.40	90
Okaihau gravelly clay	0	0.84	–	3.10	–
	1.00	1.10	102	4.10	100
	2.50	2.65	103	5.40	92

Source: Royal Society of Chemistry [49]

methods. In some cases, however, its use may make determination of an entire adsorption isotherm unnecessary, since adsorption may be determined in response to one or more predetermined metal activities. It could also be used to evaluate possible mechanisms of metal adsorption.

Roberts et al. [49] have discussed the simultaneous extraction and concentration of cadmium and zinc from soil extracts. Extractions were conducted with calcium chloride adjusted to various pH values between 3 and 11. The simultaneous recovery of cadmium and zinc was essentially quantitative over the pH range 4–7 values ranging from 92 to 102%. An extraction of pH 4.5 was adopted. Adequate recoveries were obtained when the procedure was applied to spiked soils (Table 2.14). This method is discussed more fully in section 2.60.1.1.

Carlosena et al. [50] and Hirsch and Banin [246] have conducted studies on the speciation of cadmium in soil. Feng and Barrett [247] showed that microwave dissolution of soil and dust samples with nitric–hydrofluoric acid gave recoveries of cadmium (and lead) of over 90% in 30 min digestion. Various other workers [248–251] have reviewed methods for the determination of cadmium in soils.

2.11 Caesium

Caesium contamination on soil is one system for which spectroscopic information would be of great interest [252, 253]. The 134 and 137 isotopes decay by γ emission and are formed in high fission yield; the 137 isotope has a moderately long half-life (30 years). The caesium isotopes comprise one of the lasting health problems from the Chernobyl accident [254, 255]. Caesium can be highly mobile in some environments, and geochemically it has many of the same characteristics as potassium as a consequence of similar ionic radii in solution [253]. Hence, there is motivation for understanding the interaction of Cs^+ with naturally occurring mineral surfaces at the molecular level.

Caesium sorption has been extensively investigated, primarily by using sequential extractions together with γ spectroscopy (for radioisotopes) or atomic absorption for detection [252, 254–257]. This approach has been applied to the study of caesium contamination on soils [258]. Caesium was shown to prefer the mineral soil horizons in high organic soils [259]. From these studies, it has been possible to infer mechanistic details: Caesium will tenaciously adhere to adsorption sites and can be supplanted only by K^+ and NH_4^+ It appears to prefer surface 'defects', which have been termed frayed edge, and wedge sites [260–262]. However, understanding of caesium soil systems would benefit from direct spectroscopic information.

2.11.1 Spark source mass spectrometry

The determination of caesium in soil by multi-element analysis is discussed in section 2.60.6.1.

2.11.2 Stable isotope dilution

The determination of caesium is discussed under multi-metal analysis in section 2.60.9.1.

2.11.3 Imaging time of flight secondary ion mass spectrometry

Groenewald et al. [264] used an imaging time of flight secondary ion mass spectrometer (SIMS) [263] for characterisation of soil particles that had been exposed to caesium iodide solutions. SIMS is well suited to the analysis of Cs^+ because it readily forms gas phase secondary ions. The SIMS instrument utilises microfocused primary ion guns, achieving spatial resolutions of less than 1 μm. The ion optics transmit the secondary ions through three electrostatic sectors to a channel-plate detector, such that the spatial information is preserved. They also employed scanning electron microscopy/energy-dispersive X-ray spectroscopy (SEM/EDS). The results showed that Cs^+ could be detected and imaged on the surface of the soil particles readily at concentrations dow to 160 ppm, which corresponds to 0.04 monolayer. Imaging revealed that most of the soil surface consisted of aluminosilicate material.

However, some of the surface was more quartzic in composition, primarily silica with little aluminium. It was observed that adsorbed Cs^+ was associated with the presence of aluminium on the surface of the soil particles. In contrast, in high silica areas of the soil particle where little aluminium was observed, little adsorbed Cs^+ was observed on the surface of the soil particle. Using EDS, Cs^+ was observed only in the most concentrated Cs^+ soil system and Cs^+ was clearly correlated with the presence of aluminium and iodine. These results are interpreted in terms of multiple layers of caesium iodide forming over areas of the soil surface that contain substantial aluminium. These observations are consistent with the hypothesis that the insertion of aluminium into the silica lattice results in the formation of anionic sites, which are then capable of binding cations.

2.12 Calcium

2.12.1 Spectrophotometric method

Qiu Xing-chu *et al.* [51] have described a spectrophotometric method using chlorophosphonazo 30 mA for the determination of exchangeable calcium in soils. A 1M ammonium acetate extract of the soil is treated with triethanol-amine, quinolin–8–ol masking agents and the chromogenic reagent and evaluated spectrophotometrically at 630nm. Recoveries of 99.3 to 102.7% were obtained from calcium spiked soil samples. Magnesium, iron, aluminium, manganese, copper, zinc, tungsten, chromium, molybdenum and lead did not interfere at the levels normally present in soils.

Soil cation exchangeable capacity is not only an index for evaluating the nutrient and water retention ability of the soil, but also is an important basis for amelioration of soil and for applying, rationally, fertiliser. Exchangeable cations, absorbed by soil colloid include K^+, Na^+, Ca^{2+}, Mg^{2+}, Al^{3+} and H^+, K^+, Na^+, Ca^{2+} and Mg^{2+} are exchangeable bases. Al^{3+} and H^+ are exchangeable acids and the sum of these ions is known as the cation exchangeable capacity. Exchangeable Cu^{2+}, Zn^{2+} and Mn^{2+} are present at negligible concentrations.

Among the numerous methods used for determination of the total amount of exchangeable metal cations in acidic soils, the 1M ammonium acetate leaching method and 0.1M hydrochloric acid extraction – titrimetric methods are the best known and most applied. The former involves complete evaporation of the solution obtained after eluting the soil, and ignition of the residue to change all of the exchangeable metal salts into their carbonate form. The carbonates are dissolved in standard hydrochloric acid and the excess of acid back-titrated with standard sodium hydroxide solution. The drawback of this method is that the Al^{3+} and Fe^{3+} ions precipitate as hydroxides during the titration, and absorb the indicator. The end point of the titration is indistinct and no exact result can be obtained. The same problem also occurs in the 0.1 M hydrochloric acid extraction–titrimetric method. Moreover, some non-exchangeable metal also dissolves when soil is extracted with hydrochloric acid, therefore the result includes exchangeable and some non-exchangeable metal ions in the soil.

According to the pH balance method recommended by Jackson [52], acetic acid can be used to extract the total amount of exchangeable metal cations from an acidic soil, and the pH change measured carefully. This is then compared with the pH calibration graph of acetic acid so as to obtain the decrease in H⁺ in solution. However, this method requires a very precise acidity measurement (generally an accuracy of ±0.01 pH unit when the pH value is within 2.3–2.8). To overcome this difficulty Qiu Xing-chu and Zhu Ying-quan [53] have developed the bromophenol blue spectrophotometric method for the determination of exchangeable calcium in soils.

Method

A Model 72 spectrophotometer (Shanghai Analytical Instruments Factory) of wavelength range 420–700nm was employed for absorbance measurements.

Reagents

Bromophenol blue solution, 0.1%. Dissolve 1 g of BPB (Shanghai Chemical Industries Mfg) in 30 ml of 0.05M NaOH solution and dilute to 1000 ml with water. Filter if necessary.

Standard calcium acetate solution, 0.02N. Dissolve 1.7615 g of Ca $(CH_3COO)_2$ H_2O in water and dilute to one litre. Standardise with EDTA.

Procedure

Weigh out 5.00 g of soil and place in a 100 ml glass–stoppered conical flask. Add 50 ml of 1M acetic acid and shake the flask for 1 h. Centrifuge or filter the solution, using a Buchner funnel, pipette a 10 ml aliquot of the sample solution into a 50 ml calibrated flask and add 10 ml of 0.1% bromophenol blue in 0.15% sodium hydroxide solution. Dilute to volume with water and shake well.

Measure the absorbance against a reagent blank, in 2 cm cells at 580nm.

Prepare the calibration graph by taking a series of 10 ml portions of 1M acetic acid to which exactly 0, 0.50, 1.00, 2.0, 3.00 and 4.00 ml of 0.02N calcium acetate solution have been added and treat according to the above method.

Plot absorbance against the volume of calcium acetate solution added.

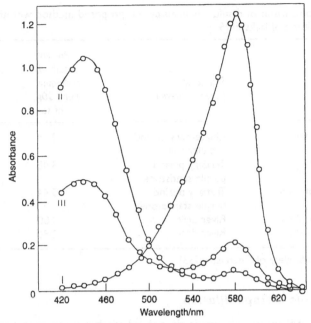

Fig. 2.14 Absorption spectra of BPB. [BPB] = 1.49 × 10^{-5} M, 1.0 cm cells. I. In sodium hydroxide solution (0.05 m); II, in acetic acid (0.2 m); and III, in hydrochloric acid (0.05 m); all against water as reference
Source: Royal Society of Chemistry [53]

The total amount of exchangeable metal cation (milliequiv per 100 g of soil) = 100 (NV/M) where N denotes the normality of the calcium acetate solution, V is the volume of calcium acetate solution found from the calibration graph (ml) and M is the mass of sample (g).

The pH range of bromophenol blue is 3.0–4.6 and its acidic and basic forms are yellow and blue, respectively. Fig. 2.14 shows the absorption spectra of the acidic and basic forms of bromophenol blue measured against water over the 420–660nm range. The absorption spectrum of the basic form has a maximum at 580nm, whereas the absorption of the acidic form is almost negligible at this wavelength.

A 98% recovery of added calcium is obtained by this procedure. Relative standard deviations are below 1.4% at the 6–5m equivalent calcium per 100 g soil level. Satisfactory agreement is obtained in a comparison of analysis performed on a variety of soils by this method and the pH balance method described by Jackson [52] (Table 2.15).

2.12.2 Inductively coupled plasma atomic emission spectrometry

The determination of calcium is discussed under multi-metal analysis in section 2.60.2.2.

Table 2.15 Comparison of results obtained by the proposed method with those of the titrimetric method of Jackson [52]

Sample No.	Soil type	Source of parent material	Present method, milliequiv. per 100g of soil	Titrimetric method, milliequiv. per 100 g of soil
1	Red soil	Quaternary period red clay soil	2.44	2.40
2	Purple soil	Tertiary period purple sandstone	3.48	3.36
3	Purple soil	Tertiary period purple standstone	5.44	5.52
4	Alluvial soil	River drift	4.00	3.92
5	Alluvial soil	River drift	2.96	2.88

Source: Royal Society of Chemistry [53]

2.12.3 Stable isotope dilution

The determination of calcium is discussed under multi-metal analysis in section 2.60.9.1.

2.12.4 Photon activation analysis

The determination of calcium is discussed under multi-metal analysis in section 2.60.4.1.

2.13 Californium

2.13.1 α spectrometry

The determination of californium in multi-element analysis is described in section 2.60.11.1.

2.14 Cerium

2.14.1 Neutron activation analysis

The determination of cerium in multi-element analysis is discussed in sections 2.60.3.1 and 2.60.3.3.

2.15 Chromium

2.15.1 Spectrophotometric methods

Qi and Zhu [54] investigated a highly sensitive method for the determination of chromium in soils. In this method, chromium VI is reacted with 0–nitrophenyl–fluorone in the presence of cetyltrimethyl ammonium bromide to form a purplish red complex at pH 4.7 to 6.6 by heating at 50°C for 10 min. The composition of the complex was determined as 1:2:2–chromium VI : NPF : CTAB. The wavelength of maximal absorbance was 582nm and the molar absorptivity was 111000 litres per mole/cm. Beers law was obeyed up to 0.2 µg per cm^3 chromium VI. Interference due to copper(II), iron(III) and aluminium(III) was eliminated by the addition of a masking reagent containing potassium fluoride, trans-1,2–diaminocyclohexanetetra-acetic acid and potassium sodium tartrate. This method was more sensitive than the diphenyl-carbazone method.

Fodor and Fischer [59] have investigated problems of chromium speciation in soils. When employing spectrophotometric detection, only a method based on the diphenylcarbazide reaction was found suitable for chromium speciation analysis.

2.15.2 Atomic absorption spectrometry

Smith and Lloyd [55] determined chromium VI in soil by a method based on complexation with sodium diethyldithiocarbamate in pH4 buffered medium followed by extraction of the complex with methylisobutyl ketone and analysis of the extract by atomic absorption spectrometry [56]. Using this method levels of chromium V between 90 and 176 mgL^{-1} were found in pastureland on which numerous cattle fatalities had occurred.

Chakraborty et al. [265] determined chromium in soils by microwave assisted sample digestion followed by atomic absorption spectrometry without the use of a chemical modifier.

The determination of chromium by atomic absorption spectrometry is also discussed under multi-metal analysis in sections 2.60.1.1, 2.60.1.3 and 2.60.1.4.

2.15.3 Inductively coupled plasma atomic emission spectrometry

The determination of chromium is discussed under multi-metal analysis in sections 2.60.2.1 and 2.60.2.2.

2.15.4 Neutron activation analysis

The determination of chromium is discussed under multi-metal analysis in section 2.60.3.1.

2.15.5 Photon activation analysis

The determination of chromium is discussed under multi-metal analysis in section 2.60.4.1.

2.15.6 Spark source mass spectrometry

The determination of chromium is discussed under multi-metal analysis in section 2.60.6.1.

2.15.7 X-ray fluorescence spectroscopy

The determination of chromium is discussed under multi-metal analysis in section 2.60.8.1.

2.15.8 Emission spectrometry

The determination of chromium is discussed under multi-metal analysis in section 2.60.5.1.

2.15.9 Miscellaneous

The sequential extraction of chromium from soils has been studied [57]. A three-step sequential extraction scheme has been proposed using acetic acid, hydroxylamine hydrochloride and ammonium acetate as extracting agents. Steps 1,2 were measured by electro-thermal atomic absorption spectrometry ETAAS. Step 3 was measured by flame atomic absorption spectrometry.

Interfering effects when measuring chromium in soils were circumvented by the use of a 1% δ-hydroxyquinoline suppressor agent [58].

Fodor et al. [59] have investigated chromium speciation in soils.

Prokisch et al. [60] have described a simple method for determining chromium speciation in soils. Separation of different chromium species was accomplished by the use of acidic activated aluminium oxide. Polarographic methods have been applied in speciation studies on chromium VI in soil extracts [61]. Milacic et al. [266] have reviewed methods for the determination of chromium VI in soils.

2.16 Cobalt

2.16.1 Spectrophotometric method

A method based on measurement of the ammonium pyrollidine dithiocarbamate complex at 240.7nm has been described for the determination of 0.5M acetic acid [62] and nitric perchloric acid soluble cobalt in soils [63].

2.16.2 Atomic absorption spectrometry

The determination of cobalt by atomic absorption spectrometry is discussed under multi-metal analysis in sections 2.60.1.1 and 2.60.1.5.

2.16.3 Inductively coupled plasma atomic emission spectrometry

The determination of cobalt is discussed under multi-metal analysis in sections 2.60.2.1 and 2.60.2.3.

2.16.4 Neutron activation analysis and photon activation analysis

The determination of cobalt is discussed under multi-metal analysis in sections 2.60.3.1 (neutron activation analysis) and 2.60.4.1 (photon activation analysis).

2.16.5 Differential pulse anodic scanning voltammetry

The determination of cobalt is discussed under multi-metal analysis in section 2.60.6.1.

2.16.6 X-ray fluorescence spectroscopy

The determination of cobalt is discussed under multi-metal analysis in section 2.60.8.1.

2.16.7 Emission spectrometry

The determination of cobalt is discussed under multi-metal analysis in section 2.60.5.1.

2.17 Copper

2.17.1 Atomic absorption spectrometry

Atomic absorption spectrometry has been used to determine 0.05M ammoniacal ethylene diamine tetracetic acid extractable [64] and nitric-perchloric acid soluble copper in soils [65]. The determination of copper by atomic absorption spectrometry is also discussed under multi-metal analysis in sections 2.60.1.1 to 2.60.1.5.

2.17.2 Inductively coupled plasma atomic emission spectrometry

The determination of copper is discussed under multi-metal analysis in section 2.60.2.1.

2.17.3 Differential pulse anodic stripping voltammetry

The determination of copper is discussed under multi-metal analysis in section 2.60.7.1.

2.17.4 X-ray fluorescence spectroscopy

The determination of copper is discussed under multi-metal analysis in section 2.60.8.1.

2.17.5 Electron probe microanalysis

The determination of copper is discussed under multi-metal analysis in section 2.60.10.1.

2.17.6 Neutron activation analysis

Fast neutron activation analysis has been studied [231] as a screening technique for copper (and zinc) in waste soils. Experiments were conducted in a sealed tube neutron generator and a germanium γ-ray detector.

2.17.7 Emission spectrometry

The determination of copper is discussed under multi-metal analysis in section 2.60.5.1.

2.17.8 Miscellaneous

Mesuere et al. [267] and Gerringa et al. [268] have reviewed methods for the determination of copper in soils. Residual copper II complexes have been determined in soil by electron spin resonance spectroscopy.

2.18 Curium

2.18.1 α spectrometry

The determination of curium in soil is discussed under multi-metal analysis in section 2.60.11.1.

2.19 Europium

2.19.1 Neutron activation analysis

The determination of europium in soil is discussed under multi-metal analysis in sections 2.60.3.1 and 2.60.3.3.

2.20 Hafnium

2.20.1 Neutron activation analysis

The determination of hafnium in soil is discussed under multi-metal analysis in section 2.60.3.1.

2.21 Indium

2.21.1 Atomic absorption spectrometry

The determination of indium in soil is discussed under multi-metal analysis in section 2.60.1.5.

2.22 Iridium

2.22.1 Neutron activation analysis

Stefanov and Daieva [67] have described a neutron activation analysis procedure for the determination of down to 30 ng of iridium in soil. Short activation times were used to avoid activation of other trace impurities.

2.23 Iron

2.23.1 Spectrophotometric method

Jayman et al. [68] have pointed out that the iron and the aluminium complexes of phenanthroline exhibit identical absorption characteristics. Attempts to mask aluminium to facilitate the determination of iron were unsuccessful. These workers have described an alternate procedure in which aluminium and phosphates are separated from iron and then the iron can be simply determined without interference.

To 2–5 g of a soil sample placed in the digestion tubes, 10 ml of 10 : 1 digestion mixture (100 ml of 60% perchloric acid mixed with 10 ml of concentrated sulphuric acid) are added. The tubes are slowly heated in an orthophosphoric acid bath until the perchloric acid has boiled off and only about 1–2 ml of solution remain. If the samples are not digested to a clear white residue, a few more millilitres of the digestion mixture are added and the digestion is continued. When the tubes have cooled they are removed from the bath and the digested samples are diluted to about 15 ml with distilled water. The solutions are then filtered into 250 ml calibrated flasks using filter paper No 542. The residues in the digestion tubes are washed repeatedly with hot distilled water and the washings passed through the filter papers into the flasks. Finally, the soil extracts in the flasks are diluted to the mark with distilled water, mixed well and suitable aliquots used for the determination of iron.

Calibration

Transfer, by pipette, into 25 ml calibrated flasks aliquots of solutions containing from 0 to 90 µg of iron; adjust the volumes to approximately 15 ml with distilled water. Add, in order, 5 ml of pH4 sodium acetate buffer solution, 2 ml of 1% WIV hydroquinine solution and finally 1 ml of 0.5% w/v phenanthroline reagent. Mix the solutions and dilute to the mark with distilled water, stopper the flasks and mix again. A reference blank is prepared in the same manner except that the iron is omitted. By use of a colorimeter read the values of the colours produced against the reference blank in 2 cm cells at a wavelength of 490 nm. The resulting graph of instrument reading *versus* concentration of iron should be linear, passing through the origin. Any suitable colorimeter can be used.

To centrifuge tubes marked with a 50 ml calibration, transfer, by pipette, suitable portions (say 10–20 ml) of solutions. Add 10 drops of bromocresol purple indicator to each and place the tubes in a boiling water bath for 3 min. Next add ammonia solution, sp gr 0.92, dropwise until precipitation just begins. Thereafter, add 4N ammonia solution dropwise until the indicator turns purple; then add a further 2 ml of the 4N ammonia solution. Stir the solution well with an air jet and place the tubes in a boiling water bath for 5 min in order to complete the precipitation. When the solution has cooled, adjust the volume to 50 ml, stir it well with the air jet and centrifuge at 200 revs min^{-1} for 5 min. Discard the supernatant liquid. To the residue in the tube add 3 ml of hot 6N hydrochloric acid and place the tubes in a boiling water bath for 2 min. Thereafter add 10 ml of hot 25% m/V sodium hydroxide solution and again place the tubes in a boiling water bath for 5 min. After cooling, adjust the volume of solution to 50 ml with distilled water, stir well with the air jet and centrifuge at 2000 rev min^{-1} for five minutes. Then discard the supernatant solution.

To the residues in the centrifuge tubes add 3 ml of hot 6N hydrochloric acid and warm them in a boiling water bath for 2 min in order to effect complete dissolution. Transfer the contents quantitatively into 50 ml calibrated flasks, washing the tubes with several small portions of water. Next make the contents of the flasks up to the mark with distilled water, stopper and mix well. The solutions in the 50 ml calibrated flasks contain iron free from interfering ions. Transfer, by pipette, portions containing 0–90 µg of iron from the separated solutions into 25 ml calibrated flasks, develop the iron colour and proceed as described above under Calibration.

2.23.2 Atomic absorption spectrometry

The determination of iron is discussed under multi-metal analysis in sections 2.60.1.1 and 2.60.1.2.

2.23.3 Inductively coupled plasma atomic emission spectrometry

The determination of iron is discussed under multi-metal analysis in sections 2.60.2.1 and 2.60.2.2.

2.23.4 Neutron activation analysis and photon activation analysis

The determination of iron is discussed under multi-metal analysis in sections 2.60.3.1 (neutron activation analysis) and 2.60.5.1 (photon activation analysis).

2.23.5 Differential pulse anodic stripping voltammetry

The determination of iron is discussed under multi-metal analysis in section 2.60.7.1.

2.23.6 X-ray fluorescence spectroscopy

The determination of iron is discussed under multi-metal analysis in section 2.60.8.1.

2.23.7 Emission spectrometry

The determination of iron is discussed under multi-metal analysis in section 2.60.5.1.

2.24 Lanthanum

2.24.1 Neutron activation analysis

The determination of lanthanum in soil is discussed under multi-metal analysis in sections 2.60.3.1 and 2.60.3.3.

2.25 Lead

Most of the lead in soil exists in sparingly soluble forms. When 2784 ppm of lead nitrate were added to soil, it was found that after three days the soluble lead content was only 17 ppm [69]. It is to be expected that all ions in nature will accumulate as their less soluble compounds, such as oxides, carbonates, silicates and sulphates, the relative proportions of each depending on the nature of the soil and on solubility.

Several acids and acid mixtures have been used for the digestion of soil samples prior to the analysis of lead, including nitric acid – perchloric acid (1 + 1) [70], hydrochloric acid [71], perchloric acid [72], nitric acid – hydrofluoric acid (1 + 1) [73, 74] and aqua regia [75].

2.25.1 Spectrophotometric method

A spectrophotometric method employing ammonium pyrolidine dithyldithiocarbamate [76] as chromogenic reagent has been applied to the determination of 0.5M acetic acid extractable lead in soils. Savvin et al. [269] have discussed a spectrophotometric method for the determination of lead in soils.

2.25.2 Atomic absorption spectrometry

Standard official methods have been described for the determination of nitric-perchloric acid soluble lead in soil [77]. Atomic absorption measurements were performed by measuring at 217nm emission from a lead hollow cathode lamp at a spectral hand width of 1.0nm. Tills and Alloway [80] investigated the speciation of lead in soil solution using a fractionation scheme, ion exchange chromatography and graphite furnace atomic absorption spectrophotometry. Soils from four sites were selected (Snertingdal in Norway, Pen Craig-ddu in Dyfed, Wales, Velvet Bottom in Somerset and Beaumont Leys Sewage Farm, Leicester, England). The sources of contamination of each soil and its chemical properties are described. The percentages of lead in cationic, anionic, neutral and less polar organic complexes were determined and discussed with respect to organic matter content and pH. No direct relationship was established between total lead content of the soil and total lead content in soil solution.

Using palladium–magnesium nitrate mixture as chemical modifiers, Hinds and Jackson [270] effectively delayed the atomisation of lead until atomic absorption spectrometer furnace conditions were nearly iso-thermal. This technique was used to determine lead in soil slurries. Zhang et al. [271] investigated the application of low pressure electrothermal atomic absorption spectrometry to the determination of lead in soils.

Hinds et al. [272, 273] investigated the application of slurry electrothermal atomic absorption spectrometry to the determination of lead in soils. Hinds and Jackson also investigated the application of vortex mixing slurry graphite furnace atomic absorption spectrometry to the determination of lead in soils. The determination of lead is also discussed under multi-metal analysis in sections 2.60.1.1 and 2.60.1.3 to 2.60.1.5.

2.25.3 Atomic fluorescence spectroscopy

Rigin and Rigina [81] determined lead in soil using flameless atomic fluorescence spectroscopy on an extract of the sample which had been preconcentrated by electrolysis on a silanised graphite rod. The limit of detection was 15 pg lead and relative standard deviation 0.4.

2.25.4 Inductively coupled plasma atomic emission spectrometry and inductively coupled plasma mass spectrometry

The determination of lead is discussed under multi-metal analysis in section 2.60.2.1 (inductively coupled plasma atomic emission spectrometry) and section 2.60.2.2 (inductively coupled plasma mass spectrometry).

2.25.5 Anodic stripping voltammetry

Somer and Aydin [78] determined the lead content of soil adjacent to roads in Turkey using anodic stripping voltammetry. These workers found that aqua regia was the most suitable acid for extracting lead from roadside soil. The lead salt that may be trapped in the silicate crystal lattice of the soil was brought into solution by keeping the soil in acid overnight. To avoid the possibility of the presence of undissolved lead salts even after digestion, EDTA was added to the digested sample in order to ensure quantitative dissolution of lead. The lead content of this solution was determined by anodic stripping voltammetry.

Apparatus

An ESA Model 2014 multiple anodic stripping analyser was used for the plating procedure. For stripping, a polarographic system was used similar to a Heath Model EUW–198 instrument with IC operational amplifiers in place of the vacuum tube types. With this instrument fast potential scanning is also possible. The cell was fitted with an Ag–AgCl reference electrode, a counter electrode and an impregnated graphite rod electrode. The graphite electrode was covered with wax and its end was polished with a fine abrasive and plated with mercury. Nitrogen was passed through the cell at a flow rate of 20 ml min $^{-1}$ to purge oxygen from the system and to stir the solution.

To plate the graphite wax test electrode with mercury a 0.50 ml volume of 8.1×10^{-3}M mercury (II) chloride solution was added to the cell, which contained 5 ml of triply distilled water, then nitrogen was passed through for several minutes. The electrode was plated for 30 min at –500mV without passing nitrogen, and was then plated for a further 30 min while nitrogen was passed. During plating the cell head was tapped in order to destroy any gas bubbles on the electrode surface.

The electrode was plated with mercury for a short period after every five or six strippings. For short platings, 0.25 ml of 8.1×10^{-3}M mercury (II) chloride solution was added to 5 ml of triply distilled water, followed by plating for 10–15 min.

Sampling

The samples were dried at 60°C and sieved through a 170 mesh sieve. Sub-samples were then taken by the coning and quartering method.

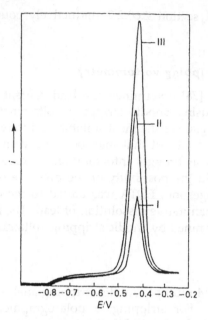

Fig. 2.15 Anodic stripping voltammogram of lead in roadside soil. I, Sample; II, sample + 100 μL of 10⁻⁵M Pb(NO₃)₂ solution; and III sample + 200 μL of 10⁻⁵M Pb(NO₃)₂ solution Source: Royal Society of Chemistry [78]

Procedure

A 100 mg soil sample was kept overnight in 2 ml of aqua regia, then evaporated nearly to dryness at 100–100°C, 2 ml of 0.1 M EDTA solution were added, the pH was adjusted to 5–6 and the solution was allowed to stand for 2h. The digested sample was diluted to 10.0 ml with water, then 100 μl of this solution and 20 μl of perchloric acid were added to 5 ml of water in the ASV cell. Nitrogen was passed through the cell for 10 min and the lead was then plated at −800 mV for 5 min. The nitrogen flow rate was adjusted to 20 ml min⁻¹ during plating. After a rest period of 45 s, a potential sweep of 60 mV s⁻¹ was applied in a positive direction and the current was recorded with a chart speed of 10 in min⁻¹. Standard additions were made by using 10⁻⁵M lead(II) nitrate solution. See Fig. 2.15.

Using this procedure, the lead content of samples of soil adjacent to some Ankara roads was determined and the results are given in Table 2.16. As expected, the lead content of the soil increased with increasing traffic volume.

Differential pulse anodic scanning voltammetry has been applied to the determination of lead in soils [274]. The determination of lead in soil by this technique is also discussed under multi-metal analysis in section 2.60.7.1.

Table 2.16 Lead contents of soil samples taken from some Ankara roads (September 1981)

Location	No of vehicles, 7.00–9.00 h plus 16.00–1800 h	No. of obser- vations	Lead content, ppm (x ± (t*s)/√N)]
Dikimevi Square	13000	7	568 ± 35
Dikimevi Uzgöreb St,†	498	9	356 ± 21
Dikimevi, Selvi St	–	8	193 ± 16
Kizialy Square ‡	22760	7	710 ± 25
Kizialy Kumrular St	5862	7	244 ± 8
Kizialy Karanfil St	1077	7	190 ± 9
Ulus Square	13430	6	610 ± 12
Ulus, Sanayi St.	4582	6	281 ± 9
Ulus, Sinasi St	–	9	356 ± 24

*90% confidence interval
†Very short street opening on to the square
‡Next to this square there is a large park

Source: Royal Society of Chemistry [78]

2.25.6 Polarography

Sakharov [79] determined lead in soil polarographically by digesting the sample with sodium carbonate, followed by dissolution in hydrochloric acid. He found that when hydrochloric, sulphuric or nitric acids were used as a digestion medium instead of sodium carbonate, no lead could be detected in the resulting solution.

Lead was determined in the digest by anodic scanning voltammetry.

2.25.7 X-ray fluorescence spectroscopy

Wegrzynek and Holynska [275] have developed a method for the determination of lead in arsenic containing soils by energy dispersive X-ray fluorescence spectroscopy. Correction for arsenic interference is based on the use of an arsenic-free reference sample. The determination of lead is discussed under multi-metal analysis in section 2.60.8.1.

2.25.8 Photon activation analysis

The determination of lead is discussed under multi-metal analysis in section 2.60.4.1.

2.25.9 Emission spectrometry

The determination of lead is discussed under multimetal analysis in section 2.60.5.1.

2.25.10 Miscellaneous

Several investigations have reviewed the determination of lead in soils [202, 209–212]. Lead has been determined in soil by using a slurry sampling technique with lead nitrate and magnesium nitrate as a chemical modifier [276]. Results were in good agreement with known concentrations of a standard reference material. Feng and Barrett [247] showed that microwave dissolution of soil and dust samples with nitric-hydrofluoric acid gave recoveries of lead (and cadmium) of over 90% in 30 min digestion.

Chen and Hong [209] found that 5-carboxy methyl-L-cysteine was especially effective for the chelating extraction of lead from contaminated soils. The chelator could be recovered and reused over consecutive runs with no loss in performance.

2.26 Magnesium

2.26.1 Atomic absorption spectrometry

This technique used to determine magnesium in 1M ammonium nitrate extracts of soil [82]. The determination of magnesium in soils is also discussed under multi-metal analysis in section 2.60.1.2.

2.26.2 Inductively coupled plasma atomic emission spectrometry

The determination of magnesium is discussed under multi-metal analysis in section 2.60.2.2.

2.26.3 Photon activation analysis

The determination of magnesium is discussed under multi-metal analysis in section 2.60.4.1.

2.27 Manganese

2.27.1 Spectrophotometric method

Alekseeva and Davydova [83] determined microamounts of manganese II in sulphuric and hydrofluoric extracts of clays by a kinetic spectrophotometric method involving the oxidation of o-dianisidine by potassium periodate. The reaction between o-dianisidine and potassium periodate is catalysed by manganese II. Spectrophotometric measurements were conducted at 460nm.

2.27.2 Atomic absorption spectroscopy

Atomic absorption spectrometry utilising the 403 nm emission has been employed to determine exchangeable and easily reduced manganese in M ammonium acetate pH7 extracts of soils [84]. Atomic absorption spectrometry has been extensively used in the multi–metal methods for the determination of manganese in soils – see section 2.60.1.1 and 2.60.1.2.

2.27.3 Inductively coupled plasma atomic emission spectrometry

The determination of manganese is discussed under multi-metal analysis in sections 2.60.2.1 and 2.60.2.2.

2.27.4 Differential pulse anodic stripping voltammetry

The determination of manganese is discussed under multi-metal analysis in section 2.60.7.1.

2.27.5 X-ray fluorescence spectroscopy

The determination of manganese is discussed under multi-metal analysis in section 2.60.8.1.

2.27.6 Emission spectrometry

The determination of manganese is discussed under multi-metal analysis in section 2.60.5.1.

2.28 Mercury

2.28.1 Spectrophotometric method

Kimura and Miller [85] have shown that at room temperature, reduction with tin(II) and aeration are suitable for quantitative separation of microgram quantities of mercury(II) from sulphuric and nitric acid extracts of soil over wide ranges of concentrations. Mercury is concentrated during the separation and is determined by a direct photometric dithizone procedure. This technique is applicable for 0.10 µg mercury with a standard deviation for a single determination of 0.05 µg mercury in the 0 to 0.5 µg range.

Kamburova [277] has reported a spectrophotometric method based on the formation of the mercury–triphenyltetrazolium chloride complex for the determination of mercury in soils.

2.28.2 Atomic absorption spectrometry

Cold vapour (or flameless) atomic absorption spectrometry is the method of choice for the determination of mercury in soils [86–97]. Ure and Shand [88] investigated various procedures for the digestion of soil samples prior to analysis by cold vapour atomic absorption spectrometry. They

Table 2.17 Comparative analysis of soils for mercury with different soil preparation methods

Soil no.	Loss on ignition (%)	Method I[a] ($\mu g\ g^{-1}$)	Method II[b] ($\mu g\ g^{-1}$)	Method III[c] ($\mu g\ g^{-1}$)
1	77	0.10	0.11	0.11
2	63	0.14	0.14	0.13
3	53	0.11	0.11	0.12
4	40	0.20	0.19	0.20
5	31	0.13	0.12	0.13
6	20	0.15	0.14	0.15
7	9	0.26	0.25	0.25
8	11	0.11	0.10	0.11
9	10	0.09	0.08	0.08
10	15	0.15	0.14	0.14
11	14	0.12	0.12	0.12
12	10	0.14	0.13	0.14
13	–	0.09	0.10	0.09

[a]$HNO_3/N_2SO_4/KMnO_4$ digestion.
[b]$HNO_3/H_2SO_4/KMnO_4$ digestion, volatisation and collection.
[c]Oxygen flask, $KMnO_4$ collecting solution.

Source: Royal Society of Chemistry [88]

found good agreement (Table 2.17) between two digestion methods involving digestion of the soil with the mixture of nitric and sulphuric acids and potassium permanganate (methods Ia and IIb) and oxygen flask combustion over acid potassium permanganate solution (method IIIc).

Method Ia and IIb

Sample preparation: The required weight of finely ground (<150 μm) air dried (<30°C) soil or peat is placed in a 100 ml Kjeldahl flask. 10 ml of a mixture of redistilled nitric acid and concentrated sulphuric (1 + 1) are added and the contents are swirled to wet the sample thoroughly. The weight taken depends on the expected mercury content and amount of oxidizable matter. The sample is digested unstoppered for 2h in a water bath at 60°C, with swirling at intervals, only the bulb of the flask being immersed so that its long neck acts as an air condenser. The flask is cooled, 20 ml of glass distilled water is added, the flask is again cooled and 5 ml aliquots of 6% potassium permanganate solution are added until the purple colour is retained for at least 1h. The sample is then left glass-stoppered at room temperature overnight, and 20% (w/v) hydroxyl-ammonium chloride is added slowly until the brown hydrated manganese oxides and the excess of potassium permanganate are reduced, giving a clear solution with a residue of insoluble material. The unstoppered flask is then left for 1h to allow the evolved gases to be liberated; no loss of mercury is observed at this stage. This solution is used for the determination of mercury in Method Ia.

Method IIIc

Initial trials of the oxygen flask combustion technique, which involved burning a pelleted sample in a platinum basket in an oxygen-filled flask containing a collecting solution to absorb the mercury evolved, gave low mercury recoveries caused by the depressive interference of platinum volatilised from the basket. Neither tantalum nor tungsten interfered in the mercury determination and recoveries of 95–100% were obtained.

Sample preparation: 1000g of finely ground (<150 µm) air dried soil or peat (dried at 30°C) is mixed with 0.500g of cellulose powder, in a polystyrene tube by a perspex ball in a vibrating ballmill, for 1 min and pressed into a 12.5 mm diameter pellet at a pressure of 600 MN m $^{-2}$ (4 tons/in^2). The pellet is wrapped in an 80 × 10 mm strip of filter paper (Whatman 540) and placed in the tantalum cup with a 30 mm paper tail to act as a fuse. A clean 5–1 round-bottomed heavy-walled combustion flask is prepared by adding to it 50 ml of the acidic potassium permanganate collecting solution and a 5 cm polypropylene encapsulated magnetic follower. The flask is flushed with oxygen for 1 min and the tantalum cup with the sample is inserted into the neck of the flask with tail fuse protruding. The fuse is ignited with a filter paper spill, the flask closed quickly and placed behind a perspex safety shield. After burning is complete, the flask is allowed to cool, first in air and then under a running cold water for 5 min. It proved essential to cool the flask to room temperature to prevent losses of mercury on removing the stopper. The sample holder with its ash is removed carefully and the flask stoppered quickly. The collecting solution is then stirred magnetically with maximum splashing for 15 min. After stirring, 2 ml of 20% hydroxylammonium chloride solution is added to reduce the excess of permanganate and any insoluble manganese oxides, the flask is swirled a few times and allowed to drain for 2 min into a 250 ml conical flask which is then stoppered ready for mercury determination by atomic absorption spectrometry.

The atomic absorption spectrometry procedure involved an agitation procedure in which the mercury in the reduced sample solution is partitioned between a fixed volume of air and the liquid phase in a closed vessel by hand-shaking or magnetic stirring. The mercury-laden air is then blown by air at a flow rate of 3 L min $^{-1}$ directly, without a drier, through the absorption cell for the atomic absorption measurement.

The apparatus is shown schematically in Fig. 2.16 where the sample vessel is the long-necked 100 ml ground glass stoppered Kjeldahl flask

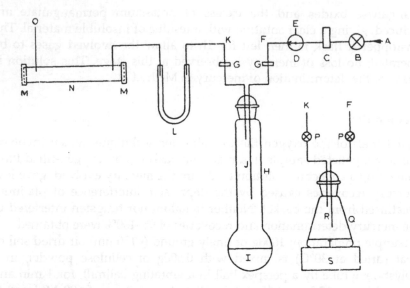

Fig. 2.16 Apparatus for the determination of mercury by cold-vapour atomic absorption. Compressed air line 105 kN m $^{-2}$ (15 lb/in^2) A; needle valve B; flow meter C; two-way tap D; vent E; connection points F and K; silicone rubber tubing with Mohr spring-clips G to Kjeldahl flask H; sample I; plain glass tubes, o.d. 0.5 cm J; empty U-tube L; 25 cm pyrex absorption tube N, od 1.9 cm, with removable fused-silica end windows; O-ring seals M and exhaust O; alternative conical flask Q; plain glass tubes R; nylon stop-cocks P; sample T; and magnetic stirrer S.

Source: Elsevier Science Publishers BV, Netherlands [88]

(total volume including the long neck about 200 ml) in which the acid digestion and wet oxidation are carried out. The flask is connected to the air train by flexible silicone rubber tubing fitted with Mohr spring clips to isolate the sample and the fixed volume of air with which mercury is partitioned by manual shaking of the flask and contents. An empty U–tube is used as a safety trap but no drier is necessary. Sample solutions prepared by the oxygen flask combustion method are decanted into the alternative vessel, a conical flask whose flat bottom makes it convenient to carry out the partitioning operation by magnetic stirring. Isolation of the contents is achieved in this case by nylon stopcocks.

The apparatus is cleaned before use and between samples by attaching the Kjeldahl flask, containing 10–20 ml of distilled water, to the air train. The clips are attached and the flask is shaken manually. After repeating with a fresh aliquot of distilled water, the clips are removed, the two-way tap is opened and air is passed at 10 L min $^{-1}$ through the flask and measuring system until no absorption is obtained. The air flow is then reduced to 3 L min $^{-1}$, the two-way tap is set to vent, and the clips are replaced.

The mercury content of a sample is then measured by adding 5 ml of 20% stannous chloride (SnC_22H_2O) in 6M hydrochloric acid to the sample solution prepared as described above (methods I, II or III) in a Kjeldahl or conical flask, which is immediately fitted to the apparatus to prevent loss of mercury. After agitation for 2 min (manual shaking for Kjeldahl, magnetic stirring for conical flask), the clips are removed (or the stopcocks opened) and air is blown through at 3 L min $^{-1}$. The absorption is measured at 253.7 nm with a single beam laboratory-built atomic absorption apparatus of conventional design fitted with a pen recorder with a chart speed of 2.5 mm min $^{-1}$. Linear calibration curves of absorbance versus mercury content are obtained up to about 0.3 μg, with slight curvature from 0.3 to 0.5 μg. For a 1 g sample, the analytical range of 0.01–0.5 μg of mercury corresponds to 0.0–0.5 ppm mercury in the soil.

Floyd and Sommers [89] evaluated a simple one step digestion procedure for extracting total mercury from soils. The sample was digested with concentrated nitric acid and 4N potassium dichromate for 4 h at 55°C and the mercury in the extract determined by flameless atomic absorption spectrometry. The method can be applied to soils containing up to 20% organic matter.

Agemian and Chau [90] have reported an improved digestion method for the extraction of mercury from soils and clays in which the sample is digested at 60°C with sulphuric acid – nitric acid (2 + 1) containing a trace amount of hydrochloric acid, and subsequently oxidised with permanganate and persulphate solutions. With this procedure mercury is successfully recovered from organic matter and resistant inorganic forms such as mercury(II) sulphide. Unlike digestion with aqua regia, this procedure is simple and safe, and is applicable to the digestion of a large number of samples simultaneously. The method can be adapted to the automated cold vapour and flame atomic absorption techniques and is therefore ideal for routine monitoring.

Method

Samples were digested in 100 ml calibrated flasks in a temperature-controlled shaker bath. The equipment used for the automated cold vapour analysis consisted of (a) an automatic sampler (Technicon AutoAnalyzer II sampler with 20–1/5 cam); (b) a proportionating pump (Carlo Erba, Model 08–59–10202); (c) Technicon AutoAnalyzer tubing of specified dimensions; (d) a gas separator as used by Agemian and Chow [90]; (e) a mercury monitor (Pharmacia Fine Chemicals); and (f) a strip chart recorder (Hewlett-Packard, Model 7101B).

For high levels of mercury, a Perkin–Elmer 503 atomic absorption spectrophotometer, equipped with an Intensitron lamp, was used with an air-acetylene flame. Determine the water content of the wet sediment by drying it overnight to constant mass at 100°C. Weigh a representative sample of wet sediment, equivalent to 0.1–2g of the dry mass, into a 100 ml calibrated flask. Wash the sediment down to the bottom of the flask with mercury-free distilled water, place the flask in an ice water bath and slowly add 15 ml of concentrated sulphuric acid–nitric acid (2 + 1). After cooling, add 2 ml of hydrochloric acid. Shake the mixture and let it stand for 5 min.

After expelling the acid fumes from the flask, place it in a shaking water bath at a temperature of 50–60°C and digest for 2 h. Then allow the flask to cool for 30 min and carefully add 10 ml of potassium permanganate 6% w/v solution while cooling in an ice water bath. If the colour does not persist for fifteen minutes, add a further amount of potassium permanganate solution. After 30 min, add 5 ml of 5% w/v potassium persulphate solution, with gentle stirring, and allow the mixture to stand overnight. If all of the permanganate is reduced, as witnessed by the absence of the purple colour, add potassium permanganate solution until the colour persists.

Add 10 ml of hydroxylammonium sulphate (6% w/v) – sodium chloride 6% w/v solution and stir the mixture gently until the solution becomes clear and all of the precipitated manganese(IV) oxide has dissolved. Make the solution up to volume and centrifuge an aliquot at 2500 rev min $^{-1}$ for 5 min. Transfer an aliquot of the clear supernatant liquid into a glass sample cup and place it in the automatic sampler for analysis. Use a cam designed for 20 samples per hour and a sample to wash ratio of 1:5, corresponding to a sampling time of 30 s and a wash time of 2.5 min. For concentrated solutions of mercury, use the flame technique. Mercury salts are reduced to metallic mercury by the addition of 10% stannous chloride in 2N sulphuric acid and the mercury swept out of the solution and estimated by atomic absorption spectrometry. The linear concentration range in solution is 0.0002–0.006 mg L $^{-1}$ for the non-flame detection system and 10–300 mg L $^{-1}$ for the flame detection system. For sample concentrations of 0.1–2 g per 100 ml, the non-flame detection system has a range of 0.01–6 mg kg $^{-1}$ and the flame method 5–300 mg kg $^{-1}$ of mercury in the sediment. Further dilution of the extracts could extend the upper limit of both detection systems. Treat all standards exactly as for the above samples.

Recent work has been focused on the methods used for sample digestion rather than the cold vapour atomic absorption spectrometric finish. Thustuwae et al. [91] used high frequency induction heating of the sample for the rapid release of mercury from soil samples prior to absorption in 0.5% potassium permanganate in 0.5M sulphuric acid and cold vapour atomic absorption spectrometry. 98–99% mercury recoveries were obtained following a 5 min induction heating period which is considerably more rapid than wet digestion procedures.

Nicholson [92] has described a rapid thermal decomposition technique for the determination of mercury in soils (and sediments). The method is applicable to samples rich in organic matter.

Table 2.18 Mercury content of three soil samples

Soil	Organic matter content, %	Mercury content by established method [93] ppm	Mercury content by Nicholson [92] method, ppm	Coefficient of variation, % of variation
Sample 1	6.8	0.33	0.37	3.8
Sample 2	25.9	0.05	0.06	9.1
Sample 3	52.0	0.06	0.08	6.7

Source: Royal Society of Chemistry [92]

The sample is heated in a nickel boat at 650°C and the products collected on gold. The gold is then heated to 750°C to release mercury which is swept into a cold vapour atomic absorption spectrophotometric cell. Three samples of soil, representing a range of organic matter contents, were chosen for statistical evaluation of the proposed method. Table 2.18 gives results for these three samples for comparison with results obtained by means of wet oxidation. Thirty replicate determinations were made on each sample.

Other workers [94] have used flameless atomic absorption spectrometry or atomic fluorescence spectrometry to determine mercury in amounts down to 10 µg kg^{-1} in soil.

Cold vapour atomic absorption spectrometry and atomic fluorescence spectrometry (253nm emission) have been applied to the determination of down to 0.01 mg kg^{-1} of mercury in soils and sediments [94].

Sakamoto et al. [278] have shown that the differential determination of different forms of mercury in soil can be accomplished by successive extraction and cold vapour atomic absorption spectrometry.

Azzaria and Aftabi [279] showed that stepwise, compared to continuous heating of soil samples before determination of mercury by flameless atomic absorption spectrometry gives increased resolution of the different phases of mercury. A gold coated graphite furnace atomic absorption spectrometer has been used to determine mercury in soils [280].

Bandyopadhyay and Das [281] extracted mercury from soils with the liquid anion-exchanger Aliquat–336 prior to determination by cold vapour atomic absorption spectrometry. The application of atomic absorption spectrometry to the determination of mercury in soils is also discussed under multi-metal analysis in sections 2.60.1.4 and 2.60.1.5.

2.28.3 Anodic stripping voltammetry

Voltammetric methods have been used to determine mercury in soil composts. The amount of mercury leaching from composts was very low [282].

2.28.4 Neutron activation analysis

Neutron activation analysis has been used to determine mercury in soils [283].

2.28.5 Miscellaneous

A study by Rasemann demonstrated by example of a non uniformly contaminated site to what extent mercury concentrations depend on the method of handling soil samples between sampling and chemical analysis [98]. Sample pretreatment contributed substantially to the variance in results and was of the same order as the contribution from sample inhomogeneity. Welz *et al.* [284] and Baxter [285] have conducted speciation studies on mercury in soils. Lexa and Stulik [286] employed a gold-film electrode modified by a film of tri-n-octylphosphine oxide in a PVC matrix to determine mercury in soils. Concentrations of mercury as low as 0.02 ppm were determined.

Cherian and Gupta [287] have described a simple field test for the determination of mercury in soil. Saouter *et al.* [288] showed that the use of hydrogen peroxide as an oxidising agent for organics in soils can result in loss of mercury. This is because hydrogen peroxide can act as a reducing agent for mercury compounds.

2.29 Molybdenum

2.29.1 Spectrophotometric methods

Various chromogenic reagents have been used for the determination of molybdenum in soil. These include toluene 3.4 dithiol [99, 100] and iron thiocyanate [101, 102]. The complexes formed with toluene 3.4 dithiol are extracted with iso-amyl acetate prior to spectrophotometric evaluation at 680 nm. Down to 0.1 µg molybdenum can be determined in the soil sample. The iron thiocyanate method determines total molybdenum [101] or ammonium oxalate–oxalic acid soluble molybdenum [102] in soils utilising the 470 nm absorption of the complex.

2.29.2 Atomic absorption spectrometry

Earlier atomic absorption methods [104–106] from the determination of molybdenum in soils employed a preliminary solvent extraction step to improve sensitivity in view of the low concentrations of molybdenum occurring in most soils [40, 107]. Baucells *et al.* [44] developed a graphite furnace atomic absorption procedure which was capable of determining down to 8.4 pg of molybdenum in a soil matrix solution with a precision of 4% for 100 µg L $^{-1}$ molybdenum. These workers showed that a char temperature of 1500°C and an atomisation temperature of 2400°C are

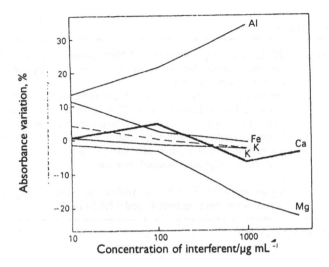

Fig. 2.17 Variation of absorbance of Mo (200 µg L⁻¹) as a function of interferent metal concentration
Source: Royal Society of Chemistry [44]

optimum for molybdenum. Under these conditions, the background absorbance is 0.015. However, the use of a char temperature of 700°C in part prevents attack of the graphite, gives a better precision and the background absorbance in the atomisation step is only 0.030.

During the extraction method described by Baucells *et al.* [44] for the determination of molybdenum, the dry residue was solubilised with nitric acid. To observe the influence of nitric acid concentration on the absorbance signal of molybdenum, different acid concentrations were used. There was a decrease of 22.86% in the peak height when 10% nitric acid was present compared with no concentrated nitric acid.

The interference study was carried out with 200 µg L⁻¹ of molybdenum at various interferent concentrations (10, 100, 1000 and 4000 µg L⁻¹). The results are shown in Fig 2.17. The most important interferences were given by aluminium, iron and magnesium.

Five replicate determinations of molybdenum in a siliceous soil sample by this proposed method gave a precision of 17.1% with a mean concentration of 35 µg L⁻¹. The main problem concerning the determination of molybdenum is the corrosion of the pyrolytic layer in the graphite tubes by the acid and the atomisation temperature used. After 30 firings (and the corresponding cleanings) the sensitivity decreased by 50%, so the minimum acceptable atomisation temperature and time must be used. Re-calibration must be carried out frequently, and a computer program was developed in order to correct for the possible variations in the readings on standards and samples during the analysis.

Table 2.19 Instrumentation

Plasma power supply	SC15: operating frequency 36 Msz: power output 0–25 kW, continuously variable; load coil 2½ turns, 6.3 mm o.d., internal diameter of coil 32 mm
Plasma torch	Demountable silica torch with brass base; coolant tube 25 mm bore, 1.5 mm wall; plasma tube 22 mm bore, 1 mm wall; injector tube 11 mm o.d., 9 mm i.d., and 2 mm orifice diameter.
Nebuliser	Pneumatic nebuliser and spray chamber, uptake rate 1.5 ml/min at argon flow-rate of 2 L/min, efficiency ~ 7%
Spectrometer	Techtron AA4 grating monochromator
Slit-width	25 μm
Read-out	Signal from PMT, after amplification, was displayed on a Servoscribe chart-recorded model RE 511–20

Source: Elsevier Science UK [40]

The application of atomic absorption spectrometry to the determination of molybdenum is also discussed under multi-metal analysis in sections 2.60.1.3, 2.60.1.4 and 2.60.1.5.

2.29.3 Inductively coupled plasma atomic emission spectrometry

Atomic emission spectrometry is not sufficiently sensitive to determine molybdenum at the levels at which it occurs in soils. Due to its greater intrinsic sensitivity, inductively coupled plasma atomic emission spectrometry is capable of achieving the required sensitivity.

Manzoori [40] has utilised inductively coupled plasma optical emission spectrometry to determine down to 0.01 mg L^{-1} molybdenum in 1M ammonium acetate extracts of soils.

Method

The ICP system used in this work utilised a Radyne 2.5 kW free-running valve oscillator operating at 36 MHz and a Techtron grating scanning monochromator. The plasma torch was a demountable Greenfield type. A dual tube aerosol chamber combined with a pneumatic nebuliser was employed for sample introduction. Details of the instrumental system are presented in Table 2.19.

A 50 g sample of soil was shaken overnight with 800 ml of neutral 1M ammonium acetate. The whole extract was filtered through an 18.5 cm Whatman No 540 filter paper, as much of the soil as possible being transferred to the paper in order to minimise the passage of clay particles. The paper was washed several times with distilled water, and the filtrate evaporated to dryness, partially on a hot plate and then to completion on a steam bath. The residue was transferred to a 100 ml standard flask and

Table 2.20 Plasma operating conditions

Net forward radiofrequency power	1200 W
Spectrometer slit	25 µm (entrance)
Argon coolant-gass flow-rate	14 L/min
Argon plasma-gas flow-rate	5 L/min
Injector gas flow-rate	1.6 L/min
Height of observation	25 mm above work coil

Source: Elsevier Science UK [40]

made up to the mark. No oxidation with nitric acid was applied, as organic matter at the concentration levels generally occurring in the soil extracts did not cause any disturbance in the plasma. However, some extracted solutions contained solid particles which made the nebulisation process unstable. In these cases a single-step oxidation with nitric acid was used, in order to obtain clear solutions (1 ml of concentrated nitric acid was added to 10 ml of solution, and the mixture was boiled to low volume and then diluted to 10 ml again).

The most sensitive molybdenum line (379.8nm) was used. There is some interference from the iron line at 379.8nm. Plasma operating conditions are given in Table 2.20. Between 0.2 and 3.1 mg L^{-1} molybdenum was found in soil extracts equivalent to 1 to 15.5 mg kg^{-1} in soils.

Thompson and Zao [103] carried out rapid determinations of 0.4–40.7 mg kg^{-1} molybdenum in soils by solvent extraction, followed by inductively coupled plasma atomic emission spectrometry. In soils, molybdenum is completely oxidised and largely associated as the MoO_4^{2-} ion with iron III oxide materials, although it may also be associated with organic matter.

The molybdenum is solubilised by treatment of the sample with 6M hydrochloric acid in capped tubes at 120°C. It is then extracted from the same medium into heptan-2-one, in which it is determined directly by inductively coupled plasma atomic emission spectrometry. Extraction of iron, which can otherwise interfere with the determination, is minimised by the presence of a reducing agent.

The layered heating block used in this procedure consisted of a lower heated section. An ARL 34000 inductively coupled plasma instrument equipped with lines for 36 elements including molybdenum (281.62nm) and iron (259.94nm), both in the second order, was used. Heptan-2-one solutions were introduced into the plasma by means of a Meinhard (Type TR–30–A3) nebuliser modified by a platinum-iridium capillary sealed into the rear end with Araldite epoxy resin. The uptake rate to the nebuliser was restricted to a 0.3 ml min^{-1} by means of a PTFE tube, 310 mm long and 0.31 mm id. No pump was used. The nebuliser discharged into a small (ca 100 cm^3) Scott-type double-pass spray chamber connected directly to the plasma torch. A stabilisation time of 20 s was followed by three 5 s integration periods, the output being the arithmetic mean of these three readings. The flow rates of argon gas to the plasma torch were as follows: coolant gas, 12 L min^{-1}; injector gas (humidified by means of a bubbler system) 1 L min^{-1}; and auxiliary gas, 0.5 L min^{-1}. Viewing height was 14 mm above the load coil and forward power was set at 1.70 kW. The reagents used were:

- Heptan-2-one. Spectrosol grade.
- L-Ascorbic acid. AnalaR grade.
- Bromine. AnalaR grade.
- Phenol.
- Ammonium molybdate standard solution, 1000 µg mL^{-1} Mo.

Weigh each sample (0.250 g of dry powder) into a clean dry thick walled 160 v 16 mm tube. Add 6M hydrochloric acid (5.00 ml) from a dispenser. (For samples containing carbonate minerals, the acid should be added cautiously in small portions, and the tubes set aside until evolution of gas has ceased.) For rock samples or others suspected of containing a small proportion of sulphide minerals, add two drops of bromine. Cap the tubes firmly with PTFE lined caps and place them in the block at 120°C. After the decomposition period (2 h for soils, 3.5 h for rocks) remove the tubes from the block and, when they are cool, break the seal momentarily to equalise the pressure. Shake the tubes vigorously to mix the contents. If bromine has been used, add two drops of phenol solution to precipitate the bromine as tribromophenol. Centrifuge the tubes and transfer 4.00 ml of the centrifugate into a test-tube equipped with a clear silicone-rubber stopper. Add concentrated hydrochloric acid (0.50 ml), ammonium iodide solution (0.5 ml of 10% w/v in 6 N hydrochloric acid) and ascorbic acid (0.2g) to each tube, stopper the tubes and shake them until the ascorbic acid has almost dissolved. Set the tubes aside for 2 h or more. Add to each tube heptan-2-one (3.00 ml, by dispenser) and shake well for 6 min on a mechanical shaker. Centrifuge the tubes briefly (or allow them to stand for 20 min) to separate the phases, and remove some of the top layer for nebulisation into the inductively coupled plasma atomic emission instrument.

Calibration solutions were prepared by following the solvent extraction procedure on tubes containing 4.00 ml of 6M hydrochloric acid and on similar tubes to which had also been added 10 µg of molybdenum as a 0.1 ml spike of ammonium molybdate solution (1000 µg). The normal computer correction was made for the spectral interference caused by the small concentration of iron also extracted. Some results obtained by this method are presented in Table 2.21.

Triplicate analyses of a series of twenty soil samples shows that the standard deviation(s) can be expressed approximately at a concentration

Table 2.21 Molybdenum results obtained by the ICPAES

| Sample | Results obtained by ICPAES µg g^{-1} | |
	With bromide	Without bromine
SO–1*	–	0,42 0.44, 0.40
SO–2*	–	1.54, 1.56, 1.48
SO–3*	–	0.74, 0.72, 0.67
SO–4*	–	0.53, 0.52, 0.53
GXR2*	1.00, 1.02	0.95, 0.95, 1.14
GXR5*	40.7, 39.6	39.6, 38.5, 38.5
GXR6*	1.93, 2.02	1.79, 2.42, 2.38

Source: Royal Society of Chemistry [103]

c as $s = 0.03 \pm 0.03c$ within the concentration range 0–100 mg kg^{-1}. This result is confirmed by two further replicate analyses of soil samples, giving, respectively: $m = 0.32$, $s = 0.02$ ($n = 11$); and $m = 33.7$, $s = 0.83$ ($n = 7$). All of these results were obtained by performing the complete procedure on separate sub-samples, with no exclusion of outlying data.

Thus the procedure seems capable of providing, under practical conditions, a detection limit of about 0.06 µg g^{-1} of molybdenum, and a coefficient of variation of 3% at concentrations well above the detection limit. The coefficient of variation of 3% is satisfactory for low trace analyses, and its magnitude seems to be due mainly to the uncontrolled final volume of the extract rather than variations in instrumental sensitivity.

Baucells et al. [44] applied ICPAES to the determination of very low levels of molybdenum (and cadmium) in soils. Of the most sensitive molybdenum lines, 202.030nm proved to be an excellent analytical line; there are two iron lines, at 201.99 and 202.074nm, but it is free from interference (Fig. 2.18). The other molybdenum-sensitive lines are subject to interference from iron, chromium and vanadium. Background measurement was carried out at –0.084nm.

Interference effects in this method are illustrated in Fig. 2.19. Calcium and magnesium are the most serious interferents when they are present at high levels (more than 1000 µg mL^{-1}). In both instances a depressant effect is observed. The effect of the calcium/molybdenum combination was studied with the molybdenum line; their joint effect was virtually identical with the sum of their separate effects (95% at the 1000 µg mL^{-1} level and 99% at the 4000 µg mL^{-1} level). These results agree with those of Maessen et al. [108]. A very good background correction is necessary in the determination of molybdenum when aluminium is present in the samples because of the enhancement of the background (Fig. 2.18).

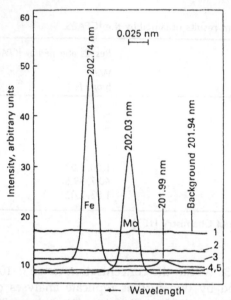

Fig. 2.18 Wavelength scans in the vicinity of the Mo II 202.030nm line. Mo, 1 μg mL $^{-1}$; and Fe, 500 μg mL $^{-1}$. Other interferents: 1, 1000 μg mL $^{-1}$ Al; 2, 100 μg mL $^{-1}$ Ca; 3, 1000 μg mL $^{-1}$ Mg; 4, 1000 μg mL $^{-1}$ Na; and 5, 1000 μg mL $^{-1}$ K. Slit widths, 300 μm entrance and 40 μm exit.
Source: Royal Society of Chemistry [44]

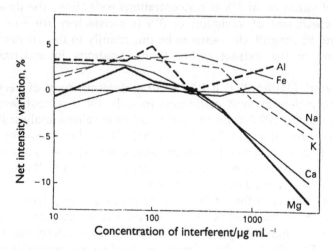

Fig. 2.19 Net line intensity variation (%) of Mo (1 μg mL $^{-1}$) as a function of interferent metal concentration
Source: Royal Society of Chemistry [44]

The application of inductively coupled plasma atomic absorption spectrometry to the determination of molybdenum in soils is also discussed under mult-metal analysis in section 2.60.2.3.

2.29.4 Neutron activation analysis

Molybdenum has been determined as discussed under multi-metal analysis in section 2.60.3.1.

2.29.5 Emission spectrometry

Molybdenum has been determined in soils by this technique [112–114] also as discussed under multi-metal analysis in section 2.60.5.1.

2.29.6 Miscellaneous

The determination of molybdenum in soil is of interest because molybdenum is necessary for normal crop growth, but an excess in forage has a toxic effect on ruminants. The absorption of molybdenum by plants is influenced by other soil components, especially extractable iron, pH and organic matter. The average abundance of molybdenum in soils is about 2 ppm, but deficient soils can have much less than 1 ppm [112].

Jiao *et al.* [289] and Rowbottom [290] have reviewed methods for the determination of molybdenum in soils.

2.30 Neptunium

2.30.1 Miscellaneous

Kim *et al.* [291] have demonstrated good agreement between methods for determining Neptunium-237 in soils, based on inductively coupled plasma mass spectrometry, neutron activation analysis and α-spectrometry.

2.31 Nickel

2.31.1 Spectrophotometric method

A spectrophotometric method has been used to determine 0.5M acetic acid extractable nickel in soil [122].

2.31.2 Atomic absorption spectrometry

This technique has been applied to the determination of nitric perchloric acid soluble nickel in soils [123]. The determination of nickel by atomic absorption spectrometry is also discussed under multi-metal analysis in sections 2.60.1.1 and 2.60.1.3 to 2.60.1.5.

2.31.3 Inductively coupled plasma atomic emission spectrometry

The determination of nickel is discussed under multi–metal analysis in section 2.60.2.1.

2.31.4 Spark source mass spectrometry

The determination of nickel is discussed under multi-metal analysis in section 2.60.6.1.

2.31.5 Differential pulse anodic stripping voltammetry

The determination of nickel is discussed under multi-metal analysis in section 2.60.7.1.

2.31.6 X-ray fluorescence spectroscopy

The determination of nickel is discussed under multi-metal analysis in section 2.60.8.1.

2.31.7 Photon activation analysis

The determination of nickel is discussed under multi-metal analysis in section 2.60.4.1.

2.31.8 Emission spectrometry

The determination of nickel is discussed under multi-metal analysis in section 2.60.5.1.

2.32 Palladium

2.32.1 X-ray absorption fine structure analysis

Manceau *et al.* [212] studied the application of X-ray absorption fine structure analysis (EXAES) to the speciation and quantification of the forms of trace metals in solid materials. Palladium was studied in particular.

2.33 Platinum

2.33.1 Inductively coupled plasma atomic emission spectrometry

Platinum has been determined by this technique as discussed in section 3.37.1 and 3.37.3.

2.33.2 Neutron activation analysis

Platinum has been determined by this technique as discussed in section 3.37.2.

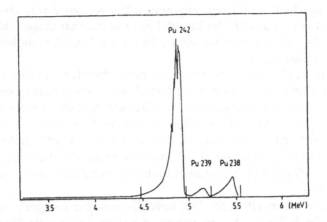

Fig. 2.20 α spectrum of vapour-deposited plutonium chloride containing 3.6 pCi ²⁴²Pu, 0.18 pCi ²³⁹Pu, and 0.43 pcl ²³⁸Pu
Source: American Chemical Society [125]

2.34 Plutonium

2.34.1 Inductively coupled plasma mass spectrometry

Kim *et al.* [292] determined the plutonium-240 to plutonium-239 ratio in soils using the fission track method and inductively coupled plasma mass spectrometry.

2.34.2 Gas chromatography

Packed column gas chromatography has been used to determine various plutonium isotopes in soils [293].

2.34.3 α spectrometry

Dienstbach and Bachmann [125] have determined plutonium in amounts down to 20 fCiPug⁻¹ soil in sandy soils by an automated method based on gas chromatographic separation and α spectrometry. In Fig. 2.20 is shown an α spectrum of vapour deposited plutonium chloride containing 3.6 pCi ²⁴²Pu, 0.18 pCi ²³⁹Pu and 0.43 pCi ²³⁸Pu. In this procedure, the sample is decomposed completely by hydrogen fluoride. The hydrogen fluoride is evaporated and the residue is chlorinated. Plutonium is separated from the sample by volatilization and separation of the chlorides in the gas phase. The plutonium is deposited on a glass disk by condensation of volatilised plutonium chloride. The concentration of plutonium is then determined by α spectroscopy.

Sekine *et al.* [126] used alpha spectrometry to determine plutonium (and americium) in soil. The chemical recovery of plutonium was 51–99%

and averaged 81% while for americium the recovery was 60–70%. The method is coupled with the liquid–liquid extraction stage taking about two days less than would the ion exchange method. A complete analysis takes about one week.

Talvitie [124] has described a radiochemical method for the determination of plutonium in soil based on chromatography on an anion exchange resin of a 9M hydrochloric acid extract of the sample. Following clean-up plutonium is desorbed by reductive elution with 1.2M hydrochloric acid, 30% hydrogen peroxide (50:1) at pH2 followed by α particle counting. The lowest detectable activity for 1000 m counts was 0.02 pCi of ^{239}Pu which is sufficient to detect global nuclear contamination in 1 g of soil.

Various workers [294, 295, 383] have discussed mass spectrometric and other methods for the determination of plutonium in soils. Plutonium in soils has been quantified using plutonium-238 as a yield tracer [296]. Hollenbach et al. [217] used flow injection preconcentration for the determination of ^{230}Th, ^{234}U, ^{239}Pu and ^{240}Pu in soils. Detection limits were improved by a factor of about 20 and greater freedom from interferences were observed with the flow injection system compared to direct aspiration.

2.35 Polonium

2.35.1 Miscellaneous

Various workers [297, 298] have reviewed methods for the determination of polonium-210 in soils.

2.36 Potassium

2.36.1 Flame photometry

Potassium has been determined in 1M ammonium nitrate extracts of soil by flame photometry [127].

2.36.2 Inductively coupled plasma atomic emission spectrometry

This technique has been used to determine potassium in soils as discussed under multi-metal analysis in section 2.60.2.2.

2.36.3 Stable isotope dilution

This technique has been used to determine potassium as discussed under multi-metal analysis in section 2.60.9.1.

2.37 Radium

2.37.1 Miscellaneous

Various workers have discussed methods for the determination of radium-226 in soils [299, 300].

2.38 Rubidium

2.38.1 Spark source mass spectrometry

This technique has been used to determine rubidium as discussed under multi-metal analysis in section 2.60.6.1.

2.38.2 Stable isotope dilution

This technique has been used to determine rubidium as discussed under multi-metal analysis in section 2.60.9.1.

2.39 Scandium

2.39.1 Neutron activation analysis

This technique has been used to determine scandium as discussed under multi-metal analysis in section 2.60.3.1.

2.40 Selenium

The fate of selenium in natural environments such as soil and sediments is affected by a variety of physical, chemical and biological factors which are associated with changes in its oxidation state. Selenium can exist in four different oxidation states (–II, 0, IV and VI) and as a variety of organic compounds. The different chemical forms of selenium can control selenium solubility and availability to organisms. Selenate (Se(VI)) is the most oxidised form of selenium, is highly soluble in water and is generally considered to be the most toxic form. Selenite (Se(IV)) occurs in oxic to suboxic environments and is less available to organisms because of its affinity to sorption sites of sediment and soil constituents. Under anoxic conditions, elemental selenium and selenide(–II) are the thermodynamically stable forms. Elemental selenium is relatively insoluble, and selenide(–II) precipitates as metal selenides(–II) of very low solubility. Organic selenium(–II) compounds such as selenomethionine and selenocystine can accumulate in soil and sediments or mineralise to inorganic selenium. Therefore, Se(VI), Se(IV) and organic selenium(–II) are the most important soluble forms of selenium in natural environments.

2.40.1 Spectrofluorimetric methods

A widely used method for the routine determination of selenium in soils is based on the reaction between selenium and 2,31 diaminonaphthalene to form a fluorescent piazoselenol product [142–147]. While methods based on this principle give satisfactory results, they require very careful technique with strict attention to detail, especially in the sample dissolution stage. Complete destruction of organic matter is necessary in order to avoid the possibility of fluorescent interference from these amounts of residual material, and is achieved by treatment of the sample with hot acidic oxidising mixtures. However, excessive temperature or prolonged heating bring about the loss of selenium by volatilization and considerable ingenuity is required to devise methods that will satisfy these conflicting requirements. These considerations and problems associated with the purity of the 2,3 diaminonaphthalene reagent are now tending to preclude recommendation of fluorometric methods.

2.40.2 Atomic absorption spectrometry

Hydride generation methods are finding increasing favour for the determination of selenium in soils and sediments. This method consists of measuring the atomic absorption of selenium hydride formed as a result of reduction of selenium and its compounds with different reducing mixtures such as sodium borohydride or, occasionally, zinc-stannous chloride-potassium iodide. Hydride generation techniques are about three orders of magnitude more sensitive for determining selenium than are classical flame ionisation techniques, a detection limit of 0.2 ngg^{-1} being achievable. They have an additional advantage of separating selenium from the matrix before atomisation thus avoiding interferences inherent to the conventional atomic absorption technique. Practical working ranges for selenium are 3–250 µg mL^{-1}, 0.03–0.3 µg mL^{-1} and up to 0.12 µg mL^{-1} respectively, for flame atomic absorption, atomic absorption and vapour generation methods [128–135].

The most intense resonance line of selenium (196.03nm) corresponds to a range near to the vacuum ultraviolet. Moreover, the most frequently applied air acetylene flame absorbs about 55% of radiation intensity of the light source. When using electrodeless discharge lamps and air-acetylene flame, appreciably lower detection limits can be achieved by application of a deuterium lamp for background correction. The argon-hydrogen flame is often used for augmentation of sensitivity but it increases interferences too. Extraction has also been attempted [136] as a means of improving sensitivity but in selenium determinations a re-extraction to a water solution is necessary.

Flameless atomic absorption spectrometric techniques offer a high sensitivity (5 × 10^{-11} gSe) but are not simple nor free from interference, due to the high volatility of selenium. This technique is suitable specially for direct analysis of samples and its additional advantage lies in possibilities of 'chemical treatment' of samples in the graphite cell in order to diminish chemical interferences.

The addition of nickel enhances significantly the sensitivity for selenium by about 30% and allows higher ashing temperatures (1000°C) without losses [137–139]. Other elements capable of forming selenides (ie barium, copper, iron, magnesium and zinc) did not interfere and arsenic interference was minimised. A detection limit of 10–12 µg kg^{-1} selenium has been achieved using a graphite electrothermal furnace and background correction with a deuterium lamp [140].

A method has recently been reported [31] for determining total arsenic (and selenium) in soils based on atomic absorption spectrometry and flow injection analysis. The method exhibits good recoveries and detection limits below 1 µg L^{-1} for an injection volume of 160 µL.

Hydride generation atomic absorption spectrometry is widely used to determine the speciation of selenium in natural water, and soil-sediment extracts because of its low detection limits. Speciation of selenium is determined by subdividing sample solutions into selective treatments. Selenite is determined by directly analysing aliquots of samples without any treatments or by analysing samples acidified to pH 2 with concentrated hydrochloric acid or samples in 4–7 N hydrochloric acid solutions. Selenate plus Se(IV) are determined after reduction of Se(VI) to Se(IV) in 4–7N hydrochloric acid at high temperatures (80–100°C) and analysis for selenium to obtain Se(VI+IV) concentrations. Selenate is determined by the difference between a determination of Se(VI+IV) and a determination of Se(IV) in another subsample. Total selenium is determined by oxidising all selenium species (organic Se(–II) and Se(IV) to Se(VI) with hydrogen peroxide or persulphate ($K_2S_2O_8$ or $(NH_4)_2S_2O_8$) then reducing Se(VI) to Se(IV) with 4–7N hydrochloric acid at a high temperature (80–100°C) and analysing for total selenium in the samples. Determination of organic Se(–II) is calculated as the difference between the Se(VI+IV) and total selenium analyses. To separate organic Se(–II) from inorganic selenium, a technique was developed by passing an acidified sample (pH 1.6–2.2) through an XAD–8 resin column to remove hydrophobic and neutral organic Se(–II) compounds before selenium species analysis. These methods have provided valuable information about selenium speciation in natural water and soil-sediment extracts.

Some drawbacks for the speciation of selenium using hydride generation atomic absorption spectrometry have been found by some researchers. Thus Se(VI) is recovered poorly from many samples after a reduction with 6N hydrochloric acid at 100°C. The addition of ammonium persulphate increased the recovery of Se(VI). However, part of the organic Se(–II) was included in the value reported for Se(VI) due to the oxidation of organic Se(–II) by persulphate. The reduction of Se(VI) to

Se(IV) in soil extracts with 6N hydrochloric acid oxidised organic Se(–II) present in the sample resulting in an overestimation of Se(VI) concentration. XAD–resin has been used to separate hydrophobic and neutral organic Se(–II) compounds. However, hydrophilic organic Se(–II) compounds in solution, such as selenomethionine which are found in soil extracts, will be detected as part of Se(VI). Also, a considerable fraction of Se(IV) is removed due to a complexion of Se(IV) with humic substances when an acidified sample is passed through an XAD–8 column, thus resulting in overestimation of organic Se(–II) or Se(VI) concentration in the samples. The net consensus view is that many of the published methods for determination of selenium speciation by using hydride generation atomic absorption spectrometry may be possible only in solutions with little or no organic Se(–II) but this situation is rarely found in natural environments.

To overcome these drawbacks Zhang *et al.* [301] developed a new method to determine organic selenium(–II) in soils and sediments. In this method persulphate is used to oxidise organic selenium(–II) and manganese oxide is used as an indicator for oxidation completion. This method was used to determine selenium speciation in soil-sediments and agricultural drainage water samples collected from the western United States. Results showed that organic selenium can be quantitatively oxidised to selenite without changing the selenate concentration in the soil–sediment extract and agricultural drainage water and then quantified by hydride generation atomic absorption spectrometry. Recoveries of spiked organic selenium and selenite were 96–105% in the soil-sediment extracts and 96–103% in the agricultural drainage water. Concentrations of soluble selenium in the soil–sediment extracts were 0.0534–2.45 µg/g of which organic selenium accounted for 4.5–59.1%. Selenate is the dominant form of selenium in agricultural drainage water, accounting for about 90% of the total selenium. In contrast, organic selenium(–II) was an important form of selenium in the wetlands. These results showed that wetland sediments are more active in reducing selenate compared to evaporation pond sediments.

The application of atomic absorption spectrometry to the determination of selenium in soils is also discussed under multi-metal analysis in sections 2.60.1.4, 2.60.1.5 and 2.55.1.6.

2.40.3 Non-dispersive atomic fluorescence spectrometry

Azad *et al.* [141] determined selenium in amounts down to 10 ng mL $^{-1}$ in soil digests by non-dispersive atomic fluorescence spectrometry using an argon-hydrogen flame and the hydride generation technique utilising sodium tetrahydroborate(III).

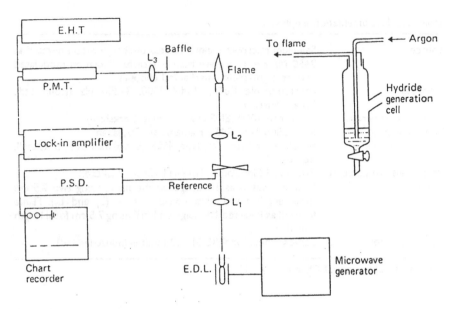

Fig. 2.21 Schematic diagram of atomic fluorescence equipment employed for the determination of selenium
Source: Royal Society of Chemistry [141]

Method

The instrumentation employed in this work was a purpose-built non-dispersive atomic fluorescence spectrometer and a simple hydride generation apparatus. A schematic diagram of the equipment employed is shown in Fig 2.21 and the details of the components employed are listed in Table 2.22. Radiation from a microwave-excited selenium electrodeless discharge lamp was focused on to a rotating sector and then refocused into the argon-hydrogen flame. The atomic fluorescence radiation stimulated from selenium atoms in the flame was then observed at 90° to the incident radiation by passage through a focusing lens to the solar-blind end-window photomultiplier. The output from the photomultiplier was taken to a lock-in amplifier whose reference signal was provided by the rotating sector in the incident radiation beam from the electrodeless discharge lamp. The analytical atomic fluorescence signals for selenium observed at the output from the lock-in amplifier were displayed at the potentiometric chart recorder.

With the flame ignited and argon passing through the hydride generation cell, sufficient time (approximately 20 s) was allowed for the replacement of any air in the apparatus. A 2 cm³ volume of sodium tetrahydroborate(III) reagent solution (5% w/v in 1% sodium hydroxide) was then transferred into the generation cell through the side arm. Acidified selenium standard solution (prepared by dissolving selenium in

Table 2.22 Instrumentation employed

Source	Selenium microwave electrodeless discharge lamp operated at 2450 MHz in a ¾-wave resonant cavity. Radiation modulated with an eight-sector mechanical chopper
Chopper	Programmable Rofin, Model 7500, 3–800 Hz (Rofin Ltd., Egham, Surrey)
Microwave generator	Microtron 2000 (EMS Ltd., Wantage, Berkshire)
Photomultiplier	Solar blind, Type R431, Hamamatsu Co., Japan
Lock-in amplifier	Brookdeal Electronics, Type 450S (Brookdeal Ltd., Bracknell, Berkshire)
Phase sensitive detector	Brookdeal Electronics, Type 411 (Brookdeal Ltd.)
Optics	Source focused as 1:1 image on the flame using two 7.5 cm focal length fused silica convex lesnes (L_1 and L_2). Flame focused as inverted 1:1 image on PMT using 7.5 cm focal length lens (L_3)
Chart recorder	Servoscribe, Model RE 511.20 (Smiths Industries Ltd).

Source: Royal Society of Chemistry [141]

Table 2.23 Optimum operating conditions for determination of selenium

Microwave power to source	50 W
Reflected power from cavity	12 W
Applied voltage to PMT	600 V
Hydrogen flow-rate	3.3 dm^3 min^{-1}
Argon flow-rate	6.0 dm^3 min^{-1}
Hydride generation cell volume	46 cm^3
Sodium tetrahydroborate(III) reagent volume (5% m/V)	2 cm^3
Selenium sample solution volume	1 cm^3

Source: Royal Society of Chemistry [141]

a minimum volume of concentrated nitric acid and making up to 1 L with 5M hydrochloric acid) (or sample solution) (1 cm^3) was then pipetted into the sodium tetrahydroborate(III) solution using a syringe pipette whose tip was fitted with a rubber sleeve to ensure a gas-tight fit with the side arm of the cell during sample introduction. The hydrogen selenide generated was then swept into the argon-hydrogen flame by the argon supply to the flame. The selenium atomic fluorescence signal was recorded at the potentiometric chart recorder; the signal duration observed was approximately 8 s for a 5 µg cm^3 selenium standard solution. The optimum operating conditions established for the procedure, with the particular instrumental arrangement employed, are summarised in Table 2.23.

Determination of selenium in soil digests

One gram amounts of soil samples were weighed into a series of test-tubes, 3.5 cm³ of concentrated nitric acid were added to each sample and the test-tubes were covered and allowed to stand overnight. A few glass boiling beads were added to each tube and then 1.5 cm³ of concentrated perchloric acid (72% m/V) were added to each. The tubes were then transferred into a cold aluminium digestion block, the temperature of which was increased steadily to 100°C over a period of 30 min. The block was maintained at this temperature for 30 min and then the temperature was increased to between 190 and 200°C and maintained at this temperature until digestion of the soil was complete. The final temperature of 200°C should not be exceeded if charring and the loss of selenium by volatilisation are to be avoided. The test tubes were then removed from the digestion block and allowed to cool. A 2 cm³ volume of potassium bromide solution (2% m/V) was added to each and the test tubes were allowed to stand in boiling water for 15 min to ensure complete reduction of selenium(VI) to selenium(IV). The solutions were then centrifuged and the residues rejected. The supernatant solution was taken for analysis; either the lanthanum nitrate-ammonia or the tellurium(IV) addition procedure was applied in order to eliminate interference from copper. The solutions were then made 5M with respect to hydrochloric acid and analysed by the hydride generation technique using the atomic fluorescence spectrometer.

Lanthanum nitrate coprecipitation procedure

Lanthanum nitrate (0.5 cm³ of a 5% m/V solution) was added to each solution prepared for analysis using the digestion procedure described above, 2 cm³ of ammonia solution were then added and the solutions were mixed. After standing for 1 min, the solutions were centrifuged and the liquid discarded. The precipitate was then dissolved in the appropriate amount of 5M hydrochloric acid.

Tellurium(IV) procedure

Tellurium(IV) oxide (0.3 cm³ of a 0.1M solution) was added to each solution prepared using the digestion procedure described above and then diluted to 5 cm³ with 5M hydrochloric acid.

Azad *et al.* [141] evaluated sample digestion recoveries obtained in the above procedure by adding to 1 g soil samples a known amount of selenium prior to their digestion. Recoveries were between 94 and 107% (Table 2.24). Nine soil samples were digested using the procedure described above. Each sample was then analysed by both the lanthanum coprecipitation and the tellurium(IV) methods of interference

Table 2.24 Recovery of selenium added to soil sample no. 4

Selenium concentration in sample/μbg/g^{-1}	Selenium added/μg	Selenium determined/μg	Recovery, %
0.7 ± 0.014	0.1	0.83	104
0.7 ± 0.014	0.2	0.87	97
0.7 ± 0.014	0.3	0.94	94
0.7 ± 0.014	0.4	1.18	107

Source: Royal Society of Chemistry [141]

Table 2.25 Comparison of results for the selenium content of soil digests

	Selenium found						
	Lanthanum nitrate method			Tellurium(IV) method			
Soil sample	Mean, ppm	SD, ppm	RSD, %	Mean, ppm	SD, ppm	RSD, %	ICP* method, ppm
1	0.37	0.017	4.5	0.35	0.009	2.6	0.38
2	0.36	0.016	4.4	0.35	0.012	3.4	0.33
3	0.24	0.010	4.1	0.23	0.015	6.5	0.23
4	0.70	0.014	2.0	0.68	0.015	2.2	0.69
5	18.7	0.64	3.4	18.6	0.49	2.6	19.2
6	111	2.93	2.6	110	2.45	2.2	–
7	0.29	0.014	4.8	0.28	0.012	4.2	0.28
8	0.31	0.020	6.4	0.30	0.015	5.0	–
9	0.30	0.023	7.6	0.29	0.018	6.2	–

*ICP = optical-emission spectrometry using an inductively coupled argon plasma source

Source: Royal Society of Chemistry [141]

suppression. The results obtained for the selenium content of the soils analysed are shown in Table 2.25 for both methods of interference suppression. As can be seen from the table there is no appreciable quantitative difference between the results obtained by both methods. These results also show extremely good agreement with those obtained by the hydride generation technique and optical-emission spectrometry using an inductively coupled argon plasma source.

2.40.4 Inductively coupled plasma atomic emission spectrometry

Pahlavanpour et al. [148] determined traces of selenium in soils by the introduction of hydrogen selenide into an inductively coupled plasma source (in emission spectrometry). This method, discussed below, is adaptable to the rapid analysis of large batches of samples. The soil digest containing the selenium is reduced with sodium tetrahydroborate(III) and the hydrogen selenide formed is swept into an inductively coupled plasma source for determination by emission spectrometry. Calibrations are linear from the detection limit (1 ng mL $^{-1}$) to about 1000 ng mL $^{-1}$ of selenium. The detection limit obtained is comparable to most of the values reported for selenium by atomic absorption-hydride systems, and the linear calibration range is much greater.

Although the method is not subject to interference from most constituents of soils, small traces of copper inhibit the release of hydrogen selenide so the selenium is separated by co-precipitation with lanthanum hydroxide. Treatment of the sample with hot oxidising acids at a controlled temperature is still required, but need not be prolonged as trace amounts of unoxidised organic matter do not interfere as they do in the fluorometric method. In the final solution selenium must be present as selenium(IV), as the efficiency of the sodium tetrahydroborate(III) reduction depends on the oxidation state. This is achieved by adding 4% of potassium bromide to the final solution and heating at 50°C.

Method

The hydride generator – ICP – spectrometer system used was that described by Thompson et al. [149, 382]. The hydride generator consisted of a separation cell and a peristaltic pump. The glass cell was designed to provide mixing of the sample and the sodium tetrahydroborate solution, separation of the hydrogen selenide plus hydrogen from the spent liquids and mixing of the gaseous products with the argon carrier gas. The pump was a Watson–Marlow type MHRE200 (ie a fast pump) fitted with narrow bore tubing to obtain fast linear flow rates, and hence rapid sample transfer and low dead space.

Plasma generator

A Radyne Model R50 cavity-controlled oscillator with a maximum forward power of 8 kW, fitted with a plasma torch, was used.

Spectrometer

An Applied Research Laboratories 29000B Quantometer fitted with 40 channels was used. Light from the tail flame of the plasma was focused on to the slit of the spectrometer in such a way that the effective viewing height of the plasma could be varied.

Operating conditions

The operating conditions were as follows: forward power, 2.7 kW; viewing height, 6 mm above load coil; coolant gas flow rate, 17 L min $^{-1}$ of argon; integration time, 20 s; pre-integration time, 15 s; selenium wavelength, 196.10nm; sample acidity, 5M in hydrochloric acid; sample flow rate, 9.2 ml min $^{-1}$; reagent composition, 1% m/V sodium tetrahydroborate(III) in 0.1M sodium hydroxide solution; reagent flow rate, 4.5 ml min $^{-1}$; and carrier gas flow rate, 0.8 L min $^{-1}$ of argon.

Heating block

A drilled-out thermostatically controlled aluminium block made by Scienco–Western Ltd which could accommodate 252 test tubes (150 × 19 mm) to a depth of 120 mm, was used. The temperature could be controlled to within 1°C.

Reagents

Analytical reagent grade reagents and de-ionised water were used throughout.

- *Potassium bromide solution.* Dissolve 400 g of potassium bromide in 1 L of water.
- *Sodium tetrahydroborate(III) solution.* Dissolve 10 g of the compound in 1 L of 0.1M sodium hydroxide solution.

Digestion of soil

Weigh 0.500 g of air dried sample, ground to pass through a 200 μm sieve, into a 19 × 150 mm borosilicate test tube. Add 2 ml of concentrated nitric acid and some crushed glass, gently swirl the mixture and leave it overnight at 50°C in a hot block. Remove the test tubes and, when cool, add 1 ml of 72% w/V perchloric acid, re-mix and replace in the hot block. Raise the temperature of the hot block to 100°C until the brown fumes cease to be evolved. Raise the temperature to 170°C and remove the test tubes when the samples are bleached but not completely dry. Add four drops of 5M hydrochloric acid to the cooled test tubes and transfer the contents with about 5 ml of water into centrifuge tubes. Spin the test tubes at 4000 rev min $^{-1}$ for 30 s, and decant the centrifuge into a graduated 10 ml centrifuge tube.

To each tube add 0.5 ml of 9% w/V lanthanum nitrate solution and 2 ml of ammonia solution, swirl the resulting suspension and centrifuge at 4000 rev min $^{-1}$ for 30 s. Discard the centrifugate and wash the residue with 1 ml of ammonia solution and 3 ml of water. Centrifuge the tube again and discard the washings. Dissolve the residue in 5 ml of 5M

Table 2.26 Comparison of mean values (μg g⁻¹) obtained using hydride generation – ICP with those obtained from other laboratories

Sample	ADAS*	ADAS*	Analysed by CCRMP†	AGRG‡ m	s	n
SO–1			0.10	0.08	0.01	7
SO–2			0.30	0.36	0.03	6
SO–3			0.05	0.18	0.02	7
SO–4			0.40	0.52	0.03	7
S778		0.20		0.30	0.04	6
S779		0.40		0.43	0.08	7
S783		0.50		0.27	0.03	5
S784		0.90		0.92	0.07	6
S788		1.10		1.08	0.08	6
S789		1.20		1.08	0.09	6
63/35	0.30			0.26	–	1
95/16	0.75			0.63	–	1
63/38	1.52			1.37	–	1

*Mr R J Hall, Ministry of Agriculture, Fisheries & Food, Newcastle upon Tyne (method: fluorimetry)
†Canadian Certified Reference Materials Project (method: fluorimetry)
‡Applied Geochemistry Research Group, Imperial College, London (method: this work). m = arithmetic mean value; s = between-batch standard deviation; n = number of observations

Source: Royal Society of Chemistry [148]

hydrochloric acid, add 1 ml of 40% w/V potassium bromide solution, heat in a water bath at 50°C for 50 min and dilute to 10 ml with 5M hydrochloric acid. (The precipitation of ammonium perchlorate during this procedure has no adverse effect.) The potassium bromide reducing agent also reduces arsenic V and antimony V to their trivalent states.

Reduce the samples sequentially with sodium tetrahydroborate(III) solution, (10% w/V in 0.1 m sodium hydroxide), using the procedure and instrumental conditions described above. Use 5M hydrochloric acid to establish the base line and a solution of sodium selenate (0.1 μg mL⁻¹ of selenium) in 5M hydrochloric acid as a calibrator, checking the readings after every ten samples.

With few exceptions, the results given by fluorometry, chromatography and neutron activation analysis compare well with those obtained by the method described above (Tables 2.26 and 2.27). The precisions obtained for the various samples were very good for between and within batch samples.

The application of hydride generation inductively coupled plasma mass spectrometry to determine selenium in soils has also been discussed by McCurdy *et al.* [302]. Selenium is discussed further under multi-metal analysis in section 2.60.2.4.

Table 2.27 Comparison of selenium concentrations (μg g⁻¹) in two standard soils obtained using gas–liquid chromatography (GLC), high-performance liquid chromatography (HPLC), neutron-activation analysis (NAA) and an inductively coupled plasma source (ICP)

Sample	GLC*	HPLC*	NAA†	ICP‡		
				m	s	n
Weald loam	0.22	0.17	0.56	0.22	0.04	14
Lower lias	0.59	0.70	0.76	0.84	0.05	14

*Dr G Nickless, Department of Inorganic Chemistry, University of Bristol
†Universities Reactor Centre, Warrington
‡Applied Geochemistry Research Group, Imperial College, London (this work). m = arithmetic mean value; s = between-batch standard deviation; n = number of observations

Source: Royal Society of Chemistry [148]

2.40.5 Neutron activation analysis

Neutron activation analysis has been used to determine selenium in soil [150–154]. The determination of selenium in soil by neutron activation analysis has also been discussed under multi-metal analysis in sections 2.60.3.1 and 2.60.3.2.

2.40.6 Miscellaneous

Agemian and Bedek [155] have described a semi-automated method for the determination of total selenium in soils.

Dong *et al.* [179] used mixtures of phosphoric acid, nitric acid and hydrogen peroxide in the digestion of soils prior to the determination of selenium.

Bem [156] has reviewed methods developed up to 1981 for the determination of selenium in soil. These methods include neutron activation analysis, atomic absorption spectrometry, gas chromatography and spectrophotometric methods. Square wave cathodic stripping voltammetry has been used to determine selenium in soils [303]. Selenium has been directly determined in soils by PIXE [304].

2.41 Silicon

2.41.1 Atomic absorption spectrometry

Pellenberg [157] analysed soils and river sediment for silicon content by nitrous oxide–acetylene flame atomic absorption spectrophotometry. He showed that total carbon and total carbohydrates both correlate well with silicon content and the correlation between sedimentary silicon and presumed sewage material is good enough to suggest silicon as a totally synthetic specific tracer for sewage in the aquatic environment.

2.41.2 Inductively coupled plasma atomic emission spectrometry

Que-Hee and Boyle [158] analysed soils for total silicon using Parr bomb digestion with hydrofluoric-nitric-perchloric acids followed by inductively coupled plasma atomic absorption spectrometry. The determination of silicon by this technique is also discussed under multi-metal analysis in section 2.60.2.2.

2.42 Silver

2.42.1 Atomic absorption spectrometry

The determination of silver by this technique is discussed under multi-element analysis in section 2.60.1.5.

2.43 Sodium

2.43.1 Atomic absorption spectrometry

Sodium has been determined in 1M ammonium extracts of soil by this method [159].

2.43.2 Inductively coupled plasma atomic emission spectrometry

Sodium has been determined by this technique as discussed under multi-metal analysis in section 2.60.2.2.

2.44 Strontium

2.44.1 Inductively coupled plasma atomic emission spectrometry

This technique has been used to determine strontium as discussed under multi-metal analysis in section 2.60.2.2.

2.44.2 Emission spectrometry

This technique has been used to determine strontium as discussed under multi-metal analysis in section 2.60.5.1.

2.44.3 Stable isotope dilution

This technique has been used to determine strontium as discussed under multi-metal analysis in section 2.60.9.1.

Table 2.28 Strontium-90 precision

DPM added × 10⁻²	DPM found × 10⁻²	DPM found/ DPM added
13.8 ± 0.2	14.0 ± 0.6	1.01
13.8 ± 0.2	14.2 ± 0.6	1.03
138 ± 2	139 ± 6	1.01
138 ± 2	138 ± 6	1.00
1331 ± 2	1310 ± 50	0.98
1331 ± 3	1310 ± 50	0.98

Source: American Chemical Society, Washington [160]

Table 2.29 Strontium-89 and -90 precision

DPM added × 10⁻²		DPM found × 10⁻²		DPM found/ DPM added	
^{90}Sr	^{89}Sr	^{90}Sr	^{89}SR	^{90}Sr	^{89}Sr
138 ± 2	17.8 ± 0.2	138 ± 6	16.2 ± 3.9	1.00	0.9
138 ± 2	17.8 ± 0.2	139 ± 6	16.5 ± 3.9	1.01	0.9
138 ± 2	55.9 ± 0.2	139 ± 6	55.5 ± 4	1.01	0.99
138 ± 2	55.9 ± 0.2	139 ± 6	53.4 ± 4	1.01	0.96
138 ± 2	1661 ± 3	143 ± 6	1740 ± 70	1.04	1.05
138 ± 2	1661 ± 3	139 ± 6	1730 ± 70	1.01	1.04

Source: American Chemical Society, Washington [160]

2.44.4 Beta spectrometry

Martin [160] has described a β counting technique for the determination of 1×10^{-1} Ci g^{-1} strontium 89 and 90 in soils. In this method fusion of the soil with potassium fluoride and potassium pyrosulphate converts strontium to the ionic state. After dissolving the fusion cake, strontium sulphate is coprecipitated with lead and calcium which are preferentially dissolved into EDTA and discarded. Subsequently, strontium sulphate is dissolved in EDTA and hydrolytic elements are separated on ferric hydroxide. Strontium sulphate is precipitated from EDTA at a lower pH, dissolved in DTPA, and set aside for ^{90}Y ingrowth. After ingrowth the strontium sulphate is reprecipitated, and the ^{90}Y is extracted from the supernate into bis(2-ethyhexyl) phosphoric acid. When both ^{89}Sr and ^{90}Sr are being determined, barium chromate is separated from a DTPA solution of strontium. Strontium sulphate is precipitated from the barium chromate supernate and counted for ^{89}Sr plus ^{90}Sr. Yttrium-90 is counted on yttrium oxalate to determine ^{90}Sr, and ^{89}Sr is determined by difference.

Accuracy and precision were studied by analysing soil samples which were spiked with either ^{90}Sr or ^{89}Sr and ^{90}Sr. The results are summarised in Tables 2.28 and 2.29. As was expected, the precision and accuracy for ^{90}Sr are about the same whether ^{90}Sr is determined in the presence of ^{89}Sr or not. Since ^{89}Sr is determined by the difference between the total strontium and the ^{90}Sr activity, the uncertainty of individual results and precision in the replicate results for strontium-89 are determined largely by the relative activities of ^{89}Sr and ^{90}Sr in the sample. The ratios of the activity found to the activity added indicate an accuracy of 2% for strontium-90 and 5 to 10% for strontium-89. In terms of relative uncertainties of individual results about the mean, the precision was 1–2% for ^{90}Sr.

2.44.5 Photon activation analysis

The determination of strontium is discussed under multi-metal analysis in section 2.60.4.1.

2.44.6 Miscellaneous

Akcay et al. [305] have shown that extraction of total strontium using an ultrasonic extraction procedure were not as good as was achieved by using conventional extraction methods. Stella et al. [306] and Ryabukhin et al. [307] have discussed the determination of radiostrontium in soils.

2.45 Tamerium

2.45.1 Emission spectrometry

The determination of tamerium is discussed under multi-metal analysis in section 2.60.5.1.

2.45.2 Neutron activation analysis

The determination of tamerium is discussed under multi-metal analysis in section 2.60.3.3.

2.46 Tantalum

2.46.1 Neutron activation analysis

The determination of tantalum is discussed under multi-metal analysis in section 2.60.3.1.

2.47 Technetium

2.47.1 Inductively coupled plasma mass spectrometry

Tagami and Uchida [308] have reported a simple method for the determination of technetium-99 in soil by inductively coupled plasma mass spectrometry.

2.47.2 Miscellaneous

Morita et al. [309] and Harvey et al. [310] have reviewed methods for the determination of technetium in soils.

2.48 Tellurium

2.48.1 Emission spectrometry

The determination of tellurium is discussed under multi-metal analysis in section 2.60.5.1.

2.49 Terbium

2.49.1 Emission spectrometry

The determination of terbium is discussed under multi-metal analysis in section 2.60.5.1.

2.49.2 Neutron activation analysis

The determination of terbium is discussed under multi-metal analysis in section 2.60.3.3.

2.50 Thallium

2.50.1 Atomic absorption spectrometry

This technique has been used to determine thallium in soil [311]. The determination of thallium is also discussed under multi-metal analysis in section 2.60.1.5.

2.50.2 Anodic stripping voltammetry

Opydo [312] used anodic stripping voltammetry to determine thallium in soil extracts in the presence of a large excess of lead.

2.50.3 Emission spectrometry

The determination of thallium is discussed under multi-metal analysis in section 2.60.5.1.

2.50.4 Miscellaneous

Chikhalikar *et al.* [313] have studied the speciation of thallium (and antimony) in soil. Lukaszewski and Zembrzuski [314] and Sagar [315] have discussed the determination of thallium in soils.

2.51 Thorium

2.51.1 Inductively coupled plasma mass spectrometry

Toole *et al.* [316] and Shaw and Francois [317] determined thorium (and uranium) in soils by inductively coupled plasma mass spectrometry.

2.51.2 Neutron activation analysis

The application of this technique to the determination of thorium is discussed under multi-metal analysis in section 2.60.3.1.

2.51.3 Emission spectrometry

The determination of thorium is discussed under multi-metal analysis in section 2.60.5.1.

2.51.4 Miscellaneous

Parsa [318] has described a sequential radiochemical method for the determination of thorium (and uranium) in soils. Mukhtar *et al.* [328] have described a laser fluorometric method for the determination of thorium (and uranium) in soils. Steam digestion has been employed in the preparation of soil samples for the determination of thorium (and uranium) [320]. To determine thorium (and uranium) in soils fluorescent X-rays were measured by the use of a germanium planar detector and chemometric techniques. No sample preparation was required in this method [321]. Various other workers have been reported for the determination of thorium (and uranium) in soils [322, 323].

2.52 Tin

2.52.1 Spark source mass spectrometry

The determination of tin in soil by this technique is discussed under multi-metal analysis in section 2.60.6.1.

2.52.2 Miscellaneous

Li et al. [324] have reviewed methods for the determination of tin in soils.

2.53 Titanium

2.53.1 Spectrophotometric method

Abbasi [161] has described a spectrophotometric method using N-p-methoxyphenyl-2-furohydroxamine acid chromogenic reagent for the determination of down to 0.007 mg L^{-1} titanium in soils. The soil was brought into solution by alkali fusion [125]. The extract was acidified to pH $\approx$ 0 with nitric acid, boiled for 5 min and filtered through a 0.45 mm membrane filter. The acidity of the extract was adjusted to 10–12M in hydrochloric acid prior to colour development and spectrophotometric evaluation at 385nm. About 60 mg kg^{-1} titanium was found in a soil sample.

2.53.2 Inductively coupled plasma atomic emission spectrometry

The determination of titanium is discussed under multi-metal analysis in section 2.60.2.2.

2.53.3 Emission spectrometry

The determination of titanium is discussed under multi-metal analysis in section 2.60.5.1.

2.53.4 Photon activation analysis

The determination of titanium is discussed under multi-metal analysis in section 2.60.4.1.

2.54 Tungsten

2.54.1 Spectrophotometric method

Quin and Brooks [162] have described a rapid method for the determination of down to 0.01 ppm of tungsten in soils. The sample is fused with potassium hydrogen sulphate and the melt leached with 10 M hydrochloric acid, then heated with stannous chloride. This solution is heated with a solution of dithiol in isoamyl acetate, then dissolved in petroleum ether prior to spectrophotometric evaluation at 630nm.

2.54.2 Emission spectrometry

Tungsten has been determined by this technique as discussed under multi-metal analysis in section 2.60.5.1.

2.55 Uranium

2.55.1 Spectrophotometric method

Spectrophotometry at 655 nm using arsenazo 111 as chromogenic reagent has been used to determine down to 0.01 mg uranium in nitric acid extracts of soils [164].

2.55.2 Inductively coupled plasma mass spectrometry

Inductively coupled plasma mass spectrometry has been used for the analysis of uranium. However, the technique suffers from spectral interferences and it has relatively poor detection limits.

Inductively coupled plasma mass spectrometry is a relatively new technique for elemental analysis and has superior limits of detection over optical methods. Also this technique has an order of magnitude better detection limit than that obtained for the conventional fluorometric method. Uranium has many stable and unstable isotopes but ^{238}U has the largest percentage abundance (99.274%). ^{238}U is free from interference from other elements and it is therefore possible to detect lower concentrations.

Boomer and Powell [325] have developed an analytical technique using inductively coupled plasma mass spectrometry to estimate the concentration of uranium in a variety of environmental samples including soil. The lower limit for quantitation is 0.1 ng mL^{-1}. Calibration is linear from the low limit to 100 ng mL^{-1}. Precision, accuracy and a quality control protocol were established. Results are compared with those obtained by the conventional fluorometric method.

In this method the soil sample is dried overnight at 85°C and ground into an homogeneous mixture. A 1 g soil sample is placed into a beaker and 10 mL of concentrated nitric acid added. The solution is heated to dryness and 5 mL of concentrated nitric acid is added. The uranium is redissolved in 5 mL of 8N nitric acid and diluted to 25 ml with distilled water. The inductively coupled plasma mass spectrometry system used was an ELAN Model 250. The ion source consists of a modified Plasma Thermal Model 2500 control box with a conventional 27 MHz rf generator. The mass spectrometer contains a quadrupole mass filter capable of a mass range to m/z 300 with a pulse counting channel electron multiplier for ion detection.

The plasma is ignited and the instrument allowed to equilibrate for a 30 min time period. The plasma and ion lenses were set to conditions previously determined by a univariate search. The forward power was set at 1200 W with the plasma flow, auxiliary flow and nebuliser pressure set at 13 L/min, 1.0 L/min and 39 psi respectively. The focusing lenses B, E1, P and S2 are set at +5.3V, –12.5V, –18.0V and –7.6V respectively.

The m/z 238 ion was monitored for two seconds with five replicates of this measurement carried out for each determination.

Toole *et al.* [316] and Shaw and Francois [317] determined uranium (and thorium) in soils by inductively coupled plasma mass spectrometry.

2.55.3 Neutron activation analysis and photon activation analysis

The application of this technique to the determination of uranium is discussed under multi-metal analysis in section 2.60.3.1 (neutron activation analysis) and section 2.60.4.1 (photon activation analysis).

2.55.4 Emission spectrometry

Uranium has been determined in soils by this technique as discussed under multi-metal analysis in section 2.60.5.1.

2.55.5 Miscellaneous

Nass *et al.* [165] used a delayed neutron counting technique to determine down to 50 ng of uranium-235 in soils. Other workers have reported the determination of uranium (and thorium) in soils [322, 323, 326]. Steam digestion has been employed in the preparation of soil samples for the determination of uranium (and thorium) [320]. Parsa [318] has described a sequential radiochemical method for the determination of uranium (and thorium) in soils.

Two methods involving dissolution in hydrogen chloride gas and microwave dissolution have been compared for the remote dissolution of uranium in soil [327]. Mukhtar *et al.* [328] have described a laser fluorometric method for the determination of uranium (and thorium) in soils. To determine uranium (and thorium) in soils, fluorescent X–rays were measured by the use of a germanium planar detector and chromometric techniques [321]. No sample preparation was required in this method.

2.56 Vanadium

Vanadium leaches soil from a large number of diverse sources, including waste effluents of iron and steel industries and chemical industries. Phosphate industries are also a major source of vanadium pollution because vanadium becomes soluble along with phosphoric acids when rock phosphates are leached with sulphuric acid. Vanadium is present in all the subsequent phosphoric acid preparations, including ammonium phosphate fertilisers, and is released into the environment along with them. Other sources of vanadium pollution are fossil fuels, such as crude petroleum, coal and lignite. Burning these fuels releases vanadium into the air which then settles in the soils.

2.56.1 Spectrophotometric method

Abbasi [163] has described a spectrophotometric method employing the violet coloured N(pN,N dimethylanilo-3-methoxy-2-naphtho) hydroxamic acid vanadium V complex and for the determination of down to 0.05 µg vanadium in soils. In this method an extract of the soil was rendered 8N in hydrochloric acid, then 0.1 ml 0.001 M potassium permanganate added to convert vanadium to the pentavalent state. The chromogenic reagent is then added prior to spectrophotometric evaluation.

2.56.2 Inductively coupled plasma atomic emission spectrometry

Vanadium has been determined in soil by this method as discussed under multi-metal analysis in section 2.60.2.2.

2.56.3 Emission spectrometry

Vanadium has been determined in soil by this method as discussed under multi-metal analysis in section 2.60.5.1.

2.57 Yttrium

2.57.1 Emission spectrometry

Yttrium has been determined in soils by this method as discussed under multi-element analysis in section 2.60.5.1.

2.58 Zinc

2.58.1 Atomic absorption spectrometry

Atomic absorption spectrometry has been used to determine 0.5M acetic and extractable [166] and nitric–perchloric acid soluble [167] zinc in soils. The application of this technique to the determination of zinc is also discussed under multi-metal analysis in Sections 2.60.1.1 and 2.60.1.3 to 2.60.1.5.

2.58.2 Inductively coupled plasma atomic emission spectrometry and inductively coupled plasma mass spectrometry

This technique has been applied to the determination of zinc as discussed under multi-metal analysis in sections 2.60.2.1 and 2.60.2.2 (inductively coupled plasma atomic emission spectrometry) and section 2.60.2.2 (inductively coupled plasma mass spectrometry).

2.58.3 Differential pulse anodic scanning voltammetry

This technique has been applied to the determination of zinc as discussed under multi-metal analysis in section 2.60.7.1.

2.58.4 X-ray fluorescence spectroscopy

This technique has been applied to the determination of zinc as discussed under multi-metal analysis in section 2.60.8.1.

2.58.5 Electron probe microanalysis

This technique has been applied to the determination of zinc as discussed under multi-metal analysis in section 2.60.10.1.

2.58.6 Photon activation analysis

The determination of zinc is discussed under multi-metal analysis in section 2.60.4.1.

2.58.7 Emission spectrometry

The determination of zinc is discussed under multi-metal analysis in section 2.60.5.1.

2.58.8 Miscellaneous

External beam photon induced X-ray emission spectrometry has been used to determine total zinc in soils [168].

Roberts *et al.* [49] have discussed the simultaneous extraction and concentration of zinc and cadmium from calcium chloride soil extracts as discussed in section 2.10.6. Adequate recoveries of zinc were obtained when the pH of the extractant was adjusted to the range 4–7. This method is discussed more fully in section 2.60.1.1.

Fast neutron activation analysis has been studied as a screening technique for zinc (and copper) in waste soils [231]. Experiments were conducted in a sealed tube neutron generator and a germanium X-ray detector.

2.59 Zirconium

2.59.1 Emission spectrometry

This technique has been applied to the determination of zirconium in soils as discussed under multi-metal analysis in section 2.60.5.1.

2.59.2 Photon activation analysis

The determination of zirconium is discussed under multi-metal analysis in section 2.60.4.1.

2.60 Multi-metal analysis

Atomic absorption spectrometry, anodic stripping voltammetry or voltammetry are the main methods that have been employed to determine heavy metals in soils. Heavy metals discussed in this section include copper, nickel, cobalt, iron, manganese, zinc, cadmium, chromium and lead.

2.60.1 Atomic absorption spectrometry

2.60.1.1 Heavy metals (cadmium, chromium, iron, manganese, nickel, copper, cobalt, zinc and lead)

Several workers have discussed the application of atomic absorption spectrometry to the determination of heavy metals in soil [39, 49, 75, 88–92, 118, 167–178]. Chao and Sanzolone [169] used atomic absorption spectrometry to determine microgram levels of cobalt, nickel, copper, lead and zinc in soil extracts containing large amounts of manganese and iron. After treating the sample with 0.1M hydroxylammonium chloride in nitric acid, manganese, iron and zinc were determined by direct atomic absorption measurements on the solution. Cobalt, nickel, copper and lead were extracted from the aqueous solution (as complexes with pyrrolidine-1-carbodithioate) into isobutyl methyl ketone before their determination. To avoid interference in the determination of cobalt, nickel, copper and lead, the 10 ml aliquot used for analysis should not contain > 20 mg of manganese; recoveries of 0.5 µg of cobalt, 1 µg of manganese or lead and 5 µg of copper were unaffected by up to 500 µg of iron in the 10 ml of solution. The procedure could be used to determine 500 to 5000 ppm of manganese or iron, 10 to 200 ppm of zinc or copper, 1 to 20 ppm of cobalt or nickel and 5 to 120 ppm of lead.

Roberts *et al.* [49] have described a method for the simultaneous extraction and concentration of cadmium and zinc from soil extracts. The procedure, using dithizone-carbon tetrachloride extraction at pH 4.5, is simple and reliable, giving an essentially quantitative recovery of cadmium and zinc added to calcium chloride extracts of several contrasting soils.

The method was used to evaluate the effect of long-term superphosphate fertiliser addition on the cadmium and zinc contents of soil.

This procedure is designed to make a ten-fold concentration of cadmium and zinc, simultaneously, from aqueous solutions or soil extracts, ie 0.01M calcium chloride (1:10 soil extraction ratio).

To a 50 ml aliquot of sample extract in a 100 ml beaker, add 10 ml of 0.5 M sodium acetate buffer (0.1M in hydrochloric acid) and 1 ml of 10% w/v hydroxylammonium chloride solution. The hydroxylammonium chloride is used to prevent the oxidation of dithizone during extraction. Adjust the pH to 4.5 with redistilled 3.1M hydrochloric acid or 25% $w/^v$ ammonia solution using a pH meter. Rinse the pH electrodes with distilled de-ionised water and transfer the solution quantitatively into a 250 ml separating funnel. Extract twice with 5 ml aliquots of dithizone solution (0.01% w/v in carbon tetrachloride). During each extraction, shake the separating funnel vigorously by hand for 1 min. Allow the aqueous and organic phases to separate and transfer the organic phase to a second separating funnel. To the combined organic phases add 5 ml of 0.01M hydrochloric acid to back-extract cadmium and zinc into the aqueous phase. Shake for 1 min, discard the organic phase and determine the cadmium and zinc concentrations in the aqueous phase using atomic absorption spectrophotometry. Prepare appropriate standards of cadmium and zinc by using the same procedure.

In this procedure recoveries of cadmium and zinc from de-ionised water ranged between 90 and 100%. Recoveries in 0.01M calcium chloride extracts ranged between 100 and 110% (cadmium) and 90 to 110% (zinc).

Pedersen et al. [118] determined copper, lead, cadmium, nickel and cobalt in EDTA extracts of soil by solvent extraction and graphite furnace atomic absorption spectrometry. Diethyl-ammonium diethyldithiocarbamate or ammonium tetramethylene dithiocarbamate metal complexes are extracted into xylene from the EDTA extracts, and metals are determined in the xylene phase by atomic absorption spectrophotometry using a graphite furnace atomiser. The detection limits (concentrations in soil) are approximately copper 0.8, lead 0.3, cadmium 0.07, nickel 2.5 and cobalt 0.8 µg g $^{-1}$. Iron, manganese, aluminium, calcium and zinc do not interfere in amounts likely to be found in extracts of natural or contaminated soils.

Reagents

- *Xylene*. BDH; laboratory reagent, sulphur free.
- *Diethylammonium diethyldithiocarbamate (DDDC), 1.0% m/w solution in xylene.*
- *Ammonium tetramethylene dithiocarbamate (ammonium pyrrolidine dithiocarbamate, APDC), 5.0% m/v solution in water.* This solution is prepared daily and filtered before use.

- *Ethylenediaminetetraacetic acid, disodium salt (EDTA)*, 0.20M solution.
- *Acetate buffer*, pH 4.6. Acetic acid (1 mol) plus sodium acetate (1 mol) are dissolved in water and diluted to 1 l with water. Before use, 1 L of this solution is purified by extraction five times with 25 ml of 1% DDDC in xylene solution and five times with 25 ml of xylene.
- *Sodium acetate solution*, 2 M.
- *Standard solutions*. Stock solutions of copper, lead, cadmium, nickel and cobalt, each containing 1000 µg mL $^{-1}$ of one of these elements, are prepared from $CuSO_4$, $5H_2O$, $Pb(NO_3)_2$, $3CdSO_48H_2O$, $NiSO_4$, $(NH_4)_2SO_46H_2O$ and $CoSO_47H_2O$ respectively, and diluted to volume with 0.2M nitric acid.

  Stock solutions of iron, manganese, aluminium, calcium and zinc, each containing 1000 µg mL $^{-1}$ of one of these elements are prepared from metallic iron, $MnSO_4$ H_2O, metallic aluminium, $CaCO_3$ and $ZnSO_47H_2O$ respectively, and diluted to volume with hydrochloric acid so that the final solutions are 0.1M in hydrochloric acid.

The atomic absorption spectrophotometer used was a Perkin–Elmer Model 303 instrument (without background corrector) equipped with Perkin–Elmer HGA–74 graphite cell, HGA–2100 controller and Model 56 recorder. Single element hollow cathode lamps were used as radiation sources. The argon flow rate through the graphite cell was 50 ml min $^{-1}$. When lead was determined the argon flow was interrupted during the atomisation stage.

Samples were injected with a Finnpipette (5–50 µl) or by means of the Perkin–Elmer AS–1 Auto-sampling system (20 µl) using polyethylene sample cups. The atomisation conditions utilised are shown in Table 2.30. For cadmium a relatively high charring temperature is necessary and for all of the elements determined a relatively long charring time is required in order to destroy the metal complexes.

Table 2.30 Atomisation conditions

| | Metal | | | | |
Condition	Cu	Pb	Cd	Ni	Co
Drying temperature/°C	175	175	175	175	175
Drying time/s	20	20	20	20	20/30
Charring temperature/°C	1000	500	500	1000	1100
Charring time/s	40	60	60	60	60
Atomisation temperature/°C	2500	2200	1800	2700	2700
Atomisation time/s	6	6	5	8	10
Wavelength/nm	324.7	283.3	228.8	232.0	240.7

Source: Royal Society of Chemistry [118]

EDTA extraction

Before extraction the material is air dried at room temperature (20–24°C), passed through a 2 mm sieve and further homogenised by hand grinding in an agate mortar. Then a 0.54 g amount is shaken for 24 h in an end-over-end shaker (30–40 rotations per minute) with 15.0 ml of acetate buffer, 15.0 ml of 0.2M EDTA solution and 60.0 ml of water, and then filtered. The pH of this solution is about 4.6.

Solvent extraction and atomic absorption spectrophotometric determination

Extraction is performed in 25 ml borosilicate test tubes with standard ground-glass stoppers. All volumes are dispensed by means of Finn pipettes with disposable polyethylene tips. A 5–7 ml aliquot of aqueous phase and 1.00 ml of xylene phase are used throughout and the stoppered test tube is shaken vigorously for 2 to 3 min by hand or by means of a reciprocating shaker. The two phases separate very rapidly. The metal content of the xylene phase is determined by graphite furnace atomic absorption spectrophotometry on the same day (for cadmium, within 2 h). Samples should not be left in the polyethylene cups of the Auto-Sampler for more than two hours and unknowns should be interspaced with standards in order to compensate for evaporation losses. It was found that 4% of the xylene would evaporate in two hours.

For the determination of 15–150 ng of copper, 1–10 ng of cadmium or 5–100 ng of lead, 5.0 ml of EDTA extract (if less than 5 ml has to be used, 2 M sodium acetate buffer) is added until the total volume of the aqueous phase is 5 ml) plus 1.00 ml of 1% diethyl ammonium diethylthiocarbamate in xylene are shaken for two minutes. For the determination of 50–800 ng of nickel, 5.0 ml of EDTA extract (or EDTA extract plus acetate buffer), 0.23 ml of nitric acid (1 + 3) decreasing the pH to about 3.8, 1.00 ml of 5% ammonia tetramethylene dithrocarbamate (ammonium pyrrolidine dithiocarbamate) in water and 1.00 ml of xylene are shaken for 3 min.

For the determination of 10–200 ng of cobalt, the same procedure as for nickel is used, except that nitric acid is omitted.

Calibration procedure

In 25 ml test tubes are prepared 5 ml volumes containing suitable amounts of metals within the ranges given above and acetate buffer, EDTA and nitric acid corresponding to unknowns. To these volumes are added 1.00 ml of 1% DDDC in xylene (copper, cadmium and lead) or 1.00 ml of 5% APDC in water plus 1.00 ml of xylene (cobalt and nickel). After shaking, the determination is carried out under the same conditions as for unknowns.

The results of the interference tests are demonstrated in Table 2.31. It should be noted that only in a few instances are the maximum amounts

Table 2.31 Interference tests

Amounts of common elements in the aqueous phase that did not interfere in the presence of the stated amounts of elements to be determined in 1 ml of xylene phase

Amount of metal to be determined/µg	Amount of metal tolerated/µg				
	Fe	Mn	Al	Ca	Zn
Cu, 0.100	100	100	100	100	100
Pb, 0.050	100	100	100	100	100
Cd, 0.004	1000	100	100	1000	100
Ni, 0.200	100	100	1000	2000	100
Co, 0.100	50	200	1000	2000	50

Source: Royal Society of Chemistry [118]

Table 2.32 Representative results

Metal	Determination*	Standard-additions tests		
		Soil/ µg g^{-1}	Amount added to EDTA soil extract/ng	Recovery %
Cu	A	9.79, 9.12	50	101
	B	8.21, 8.70		
Pb	A	8.44, 8.78	25	104
	B	8.73, 8.21		
Cd	A	0.122, 0.098	4	94
	B	0.093, 0.086		
Ni	A			
	B			
Co	A	0.32, 0.19	10	99
	B	0.27		

*For each metal, A represents replicate determinations on the same ETDA extracts and B represents replicate determinations on EDTA extracts of an identical sub-sample.

Source: Royal Society of Chemistry [118]

that can be tolerated shown. However, larger amounts than those tested are rarely encountered in EDTA extracts of soils. When about 1 mg of iron is present a precipitate is formed in the xylene phase, but even this amount did not interfere in the determination of cadmium. However, when determining nickel and cobalt the iron content may become close to the interference limit. With cobalt, up to 150 µg of iron may be present without interfering if the pH is decreased to 3.5.

Representative results from duplicate determinations on soil are shown in Table 2.32. A relative deviation of about 10% or less between duplicate

EDTA extracts of solid sub-samples was considered satisfactory. Standard additions tests demonstrated the absence of matrix interferences.

Weitz et al. [170] have described an atomic absorption spectrometric method for the determination of cadmium and lead in the ppb range in soils after separation by volatilization. The sample was dried at 85°C and then oxidised by heating to 300°C in an argon–oxygen gas stream, the temperature being increased at a controlled rate and held at 300°C for 30 min. Following this, the apparatus was purged with argon and the sample reheated to 870°C in an argon–hydrogen gas stream, when both cadmium and lead are volatilised. The metals are condensed on a cold finger, dissolved in nitric acid and then determined by atomic absorption spectrometry. A recovery of 87% of the cadmium present in the soil is reported.

Berrow and Stein [75] carried out experiments discussed below on the extraction with aqua regia of chromium, copper, lead, manganese, cadmium, iron and zinc from soils prior to the determination of these elements by atomic absorption spectrometry.

All digestions in open vessels were carried out in 100 ml Pyrex beakers on a polypropylene steam bath. Digestions with aqua regia under reflux were carried out in 100 ml round-bottomed flasks equipped with a 30 cm condenser on an Electrothermal Extraction apparatus (E30–220, Macfarlane Robson Ltd, Tyne and Wear). Bomb digestions were carried out in 25 ml capacity PTFE–lined general purpose acid digestion bombs (Parr 4745, Parr Instrument Co, Moline, IL, USA).

All determinations were made using an IL 751 or a PE 560 atomic absorption spectrophotometer, equipped with a single slot burner and an air–acetylene flame except for aluminium, chromium and iron, which were determined in a dinitrogen oxide–acetylene flame. Hollow cathode lamps and the following wavelengths were used: Al 396.3, Cd 228.8, Cu 324.7, Cr 357.9, Fe 248.3, Pb 217.0, Mn 279.5, Ni 232.0 and Zn 213.9 nm. For all elements determined at a wavelength of less than 300nm a deuterium lamp background correction was applied. For the determination of aluminium 1000 mg L^{-1} of sodium were added to all solutions as an ionisation suppressor. In some of the soil digests where the levels of cadmium were below the limit of detection of the normal flame technique the improved sensitivity of an atom-trapping device attached to a Techtron AA6 instrument was used. Soil samples were dried in air at a temperature not exceeding 30°C and sieved through a 2 mm sieve. The sieved fine soil was thoroughly mixed, coned and quartered and about 30 g ground in an agate planetary ball mill (Fritsch Pulverisette No 5, Alfred Fritsch OGH, Oberstein, West Germany) for 30 min to 150 μm in size. Ground soils and sludges were stored in glass bottles.

Seven soils (A–G), which included both uncontaminated (samples E and F) and metal-contaminated soils, were each analysed five times and the results are reported in Table 2.33. RSD values again varied according

Table 2.33 Mean and relative standard deviation values for five analyses of seven soils using aqua regia reflux digestion

Mean expressed as milligrams per kilogram in dry soil and relative standard deviation (RSD) in %

Soil	Cadmium Mean	RSD	Chromium Mean	RSD	Copper Mean	RSD	Lead Mean	RSD	Manganese Mean	RSD	Nickel Mean	RSD	Zinc Mean	RSD
A	6.2	3.4	94	1.1	71	1.3	69	3.7	310	3.8	30	8.0	700	2.2
B	5.0	3.8	88	2.7	612	0.9	57	5.3	300	4.7	34	2.6	570	2.0
C	<1.7	—	40	5.5	24	1.5	27	6.6	320	3.9	16	11	380	1.6
D	<1.7	—	34	5.9	21	2.3	26	8.8	310	6.3	13	9.6	300	5.4
E	<1.7	—	58	3.3	31	3.5	20	4.9	440	3.3	27	4.0	81	2.8
F	<1.7	—	30	1.2	26	3.5	27	1.9	470	2.8	27	7.9	95	1.2
G	29	3.2	240	2.4	230	2.6	1220	840	2.7	94	1.0	1420	3.2	1.2

Source: Royal Society of Chemistry [75]

Table 2.34 Metal contents of soils extracted by refluxing with aqua regia, in acid-insoluble residues and in undigested soil

Contents expressed as milligrams per kilogram in dry soil

Soil content	Chromium	Copper	Lead	Manganese	Nickel
A–					
AR*	94	71	69	310	30
AIR†	23	6	3	37	<9
Total‡	80	70	80	350	40
B–					
AR	88	61	57	300	34
AIR	21	5	3	50	<8
Total	100	60	80	350	40
C–					
AR	40	24	27	320	16
AIR	8	3	3	35	<8
Total	30	25	30	400	30
D–					
AR	34	21	26	310	13
AIR	8	3	3	35	<8
Total	35	25	30	400	25
E–					
AR	58	31	20	440	27
AIR	20	4	2	40	13
Total	60	30	30	500	40
F–					
AR	50	26	27	470	27
AIR	20	3	2	50	17
Total	70	21	35	500	45
G–					
AR	240	230	1220	840	94
AIR	27	5	5	60	13
Total	160	300	1400	800	125

*AR, aqua regia
†AIR, acid-insoluble residues
‡Total, undigested soil

Source: Royal Society of Chemistry [75]

to element determined and soil sample and tended to be better for copper (mean 2.2) and zinc (mean 2.6) than for the other elements but overall averaged 3.9%. The results of analysis of the residues remaining after extraction and of the original soils by dc arc emission spectrography are reported in Table 2.34. The aqua regia extractable values and the contents of the residues, adjusted for the losses in mass during digestion, should sum to the total content of the soil within the limits of error of the semi-quantitative method used, which is generally so and indicates the degree of retention of metals by the siliceous residues. Expressing the aqua regia extractable values as percentages of the total contents shows that on average some 70% of the nickel, 80% of the lead and 90% of the chromium, copper and manganese have been extracted.

Because of the poor limits of detection for cadmium and zinc obtained using emission spectrography, it has not been possible previously to determine the amounts of these elements in either of the soils or their aqua regia insoluble residues. Data on the efficiency of extraction of these two elements have been obtained, however, by analysis of the Canadian Reference soils for which there are established recommended values [175]. Table 2.35 reports the mean RSD recommended (RV) and percentage solubility (S) values for the four soils. The soils are as follows: SO-1 a clay soil C-horizon, SO-2 a podzol B-horizon, SO-3 a calcareous soil, and SO-4 a chernozem A-horizon. The recommended cadmium values are overall means after rejection of outliers and are not yet certified values because of analytical variability. The values are therefore reported in parentheses, as is the value for nickel in SO-2. All values for cadmium are at the lower end of the range for cadmium in uncontaminated soils, which ranges from about 0.02-2, mean 0.6 mg kg^{-1}. The cadmium concentrations in the aqua regia digests were measured using the improved sensitivity of the atom-trapping technique in flame atomic absorption. The results show that in three of the soils all the zinc and in SO-4, 90% of the total cadmium has been extracted. The recommended values for chromium, copper, lead and nickel in SO-2 and SO-3 are also all at the low end of the ranges for these elements in soils and the relatively poor RSD values for these elements, in soil SO-2 in particular, probably reflects this. The proportions of the recommended total contents extracted by aqua regia for the nine elements determined, expressed as a mean percentage solubility, are aluminium 35, cadmium 70, chromium 67, copper 78, iron 85, lead 46, manganese 64, nickel 58 and zinc 90%. The low percentage solubilities for aluminium and lead suggest that the bulk of the generally low lead contents are held in aluminosilicates such as the potassium feldspars, which are moderately resistant to acid dissolution.

Electrothermal atomisation atomic absorption spectrometry has been used to directly determine heavy metals in soil slurries [172]. Lead and cadmium are determined directly in soil by pipetting an aqueous slurry into a graphite furnace electrothermal atomiser. Aqueous calibration standards are used, and analyte recovery is complete provided that the soil is finely ground. The precision (relative standard deviation) at typical concentrations in soil is approximately 2-4% for cadmium and 8-9% for lead. The importance of soil particle size has been studied. Insufficient grinding leads to poor recovery from larger particles. The major error is in the atomisation process, and analyte losses during pipetting of slurries into the atomiser are relatively small. In this method both mills were used to grind the soils [176] and an automatic agate mortar and pestle (Mortar Grinder, Model 155, Fisher Scientific Co) was used for the coarse grinding

Table 2.35 Analysis of Canadian reference soils SO–1–SO–4 using aqua regia reflux digestion

Mean and recommended values (RV) expressed in milligrams per kilogram (except for Al and Fe, %) and relative standard deviation (RSD) and solubility values (SOl) in % on an oven-dry basis (110°C), %

Soil	Aluminium, %				Cadmium				Chromium			
	Mean	RSD	RV	Sol	Mean	RSD	RV	Sol	Mean	RSD	RV	Sol
SO–1	4.38	5.0	9.38	47	0.12	33	(0.15)	80	171	3.5	160	107
SO–2	2.39	2.3	8.07	30	0.11	32	(0.18)	61	9.2	13.5	16	58
SO–3	0.99	3.7	3.05	33	0.07	4.2	(0.14)	50	16.6	3.4	26	64
SO–4	1.70	2.6	5.46	31	0.38	8.3	(0.42)	90	27	5.0	61	44

Soil	Copper				Iron, %				Lead			
	Mean	RSD	RV	Sol	Mean	RSD	RV	Sol	Mean	RSD	RV	Sol
SO–1	56	2.1	61	92	5.55	1.7	6.00	93	9.8	5.8	21	47
SO–2	4.3	3.5	7	61	3.58	4.1	5.56	64	4.2	18	21	20
SO–3	14	1.8	17	82	1.44	1.9	1.51	95	8.7	7.2	14	62
SO–4	17	3.3	22	77	2.11	1.3	2.37	89	9.0	1.7	16	56

Soil	Aluminium, %				Cadmium				Chromium			
	Mean	RSD	RV	Sol	Mean	RSD	RV	Sol	Mean	RSD	RV	Sol
SO–1	676	2.1	890	76	74	2.0	94	79	157	2.9	146	108
SO–2	189	1.9	720	26	<3.4	–	(12)	<28	75	3.9	124	60
SO–3	370	3.6	520	71	9.0	5.6	16	56	52	1.1	52	100
SO–4	494	1.3	600	82	17	2.6	26	67	98	1.6	94	104

Source: Royal Society of Chemistry [75]

of the samples [185]. Electrothermal atomic absorption spectrometry was carried out on two atomic absorption spectrometers, a Model 157 Instrumentation Laboratory spectrometer equipped with a Model 555 CTF electrothermal atomiser, and a Perkin–Elmer Model 2280 spectrometer with a Model HGA 500 electrothermal atomiser. Both electrothermal atomisers were used with platform atomisation, the Instrumentation Laboratory instrument using graphite tubes with a square cross–section and the Perkin–Elmer instrument using conventional cylindrical tubes with L'vov-type platforms. Deuterium-arc background correction was used for all determinations. Analyses by flame atomic absorption spectrometry were performed on a Varian AA 1275 spectrometer and a Perkin–Elmer 2280 spectrometer, both of which were equipped with deuterium-arc background correction.

Procedure

Soil pre-treatment

Samples included sandy soils and those of a high clay content. All soils were dried at 105°C for 24 h, and those to be used for the comparative electrothermal and flame studies were coarse ground in a porcelain ball mill, followed by fine grinding in a miniature ball mill [94]. This ensured that at least 90% of the particles were ≤11 μm in diameter, and the remainder ≤30 μm in diameter.

Comparative soil analysis

Slurries for electrothermal analysis were prepared by weighing an appropriate amount of finely ground soil into a beaker, adding 25 ml of distilled or de-ionised water and stirring magnetically for 3 min. The amount of soil was chosen to give a concentration in the linear range of the appropriate calibration graph. Typically 5–20 mg of soil were slurried for lead determinations and 10–500 mg for cadmium. Either 20 or 50 μL of the stirred slurry were pipetted, in triplicate, into the electrothermal instrument and the absorption peak height or area was measured. The heating programmes shown in Table 2.36 were used for the two graphite furnaces. Aqueous calibration standards were used. Owing to the high concentration range of lead in soil it was convenient to use two analytical wavelengths of different sensitivity. For work at 283.3nm the standards were in the range 0–0.1 mg L $^{-1}$, and the line at 261.4nm had a linear range of 0–10 mg L $^{-1}$. Digestion of soil with nitric acid, followed by flame atomic absorption spectrometry measurement, leads to incomplete recoveries of lead compared with the slurry electrothermal procedure. Digestion with mixtures containing hydrofluoric acid, which also dissolves the siliceous material in the soil, gives a more complete recovery of the trace metals than do other acid mixtures. Therefore, the following total digestion procedure was used in the comparative flame atomic absorption spectrometric

Table 2.36 Furnace operating programmes for the two atomic-absorption spectrometers

(a) *Instrumentation Laboratory 555 CTF–*

Element	Dry		Ash			Atomise	
Lead:							
Time/s	5	15	35	5	0		10
Temperature/°C	70	125	450	450	2000		2000
Cadmium:							
Time/s	15	25	25	5	0		15
Temperature/°C	70	125	200	250	1800		1800

(b) *Perkin–Elmer HGA 500–*

	Dry	Ash	Atomise	Clean
Lead:				
Ramp time/s	20	30	0	2
Hold time/s	30	60	10	8
Final temperature/°C	110	700	1800	2400
Cadmium:				
Ramp time/s	20	30	0	2
Hold time/s	30	60	10	8
Final temperature/°C	200	200	1600	2400

Source: Royal Society of Chemistry [172]

method. Concentrated nitric acid (20 ml) and 40% hydrofluoric acid (10 ml) were added to 1 g of soil and the mixture was evaporated slowly to near dryness at 120°C. A further 40 ml of nitric acid were added and evaporated almost to dryness. This was followed by 20 ml of perchloric acid (35%) diluted 1+1 with distilled water, which was also evaporated almost to dryness. Throughout this procedure, care was taken to ensure that the sample was not evaporated to *complete* dryness. Finally, the residue was dissolved in a minimum volume of 10% hydrochloric acid and diluted to a known volume with distilled or de-ionised water. The final volume of solution depended on the concentration of lead or cadmium in the soil, and in order to obtain adequate sensitivity for cadmium, a small volume (eg 10 ml) was desirable. The upper concentrations of the calibration ranges are shown in Table 2.37. The slightly lower calibration range for the Perkin–Elmer HGA 500 electrothermal atomiser gives an indication of its better sensitivity, compared with the Instrumentation Laboratory electrothermal atomiser. For this application, however, high sensitivity is not required and a more meaningful parameter is the precision when a typical slurry is analysed. Also in Table 2.37 are the precision data for replicate analyses of slurries whose lead and cadmium concentrations are within the optimum concentration ranges. It is seen that cadmium is determined more precisely than lead in both electrothermal atomisers. Nevertheless, the lead precision is considered adequate for this application.

Table 2.37 Calibration ranges and precision of slurry – ETA–AAS determinations at typical concentrations in soil

Spectrometer	Element and wavelength/ nm	Calibration range/ mg L⁻¹	Concentration slurry/mg L⁻¹	Concentration soil/µg g⁻¹	R.s.d.,*%
555 CTF	Pb, 283.3	0.1	0.065	81	8
	Cd, 228.8	0.04	0.006	1.4	1.7
HGA 500	Pb, 283.3	0.05	0.024	11	9
	Cd, 228.8	0.01	0.002	0.8	3.7

*n = 10

Source: Royal Society of Chemistry [172]

Table 2.38 Measured lead and cadmium concentrations

Lead/µg g⁻¹		Cadmium/µg g⁻¹	
ETA–AAS	Flame AAS	ETA–AAS	Flame AAS
50	53	1.4	1.3
81	81	0.47	0.45
156	148	99	105
11	13		

Source: Royal Society of Chemistry [172]

Table 2.38 shows comparative results for several soils analysed by the slurry – electrothermal and nitric acid-hydrofluoric acid–perchloric acid digestion – flame AAS procedures. Good agreement for both lead and cadmium over a wide concentration range confirms complete recovery by the slurry – electrothermal procedure when aqueous calibration standards are used. Both electrothermal atomisers gave equivalent results for the slurries.

Two procedures for determining extractable metals (cadmium, copper, lead, manganese, nickel and zinc) in air dried soils have been described [171]. 0.5M EDTA was used to extract these elements while 0.5M acetic acid was used to extract nickel and zinc. In both cases atomic absorption spectrometry was used to determine the extracted metals. Chester and Hughes [177] and Chowdbury and Bose [178] have studied the efficiency of extraction of copper, cadmium, lead, zinc, nickel and chromium from soils by various extraction agents including 4N nitric acid plus 0.7N hydrochloric acid, 0.5N hydrochloric acid, 1N hydroxylamine, hydrochloride acid plus 25% acetic acid, 0.05N EDTA. It was shown that copper complexed with humic components of the soil is readily liberated

Table 2.39 Extraction of heavy metal from soil

Site	Cd				Cu				Pb			
					Method Used[a]							
	1	2	3	4	1	2	3	4	1	2	3	4
B	–	4.7	–	9	9.3	9.1	7.5	8.1	12	16	16	17
T	–	45	39	78	4.4	4.2	1.5	5.3	6.5	8.2	7.7	8.9
W	–	6.0	5.0	9	4.2	4.1	1.5	5.3	6.5	7.9	7.4	8.7
X	–	3.3	4.0	6	2.8	3.0	1.5	3.6	13	17	16	17

Site	Zn				Ni				Cr			
					Method Used[a]							
	1	2	3	4	1	2	3	4	1	2	3	4
B	10	18	32	34	–	–	–	2	–	2.1	3.6	4.4
T	8.5	16	28	34	5.9	13	18	32	3.7	6.6	15	18
W	5.1	7.3	13	15	–	–	2.7	–	–	–	3.7	6.4
X	4.9	7.8	14	15	–	–	–	–	–	–	–	–

a 1 = 4.0 N HNO_3 + 0.7 N HCl; 2 = 0.5 N HCl; 3 - 1 N NH_2OH_2Hcl + 25% acetic acid; 4 = 0.05 N EDTA

Source: Elsevier Science BV, Netherlands [177]

from organic matter with 0.5N hydrochloric acid (pH 1.0). Thus, unlike 1N hydroxylamine hydrochloride plus 25% acetic acid, extraction with 0.5N hydrochloric acid is suitable for copper extraction. Results obtained from copper and the five other elements are reviewed in Table 2.39.

Official methods [171] have been described for determining extractable metals in soils. The first concerning the use of 0.05 M EDTA for separation of extractable amounts of cadmium, copper, lead, manganese, nickel and zinc, and the second using 0.5 M acetic acid for nickel and zinc. In both cases an atomic absorption spectrometry method is used for the final determination. All the determinations are performed on air–dried samples which have been ground and passed through a 2 mm sieve.

2.60.1.2 Heavy metals (iron, manganese, copper and magnesium)

Stupar and Ajlec [180] have described a flame atomic absorption spectrometric method for the determination of magnesium, iron, manganese and copper in soil suspensions. An investigation was made of the factors influencing the atomisation efficiency of these elements when suspensions of soil sample were aspirated into the flame. Particle size, flame temperature and position in the flame were found to be critical in

Table 2.40 Instrumental parameters employed in atomc-absorption spectrometric measurements of soil suspensions

Element	Wavelength/ nm	Spectral band pass/	Lamp current/	Flame	Working range/ μg mL $^{-1}$
Iron	372.0*	0.33	9	Air–acetylene	0–100
	372.0	0.33	9	Air–N$_2$O–acetylene	0–60
Manganese	279.5	0.17	5	Air–acetylene	0–5
	279.5	0.17	5	Air–N$_2$O–acetylene	0–3
Copper	324.7*	0.33	3	Air–acetylene	0–2
	324.7	0.33	3	Air–acetylene	0–1
Mangesium	285.2*	0.17	3	N$_2$O–acetylene	0–20
	285.2	0.17	3	Air–N$_2$O–acetylene	0–15

*Solutions of soil samples obtained by conventional decomposition

Source: Royal Society of Chemistry [180]

determining the fractions of particular elements atomised. Special emphasis was given to the preparation of the soil suspensions, which is the most critical step in the whole analytical procedure. Magnetic and ultrasonic devices were used for stirring purposes. The latter proved to be more efficient, particularly when suspensions of high clay content soils are being prepared.

Method

Varian–Techtron Model AA–5 atomic absorption spectrometer was used in this work. A standard nebuliser–spray chamber system and the burner heads for air–acetylene (10 cm slot) and dinitrogen oxide-acetylene flames (5 cm slot) were employed. A supplementary system was designed for mixing air and dinitrogen oxide in order to prepare various oxidant gas mixtures. This was connected directly to the nebuliser. Suspensions of soil samples were prepared in 100 ml Erlenmeyer flasks and were stirred with a magnetic device during aspiration into the flame.

Instrumental parameters employed in the atomic absorption spectrometric measurements of iron, manganese, copper and magnesium are presented in Table 2.40. In general, the same absorption line, spectral band pass and lamp current were used for both suspension and solution measurements. A hydrogen lamp was employed for background absorption measurements. An oxidant mixture of dinitrogen oxide (30%) – air (70%) was used. Air-acetylene or dinitrogen oxide-acetylene flames were used for atomisation of solutions. The 372.0 nm iron line was selected for absorption measurement because of its suitable sensitivity and linear response.

Table 2.41 Particulate composition of soil samples

Sample	Organic matter, clay, % (<2 µm)	Silt, % (2–20 µm)	Sand, % (20–2000 µm)
P4B	53.9	39.6	4.5
P13B	42.3	47.4	10.3
P2B	47.9	46.0	6.1
P63G	20.5	43.8	35.7
P73G	22.7	42.0	35.8
P71G	34.9	51.9	13.2
T3	60.4	16.6	23.0
T5	60.0	17.8	22.2
T7	60.6	17.0	22.4
T9	75.8	10.2	14.0
T11	62.0	15.4	22.6
T13	59.0	18.4	22.6

*P = Peat soil and T = terra rossa soil

Source: Royal Society of Chemistry [180]

Preparation of samples is an essential stage in the direct atomic absorption spectrometric analysis of solids in flames, particularly in dealing with soils. These differ significantly in their chemical (mineral) and physical (particle size) composition. Soils are generally characterised by three different particle size fractions: less than 2 µm, organic matter, clay; 2–20 µm, silt; and 20–2000 µm, sand. Clay, silt and said contain a great variety of minerals whose proportions vary tremendously between soils of different types. Under dry conditions clay minerals are liable to form aggregates (silica chains of various lengths) that disintegrate in the presence of water and certain electrolytes to small segments. An illustration of the typical grain size composition of the soils is given in Table 2.41. Stupar and Ajlec [180] studied two soil types, peat soil and terra rossa in this study.

Grinding is the first step involved in the preparation of soil samples after their collection. The purpose of this is to reduce the particle size of the sand fraction, which contains particles up to 2 mm in diameter. A wet method of grinding (using ethanol) was performed in an agate ball mill (Fritsch, 25 balls, 14 mm in diameter). When the procedure was completed, samples were dried in an oven at 105°C and crushed with an agate pestle and mortar. Particle size analysis of soil samples after grinding for 0.5 h showed that the largest particles present in the samples did not exceed 40 µm in diameter. The absorbances measured after 0.5 h grinding are on average 6–15% lower in comparison with those after grinding for 1 h, but the precision of the measurement (relative standard deviation) of major elements (iron, magnesium) is not seriously affected. On the other hand, the precision of the measurement of some minor (manganese) and trace elements may vary considerably with the sample grinding time.

On the basis of the above experimental evidence, an attempt was made to use the suspension technique in routine soil analysis. Two sets of soil samples of different geochemical origin (peat soil, terra rossa) were selected to test the precision and accuracy of the suspension technique. Samples were ground and dried at 105°C prior to analysis. The true concentrations of copper, iron, manganese and magnesium were determined by wet dissolution and a conventional atomic absorption spectrometric procedure. Direct analyses were performed on suspensions prepared in 20% and 50% propan-2-ol-water mixtures. Mixing was carried out for 15 min using a magnetic stirrer or preferably ultrasonic agitation for 2 min. Generally 0.25% m/V suspensions were employed for iron and magnesium determinations whereas 1–2% m/V suspension concentrations were found to be the most convenient for measurement of manganese and copper. The latter was measured in an air-acetylene flame, whereas for the other three elements a [dinitrogen oxide (30%) - air (70%)]-acetylene flame was preferred.

Significant variations in mineral, chemical and particle size composition between soils of different origin entail the use of a separate set of standards for each particular soil type if reasonable accuracy is to be expected. Our approach was to prepare for each soil type one average sample from which a standard suspension for calibration could be made. The average sample was prepared from equal portions of all the samples in a particular set (six samples of peat soil and 15 samples of terra rossa). The elemental content of the standard was determined by conventional atomic absorption spectrometric and other analytical methods. The same preparation procedure (grinding, mixing of the suspension, etc.) was applied for the standard as for the other samples in the series. The standard suspension was measured with each set of identical samples. The concentration of a particular element in the sample was calculated taking a simple linear relationship between the absorbances of the standard and of the sample. Background absorption was substracted from the reading if necessary but no allowance was made for the solvent-soluble fraction in 20% propan-2-ol-water.

The standard suspension was assumed to have an average chemical mineral and grain-size composition of the whole series of similar samples. Thus an approximately equal number of the results should be biased positively and negatively with respect to the true value. The concept of an average sample as a standard was tested on a set of six peat soil samples. Suspensions in 20% propan-2-ol-water were mixed for 15 min on a magnetic stirrer, left overnight and stirred for approximately 1 min before the measurement. No difficulties such as capillary clogging, drifting of

Table 2.42 Concentrations of iron, manganese, magnesium and copper in peat soil determined by atomic-absorption spectrometry without decomposition of the samples and with magnetic stirring

Sample*	Fe Concentration ppm	Δ, %	$\varepsilon_{a,rel}$	Mn Concentration ppm	Δ, %	$\varepsilon_{a,rel}$
B2	55,800	+12.5	0.38	2750	+24.5	0.38
B4	34,100	+6.1	0.38	240	+35.5	0.37
B13	20,100	−13.6	0.28	200	−9.3	0.25
G67	30,100	−3.9	0.34	210	+17.0	0.34
G71	34,700	−1.1	0.34	450	−20.0	0.22
G73	21,800	−16.7	0.26	590	−5.3	0.26
Standard G	28,600	0	0.34	714	0	0.27

Sample*	Fe Concentration ppm	Δ, %	$\varepsilon_{a,rel}$	Mn Concentration ppm	Δ, %	$\varepsilon_{a,rel}$
B2	14,650	+13.2	0.30	40	−6.7	0.41
B4	15,650	+13.9	0.30	51	−1.9	0.42
B13	11,600	−4.6	0.25	33	−25.4	0.33
G67	10,100	+1.1	0.27	42	+7.5	0.46
G71	10,250	+22.0	0.33	32	−33.0	0.30
G73	5,500	−5.5	0.25	28	−9.5	0.39
Standard G	6,980	0	0.26	34.7	0	0.40

*Suspensions in 20% propan-2-ol-water

Source: Royal Society of Chemistry [180]

the absorption signals, etc, were observed during the measurements. The results of this trial are collected in Table 2.42 where deviations from the true values [Δ(%)] and $\varepsilon_{a,rel}$ are also given. It is evident that the concept of the standard is fairly well demonstrated and close correlation between $\varepsilon_{a,rel}$ and Δ can be observed.

A similar test was made with a series of 15 terra rossa soil samples. Using the same procedure some difficulties arose in obtaining reproducible results. Anomalous behaviour of some particular samples in the series was observed. If such samples were mixed for different times large variations in the results were noted. In Fig. 2.22 the variation of the analytical error is plotted as a function of the time of stirring the suspension. With these samples rather noisy absorption signals were observed, showing a continuous rise during prolonged aspiration into the flame. When stirring was interrupted the absorption signal dropped rapidly, the remaining signal being substantially improved in stability.

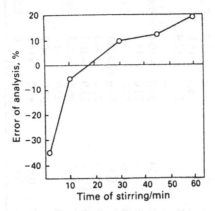

Fig.2.22 Variation of the analytical error (deviation from the true value) with time of stirring suspension. Sample, terra rossa No. 9, 0.25% m/V suspension in 20% V/V propan-2-ol-water; determination of iron; flame, [air (70%) − dinitrogen oxide (30%)] − acetylene.

Source: Royal Society of Chemistry [180]

Ultrasonic agitation was found to be effective in solving this problem. Therefore, the analysis of terra rossa samples was performed on suspensions treated (for 2 min) with an ultrasonic vibrator. After approximately 24 h, the suspensions were measured under continuous stirring employing a magnetic device.

Although the variability of the measured signals was acceptable, the results of the analysis given in Table 2.43 shows a definite negative bias. The reason for the systematic error was found to be in the relatively large proportions of the elements in the soluble form up to 40% of the total amount). The solvent-soluble fraction of the elements would have to be measured for each sample and the standard if an efficient correction were to be performed. To avoid this time-consuming operation, 50% V/V propan-2-ol-water suspensions were prepared with the same set of samples. In this instance the solvent-soluble fraction could be neglected and the stability of the absorption signals was slightly improved.

The results of this trial are summarised in Table 2.44 where the negative and positive deviations are represented in an almost equal proportion for manganese and magnesium. This confirmed the validity of the average standard suspension for calibration purposes for manganese and magnesium; however, for copper and iron a considerable negative bias was observed on average.

In conclusion, it can only be stated that the majority (80–90%) of samples can be analysed with an accuracy of ±20%. This should be acceptable in most applications where a large number of samples are to be analysed.

Table 2.43 Concentrations of iron, manganese, magnesium and copper in terra rossa soil determined by atomic-absorption spectrometry without decomposition of the samples and with ultrasonic agitation

Sample*	Fe Concentration ppm	Δ, %	$\varepsilon_{a,rel}$	Mn Concentration ppm	Δ, %	$\varepsilon_{a,rel}$	Mg Concentration ppm	Δ, %	$\varepsilon_{a,rel}$	Cu Concentration ppm	Δ, %	$\varepsilon_{a,rel}$
1	25,400	−37.4	0.36	710	−26.8	0.31	2950	−46.8	0.21	29.0	−30.4	0.39
2	56,500	−4.7	0.60	1050	−4.5	0.39	5950	−6.5	0.35	41.0	−2.6	0.58
3	53,000	−5.0	–	850	−1.5	–	5,800	−7.5	–	36.5	+5.8	–
4	38,500	−33.1	–	660	°13.9	–	4200	−34.3	–	33.0	−5.4	–
5	55,000	−4.4	–	900	−13.5	–	6100	0.0	–	–	–	–
6	50,300	−13.4	0.47	900	−13.5	0.34	6350	−13.4	0.34	37.0	−10.4	0.48
7	46,400	−25.3	–	680	−15.4	–	4600	−17.6	–	33.0	−15.3	–
8	77,500	+5.6	0.43	840	−14.3	0.47	6150	+13.5	0.40	55.0	+15.6	0.68
9	67,900	−10.5	0.51	550	9.5	0.48	8650	+23.7	0.41	51.0	+2.3	0.61
10	49,600	−11.9	–	1100	−9.1	–	6400	−8.9	–	38.5	−8.5	–
11	39,900	−13.8	–	1500	−0.2	–	5150	−17.5	–	33.0	−36.2	–
12	60,200	+3.6	0.52	1500	−13.8	0.35	5400	+11.7	0.41	39.5	+10.3	0.66
13	44,200	−25.3	–	1330	+0.5	–	4700	−29.2	–	40.0	−1.4	–
14	41,400	−12.1	–	1700	+9.1	–	3850	−9.1	–	37.5	+0.7	–
15	72,100	+5.6	0.50	860	+3.6	0.51	7300	+26.7	0.43	49.0	+14.1	0.64
Standard	56,260	0	0.53	1092	0	0.40	5980	0	0.34	41.2	0	0.56

*Suspensions in 20% propan-2-ol-water

Source: Royal Society of Chemistry [180]

Table 2.44 Concentration of iron, manganese, magnesium and copper in terra rossa soil determined by atomic-absorption spectrometry without decomposition of the samples and with ultrasonic agitation

Sample*	Fe		Mn		Mg		Cu	
	Concentration, ppm	Δ.%	Concentration, ppm	Δ.%	Concentration, ppm	Δ.%	Concentration, ppm	Δ.%
1	38,600	−4.9	840	−13.3	3900	−30.1	34.0	−18.3
2	58,200	−1.8	1240	+12.3	6900	+7.9	48.6	+14.8
3	51,700	−7.3	890	+2.8	5700	−8.8	33.0	−4.3
4	42,700	−25.6	580	−24.7	4100	−35.9	24.5	−29.6
5	55,200	−3.9	970	−5.3	6250	+1.9	—	—
6	51,700	−11.0	900	−13.2	6350	−13.5	33.0	−19.7
7	47,800	−22.9	670	−17.1	4850	−12.8	28.0	−28.5
8	79,200	+7.9	900	+23.1	6700	+23.5	52.0	+9.3
9	64,400	−15.1	470	−7.3	6400	−2.6	42.0	−15.9
10	56,900	+0.9	1210	+0.2	7400	+5.2	41.0	−2.7
11	47,500	+2.6	1610	+6.5	6150	+1.6	34.5	−33.5
12	54,100	−7.0	1480	−15.3	5200	+7.1	39.5	+10.1
13	53,300	−9.9	1270	−17.5	6000	−10.3	41.0	0.0
14	50,100	+6.3	1610	+4.1	4750	+12.4	36.0	−3.1
15	61,700	−9.7	980	+7.0	6600	+14.4	47.0	+9.1
Standard	56,260	0	1092	0	5980	0	41.2	0

*Suspension in 50% propan-2-ol-water

Source: Royal Society of Chemistry [180]

2.60.1.3 Heavy metals (cadmium, chromium, copper, molybdenum, nickel, lead and zinc)

Davis et al. [181, 182] used atomic absorption spectrometry to study the distribution of molybdenum, cadmium, chromium, copper, nickel, lead and zinc in grassland soils following application of sewage sludge. Baucells et al. [44] have described methods for the direct determination of cadmium and molybdenum in soil extracts by graphite furnace atomic absorption spectrometry (GFAAS) and inductively coupled plasma (ICP) spectrometry. Optimum operating conditions, analyte modifiers and matrix interferences were investigated. The use of a mini-furnace allowing an atomisation temperature of 1200°C and the addition of molybdenum plus hydrogen peroxide plus nitric acid as a matrix modifier are the most important features for the successful direct determination of cadmium by graphite furnace atomic absorption spectrometry. The characteristic amounts are 0.164 pg for cadmium and 8.39 pg for molybdenum, with relative standard deviations (RSD) of 7.5% and 4% respectively, based on 10 replicate measurements on a soil extract. The ICP detection limits are 1.49 µg L $^{-1}$ for cadmium and 4.09 µg L $^{-1}$ for molybdenum, with RSDs better than 1%. Matrix interferences in ICP are negligible compared with GFAAS and the precision of the former technique is better, but the sensitivity of GFAAS is clearly superior.

A Varian Model 875 atomic absorption spectrometer, equipped with a CRA–90 graphic furnace and an LBK 2210 recorder, was used in this work. Pyrolytically coated graphite tubes (Ringsdorff–Werke RWO 278/PYC) were used. Background correction was effected using a deuterium lamp. Manual injection with an Eppendorf micropipette was used because the ASD–53 autosampler connected to the graphite furnace did not work properly with the samples studied, not only because of the poor precision but also because the solution remained in the hole of the mini-furnace. The amounts of sample injected were 5 µL in all instances. Recommended spectrometer settings and furnace temperatures and time programmes are given in Tables 2.45 and 2.46. The sequential ICP system used and the operating parameters are given in Table 2.47. To smooth sample aspiration effects, a peristaltic pump was used. Nebuliser clogging was avoided by using a suitable uptake and rinsing cycle. A 30 s pre-observation aspiration, a 30 s measurement time and a 30 s rinse with triply distilled water were applied in order to avoid nebuliser drift, which is critical at high salt contents.

Cadmium extraction was performed by the method of Lindsay and Norvell [183], using diethylenetriaminepentaacetic acid (DTPA). A 10 g amount of air-dried soil was shaken with 20 ml of a mixture of 0.005 M DTPA, 0.1M triethanolamine and 0.01M $CaCl_2$ solution (pH 7.3) for 2 h.

Molybdenum extraction was performed with 1M ammonium acetate solution (pH 7). A 12.5 g amount of air-dried soil was shaken for 15 h with

Table 2.45 Recommended atomic-absorption spectrometer settings

Element	Wavelength/nm	Spectral band width/nm	Lamp current/mA
Cd	228.8	0.5	3.5
Mo	313.3	0.5	7

Source: Royal Society of Chemistry [44]

Table 2.46 Recommended furnace programmer settings. Peak height mode; deuterium arc; interal gas (argon) flow-rate, 5 L min $^{-1}$

Element	Dry	Char	Atomise ramp	Atomisation	Clean-out
Cd	100°C, 35 s	400°C, 20 s	1200°C, 2 s	700°C, s $^{-1}$	3000°C, 1.5 s
Mo	100°C, 35 s	500°C, 20 s	2400°C, 1.5 s	700°C, s $^{-1}$	3000°C, 1.5 s

Source: Royal Society of Chemistry [44]

Table 2.47 ICP instrumental and operational parameters

ICP equipment:

Generator	Plasma–Therm HFS 200D; frequency 27.12 MHz; maximum power output 2.5 kW, with power stabilisation
Torch	Quartz, 18 mm o.d., three concentric tubes design; outer gas (argon) flow-rate 20 L min $^{-1}$; no intermediate gas flow
Nebulisation	Fixed cross-flow nebuliser (Jarrell–Ash) operated at 15–50 lb in $^{-2}$ [aerosol gas (argon) flow-rate, 0.35 L min $^{-1}$]; 'Scott'-type nebulisation chamber; forced sample aspiration with a Minipuls II peristaltic pump with 2 L min $^{-1}$ sample delivery rate; humidified carrier gas (argon)

Spectroscopic equipment:

Spectrometer	Jobin–Yvon HR 1000 (thermoregulated), 1 m Czerny–Turner holographic grating, 2400 grooves mm $^{-1}$ with a reciprocal linear dispersion of 0.4nn mm $^{-1}$ (1st order)

Element line/nm	Observation height above r.f. coil/mm	Operating power/kW	Carrier gas (argon) pressure/lb in $^{-2}$
Mo 202.030 (II)	16	1.07	32
Cd 228.802(I)	20	1.00	30

Source: Royal Society of Chemistry [44]

200 ml of the ammonium acetate extractant solution, filtered through Whatman No 40 filter paper and evaporated to dryness on a steam bath. A later two-step oxidation was carried out, first with 5 ml of nitric acid (1+1) and then with 2.5 ml of concentrated nitric acid. Finally, the residue was dissolved in 10 ml of nitric acid (1+3) and transferred into a 25 ml calibrated flask.

The determination of cadmium by GFAAS is carried out after 25-fold sample dilution using a calibration method with standards ranging from 0.3 to 5 μg L^{-1} of cadmium in 25-fold diluted extractant solution. The addition of 1% nitric acid – 100 μg mL^{-1} molybdenum – 1.5% hydrogen peroxide to the standards and samples as a matrix modifier is necessary.

The determination of molybdenum by GFAAS is carried out directly on the samples extracted with ammonium acetate solution, evaporated to dryness and redissolved in 10% nitric acid, by calibration with standards ranging from 15 to 200 μgL^{-1} of molybdenum and containing 10% nitric acid, and 500 μg ml^{-1} of magnesium (as nitrate). The working conditions are given in Tables 2.45 and 2.46.

The determinations of cadmium and molybdenum by ICP are carried out directly on the extractant sample solution. A calibration method using external standards was employed. Standards starting at 15 μg L^{-1} were prepared with the DTPA – triethanolamine – calcium chloride extractant solution for cadmium, and standards starting at 25 μg L^{-1} with 10% nitric acid for molybdenum. The working conditions are given in Table 2.47.

The RSDs of the 10 average net intensities, running the experiment on two different days, were 1.90 and 1.00% for cadmium and 1.28 and 0.32% for molybdenum. A similar precision study with 64 μL^{-1} of cadmium and 40 μg L^{-1} of molybdenum in soil extracts gave RSDs of 0.99 and 1.4% for cadmium and 9.64 and 7.30% for molybdenum.

From these studies it is concluded that a recalibration every five samples is necessary.

The precision of the analysis, based on five replicate determinations of cadmium and molybdenum in the same extract was 3% (mean concentration 65 μg L^{-1}) and 7.12% (mean concentration 75 μg L^{-1}). Detection limits were calculated from calibration graphs obtained from 15–250 μg L^{-1} standard solutions of cadmium in the DTPA-triethanolamine – calcium chloride extractant solution and 25–500 μg L^{-1} standard solutions of molybdenum in 10% nitric acid. Net intensities were obtained for blank and standard solutions, subtracting the background; then the concentration that yielded three times the standard deviation of the blank plus the net intensity of the blank (obtained from the calibration graphs) was taken as the limit of detection. Average values were 3.41 μg L^{-1} of cadmium and 6.10 μg L^{-1} of molybdenum.

In order to determine the accuracy, samples were analysed by the GFAAS and ICP methods and with a calibration method using standards very accurately matched with the samples. The results obtained are given

Table 2.48 Summary of results of GFAAS and ICP analyses of Cd in DTPA – triethanolamine – CaCl$_2$ extratcs of soils. Means ± standard deviations for five replicate determinations

Sample	GFAAS/µg L^{-1}		ICP/µg L^{-1}	
	Proposed method	With matching	Proposed method	With matching
Sediment A	214 ± 15.8	240 ± 16.0	252 ± 8.4	260 ± 8.1
Sediment B	229 ± 16.1	230 ± 16.5	212 ± 7.9	229 ± 8.3
Lleida soil	27 ± 2.9	30 ± 3.5	34 ± 1.8	32 ± 2.0
Badajoz soil	80 ± 7.2	77 ± 6.9	72 ± 2.3	76 ± 2.5
Barcelona soil	69 ± 6.5	66 ± 7.0	65 ± 2.1	68 ± 2.3

Source: Royal Society of Chemistry [44]

Table 2.49 Summary of results of GFAAS and ICP analyses of Mo in ammonium acetate extracts of soils. Means ± standard deviations for five replicate determinations

Sample	GFAAS/µg L^{-1}		ICP/µg L^{-1}	
	Proposed method	With matching	Proposed method	With matching
Sediment A	640 ± 71.0	689 ± 74.5	679 ± 40.8	669 ± 43.5
Sediment B	298 ± 22.3	330 ± 21.9	315 ± 18.9	322 ± 17.7
Lleida soil	55 ± 8.2	55 ± 7.6	49 ± 3.,4	50 ± 4.1
Badajoz soil	35 ± 6.0	43 ± 6.4	44 ± 3.1	41 ± 2.9
Barcelona soil	73 ± 10.1	81 ± 10.8	75 ± 5.3	85 ± 5.7

Source: Royal Society of Chemistry [44]

in Tables 2.48 and 2.49. The results obtained for cadmium and molybdenum by the GFAAS technique agree closely with those obtained by ICP spectrometry. Maximum differences obtained between proposed methods and the calibration method were 10.8% for GFAAS (for sediment A) and 7% for ICP (for sediment B) for cadmium, and 18% for GFAAS (Badajoz soil) and 11.8% for ICP (Barcelona soil) for molybdenum.

The results indicated that both methods for the determination of cadmium are excellent. The GFAAS method for molybdenum is good, but could be improved if a previous sample classification (acidic or basic soils) were carried out in order to be able to 'spike' standards for calibration with amounts of magnesium similar to those in the extracts. The determination of molybdenum by the ICP method could be carried out without difficulty; samples with an magnesium plus calcium content higher than 400 µg mL^{-1} could be analysed more accurately using standards containing calcium and molybdenum.

2.60.1.4 Heavy metals (cadmium, chromium, copper, nickel, lead, zinc, selenium, arsenic, mercury and molybdenum)

Davis and Carlton-Smith [182] carried out an interlaboratory comparison on behalf of the Department of the Environment (UK) on the determination in soil of total metals (cadmium, chromium, copper, nickel, lead and zinc), the determination of extractable copper, nickel and zinc, and the determination of total molybdenum, mercury, arsenic and selenium. Heavy metals were determined by atomic absorption spectrometry. Very few of the 13 participating laboratories carried out analysis for molybdenum, mercury, arsenic and selenium and the results obtained were inconclusive and in a wide range.

If an accuracy (MPD) of ±20% is taken as an arbitrary level of acceptable error then the results (median values) of the interlaboratory comparison showed that:

1. determinations of copper and zinc were acceptable by all the digestion procedures studied
2. determinations of nickel, lead, cadmium and chromium unacceptable by all the digestion procedures studied. The digestion procedures studied were HNO_3/H_2O_2, HNO_3, $HNO_3/HClO_4$, HNO_3/H_2NO_2.

2.60.1.5 Heavy metals (cadmium, cobalt, copper, nickel, lead, zinc, iridium, silver, selenium, mercury, bismuth, thallium and molybdenum)

Eidecker and Jackwerth [184] employed a trace element preconcentration step using a small amount of iron as a collector element and collector precipitation with hexamethylene-ammonium hexamethylene dithiocarbamate (HMA–HMDTC) followed by atomic absorption spectrometry and X-ray fluorescence spectroscopy for the determination in soils of iridium, silver, selenium, mercury, bismuth, molybdenum, cadmium, cobalt, copper, nickel, lead and zinc. Detection limits were between 0.1 and 1 mg kg $^{-1}$. The relative standard deviation for the total method was 0.03. To determine lead, cadmium and thallium in soils Lopez-Garcia et al. [210] suspended the sample in water containing 5% (v/v) hydrofluoric acid before injection into an electrothermal atomic absorption spectrometer. No modifier was needed other than the hydrofluoric acid.

2.60.1.6 Arsenic, antimony, bismuth and selenium

Haring et al. [28] have used atomic absorption spectrometry and hydride generation atomic absorption spectrometry to determine arsenic and antimony in soils. In this method samples are acidified with hydrochloric acid. After the addition of a solution of sodium borohydride to the acidified sample, volatile metal hydrides will be formed.

The metal hydrides are transferred into a heated quartz cell by means of a nitrogen or argon gas flow. The metal hydrides will decompose in the heated quartz cell and the atomic absorption of the elements can be determined.

For the determination of arsenic and antimony a prereduction with potassium iodide and ascorbic acid is necessary because As(V) and Sb(V) are less reactive with respect to the formation of the metal hydrides than As(III) and Sb(III). Potassium iodide is also known to mask interferences by other metal ions.

Method I: Hydride generating technique combined with atomic absorption spectrometry

A Perkin–Elmer model 380 AAS equipped with the Mercury Hydride System MHS–1 and EDL lamps for As and Sb was used in this study. Sodium borohydride 10% $NaBH_4$, 5% sodium hydroxide prepared daily.

Arsenic standard solutions. As(V) standards in 6% hydrochloric acid were prepared from a Merck-titrisol. As(III) standards in 6% hydrochloric acid were prepared from a stock solution of 1 g of As(III) per litre (1.735 g of sodium arsenite in 1 litre of twice distilled water).

Antimony standard solutions. Sb(V) standards in 6% hydrochloric acid were prepared from a stock solution of 1 g of Sb(V) per litre (3.952 g of Na_3Sb $S_4.9H_2O$ in 1 l of twice distilled water). Sb(III) standards in 6% hydrochloric acid were prepared from a Merck-titrisol.

Potassium iodide.

Ascorbic acid.

Procedure

The recommended procedure used for the determination of arsenic and antimony involves the addition of 20 ml of concentrated hydrochloric acid (30%), 1 g of potassium iodide and 1 g of ascorbic acid to an 80 ml sample. This solution should be kept at room temperature during at least 5 h before initiation of the programmed MHS–1 system (addition of sodium borohydride to 25 ml sample and measurement of atomic absorption).

The sodium borohydride (10% w/v in 5% sodium hydroxide) is cooled with ice to prevent losses of the reducing capability of the reagent. The instrumental parameters are given in Table 2.50.

Table 2.50 Instrumental parameters for AA measurement for arsenic and antimony with the hydride generating technique

Parameters		As	Sb
Wavelength (nm)		193.7	217.6
Slit width (nm)		0.7	0.2
Random noise supression (s)		0.4	0.4
Quartz cell temperature	MHS–I (C)	1000	1000
Programme	MHS–I	I	I
Pump vollume (NaBH₄)	MHS–I (ml)	2.5	2.5
Sample volume	MHS–I (ml)	25.0	25.0

Source: Springer-Verlag Chemie GmbH, Germany [28]

Table 2.51 Instrumental parameters for graphite furnace AA determination of arsenic and antimony

Parameters	As	Sb
Wavelength (nm)	193.7	217.6
Slit width (nm)	0.7	0.2
Ashing temperature (°C)	1300	1000
Ashing time (s)	30	30
Miniflow during atomisation (ml/min)	50	50
Deuterium background correction	Yes	Yes
Atomisation temperature (°C)	2600	2500 (max. power)
Atomisation time (s)	4	4

Source: Springer Verlag Chemie GmbH, Germany [28]

Method 2: Graphite furnace atomic absorption spectrometry

A Perkin–Elmer model 5000 AAS equipped with a HGA–500 graphite furnace was used. Standard graphite tubes were used in this study.

Procedure

To each 100 ml of arsenic standard solution (and sample) 1 ml of 5% nickel solution is added (24.77 g of nickel nitrate borohydrate/100 ml). A small volume of sample or standard (usually 10, 20 or 50 µL) is injected into the graphite furnace.

The temperature programme of the graphite furnace and a number of other instrumental parameters are given in Table 2.51. Calibration curves for arsenic and antimony are used to determine the sample concentrations (prepared using arsenic and antimony standards adjusted to pH 2 with nitric acid).

Table 2.52 Result of intercomparison experiments of arsenic determinations in soil* samples

Arsenic all concentrations in µg l	Colorimetric III technique**	Hydride generation technique	Graphite furnace technique
Sample A	220	220	250
Sample A + 200 µg L⁻¹ As	420	380	400
Sample B	300	260	300
Sample B + 200 µg L⁻¹ As	540	420	500
Sample C	220	200	250
Sample C + 200 µg L⁻¹ As	440	400	450
Blank	<20	<20	<50
Standard 400 µg L⁻¹ As	400	360	350

*25,9 g of soil was refluxed (1 h) with 100 ml of HCl-HNO₃–H₂O (3:1:1) and next (45 min) with 20 ml of 50% NH₃OHCl
**Not neutralised samples – unknown concentration of residual NH₃OHcL
Neutralised samples – 2% NH₃OHCl

Source: Springer Verlag Chemie GmbH, Germany [28[

Table 2.53 Result of intercomparison experiments of antimony determinations in soil* samples

Arsenic all concentrations in µg l	Not neutralised samples		
	Colorimetric III technique**	Hydride generation technique	Graphite furnace technique
Sample A	80	80	100
Sample A + 200 µg L⁻¹ Sb	260	320	300
Sample B	800	800	750
Sample B + 200 µg L⁻¹ Sb	1020	1000	1000
Sample C	340	420	400
Sample C + 200 µg L⁻¹ Sb	520	620	600
Blank	<20	<20	<50
Standard 400 µg L⁻¹ Sb	380	400	350

*25.0 g of soil was refluxed (1 h) with 100 ml of HCl–NHO₃–H₂O (3:1:1) and next (45 min) with 20 ml of 50% NH3OHCl
**with KI addition: 3% KI

Source: Springer Verlag Chemie GmbH, Germany [28]

Excellent agreement in antimony and arsenic contents of soils were obtained between these two methods and a spectrophotometric method based on the formation of the silver diethyldithiocarbamate complexes of arsenic and bismuth [111] (see Tables 2.52 and 2.53).

2.60.2 Inductively coupled plasma atomic emission spectrometry and inductively coupled plasma mass spectrometry

Inductively coupled plasma atomic emission spectrometry is displacing atomic absorption spectrometry as the method of choice for the determination of metals in soils. It has the advantages of unattended overnight operation, high sensitivity, the ability to analyse large batches of samples in a single run, and finally the ability to determine a large number of different elements in a single run. Earlier work [10, 185–190] is not discussed here in detail.

2.60.2.1 Heavy metals (chromium, iron, nickel, copper, cobalt, cadmium, zinc, lead and manganese)

Inductively coupled plasma atomic emission spectrometry has been used to determine heavy metals and trace elements in soils [115]. The necessary equipment and sample preparation are described. The use of a 16 channel 1 CP spectrometer enabled the simultaneous determination of 16 elements to be carried out. The particular feature of this method of excitation (at temperatures of 10000°K) are outlined by these workers. A comparison of aqua regia extraction and total ashing procedures is presented. Electrothermal vaporisation has been used with inductively coupled plasma atomic emission spectrometry for the direct determination of arsenic, cadmium, lead and zinc in soils [211]. By the addition of toluene vapour to the internal furnace gas and sodium selenite to both standard solutions and solid samples, the transport losses of analytes were considerably reduced. Li [230] has reported on the development of a five step sequential extraction scheme for the analysis of soils by inductively coupled plasma mass spectrometry. Application of the method to soils contaminated by past mining and smelting activities showed distinctive partitioning patterns of heavy metals from the two sources.

2.60.2.2 Heavy metals (cadmium, chromium, iron, manganese and zinc) and aluminium barium, calcium, potassium, magnesium, sodium, silicon, strontium, titanium and vanadium

Que Hee and Boyle [158] analysed soils using Parr bomb digestion with hydrofluoric acid and nitric acid–perchloric acid followed by inductively coupled plasma atomic emission spectrometry. Zaray and Kontor [211] used electrothermal vaporization with inductively coupled plasma mass spectrometry for the direct determination of arsenic, cadmium, lead and zinc in soils. By the addition of toluene vapour to the internal furnace gas and sodium selenite to both standard solutions and soil samples, the transport losses of analytes were considerably reduced.

2.60.2.3 Molybdenum, cobalt and boron

Manzoori [40] has developed an inductively coupled radio frequency plasma method for the determination of these elements in amounts down to 0.01 ppm (molybdenum), 0.05 ppm (cobalt) and 0.05 ppm (boron) with a mean relative standard deviation of 0.9%. In this method a 20 g sample of air-dried soil was shaken overnight on an end-over-end shaker with 800 ml of 0.5% acetic acid solution. The whole extract was filtered in the same manner as the ammonium acetate extracts and evaporated to dryness. The residue was then taken up in 100 ml of 0.1M hydrochloric acid.

The most sensitive molybdenum line (379.8nm) was used although there is some interference from the iron line at 379.8nm. The molybdenum 390.3nm line was found to be as sensitive but the iron 390.3nm line, which is very intense, strongly interfered. The most sensitive cobalt line (345.3nm) was used. This line was found to be free from spectral interferences. Boron has a very simple spectrum with the sensitive doublet at 249.7 and 249.8nm being the only useful analytical lines.

Aluminium interferes in the determination of molybdenum and calcium interferes in the determination of molybdenum, cobalt and boron. Good agreement was obtained for all three elements on ammonium acetate extracts of standard soil samples supplied by the Macaulay Institute.

2.60.2.4 Arsenic, antimony, bismuth and selenium

Pahlavanpour *et al.* [191] have described a method for the simultaneous determination of traces of arsenic, antimony and bismuth in soils by an inductively coupled plasma volatile hydride method after rapid attack with concentrated hydrochloric acid in sealed tubes. The samples are treated with the acid at 150°C for 2 h in capped test tubes. After addition of potassium iodide solution, the hydrides are formed by mixing the solution with sodium tetrahydroborate(III) solution in a continuous flow system, and are swept into the plasma by a stream of argon for determination by atomic emission spectrometry. Acceptable precision and accuracy are obtained, and the detection limits for all three analytes are about 0.1 µg g^{-1}. Approximately 200 samples can be analysed by one person in a two day cycle.

Equipment

Hydride generator – plasma system: The system consisted of a continuous flow hydride generator, a Radyne R50 plasma generator and an Applied Research Laboratories 29000B quantometer. The operating conditions and procedure were as previously described [192].

Screw-capped test tubes: Sample attacks were carried out in Sovirel screw-capped borosilicate glass test tubes (160 × 1 mm) with specially made cap liners consisting of a layer of silicone rubber (Esco type SR70, 3.2 mm thickness) covered with a chemically resistant film (Du Pont Teflon FEP Type A, 0.36 mm thickness).

Heating block: Batches of tubes were heated in a thermostatically controlled aluminium block to a depth of about 65 mm. The Bakelite caps and the tops of the tubes were kept cool by means of a blast of air from a domestic vacuum cleaner.

Materials
Analytical reagent grade chemicals and purified water were used throughout.

Sample digestion
The powdered samples (0.250g) were weighed into test tubes and 5 ml of concentrated hydrochloric acid were added to each tube. The tubes were capped and placed in the heating block at 150°C for 2 h, then they were removed cautiously, cooled rapidly in a cold water bath and unsealed. A 5 ml volume of 0.2% m/V potassium iodide solution was added to each tube and the contents were mixed by thorough shaking. The solid residue was allowed to settle (for about 4 h) and the solutions were used directly for the determinations.

Determination
The arsenic, antimony and bismuth were determined simultaneously in the analyte solution. A calibrating solution was made containing the analytes (as arsenic(III)), antimony(III) and bismuth(III) at a concentration of 100 ng ml^{-1} in 1+1 hydrochloric acid and was run, together with a blank solution, after every 10th sample solution. Sample solutions containing an analyte at a concentration above the limit of linear calibration (arsenic 800 ng ml^{-1}, antimony 1500 ng ml^{-1}, bismuth 500 ng ml^{-1}) were diluted 10-fold and re-run. The above methods were applied to a range of standard Canadian Certified Soils. It is seen (Table 2.54) that the method produced comparable although generally slightly lower values than those currently recommended, but the number of laboratories returning data for arsenic, antimony and bismuth was small, and most of the results obtained for bismuth are close to the detection limit of the proposed method and apparently the CCMP method.

Goulden *et al.* [193] have described a semi-automated system for the determination of arsenic and selenium by hydride generation – inductively coupled argon plasma emission spectrometry. Soils are brought into solution by fusion with sodium hydroxide in a zirconium crucible. The automated measurement system uses conventional continuous flow equipment connected to a 'larger diameter than usual' torch in the inductively coupled plasma (ICAP) instrument.

Table 2,54 Results produced on the Canadian Certified Materials Project standard soils compared with the recommended values

SO–1, SO–2 and SO–3 are single determinations; SO–4 is the mean and standard deviation of eight determinations oni separate portions of the material

Sample	Result	As/mg kg $^{-1}$	Sb/mg kg $^{-1}$	Bi/mg kg $^{-1}$
SO–1	Found	1.94	0.12	0.24
	Recommended	1.9 ± 0.3	0.2	0.5
SO–2	Found	0.77	0.1	0.03
	Recommende	1.2 ± 0.2	0.1	0.1
SO–3	Found	2.32	0.22	0.03
	Recommended	2.6 ± 0.1	0.3	0.1
SO–4	Found	6.45 ± 0.07	0.25 ± 0.03	0.19 ± 0.02
	Recommended	7.1 ± 0.7	0.7	0.1

Source: Royal Society of Chemistry [191]

2.60.2.5 Miscellaneous elements

Kanda and Taira [197] presented results from a study on the use of a computer-controlled rapid-scanning echelle ICPAES monochromator to determine major, minor and trace elements in sediments and soils. The concentrations of 17 elements in five standard reference materials were determined by using a single set of analytical lines without any corrections for line-overlap interference. Lowest determinable concentrations and relative sensitivities for 13 metals are included.

Kheboian and Bauer [174] demonstrated in experiments on the sequential extraction of trace models in aquatic sediments using three different multiphase model sediments that element redistribution of sediment phases during extraction was a major problem in interpreting results. Evaluation of the sequential extraction process by atomic absorption and inductively coupled plasma indicated that recovery of the trace metals was incomplete and variable.

2.60.3 Neutron activation analysis

2.60.3.1 Heavy metals (chromium, iron and cobalt) and scandium, arsenic, selenium, molybdenum, antimony, lanthanum, cerium, europium, hafnium, tantalum, thorium and uranium

Randle and Hartman [194] used this technique to investigate the metal concentrations in humic compounds in organic soils.

The methods used to isolate the humic substances from the soil are based on extractions with sodium hydroxide or sodium pyrophosphate. After drying, the humic samples were analysed for scandium, chromium, iron, cobalt, arsenic, selenium, bromine, molybdenum, antimony, lanthanum, cerium, europium, hafnium, tantalum, thorium and uranium, by thermal neutron activation analysis. Concentrations ranged from 0.02 ppm (lanthanum) to 7900 ppm (iron).

Fast neutron activation analysis has been studied as a screening technique for copper and zinc, for application to waste site soils. Experiments were conducted with a sealed tube neutron generator and a Geγ-ray detector [231].

2.60.3.2 Arsenic and selenium

Kronborg and Steinnes [151] have described a neutron activation analysis method for the simultaneous determination of arsenic and selenium in soils. After fusion of the soil with alkali and removal of the insoluble hydroxides, arsenic and selenium are determined in the solution, after acidifying it with hydrochloric acid, by a simple precipitation with thioacetamide. Chemical yields are determined by re-irradiation of the sulphite precipitate. This method is discussed briefly below in order to give some idea of the experimental technique employed in neutron activation analysis.

For the irradiation air-dried samples, each of about 200 mg, were accurately weighed and wrapped in aluminium foil; 1 ml aliquots of arsenic and selenium standard solutions were transferred into quartz ampoules, which were heat–sealed. Twelve samples were irradiated together with the standards for two days in the JEEP–II reactor (Kjeller, Norway) at a thermal neutron flux of about 1.5×10^{13} neutrons cm $^{-2}$ s $^{-1}$. The irradiated samples were stored for seven days before starting the separations.

In the radiochemical procedure, 1.00 ml each of arsenic and selenium carrier solutions are transferred with a pipette, into a nickel crucible containing sufficient sodium hydroxide to make the resulting mixture alkaline, and the mixture evaporated to dryness under a heating lamp. The aluminium foil is unwrapped and the soil sample poured quantitatively into the crucible. The sample is covered with 2 g of sodium hydroxide pellets and 2 g of sodium peroxide and moistened with a few drops of water. If the solid contains appreciable amounts of organic material, a moderate reaction takes place in the crucible. The arsenic carrier solution consists of As(III) in 1 M ammonia to give a solution containing 5 mg L $^{-1}$ arsenic, then diluted 1 to 500 ml with 1M ammonia. The selenium carrier solution consists of Se IV) in water to give a solution containing 20 mg L $^{-1}$ selenium, then diluted, 1 to 1000 ml with 0.1M nitric acid.

After the first reaction has ceased, the crucible is heated carefully over an electrothermal burner at short intervals until the breakdown of organic material has been brought almost to completion, then the mixture fused for 5 min with the electrothermal burner. The fused material is released from the crucible with water and transferred into a 50 ml centrifuge tube. The liquid is centrifuged and transferred into a 250 ml beaker containing 75 ml of water, and 75 ml of concentrated hydrochloric acid is added. 100 mg of thioacetamide is added and the mixture heated until precipitation occurs. The precipitate is filtered on to a membrane filter, washed several times with 6M hydrochloric acid, and then several times with water in order to remove the hydrochloric acid. The filter plus precipitate is sealed between two sheets of polyethylene.

0.100 ml of selenium standard solution and 0.500 ml of arsenic-standard solution are transferred into a nickel crucible containing arsenic and selenium carrier in strongly alkaline solution (pH ~14). The mixture is evaporated carefully to dryness under a heating lamp. Proceed as described above for the samples.

The gamma activities of the samples and the standards were recorded by means of a 35 cm^3 germanium (lithium) detector connected to a multichannel analyser. The 265 keV peak of 120-d selenium-75 and the 559 keV peak of 26.4 h arsenic-76 were made the basis for the quantitative measurements. Waiting for seven days before the counting gave optimum conditions for the simultaneous measurement of arsenic-76 and selenium-75 at the concentration levels encountered, when an irradiation time of two days was employed.

Determination of chemical yield

To determine chemical yield, after the arsenic-76 activity of the separated samples had decayed to a negligible level, the chemical yields were determined by re-activation. The polyethylene 'envelopes' containing the precipitates were irradiated for 30 s under the same conditions as before. After a delay of 30 min in order to allow the decay of 3.9 min selenium-97m, the recovery of carriers was determined by measurement of the 103 keV γ-ray of 57 min selenium-81m and the 559 keV γ-ray of arsenic-76 using γ-spectrometry. The chemical yield was in most instances 80% or above for both elements. The reproductibility of the method is estimated to be about 5% for both elements.

In Table 2.55 are shown the results for duplicate analyses of 10 typical soils, the inorganic portion of which consisted mainly of morainic material. The results lie within the accepted ranges for arsenic (1–10 mg L^{-1}) and selenium (0.01–1 mg L^{-1}) in soils from various geological regions. It can be seen that in some instances the deviation between the duplicate values significantly exceeds the estimated 5% figure for the analytical precision and is probably associated with inhomogeneous distribution of the elements within the sample, as the soils were not extensively homogenised before analysis.

Table 2.55 Results of duplicate analyses of some air-dried soil samples

Sample No.	Loss on ignition, %	Arsenic, ppm		Selenium, ppm	
1	16.4	3.0	3.5	0.32	0.28
2	97.2	1.13	1.17	0.30	0.41
3	59.5	2.9	2.2	0.76	0.66
4	73.5	3.0	2.6	0.48	0.45
5	90.0	0.87	0.70	0.69	0.81
6	64.3	2.2	1.8	1.4	1.6
7	96.6	1.00	1.23	0.75	1.01
8	63.4	0.72	0.74	0.30	0.29
9	46.1	7.1	0.2	0.55	0.66
10	58.1	0.50	0.51	0.20	0.18

Source: Royal Society of Chemistry [151]

2.60.3.3 Lanthanides

Tsukada *et al.* [329] have discussed the application of neutron activation analysis to the determination of lanthanides in soils.

2.60.4 Photon activation analysis

2.60.4.1 Heavy metals (chromium, iron, cobalt, lead, nickel and zinc) and aluminium, magnesium, calcium, titanium, arsenic, strontium, zirconium, antimony and uranium

Randle and Hartman [194] used this technique to investigate concentrations of these metals in soils.

2.60.5 Emission spectrometry

2.60.5.1 Miscellaneous metals

This technique has been used to determine metals in soils [195, 383]. Elements determined include heavy metals, aluminium, barium, beryllium, molybdenum, strontium, vanadium, terbium, tellurium, thorium, titanium, thallium, tamerium, uranium, tungsten, yttrium and zirconium. Wisbrun *et al.* [330] has been developing new methods for heavy metals in soil based on laser plasma spectrometry. For the application of time-resolved optical emission spectrometry from laser-induced plasma to the determination of metals in solid samples, detection limits in the 10 µg g^{-1} range were obtained.

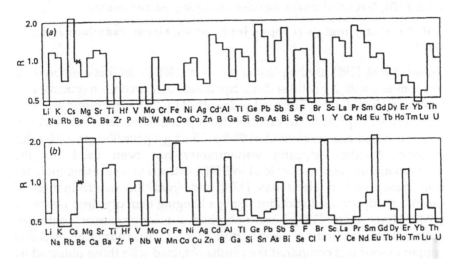

Fig. 2.23 Examples of compositional fingerprints for two soils, C and D, developed on granite/granitic gneiss (a) and on basic igneous (b) parent material. F, Ratio of element content in a particular soil to the average content of that element in a number of soils. ×, No analytical value for this element, plotted as $R = 1$.
Source: Royal Society of Chemistry [196]

2.60.6 Spark source mass spectrometry

2.60.6.1 Heavy metals (chromium and nickel) and rubidium, tin, antimony and caesium

Ure and Bacon [196] have described a technique for the quantitative analysis of soils by spark source mass spectrometry with aluminium as the conducting material. Details are given of photographic plate emulsion calibration and evaluation methods together with the interference correction procedures essential before measuring, and finally before applying, the relative sensitivity coefficients used to standardise the analysis. More than 50 elements can be determined and a typical precision of ±10–15% is attained. The techniques are illustrated by application to the analysis of soils, concentrates of acetic acid extracts of soils and US Geological Survey Standard Rocks G–2 and AGV–1.

This technique can be used to obtain element fingerprints for soil samples. Examples of such fingerprints for two Scottish soils are shown in Fig. 2.23. The ordinate R, plotted on a logarithmic scale, is the ratio of the element content in a particular soil to the average content of that element determined by these methods in a number of soils. These diagrams therefore show changes in composition with respect to an average content and present together data for major and trace element contents. For convenience in presentation, the ratio R displayed is limited to the range 0.5–2.0 with values outside this range plotted just beyond these limits.

2.60.7 Differential pulse anodic stripping voltammetry

2.60.7.1 Heavy metals (copper, lead, cobalt, nickel, cadmium and zinc)

Goulden et al. [193] used a plasma power of 1400 W. In this method 0.2 g soil is mixed with 2 g solid sodium hydroxide in a zirconium crucible and heated to 350°C for 2 h. The melt is dissolved in 2N hydrochloric acid and heated to 90°C prior to decantation from precipitated silicic acid. A soil with a certified arsenic content of 93.9 ± 7.5 µg g $^{-1}$ gave 95.1 µg g $^{-1}$ by this method. Anodic stripping voltammetry has been used for the determination of copper and lead in EDTA extracts of soils using mercury film glossy carbon electrodes [173] and from the determination of cadmium, copper, lead and zinc using a hanging mercury drop electrode [198]. Meyer et al. [199] have described a microprocessor controlled voltammetric method for the determination of zinc, cadmium, lead and copper in soils and compared the results obtained with those obtained by atomic absorption spectrometry.

The determination of zinc, cadmium, lead and copper in soils was carried out using a Metrohm VA–Processor 646. With this fully automated instrument, peak current measurements are made in a suitable sensitivity range (auto-ranging). Moreover, when presetting the working conditions for a single determination run, the voltage sweep can be divided in up to eight different segments. Thereby, each element is recorded under optimal conditions and each measuring signal is evaluated with high accuracy.

In this method the samples were dried at 105°C to constant weight and ground to a final grain size <1 mm in diameter. After repeated drying, approximately 3 g of the sample were weighed, transferred to a 250 ml round bottomed flask, wetted with little water and finally filled up with 21 ml of 25% hydrochloric acid and 7 ml of concentrated nitric acid. The bottle was then placed on an oil bath and equipped with a reflux cooler; the absorption tube on top of the reflux cooler contains 10 ml of a 0.5 mol/l nitric acid. After standing for some time at room temperature the sample is heated to the boiling point for 2 h. After cooling to room temperature, the nitric acid from the absorption tube is transferred to the round bottomed flask through the reflux cooler; the absorption tube is rinsed with 5 ml, the reflux cooler with 10 ml of 0.5 mol/l nitric acid. Now the sample is transferred into a 100 ml measuring flask. The round bottom flask is washed with 5 ml nitric acid, and the measuring flask is filled up with bidistilled water to the mark. Prior to the determination step, the undissolved part of the sample is filtered through a 0.45 µm membrane. To avoid interference by organic matter, the extract was diluted five fold with distilled water and 5 ml of M acetic acid/sodium acetate added. From this solution zinc, cadmium, lead and copper can be deposited and concentrated simultaneously at a deposition potential of –1.2V (vs Ag/AgCl at the HMDE). Because of different element concentrations the

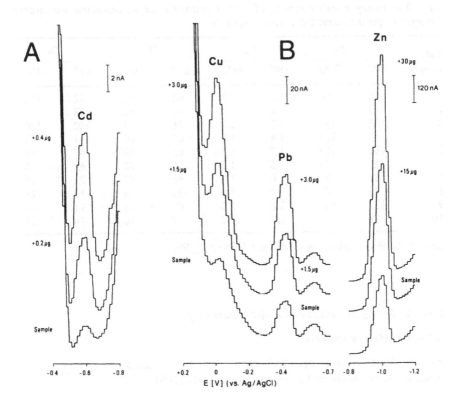

Fig. 2.24 Determination of cadmium (A) and copper, lead, zinc (B) in soil by DPASV

Source: Springer Verlag Chemie GmbH, Germany [199]

anodic dissolution peaks have to be recorded at different sensitivity ranges. As shown for a soil sample in Fig. 2.24 it is advisable to measure cadmium when present in low concentrations in a separate step. In part A of Fig. 2.24 the cadmium peak (E_p = –0.60V) registered at 1.10^{-10} A/mm is shown; the deposition was made at –0.8V, the deposition time was 20 s. After the determination of cadmium the four metals are deposited simultaneously at –1.2V. During the subsequent anodic dissolution step as shown in part B of Fig. 2.24 the zinc peak appears first at a potential of –1,02V and a registration sensitivity of 6.10^{-9} A/mm. Subsequently, the current sensitivity is changed to 1.10^{-9} A/mm for the following recording of lead (E_p = –0.43 V) and copper (E_p = –0.08V). The excellent agreement obtained between voltammetric and atomic absorption spectrometric techniques is illustrated in Table 2.56.

Table 2.56 Heavy metals contents of soils. Comparison of voltammetric and atomic absorption spectrometric techniques, (mg kg^{-1})

Soil sample	Zn		Cd		Pb		Cu	
	DPP	FAAS	DPP	FAAS	DPP	FAAS	DPP	FAAS
1	156	160	1.5	1.7	62	60	152	153
2	143	140	1.2	1.3	60	63	72	42
3	156	167	1.2	1.3	56	63	240	240
4	129	137	1.4	1.6	61	60	63	52
5	150	159	1.7	1.8	63	68	43	48
6	121	137	1.6	1.6	55	62	42	36
7	92	93	1.6	1.6	46	50	21	18
8	75	75	1.4	1.4	35	34	19	21
9	132	142	2.8	2.5	69	68	94	107
10	24	27	0.5	0.3	17	16	24	27

Source: Springer Verlag Chemie GmbH, Germany [199]

2.60.8 X-ray fluorescence spectroscopy

2.60.8.1 Heavy metals

X-ray fluorescence spectroscopy has been used to carry out a homogeneity study of heavy metals in soils [200].

2.60.8.2 Miscellaneous metals

Levinson and Pablo [201] have described a procedure for the analysis of soil samples which are placed in the sample holder of an X-ray spectrometer. Detection limits between 10 and 50 ppm can be obtained for elements heavier than iron with counting times of 20 s; and a number of elements can be determined simultaneously. For most samples, accuracies of 10–20% are achievable.

2.60.9 Stable isotope dilution mass spectrometry

2.60.9.1 Potassium, rubidium, caesium, calcium, strontium and barium

Hinchley [202] used this technique to determine a wide range of alkali metals and alkali earths (potassium, rubidium, caesium, calcium, strontium, barium) in small volumes of soil extracts. Stable isotope dilution mass spectrometric analysis is a most sensitive and accurate method for determining these six elements in soil samples (parts per 10^9 when using 0.3 g soil sample).

2.60.10 Electron probe microanalysis

2.60.10.1 Copper and zinc

Lee and Kittrick [203] carried out an electron probe study of elements associated with zinc and copper in an oxidising and an anaerobic soil environment to obtain direct evidence of elements associated with heavy metal phases in the sample. Approximately 12000 particles were scanned to locate thirty that produced more than 100 counts per second zinc or copper x-radiation. Over 50% of the copper containing particles in the soil were not associated with elements above atomic number 9, while sulphur and iron were the most important elements associated with the remaining copper particles. Zinc and copper appeared to be precipitated with the associated elements rather than adsorbed on particle surfaces.

2.60.11 Alpha spectrometry

2.60.11.1 Alpha-emitting nucleides (americium, californium, curium, radium through californium)

Sill *et al.* [8] have described a procedure for the determination of virtually all alpha-emitting nucleides in soil, either singly or in any combination on a single sample. The sample is decomposed completely by a combination of potassium fluoride and pyrosulphate fusions with simultaneous volatilization of hydrogen fluoride and silicon tetrafluoride. The cake is dissolved in dilute hydrochloric acid and all alpha emitters are precipitated with barium sulphate. An option is provided by which uranium can be obtained in a separate fraction to eliminate mutual interference with ^{237}Np. The barium sulphate is dissolved in acidic aluminium nitrate and the elements from thorium through plutonium are extracted into Aliquat-336. Acid-deficient aluminium nitrate is then added to the aqueous phase to reduce the acidity, and americium, curium and californium are extracted into fresh Aliquat-336. Thorium is removed by scrubbing the first organic extract with 10M hydrochloric acid, and protactinium, uranium, neptunium and plutonium are stripped with a mixture of perchloric and oxalic acids. Americium, curium and californium are stripped from the second organic extract with 8 M nitric acid. Each fraction is electrodeposited and analysed by alpha spectrometry with few interferences. Overall recovery excluding electro-deposition ranges from 88% for protactinium to 99% for plutonium. The rationale of this scheme of analyses is shown in Table 2.57.

2.60.12 Miscellaneous methods

2.60.12.1 General

An HMSO (UK) publication discusses three methods (gas chromatography,

Table 2.57 Soil 10 g

Fuse with 30 g KF, transpose with 35 ml H_2SO_4 and 20 g of Na_2SO_4 volatilising HF and SiF_4, and dissolve cake in 400 ml H_2O and 25 ml HCl

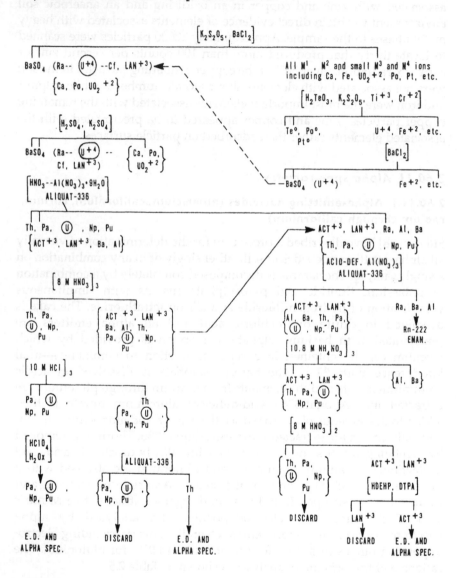

Diagram of procedure. Double vertical lines indicate precipitates or organic extracts; single vertical lines indicate aqueous solutions. Square brackets indicate reagents used; parentheses indicate elements precipitated with barium sulfate; circle around uranium and dotted line indicates option on uranium; and curved brackets indicate small quantities of certain elements that require special attention to eliminate detectable interference fluorometric method and hydride generation–atomic absorption spectrometry) for the determination of selenium and arsenic in soils [204].

Agemian and Bedak [32] have described a semi-automated method for the determination of total arsenic in soils. Jan and Schwedt [205] applied ion chromatography to the determination of magnesium, calcium, manganese and zinc in soil extracts. Separations were achieved on the low capacity cation exchange resin Dionex CS–2 and a conductivity detector used. Various organic complexing agents were studied. Detection limits in the soil digest were 20 µg L^{-1} (calcium) and 10 µg L^{-1} (magnesium). Senesi and Sposito [66] carried out an electron spin resonance study of residual copper II complexes in purified soil. Spectra were obtained for both frozen aqueous solutions and solid samples for organic matter in soil. Residual divalent copper complexes were present in all the soil samples examined.

Chowdhury and Bose [178] have shown that copper complexed with humic compounds isolated from soils is readily liberated with dilute hydrochloric acid (pH 1.0). Therefore 0.5N hydrochloric acid liberates copper from organic matter. Thus, unlike 1N hydroxylamine hydro-chloride plus 25% acetic acid, the above method is suitable for copper extraction. The mean level of copper with 0.5N hydrochloric acid was shown to be of the same order of magnitude as that of 0.05N EDTA. It is interesting to note that the mean levels of lead with all four partial extraction methods were very similar.

Table 2.58 represents the contrast of the different metals with each method for the four major anomalies at four locations. The contrast for copper and lead is essentially constant for all methods with the exception of method 3 for copper and method 1 for lead. For the other four trace elements, the contrast increases as the strength of the extractant decreases from methods 1 to 4.

From the above discussion, a few conclusive points arise. The 1N hydroxylamine hydrochloride plus 25 vol % acetic acid is satisfactory for the extraction of ferromanganese and carbonate minerals and adsorbed trace elements, although it has no effect on authigenically formed sulphide minerals and organic complexes. Therefore, it is not suitable for the simultaneous extraction of a large number of the nonsilicate authigenically formed elements. The 4.0N nitric acid plus 0.7N hydro-chloric acid solution has a considerable effect on the aluminosilicate crystal lattice. Thus it may provide misleading environmental information if the mineralogy of the area of study is not constant. It also gives rise to lower contrast of anomalous to background samples than the other methods and thus is not as preferable. The other two methods are informative because they provide high contrast of anomalous to background samples, and simultaneously extract a large number of metals from sediments without attacking the aluminosilicate minerals. The 0.5N hydrochloric acid method is the preferred method for

Table 2.58 Comparison of metal content at four different locations, determined using four diffrent extractants (mg kg^{-1})

Site	Cd				Cu				Pb			
					Method used[a]							
	1	2	3	4	1	2	3	4	1	2	3	4
B	–	4.7	–	9	9.3	9.1	7.5	8.1	12	16	16	17
T	–	45	39	78	4.4	4.2	1.5	5.3	6.5	8.2	7.7	8.9
W	–	6.0	5.0	9	4.2	4.1	1.5	5.3	6.5	7.9	7.4	8.7
X	–	3.3	4.0	6	2.8	3.0	1.5	3.6	13	17	16	17

Site	Cd				Cu				Pb			
					Method used[a]							
	1	2	3	4	1	2	3	4	1	2	3	4
B	10	18	32	34	–	–	–	2	–	2.1	3.6	4.4
T	8.5	16	28	34	5.9	13	18	32	3.7	6.6	15	18
W	5.1	7.3	13	15	–	–	2.7	–	–	–	3.7	6.4
X	4.9	7.8	14	15	–	–	–	–	–	–	–	–

[a] 1 = 4.0N HNO$_3$ + 0.7N HCl: 2 = 0.5N HCl; 3 = 1N NH$_2$OH HCl + 25% acetic acid; 4 = 0.05N EDTA

Source: Editors of Geochemical Exploration [178]

measuring authigenically formed metals in soils and sediments since, due to some natural processes, the adsorbed metals in some samples are bonded more strongly, making the 0.05N EDTA method incapable of extracting them completely.

Yamamoto *et al.* [213] have developed a portable instrument for metals on soil surfaces based on laser-induced breakdown spectroscopy. Detection limits for selected metals ranged from about 10 to 300 ppm. The potential on EXAFS spectroscopy for the speciation and quantification of the forms of trace metals in solid materials using palladium as a case study, has been examined by Manceau *et al.* [212]. Espinosa and Silva [214] described the use of nuclear track detectors for the determination of α contaminated soils. He used a stake type device containing strips of Cr–39 that can be inserted into the soil, after which the α tracks are developed by etching with potassium hydroxide. Another simple passive sampling device was reported by Burnett *et al.* [215]. For the measurement of radon flux from soil, a charcoal canister was fastened inside a PVC cylindrical container. Radon activity onto the charcoal was measured by using a sodium iodide detector, which could be related to total radon flux after determining the charcoal adsorption efficiency. Smith developed an improved analytical scheme for actinides in soil by using actinide-selective extraction chromatography and ion-exchange resins designed to satisfy the requirements of both γ-spectrometry and inductively coupled plasma atomic emission spectrometry [216]. Flow injection preconcentration was used by Hollenbach et al. [217] for the determination of ^{230}Th, ^{234}U, ^{239}Pu and ^{240}Pu in soils. Detection limits were improved by a factor of ~20, and greater freedom from interference was observed with the flow injection system, compared to direct aspiration. In another study, the performance of 24 European measurement teams for an in situ γ-spectrometry intercomparison exercise was reported [218].

A new type of pedologically based soil sampling technique (ie by soil horizon rather than incrementally in depth) has been suggested by Meriwether et al. [219]. He gives an example where classical sampling approaches would lead to erroneous conclusions about anthropogenic contamination. Considerable simplification of sample preparation for trace elements in sediments and soils has been reported by using ultrasonic slurry sampling and graphite furnace atomic absorption spectrometry [220]. Before analysis, samples were ground to a grain size of <63 μm.

Lopez Garcia et al. [210] suspended soil samples in water containing 5% (v/v) concentrated hydrofluoric acid before injection into an electrothermal atomic absorption spectrophotometric system. No modifier other than the hydrofluoric acid was required for the determination of lead, cadmium and thallium.

Jerrow et al. [331] have reviewed methods for the determination of alkaline earths in soils.

Opydo [332] used anodic stripping voltammetry to determine zinc, cadmium, lead and copper in soil extracts. Ure [333] and Spevachova and Kucerna [334] studied the determination and speciation of trace metals in soil. Sagar [335] has reviewed the chemical speciation and environmental mobility of metals in soils. Stachel et al. [227] performed comparative studies using different digestion procedures in the determination of heavy metals and arsenic in the five grain particle size fraction of suspended particulate matter. The highest metal concentrations were found for microwave heating in a closed system using an acid mixture of nitric acid–hydrofluoric acid. In a separate study, only digestion procedures using nitric acid and hydrofluoric acid with subsequent evaporation to dryness and dissolution in hydrochloric acid gave appropriate results for a wide range of elements [229].

2.60.12.2 Radionucleides

Noey [326] has described methods for detecting α emitting radionucleides in field soils. Knizhnik et al. [336] experimentally verified a formula for evaluating the α radioactivity of soils by using track detectors. Espinosa and Silva [214] have described the use of nuclear track detectors for the determination of α contaminated soils. These workers used a stake-type device containing strips of chromium-39 that can be inserted in the soil sample, after which the α tracks are developed by etching with potassium hydroxide. Smith *et al.* [216] have developed an improved analytical scheme for actinides in soil by using actinide-selective extraction chromatography and ion-exchange resins designed to satisfy the requirements of both α spectrometry and inductively coupled plasma mass spectrometry.

Korun *et al.* [337] have described an improved method for the determination of radionucleides in soil based on in situ γ.-ray measurements made using a portable spectrometer. An intercomparison study by 24 European measurement teams has been reported for an in situ γ spectrometry survey on soils [218]. Wu and Landsberger [338] have found that various conditions are needed for the determination of medium-lived radionucleides in soil by neutron activation methods. A simple passive sampling device has been reported by Burnett *et al.* [215]. For the measurement of radon flux from soil, a charcoal canister was fastened inside a PVC cylindrical container. Radon activity onto the charcoal was measured by using a sodium iodide detector, which could be related to total radon flux after determining the charcoal adsorption efficiency.

2.60.13 Extraction of metals from soil samples

2.60.13.1 Extraction with ammonium acetate

Del Castilho and Rix [339] have reviewed the suitability of the ammonium acetate extraction method for the prediction of heavy metal availability in soils.

2.60.13.2 Extraction with ethylene-diamine tetraacetic acid

Haynes and Swift [340] investigated the effects of air and oven drying on the ethylene diamine tetraacetic acid extractable concentrations of copper, zinc, iron and manganese from soil. Drying increased the amounts of extractable micronutrients and organic matter from soil samples, and oven drying had a more pronounced effect than air drying.

Crosland *et al.* [341] in a collaborative interlaboratory study evaluated an EDTA extraction procedure for determining metals in soils. All laboratories produced some extreme outlying results, but most results were in good agreement once the outliers were removed.

2.60.13.3 Acid extraction

Davis and Carlton-Smith [182] conducted an interlaboratory comparison of metal determinations in soil and sewage sludge. The elements studied included cadmium, chromium, copper, nickel, lead, zinc, molybdenum, mercury, arsenic, selenium and boron. Various acid digestion procedures were applied to the soil samples prior to analysis by atomic absorption spectrometry. These included (a) wet digestion with nitric acid–perchloric acid or nitric acid, (b) dry ashing before digestion with nitric acid–perchloric acid or, nitric acid–sulphuric acid or nitric acid–sulphuric acid–hydrogen peroxide, or nitric acid–hydrochloric acid–hydrogen peroxide or nitric acid–hydrochloric acid or hydrochloric acid–hydrogen peroxide or perchloric acid–hydrogen peroxide or hydrochloric acid.

In general the conclusion reached was that the nitric acid–perchloric acid digestion procedure tended to recover more of the heavy metals (except copper) from soil samples than other wet digestion procedures and was also slightly more precise. Regarding determinations of total molybdenum, mercury, arsenic, selenium and boron, very few of the 100 laboratories involved submitted results and those that were submitted served to underline the need for further work on method development.

Sturgeon et al. [228] showed that only digestion procedures using nitric–hydrofluoric acids with subsequent evaporation to dryness and dissolution in hydrochloric acid gave appropriate results for a wide range of elements. Schramel et al. [224, 225] used a mixture of hydrofluoric acid and nitric acid under pressure for 6 h at 170°C for the wet digestion of soils. The technique was tested for the determination of aluminium, titanium, chromium, copper, cadmium, iron, manganese, nickel, lead and zinc. Values in very close agreement with certified values were obtained for a range of standard soils.

2.60.13.4 Extraction with organic complexing agents

Ivanova et al. [342] reported on the application of the extraction system ammonium tetramethylenecarbodithioate/isobutyl methyl ketone to elements in soil digests. Good results were obtained for the simultaneous preconcentration of arsenic, cadmium and thallium, followed by determination by atomic absorption spectrometry.

Donaldson [343] extracted cobalt, nickel, lead, bismuth and iridium from soils using chloroform extraction of these xanthate complexes.

2.60.13.5 Microwave extraction

Various workers [108–111, 221, 227, 228, 265, 327, 344–355] have applied microwave-assisted acid extraction techniques to the determination of metals in soils. Kammin and Brandt [349] showed that microwave digestion had considerable promise as a high speed alternative to the Environmental Protection Agency digestion method 3050 for the determination of trace metals in soils. Kratchvil and Mamba [348] showed that all the zinc and copper were released from soils within 7 min using a commercial microwave oven. Lo and Fung [351] studied the recovery of heavy metals from soils during acid digestion with different acid mixtures by a block heater and by microwave digestion. Reynolds [352] has reviewed microwave digestion procedures for the analysis of metal contaminated soils. Kingston and Walter [353] compared microwave digestion with conventional dissolution methods for the determination of metals in soils. Feng and Barrett [247] showed that microwave dissolution of soil samples with nitric-hydrofluoric acid gave recoveries of lead and cadmium of over 90% in 30 min.

D'Silva et al. [327] compared two methods involving dissolution in hydrogen chloride gas and microwave dissolution for the remote dissolution of uranium in soil. Krishnamurti et al. [354] have shown that microwave extraction of cadmium from a soil reference sample gave results comparable to those found by conventional extraction procedures. Real et al. [355] have shown that optimising the microwave heating procedure would optimise results obtained in sequential extraction procedures. Chakraborty et al. [265] determined chromium in soils by microwave-assisted sample digestion followed by atomic absorption spectrometry without the use of a chemical modifier.

Comparative studies using different digestion procedures have been performed from the determination of heavy metals and arsenic in the fine grain particle size fraction of suspended particulate matter [253]. The highest metal concentrations were found for microwave heating in a closed system using a mixture of nitric and hydrofluoric acids.

A continuous flow microwave-assisted digestion of soils has been found to be an effective approach, giving trace metal recoveries averaging 90% with good precision [254]. Torres [221] found that a microwave-assisted robotic method for trace metals in soil decreased sample digestion times from 2 h to 3 min.

2.60.13.6 Ultrasonic extraction

Akcay et al. [305] have shown that extraction of total strontium using an ultrasonic extraction procedure was not as good as was achieved by using conventional extraction methods. Van den Akker and Van den Heuvel [356] used an ultrasonic sample to prepare soil samples for analysis.

Sanchez et al. [357] obtained acceptable accuracy and precision in the determination of metals in soil using an ultrasonic bath digestion procedure. Klemm and Bombach [220] have reported a considerable simplification of sample preparation for the determination of trace elements in soils by using ultrasonic slurry sampling and graphite furnace atomic absorption spectrometry. Before analysis, the samples were ground to a grain size of <63 μm.

2.60.13.7 Sequential extraction

Sequential extraction techniques have been widely used for the speciation analysis of major and trace elements in soils and sediments. Due to the growing interests in the development and application of the sequential extraction techniques, several reviews have appeared in the literature [358–361]. Several different sequential extraction programmes have been developed [362–375]. Most include up to five extractants: cation-exchange extractants, carbonate-dissolving extractants, acidic reducing extractants, extractants that release organic and sulphide-bound metals, and strong acidic extractants for dissolving silicates or minerals that have not been attacked by the milder reagents. The most widely recognised sequential extraction is Tessier's method [364] which has been frequently used [370].

In this scheme the metals are partitioned into five fractions:

Fraction 1: Exchangeable. Metals in this fraction are bound to the sediments by weak adsorption onto sediment particles. Changes in ionic strength of the water are likely to affect the adsorption-desorption or ion exchange processes resulting in the uptake or release of metals at the sediment/water interface. Metals are extracted from this fraction by increasing the ionic strength with 1M $MgCl_2$ (pH 7) for 1 h at room temperature.

Fraction 2: Carbonate Bound. Metals bound to carbonates are sensitive to pH changes with the lowering of pH being associated with the release of metal cations. Metals are extracted from this fraction at pH 5 with an acetate buffer for 5 h at room temperature.

Fraction 3: Iron/Manganese Oxide Bound. Metals bound to iron/manganese oxide fraction are unstable under reducing (anoxic) conditions. These conditions result in the release of metal ions to the dissolved fraction. Metals are extracted from this fraction using 0.04M hydroxylamine hydrochloride in 25% (v/v) acetic acid for 6 h at 96°C.

Fraction 4: Organic Bound. Degradation of organic matter under oxidising conditions can lead to the release of soluble metals bound to those materials. Metals are extracted from this fraction using 30% H_2O_2 in 0.02M HNO_3 for a total of 5 h at 85°C. At the end of the extraction, 3.2M ammonium acetate in 25% (v/v) nitric acid is added to prevent readsorption of metals in oxidised sediment.

Fraction 5: Residual. This fraction should contain naturally occurring minerals which may hold trace metals within their crystalline matrix. These metals would not be readily released to a soluble state in conditions encountered in nature. An $HClO_4$/HF/HCl digestion requiring about 4 h is used for the extraction of residual metals.

Miller *et al.* [370] examined the order of extraction for key steps in the sequential, extraction and proposed a nine-step sequential method to characterise trace metals in agricultural polluted and waste amended soils. In order to simplify the sequential extraction method, Chao and Sanzolone [372] divided soil selenium into immediate and potential impact categories and recommended a two-step extraction procedure.

Although various sequential extraction approaches have been recommended in the literature, these methods suffered from non-selectivity of extractants and trace element redistribution among phases during extraction. Guy et al. [376] used bentonite, manganese oxide, and humic acid spiked with copper and lead to evaluate the extraction methods, and the results indicated that the chemical extraction procedures were unable to determine unequivocally the sites of the adsorbed metals in soils and sediments. The intercomponent interferences prevented 100% recoveries of the doped metals. Tipping et al. [377] observed the direct evidence of redistribution when manganese oxide in a mine wall deposit was selectively separated from iron oxide by using hydroxylamine. Electron microscopy analysis before extraction indicated that lead was in the manganese phase, but none was found in the iron phase. After extraction, lead was present in the iron phase. The reason for the transfer of a large amount of lead from manganese to iron oxide was ascribed to the adsorption of lead on the remaining iron oxide during hydroxylamine treatment. Kheboian and Bauer [174] criticised the accuracy of Tessier's sequential extraction for metal speciation by using model aquatic sediments. Trace elements of lead, zinc, copper and nickel were doped into each phase by coprecipitation or adsorption. Then, the model aquatic sediments were successively treated with a five-step extraction procedure under the conditions recommended by Tessier et al. [364]. Generally, none of the trace elements were removed at the stage predicted according to Tessier's method. This was mainly due to the elemental redistribution.

The advantages of using purified commercially available illite, silicate gel, and humic acid and other pure chemical compounds and of using doped trace elements in each phase are the simplicity and ease of understanding the behaviour of trace elements during sequential extraction. However, soils and sediments are a complex mixture of mineral fragments and decomposition products, reflecting the nature of the original base rock, the degree of degradation and leaching introduced by weathering cycles, and the influence of external inputs such as plant debris or contamination introduced by human activities. Therefore, the properties of the model aquatic sediments are unlikely to be sufficiently representative of real sediments. Tessier and Campbell [378] argued the validity of the approach and experimental problems associated with the methods used by Kheboian and Bauer [174] and some others. The main drawback of the model soil or sediment as prepared by mixing a phase doped with a trace metal with other phases not doped with that metal is non-equilibrium distribution of trace metals among various solid phases; in other words, the model sediment is biased toward trace metal redistribution.

Xiao Quan and Bin [208] have evaluated the applicability of Tessier sequential extraction as a method for the speciation of trace metals in soils containing natural minerals and humic acids. These workers found that although the recommended sequential extractants are able to bring the individual geochemical phases into solution at the experimentally defined stage, none of the extractants is completely specific and efficient. Various proportions of metals released at exchangeable, carbonate-bound, Fe–Mn oxide-bound, organic-bound fractions are readsorbed onto the other solid geochemical phases during sequential extraction. However, it is difficult to quantify the extent of readsorption and/or redistribution for the individual element and for the individual geochemical phase.

Slavek *et al.* [206] have studied the ability of a range of electrolytes to release metal ions (copper, lead, cadmium and zinc) presorbed on two samples of humic acid. Though treatment with mineral acid or a chelating agent released a high proportion of the retained metal ion, recoveries were never total. Concentrated salt solutions displaced about 80% of the retained cadmium or zinc, and about half of any copper or lead held by the organic matter, which indicates that most of the adsorbed metal ion is exchangeable, the extraction efficiency being controlled by competing equilibria. The effect of added clay suspensions was also examined. Analytical procedures for fractionating the total metal content of soils into subgroups have been assessed against the observed extraction behaviour.

The electrolytes studied included: nitric acid, ammonium acetate, ammonium nitrate, magnesium chloride (1M), hydrochloric acid, acetic acid, acetic acid–ammonium acetate, sodium chloride (0.5M), sodium hydroxide, sodium pyrophosphate (0.1M), calcium chloride (0.05M) and EDTA, DPTA (0.005 M).

Hickey and Kittrick [207] used sequential extraction to investigate the chemical partitioning of cadmium, copper, nickel and zinc in soil which had been subject to massive additions of heavy metals over six years. The metals were partitioned into five operationally defined geochemical fractions: exchangeable, bound to carbonates, bound to iron/manganese oxides, bound to organic matter and residual. Cadmium had the highest mobility and potential bioavailability followed by zinc, copper and nickel in descending order.

Lopez-Sanchez *et al.* [379] found that significantly different results can be obtained when different sequential extraction procedures are used. The sequential extraction of calcium and magnesium, also heavy metals, strontium, vanadium and aluminium in soils containing natural minerals and humic acid has been studied [208]. Shan and Chen [380] have reported that various proportions of metals released from exchangeable carbonate-bound, iron and manganese oxide-bound, and organic-bound fractions were readsorbed onto the other solid geochemical phases during sequential extraction. Sequential extraction schemes have been studied by several investigators [223, 226, 227]. Ure *et al.* reviewed single and sequential extraction schemes for trace metal speciation in soil [222]. Magi studied the optimisation of the extraction of metal – humic acid complexes from marine sediments [223]. Li *et al.* [230] reported on the development of a five step sequential extraction scheme for the analysis of soils by inductively coupled plasma mass spectrometry. Application of the method to soils contaminated by past mining and smelting activities showed distinctive partitioning patterns of heavy metals from the two sources.

2.60.13.8 Miscellaneous

Desaules *et al.* [381] have conducted an interlaboratory study in which soil extraction techniques were specified but the measurement method was optional. This work showed a strongly differing degree of comparability for the various metals investigated also for the extraction techniques and measurement techniques used by the different laboratories.

References

1 Dodson, A. and Jennings, V.J., *Talanta*, **19** ,801 (1972).
2 Reis, B.F., Bergamin F, H., Zegatto, E.A.G. and Krug, F.J., *Analytica Chimica Acta*, **107**, 309 (1979).
3 Tecator Ltd., Application Note ASN 78–31/85, Determination of Aluminium in Soil by Flow Injection Analysis (1985).
4 Tecator Ltd., Application Note ASTN 10/84, Determination of Aluminium by Flow Injection Analysis (1984).
5 Zoltzer, D. Diplomarbeit, Inst für Anong Chemie, University of Gottingen, West Germany (1882).
6 Mitrovic, B., Milacic, R., and Pihlar, B. *Analyst* (London), **121**, 627 (1996).
7 Kozuk, N., Milacic, R., and Gorenc, B. *Ann Chim*, **86**, 99 (1996).
8 Sill, C.W., Puphal, K.W. and Hindman, F.D. *Analytical Chemistry*, **46**, 1725 (1974).
9 Keay J. and Menage, P.M.A. *Analyst* (London), **95**, 379 (1970).
10 Alder, J.F., Gunn, A.M. and Kirkbright, G.F. *Analytica Chimica Acta*, **92**, 43 (1977).
11 Bremner, J.M. *Monogram American Soc Argon*, 9 (1965).
12 Tecator Ltd., Application Note ASN 65–32/83, Determination of Ammonia Nitrogen in Soil Samples Extractable by 2 M KCl using Flow Injection Analysis (1983).
13 HMSO, London, The Analysis of Agricultural Materials, RB 427, 2nd ed, Method 60, Ammonium nitrate and nitrite nitrogen potassium chloride extractable in moist soil (1979) ISBN 011 240 352.2.
14 Waughman, G.J. *Environmental Research*, **26** , 529 (1981).
15 Chikhalikar, S., Sharma, K. and Patel, K.S. *Communications Soil Sci Plant Anal*, **26**, 621 (1995).
16 Agrawal, Y.K. and Patke, S.K. *International Journal of Environmental Analytical Chemistry*, **8**, 157 (1980).
17 Lawless, E.W., Von Rumber, R. and Ferguson, R.L. Technical Studies Dept T 500–72, Environmental Protection Agency, Washington DC (1972).
18 Woolson, E.A., Axley, J.J. and Kearney, P.C. *Soil Sciences Society Am Proc*, **35**, 101 (1971).
19 Schroeder, H.A. and Balassa, J.J. *Journal of Chronic Disorders*, **19**, 1 (1966).
20 Analytical Sub Comm, *Analyst* (London), **85** , 679 (1960).

21 Analytical Methods Committee, *Analyst* (London), **100**, 54 (1975).
22 Sandhu, S.S. *Analyst* (London), **106**, 311 (1981).
23 Forehand, T.J., Dupey, A.E. and Tai, H. *Analytical Chemistry*, **48**, 999 (1976).
24 Thompson, K.C. and Thomerson, D.R. *Analyst* (London), **99**, 595 (1974).
25 Thompson, A.J. and Thoresby, P.A. *Analyst* (London), **102**, 9 (1977).
26 Wanchange, R.D. *Atomic Absorption Newsletter* Perkin–Elmer, **15**, 64 (1976).
27 Ohta, K. and Suzuki, M. *Talanta*, **25**, 160 (1978).
28 Haring, B.J.A., van Delft, W. and Bom, C.M. *Fresenius Zeitschrift für Analytische Chemie*, **310**, 217 (1982).
29 Merry, R.H. and Zarcinas, B.A. *Analyst* (London), **109**, 998 (1980).
30 HMSO (London). Selenium and Arsenic in Sludges, Soils and Related Materials (1985). A note on the use of hydride genetator kits, (1987).
31 Jiminez de Blas, O., Mateos, N.R. and Sanchez, A.G. *Journal of American Association of Analytical Chemists (AOAC) Int*, **79**, 764 (1996).
32 Agemian, H. and Bedak, E. *Analytica Chimica Acta*, **119**, 323 (1980).
33 Wenclawiak, B.W. and Krab, M. *Fresenius Journal of Analytical Chemistry*, **351**, 134 (1995).
34 Chappell, J., Chiswell, B. and Olszowy, H. *Talanta*, **42**, 323 (1995).
35 Ross, D.S., Bartlett, R.J. and Magdoff, F.R. *Atomic Spectroscopy*, **7**, 158 (1986).
36 HMSO, The Analysis of Agricultural Materials, RB 427, 2nd edition (ISBN 011240 3522) Method 8, Boron, water soluble in soil (1979).
37 Aznarez, J., Bonilla, A. and Vidal, J.C. *Analyst* (London), **108**, 368 (1983).
38 Ducret, L. *Analytica Chimica Acta*, **17**, 213 (1957).
39 Pasztor, L., Bode, J.D. and Fernando, Q. *Analytical Chemistry*, **32**, 277 (1960).
40 Manzoori, J.L. *Talanta*, **27**, 682 (1980).
41 Zarcinas, B.A. and Cartwright, M.B. *Analyst* (London), **112**, 1107 (1987).
42 HMSO, London. The Analysis of Agricultural Materials, RB 427 (ISBN 011240 352 2), 2nd ed, Method 11, Cadmium nitric perchloric acid soluble in soil (1979).
43 HMSO, London. The Analysis of Agricultural Materials, RB 427 (ISBN 011240 352 2), 2nd ed, Method 10, Cadmium acetic acid extractable in soil (1979).
44 Baucells, M., Lacort, G. and Roura, M. *Analyst* (London), **110**, 1423 (1985).
45 Bolt, G.H. and Bruggenwert, M.G.M. in *Soil Chemistry* Part A, Elsevier, Amsterdam (1978).
46 Lewin, V.H. and Beckett, O.H.T. *Effluent and Waste Treatment Journal*, **20**, 162 (1980).
47 Christensen, T.H. and Lun, X.Z. *Water Research*, **23**, 73 (1989).
48 Turner, M.A., Hendrickson, L.L. and Corey, R.B. *Soil Science Society of America Journal*, **48**, 763 (1984).
49 Roberts, A.H.C., Turner, M.A. and Syers, J.K. *Analyst* (London), **101**, 574 1976.
50 Carlosena, A., Prada, D., Andrade, J.M., Lopez, P. and Muniategui, B. *Fresenius Journal of Analytical Chemistry*, **355**, 289 (1996).
51 Qin Xing-chu, Zhang Yu-sheng and Zhu Ying-quan, *Analyst* (London), **108**, 754 (1983).
52 Jackson, Y.L. in *Soil Chemical Analysis*, Prentice–Hall, Eaglewood Cliffs, NJ (1958).
53 Qin Xing-chu and Zhu Ying-quan, *Analyst* (London), **110**, 185 (1985).
54 Qi, W.B. and Zhu, L.Z. *Talanta*, **33**, 694 (1986).
55 Smith, G.H. and Lloyd, O.L., *Chemistry in Britain*, February, p139 (1986).
56 Evans, R. Personal communication, Dept of City Analyst, Dundee (1985).
57 Sahuquillo, A., Lopez-Sanchez, J.F., Rubio, R., Rauret, G. and Hatje, V. *Fresenius Journal of Analytical Chemistry*, **351**, 197 (1995).
58 Sahuquillo, A., Lopez-Sanchez, J.F., Rubio, R. and Rauret, G. *Mikrochimica Acta*, **119**, 251 (1995).

59 Fodor, P. and Fischer, I. *Fresenius Zeitschrift für Analytische Chemie*, **351**, 454 (1995).

60 Prokisch, J., Kovacs, T.S., Gyori, Z. and Loch, J. *Communications in Soil Science*, **26**, 2051 (1995).

61 Florez-Velez, L.M., Gutierrez–Ruiz, H.E., Reyes-Salas, O., Cram-Heydrich, S. and Baesze-Reys, A. *International Journal of Environmental Analytical Chemistry*, **61**, 177 (1995).

62 HMSO, London, The Analysis of Agricultural Materials, RB 427 (ISBN 011240 352 2) 2nd ed, Method 22, Cobalt extractable in soils (1979).

63 HMSO, London, The Analysis of Agricultural Materials, RB 427 (ISBN 011240 352 2) 2nd ed, Method 23, Cobalt, nitric-perchloric soluble, in soil (1979).

64 HMSO, London, The Analysis of Agricultural Materials RB 427 (ISBN 011240 352 2) 2nd edition, Method 26, Copper, EDTA extractable in soil (1979).

65 HMSO, London, The Analysis of Agricultural Materials RB 427 (ISBN 011240 352 2) 2nd ed, Method 27, Copper, nitric-perchloric soluble, in soil (1979).

66 Senesi, N. and Sposito, G. *Soil Science Society of America Journal*, **48**, 1247 (1984).

67 Stefanov, G. and Daieva, L. *Isotopenpraxis*, **8**, 146 (1972).

68 Jayman, T.C.Z., Sivasubramanian, S. and Wijeedasa, M.A. *Analyst* (London), **100**, 716 (1975).

69 Keaton, C.M. *Soil Science*, **43**, 40 (1937).

70 Agrawal, Y.K., Raj, K.P.S., Desai, S., Patel, S.G. and Merh, S.S. *International Journal of Environmental Science*, **14**, 313 (1980).

71 Chow, T.J. *Nature* (London), **225**, 295 (1970).

72 Motto, H.L., Daines, R.H., Chilko, D.M. and Motto, C.K. *Environmental Science and Technology*, **4**, 231 (1970).

73 Low, K.S., Lee, C.K. and Arshad, M.M. *Pertanika*, **2**, 105 (1979).

74 Ward, N.I., Reeves, R.D. and Brooks, R.R. *Environmental Pollution*, **6**, 149 (1974).

75 Berrow, M.L. and Stein, W.M. *Analyst* (London), **108**, 277 (1983).

76 HMSO, London, The Analysis of Agricultural Materials, RB 427 (ISBN 011240 352 2) 2nd ed, Method 43, Lead extractable in soil (1979).

77 HMSO, London, The Analysis of Agricultural Materials, RB 427 (ISBN 011240 352 2) 2nd ed, Method 44, Lead nitric perchloric soluble in soil (1979).

78 Somer, G. and Aydin, H. *Analyst* (London), **110**, 631 (1985).

79 Sakharov, A.A. *Pockvovedenie*, **1**, 107 (1967) .

80 Tills, A.R. and Alloway, B.J. *Environmental Technology Letters*, **4**, 529 (1983).

81 Rigin, V.I., Rigina, I.I. *Zhur Analiticheskoi Khim*, **34**, 1121 (1979).

82 HMSO, London, The Analysis of Agricultural Materials, RB 427 (ISBN 011240 352 2) 2nd ed, Method 46, Magnesium, extractable in soil (1979).

83 Alekseeva, I.I. and Davydova, Z.P. *Zhur Analit Khim*, **26**, 1786 (1971).

84 HMSO, London, The Analysis of Agricultural Materials, RB 427 (ISBN 011240 352 2) 2nd ed, Method 48, Manganese exchangeable and easily reducible in soil (1979).

85 Kimura, Y. and Miller, V.L. *Analytica Chimica Acta*, **27**, 325 (1962).

86 Hatch, A. and Ott, W.L. *Analytical Chemistry*, **40**, 2085 (1968).

87 Hoggins, F.E. and Brooks, R.R.J. *Association of Official Analytical Chemists*, **56**, 1306 (1973).

88 Ure, A.M. and Shand, C.A. *Analytica Chimica Acta*, **72**, 63 (1974).

89 Floyd, M. and Sommers, L.E. *Journal of Environmental Quality*, **4**, 323 (1975).

90 Agemian, H. and Chau, A.S.Y. *Analyst* (London), **101**, 91 (1976).

91 Kuwae, Y., Hasegawa, T. and Shono, T. *Analytica Chimica Acta*, **84**, 185 (1976).

92 Nicholson, R.A. *Analyst* (London), **102**, 399 (1977).

93 Head, P.C. and Nicholson, R.A. *Analyst* (London), **98**, 53 (1973).

94 HMSO, Methods for the Examination of Water and Associated Materials. Mercury in waters effluents soils and sediments etc, Additional Methods 40453 39 pp (1987).
95 Lutze, R.G. *Analyst* (London), **104**, 979 (1979).
96 Grantham, P.L. *Laboratory Practice*, **27**, 294 (1978).
97 HMSO, London, The Analysis of Agricultural Materials, RB 427 (ISBN 011240 352 2) 2nd ed, Method 86, Mercury in soil and plant material (1979).
98 Raseman, W., Seltmann, U. and Hempel, M., *Fresenius Journal of Analytical Chemistry*, **351**, 632 (1995).
99 Wenger, R. and Hagel, O. *Mitt Geb Lebensmittelunters U Hyg*, **62**, 1 (1971).
100 Micham, P., Maksvytis, A. and Barkus, B. *Analytical Chemistry*, **44**, 2102 (1972).
101 HMSO, London, The Analysis of Agricultural Materials, RB 427 (ISBN 011240 352 2) 2nd ed, Method 51, Molybdenum total in soil (1979).
102 HMSO, London, The Analysis of Agricultural Materials, RB 427 (ISBN 011240 352 2) 2nd ed, Method 50, Molybdenum extractable in soil (1979).
103 Thompson, M. and Zao, L. *Analyst* (London), **110**, 229 (1985).
104 Kim, C.H., Owens, C.M. and Smythe, L.E. *Talanta*, **21**, 445 (1974).
105 Kim, C.H., Alexander P.W. and Smythe, L.E. *Talanta*, **23**, 229 (1976).
106 Ni, Z.M., Chin, L.C. and Wu, T.H. *Huan Ching K'o Hsueh* 1979 No 6 25 Analytical Abstracts 40 5H65 (1981).
107 Munter, R.C. and Grande, R.A. in Barnes, R., ed. *Developments in atomic plasma spectrochemical analysis*, Heyden, London pp 653–72 (1981).
108 Maessen, F.J.M.L., Balke, J. and de Beer, J.L.M. *Spectrochimica Acta*, Part B, **37**, 517 (1982).
109 David, D.J. *Prog Analytical Atomic Spectroscopy*, **1**, 225 (1978).
110 Jones, J.B. *Journal of the Association of Official Analytical Chemists*, **58**, 764 (1975).
111 Mitchell, R.L. and Scott, R.O. *Journal of Society of Chemical Industry*, **66**, 330 (1974).
112 Reisenauer, M.H. in *Methods of Soil Analysis*, Part 2. American Society of Agronomy, Madison WI, pp 1054–57 (1965).
113 Sperling, K.R. *Zeitschrift für Analytische Chemie*, **283**, 30 (1977).
114 Ure, A.M. and Mitchell, M.C. *Analytica Chimica Acta*, **87**, 283 (1976).
115 Iu, K.L., Pulford, I.D. and Duncan, H.J. *Analytical Chimica Acta*, **106**, 319 (1979).
116 Crompton, T.R. unpublished work.
117 Ure, A.M., Hernandez, M.P. and Mitchell, M.C. *Analytica Chimica Acta*, **96**, 37 (1978).
118 Pedersen, B., Willems, M. and Storgaard, J.S. *Analyst* (London), **105**, 119 (1980).
119 Henn, E.L. in *Flameless Atomic Absorption Analysis – an update*, ASTM STP 618 American Society for Testing and Materials, Philadelphia pp 54–64 (1977).
120 Dahlquist, R.L. and Knoll, J.W. *Applied Spectroscopy*, **32**, 1 (1978).
121 Jones, J.B. *Communications Soil Science Plant Analysis*, **8**, 340 (1977).
122 HMSO, London, The Analysis of Agricultural Materials, RB 427 (ISBN 011240 352 2) 2nd ed, Method 53, Nickel extractable in soil (1979).
123 HMSO, London, The Analysis of Agricultural Materials, RB 427 (ISBN 011240 352 2) 2nd ed, Method 54, Nickel nitric perchloric soluble in soil (1979).
124 Talvitie, N.A. *Analytical Chemistry*, **43**, 1827 (1971).
125 Dienstbach, E. and Bachmann, K. *Analytical Chemistry*, **52**, 620 (1980).
126 Sekine, K., Imai, T. and Kasai, A. *Talanta*, **34**, 567 (1987).
127 HMSO, London, The Analysis of Agricultural Materials, RB 427 (ISBN 011240 352 2) 2nd ed, Method 68, Ammonium acetate extractable potassium in soil (1979).

128 Weltz, B. *Atomic Absorption Spectrometry*, Verlag Chemie, New York (1976).
129 Pinta, M. *Atomic Absorption Spectrometry Applications to Chemical Analysis*, PWN Warsaw (1977).
130 Schrenk, W.G. Modern *Analytical Chemistry Analytical Atomic Spectroscopy*, Plenum Press, New York (1975).
131 Brodie, K.G. *International Laboratory*, **40**, July/August (1979).
132 Laurakis, V., Barry, E. and Golembeski, T. *Talanta*, **22**, 547 (1975).
133 Ohta, K. and Suzuki, M. *Talanta*, **22**, 465 (1975).
134 Thompson, K.C. *Analyst* (London), **100**, 307 (1975).
135 HMSO, London. Selenium and Srsenic in sludges, soils and related materials (1985) – a note on the use of hydride generation kits (1987).
136 Chambers, J.C. and McClellan, D. *Analytical Chemist*, **48**, 2061 (1976).
137 Inhat, M. *Analytica Chimica Acta*, **82**, 292 (1976).
138 Henn, E.L. *Analytical Chemistry*, **47**, 428 (1975).
139 Ishizaka, M. *Talanta*, **25**, 167 (1978).
140 Montaser, A. and Mehrabzadeh, A.A. *Analytical Chemistry*, **50**, 1697 (1978).
141 Azad, J., Kirkbright, G.F. and Snook, R.D. *Analyst* (London), **104**, 232 (1979).
142 Parker, C.A. and Harvey, L.G. *Analyst* (London), **87**, 558 (1962).
143 Alloway, W.H. and Cary, E.E. *Analytical Chemistry*, **36**, 1359 (1964).
144 Watkinson, J.H. *Analytical Chemistry*, **32**, 98 (1960).
145 Watkinson, J.H. *Analytical Chemistry*, **38**, 92 (1966).
146 Ewan, R.C., Baumann, C.A. and Pope, A.L. *Journal of Agricultural and Food Chemistry*, **16**, 212 (1968).
147 Hall, R.J. and Gupta, P.L. *Analyst* (London), **94**, 292 (1969).
148 Pahlavanpour, B., Pullen, J.H. and Thompson, M. *Analyst* (London) **105**, 274 (1980).
149 Thompson, M., Pahlavanpour, B,. Walton, S.J. and Kirkbright, G.F. *Analyst* (London), **103**, 568 (1978).
150 Zmijewska, W. and Semkow, T. *Chem Anal* (Warsaw), **23**, 583 (1978).
151 Kronborg, O.J. and Steinnes, E. *Analyst* (London), **100**, 835 (1975).
152 Mignosin, E.P. and Roelandts, I. *Chemical Geology*, **16**, 137 (1975).
153 Van der Klugt, N., Poelstra, P. and Zwemmer, E. *Journal of Radio Analytical Chemistry*, **35**, 109 (1977).
154 Baedecker, P.A., Rowe, J.J. and Steinnes, E. *Journal of Radio Analytical Chemistry*, **40**, 115 (1977).
155 Agemian, H. and Bedek, E. *Analytical Chemistry*, **119**, 394 (1980).
156 Bem, E.M. *Environmental Health Perspectives*, **37**, 183 (1981).
157 Pellenberg, R. *Marine Pollution Bulletin*, **10**, 267 (1979).
158 Que-Hee, S.G. and Boyle, J.R. *Analytical Chemistry*, **60**, 1033 (1988).
159 HMSO, London, The Analysis of Agricultural Materials, RB 427 (ISBN 011240 352 2) 2nd ed, Method 72, Sodium extractable, in soil (1979).
160 Martin, D.B. *Analytical Chemistry*, **51**, 1968 (1979).
161 Abbasi, S.A. *International Journal of Environmental Analytical Chemistry*, **11**, 1 (1982).
162 Quin, B.F. and Brooks, R.R. *Analytica Chimica Acta*, **58**, 301 (1972).
163 Abbasi, S.A. *International Journal of Environmental Studies*, **18**, 51 (1981).
164 Prister, B.S., Zubach, S.S. *Radiokhimiya*, **10**, 743 (1968).
165 Nass, H.W., Molinski, V.J. and Kramer, H.H. *Mater Res Stand*, **12**, 24 (1972).
166 HMSO, London, The Analysis of Agricultural Materials, RB 427 (ISBN 011240 352 2) 2nd ed, Method 82, Zinc extractable, in soil (1979).
167 HMSO, London, The Analysis of Agricultural Materials, RB 427 (ISBN 011240 352 2) 2nd ed, Method 83, Zinc, nitric perchloric soluble, in soil (1979).
168 Abdullah, M., Zaman, M.B., Kaliquzzaman, M. and Khan, A.H. *Analytica Chimica Acta*, **118**, 175 (1980).

169 Chao, T.T. and Sanzolone, R.F. *J Research US Geological Survey*, **1**, 681 (1973).
170 Weitz, A., Fuchs, G. and Bachmann, K. *Fresenius Zeitschrift für Analytische Chemie*, **313**, 38 (1982).
171 HMSO, London. Department of the Environment/National Water Council Standing Technical Committee of Analysts. *Methods for the examination of water and associated materials – determination of extractable metals in soil sewage sludge treated soils and associated materials*, 17 pp (22 BC ENV) (1983).
172 Hinds, M.W., Jackson, K.W. and Newman, A.P. *Analyst* (London), **110**, 947 (1985).
173 Edmonds, T.E., Guogang, P. and West, T.S. *Analytica Chimica Acta*, **120**, 41 (1980).
174 Kheboian, C. and Bauer, C.F. *Analytical Chemistry*, **59**, 1417 (1987).
175 Bowman, W.S., Faye, G.H., Sutarno, B., McKeague, J.A. and Kodama, H. *Geostand Newsletter*, **3**, 109 (1979).
176 Jackson, K.W. and Newman, A.P. *Analyst* (London), **108**, 261 (1983).
177 Chester, R. and Hughes, M. *Chemical Geology*, **2**, 249 (1967).
178 Chowdbury, A.N. and Bose, B.B. *Geochemical Exploration*, **11**, 410 (1971).
179 Dong, A., Rendig, V.V., Burau, R.G. and Besga, G.S. *Analytical Chemistry*, **59**, 2728 (1987).
180 Stupar, J. and Ajlec, R. *Analyst* (London), **107**, 144 (1982).
181 Davis, R.P., Carlton-Smith, C.H., Stark, J.H. and Campbell, J.A. *Environmental Pollution*, **49**, 99 (1988).
182 Davis, R.P., Carlton-Smith, C.H. *Water Pollution Control*, **82**, 290 (1983).
183 Lindsay, W.L. and Norvell, W.A. *Soil Science Society of America Journal*, **42**, 421 (1978).
184 Eidecker, R. and Jackwerth, E. *Fresenius Zeitschrift für Analytische Chemie*, **328**, 469 (1987).
185 Schramel, P., Xu, L.Q., Wolf, A. and Hasse, S. *Fresenius Zeitschrift für Analytische Chemie*, **313**, 213 (1982).
186 Greenfield, S., Jones, H., McGeachin, H.D. and Smith, P.B. *Analytica Chimica Acta*, **74**, 225 (1975).
187 Butler, C.C., Kniseley, R.N. and Fassel, V.A. *Analytical Chemistry*, **47**, 825 (1975).
188 Fassel, V.A., Peterson, F.N., Abercrombie, F.N. and Kniseley, R.N. *Analytical Chemistry*, **48**, 516 (1976).
189 Gunn, A.M., Kirkbright, G.F. and Opheim, L.N. *Analytical Chemistry*, **49**, 1492 (1977).
190 Watson, E., Russel, G.M. and Balaes, S. South Africa National Institute of Metals Report No 1815, April 15 (1976).
191 Pahlavanpour, B., Thompson, M. and Thorne, L. *Analyst* (London), **105**, 756 (1980).
192 Thompson, M., Pahlavanpour, B., Walton, S.J. and Kirkbright, G.F. *Analyst* (London), **103**, 568 (1978).
193 Goulden, P.D., Anthony, D.H.J. and Keith, D. *Analytical Chemistry*, **53**, 2027 (1981).
194 Randle, K. and Hartman, E.H. *Journal of Radioanalytical and Nuclear Chemistry*, **90**, 309 (1985).
195 Ure, A.M., Ewen, G.J. and Mitchell, M.C. *Analytica Chimica Acta*, **118**, 1 (1980).
196 Ure, A.M. and Bacon, J.R. *Analyst* (London), **103**, 807 (1978).
197 Kanda, Y. and Taira, M. *Analytica Chimica Acta*, **207**, 269 (1988).
198 Reddy, S.J., Valenta, P. and Nurnberg, H.W. *Fresenius Zeitschrift für Analytische Chemie*, **313**, 390 (1982).
199 Meyer, A., de la Chevallerie-Haaf, U. and Henze, G., *Fresenius Zeitschrift für Analytische Chemie*, **328**, 565 (1987).
200 Muntau, H., Crossman, G., Schramel, P., Gallorni, M. and Orvini, E. *Fresenius Zeitschrift für Analytische Chemie*, **326**, 634 (1987).

201 Levinson, A. and Pablo, L. *Journal of Geochemical Exploration*, **4**, 339 (1975).
202 Hinkley, T. *Nature* (London), **277**, 444 (1979).
203 Lee, F.Y. and Kittrick, J.A. *Soil Science of America Journal*, **48**, 548 (1984).
204 HMSO, London. Method for the Examination of Waters and Associated Materials, Selenium in waters (1984), selenium and arsenic in sludges soils and related materials (1985), a note on hydride generator kits (1987), 42 pp (40548) (1987).
205 Jan, D. and Schwedt, G. *Fresenius Zeitschrift für Analytische Chemie*, **320**, 121 (1985).
206 Slavek, J., Wold, J. and Pickering, W.F. *Talanta*, **29**, 743 (1982).
207 Hickey, M.G. and Kittrick, J.A. *Journal of Environmental Quality*, **13**, 372 (1984).
208 Xiao-Quan Bin, S. *Analytical Chemistry*, **65**, 802 (1993).
209 Chen, T.C. and Hong, A. *J Hazardous Materials*, **41**, 147 (1995).
210 Lopez-Garcia, L., Sanchez-Merlos, M. and Hernandez-Cordoba, M. *Analytica Chimica Acta*, **328**, 19 (1996).
211 Zaray, G. and Kantor, T. *Spectrochimica Acta* Part B, **50B**, 489 (1995).
212 Manceau, A., Boisset, M.C., Sarret, G., Hazemann, J.L., Mench, M., Cambier, P. and Prost, R. *Environmental Science and Technology*, **30**, 1540 (1996).
213 Yamamoto, K.Y., Cremers, D.A., Ferris, M.J. and Foster, L.E. *Applied Spectroscopy*, **50**, 222 (1996).
214 Espinosa, G. and Silva, R. *Journal of Radioanalytical and Nuclear Chemistry*, **194**, 207 (1995).
215 Burnett, W.C., Cable, P.H. and Chanton, J.P. *Journal of Radioanalytical and Nuclear Chemistry*, **193**, 281 (1995).
216 Smith, L.L., Crain, Y.S., Yeager, J.S., Horwitz, E.P., Diamond, H. and Chiarizla, R. *Journal of Radioanalytical and Nuclear Chemistry*, **194**, 151 (1995).
217 Hollenbach, M., Grohs, J., Kraft, M. and Mamich, S., ASTM Special Technical Publication STP 11291. Applications of Inductively Coupled Plasma Mass Spectrometry to Radionucleide Determination (1995).
218 Lettner, H., Andrasi, A., Hubmer, A.K., Lavranich, E., Steger, F. and Zombori, P. *Nuclear Instrum Methods Phys Res Section A*, **369**, 547 (1996).
219 Meriwether, J.R., Burns, S.F., Thompson, R.H. and Beck, J.N. *Health Physics*, **69**, 406 (1995).
220 Klemm, W. and Bombach, G. *Fresenius Journal of Analytical Chemistry*, **353**, 12 (1995).
221 Torres, P., Ballesteros, E. and Luque de Castro, M.D. *Analytice Chimica Acta*, **308**, 371 (1995).
222 Ure, A.M., Davidson, C.M. and Thomas, R.P. *Technical Instrumentation Analytical Chemistry*, **17**, 505 (1995).
223 Magni, E., Guisto, T. and Frache, R. *Analytical Proceedings*, **32**, 267 (1995).
224 Schramel, P., Klose, B.J. and Hasse, S. *Fresenius Zeitschrift für Analytische Chemie*, **310**, 209 (1982).
225 Schramel, P., Lill, G. and Seif, R. *Fresenius Zeitschrift für Analytische Chemie*, **326**, 135 (1987).
226 Coles, B.J., Ramsey, M.H. and Thornton, I. *Chemical Geology*, **124**, 109 (1995).
227 Stachel, B., Elsholz, O. and Reincke, H. *Fresenius J Analytical Chemistry*, **353**, 21 (1995).
228 Sturgeon, R., Willie, S.N., Methven, B.A., Lam, J.W.H. and Matusiewicz, H.I. *Analytical Atomic Spectroscopy*, **10**, 981 (1995).
229 Krauss, P., Erbsloueh, L.B., Niedergasaess, R., Pepeinik, R. and Prange, A. *Fresenius J Analytical Chemistry*, **353**, 3 (1995).
230 Li, X., Coles, B.J., Ramsey, M.H. and Thornton, I. *Chemical Geology*, **124**, 109 (1995).
231 Shapiro, J.B., James, W.D. and Schweikert, E.A. *Journal of Radioanalytical and Nuclear Chemistry*, **192**, 275 (1995).

232 Gibson, J.A.E. and Willett, I.R. *Communications in Soil Science Plant Analysis,* **22,** 1303 (1991).
233 Garotti, F.V., Massaro, S. and Serrano, S.H.P. *Analusis,* **20,** 287 (1992).
234 Downard, A.J., Kipton, H., Powell, J. and Xu, S. *Analytica Chimica Acta,* **256,** 117 (1992).
235 Schmidt, S., Koerdel, W., Kloeppel, H. and Klein, W. *Journal of Chromatography,* **470,** 289 (1989).
236 Joshi, S.R. *Applied Radiation and Isotopes,* **40,** 691 (1989).
237 Livens, F.R. and Singleton, D.L. *Analyst* (London), **114,** 1097 (1989).
238 Chickhaikar, S., Sharma, K. and Patel, K.S. *Communications Soil Science Sc Plant Anal,* **26,** 625 (1995).
239 Asami, T, Kubota, M. and Salto, S. *Water Air and Soil Pollution,* **62,** 349 (1992).
240 Lasztity, A., Krushevska, A., Kotrebai, M., Barnes, R.M. and Amarasiriwardena, D. *Journal of Analytical Atomic Spectroscopy,* **10,** 505 (1995).
241 Hwang, J.D., Huxley, H.P., Diomiguardi, J.P. and Vaughn, W. *Journal of Applied Spectroscopy,* **44,** 491 (1990).
242 Rurikova, D. and Beno, A. *Chem Pap,* **46,** 23 (1992).
243 Masscheleyn, P.H., Delaune, R.D. and Patrick, W.H. *Environmental Science and Technology,* **25,** 1414 (1991).
244 McGeehan, S.L. and Naylor, D.V. *Journal of Environmental Quality,* **21,** 68 (1992).
245 Honore Hansen, G., Larsen, E.H., Pritzi, G. and Cornett, C. *Journal of Analytical Atomic Spectroscopy,* **7,** 629 (1992).
246 Hirsch, D. and Banin, A. *Journal of Analytical Quality,* **19,** 366 (1990).
247 Feng, Y. and Barratt, R.S. *Science of the Total Environment,* **143,** 157 (1992).
248 Kim, N,D. and Fergusson, J.E. *Science of the Total Environment,* **105,** 191 (1991).
249 Dubois, J.P. *Trace Microprobe Techniques,* **9,** 149 (1991).
250 Mazzucotelli, A., Soggia, F. and Cosma, B. *Applied Spectroscopy,* **4,** 504 (1991).
251 Millward, C.G. and Kluckner, P.D. *Journal of Analytical Atomic Spectroscopy,* **6,** 37 (1991).
252 Evans, D.W., Alberts, J.J. and Clark, R.A. *Geochimica and Cosmochimica Acta,* **47,** 1041 (1983).
253 Kim, C.S., Kim, S.J. and Park, S.W. *Journal of Environmental Science and Health,* **A31,** 2173 (1996).
254 Fawaris, B.H. and Johansen, K.J. *Science of the Total Environment,* **170,** 221 (1995).
255 Carbol, P., Skarnemark, G. and Skalberg, M. *Science of the Total Environment,* **131,** 129 (1993).
256 Williams, T.M. *Environmental Geology,* **21,** 62 (1993).
257 Von Gunten, H.R. and Benes, P. *Radiochimica Acta,* **69,** 1 (1995).
258 Essington, E.H., Fowler, E.B. and Polzer, W.L. *Soil Science,* **132,** 13 (1981).
259 Bunzl, K. and Schimmack, W. *Chemosphere,* **18,** 2109 (1989).
260 Wauters, J., Vidal, M., Elsen, A. and Cremers, A. *Applied Geochemistry,* **11,** 595 (1996).
261 Vidal, M., Roig, M., Rigol, A., Llarado, M., Rauret, G., Wauters, J., Elsen, A. and Cremers, A. *Analyst* (London), **120,** 1785 (1995).
262 Thiry, Y. and Myttenaere, C. *Journal of Environmental Radioactivity,* **18,** 247 (1993).
263 Scheuler, B., Sander, P. and Read, D.A. *Vacuum,* **41,** 1661 (1990).
264 Groenewold, G.S., Ingram, J.C., McLing, T. and Gianotto, A.K. *Analytical Chemistry,* **70,** 534 (1998).
265 Chakraborty, R., Das, A.K., Cervera, M.L. and De la Guardia, M. *Journal of Analytical Atomic Spectroscopy,* **10,** 353 (1995).
266 Milacic, R., Stupar, J., Kozuh, N. and Korosin, J. *Analyst* (London) **117,** 125 (1992).

267 Mesuere, K., Martin, R.E. and Fish, W. *Journal of Environmental Quality*, **20**, 114 (1991).
268 Gerringa, L.J.A., Van der Meer, J. and Cauwet, G. *Marine Chemistry*, **36**, 51 (1991).
269 Savvin, S.B., Petrova, T.V., Ozherayan, T.G. and Reikhshat, M.M. *Fresenius Journal of Analytical Chemistry*, **340**, 217 (1991).
270 Hinds, M.M. and Jackson, K. *Journal of Analytical Atomic Spectroscopy*, **5**, 199 (1989).
271 Zhang, B., Tao, K. and Feng, J. *Journal of Analytical Atomic Spectroscopy*, **7**, 171 (1992).
272 Hinds, M.W., Latimer, K.E. and Jackson, K.W. *Journal of Analytical Atomic Spectroscopy*, **6**, 473 (1991).
273 Hinds, M.W. and Jackson, K.W. *Atomic Spectroscopy*, **12**, 109 (1991).
274 Fernando, A.R. and Plambeck, J.A. *Analyst* (London), **117**, 39 (1992).
275 Wegrzynek, D. and Holynska, B. *Applied Radiation and Isotopes*, **44**, 1101 (1993).
276 Bermejo-Barrera, P., Barciel-Alonso, C., Aboal-Somaza, M. and Bermejo-Barrera, A. *Journal of Atomic Spectroscopy*, **9**, 469 (1994).
277 Kamburova, M. *Talanta*, **40**, 719 (1993).
278 Sakamoto, H., Tomlyasu, T. and Yonehara, N. *Analytical Science*, **8**, 35 (1992).
279 Azzaria, L.M. and Aftabi, A. *Water Air and Soil Pollution*, **56**, 203 (1991).
280 Lee, H.S., Jung, K.H. and Lee, D.S. *Talanta*, **36**, 999 (1989).
281 Bandyopadhyay, S. and Das, A.K. *Journal of Indian Chemical Society*, **66**, 427 (1989).
282 Golimowski, J., Orzechowska, A. and Tykarska, A. *Fresenius Journal of Analytical Chemistry*, **351**, 656 (1995).
283 Robinson, L., Dyer, F.F., Combs, D.W., Wade, W., Teasley, N.A. and Carlton, J.E. *Journal of Radioanalytical and Nuclear Chemistry*, **179**, 305 (1994).
284 Welz, B., Schlemmeer, G. and Mudakavi, J.R. *Journal of Analytical Atomic Spectroscopy*, **7**, 499 (1992).
285 Baxter, D.C., Nichol, R. and Littlejohn, D. *Spectrochimica Acta*, Part B, **47**, B 1155 (1992).
286 Lexa, J. and Stulik, K. *Talanta*, **36**, 843 (1989).
287 Cherian, L. and Gupta, U.K. *Fresenius Zeitschrift für Analytische Chemie*, **336**, 400 (1990).
288 Saouter, E., Campbell, P.G.C., Ribeyre, F. and Boudon, A. *International Journal of Environmental Analytical Chemistry*, **54**, 57 (1993).
289 Jiao, K., Jin, W. and Metzner, H. *Analytica Chimica Acta*, **260**, 35 (1992).
290 Rowbottom, W.H. *Journal of Analytical Atomic Spectroscopy*, **6**, 123 (1991).
291 Kim, C.K., Takaku, A., Yamamoto, M., Kawawamura, H. Shiraishi, K., Igarashi, Y., Igarashi, S., Takayama, H. and Ikeda, N. *Journal of Radioanalytical and Nuclear Chemistry*, **132**, 131 (1989).
292 Kim, C.K., Oura, Y., Takaku, N.H., Igarashi, V. and Ikeda, N. *Journal of Radioanalytical and Nuclear Chemistry*, **136**, 353 (1989).
293 Jia, G., Testa, C., Desideri, D. and Mell, M.A. *Analytica Chimica Acta*, **220**, 103 (1989).
294 Green, L.W., Miller, F.C., Sparling, J.A. and Joshi, S.R. *Journal of American Society of Mass Spectrometry*, **2**, 240 (1991).
295 Holgye, Z. *Journal of Radioanalytical and Nuclear Chemistry*, **149**, 275 (1991).
296 Zhu, H.M. and Tang, X.Z. *Journal of Radioanalytical and Nuclear Chemistry*, **130**, 443 (1989).
297 Nakanishi, T., Satoh, M., Takei, M., Ishikawa, A., Murata, M., Dairyah, M. and Higuchi, S. *Journal of Radioanalytical and Nuclear Chemistry*, **138**, 321 (1990).
298 El-Daoushy, F., Garcia-Tenorio, R. *Journal of Radioanalytical and Nuclear Chemistry*, **138**, 5 (1990).

299 Williams, L.R., Leggett, R.W., Espegren, M.L. and Little, C.A. *Environmental Monitoring Assessment*, **12**, 83 (1989).

300 Hafez, A.F., Moharram, B.M., El-Khatib, A.M. and Adel-Naby, A. *Isotopenproxis*, **27**, 185 (1991).

301 Zhang, Y., Moore, J.N. and Frankenberger, W.T. *Environmental Science and Technology*, **33**, 1652 (1999).

302 McCurdy, E.J., Lange, J.D. and Haygarth, P.M. *Science of the Total Environment*, **135**, 131 (1993).

303 Rojus, C.J., de Maroto, S.B. and Valenta, P. *Fresenius Journal of Analytical Chemistry*, **348**, 775 (1994).

304 Cruvinel, P.E. and Flocchini, R.G. *Nuclear Instrumental Methods of Physical Research*, Section B, **B75**, 415 (1993).

305 Akcay, M., Elik, A. and Savasci, S. *Analyst* (London), **114**, 1079 (1989).

306 Stella, R., Ganzerli, V.M.T. and Maggi, L. *Journal of Radioanalytical and Nuclear Chemistry*, **161**, 413 (1992).

307 Ryabukhin, V.A., Volynets, M.P., Myasoedov, B.F., Rodionova, I.M. and Tuzova, A.M. *Fresenius Journal of Analytical Chemistry*, **341**, 636 (1991).

308 Tagami, K. and Uchida, S., *Radiochimica Acta*, **63**, 69 (1993).

309 Morita, S., Kim, C.K., Takaku, Y., Seki, R. and Ikeda, N. *Applied Radiation and Isotopes*, **42**, 531 (1991).

310 Harvey, B.R., Williams, K.J., Lovett, M.B. and Ibbett, R.D. *Journal of Radioanalytical and Nuclear Chemistry*, **158**, 417 (1992).

311 De Ruck, A., Vandecastelle, G. and Dams, R. *Analytical Letters* (London), **22**, 469 (1989).

312 Opydo, J. *Mikrochimica Acta*, **2**, 15 (1989).

313 Chikhalikar, S., Sharma, K. and Patel, K.S. *Communications in Soil Science and Plant Analysis*, **26**, 621 (1995).

314 Lukaszewski, Z. and Zembrzuski, W. *Talanta*, **39**, 221 (1992).

315 Sagar, M. *Mikrochimica Acta*, **106**, 241 (1992).

316 Toole, J., McKay, K. and Baxter, M. *Analytica Chimica Acta*, **245**, 83 (1990).

317 Shaw, T.J. and Francois, R. *Geochimica and Cosmochimica Acta*, **55**, 2075 (1991).

318 Parsa, B. *Journal of Radioanalytical and Nuclear Chemistry*, **157**, 65 (1992).

319 Mukhtar, O.M., Grods, A. and Khangi, F.A. *Radiochimica Acta*, **54**, 201 (1991).

320 Mann, D.K., Oatis, T. and Wong, G.T.F. *Talanta*, **39**, 1199 (1992).

321 Lazo, E.N., Roessier, G.S. and Bervani, B.A. *Health Physics*, **6**, 231 (1991).

322 Shuktomova, I.I. and Kochan, I.G. *Journal of Radioanalytical and Nuclear Chemistry*, **129**, 245 (1989).

323 Lazo, E.N., Report, 1988. DOE (Department of Environment) /OR/0033–T 424, Order No DE89010612 Avail NTIS 350 pp (1988).

324 Li, Z., McIntosh, S., Carnride, G.R. and Slavin, W. *Spectrochimica Acta*, Part B, **47B**, 701 (1992).

325 Boomer, D.W. and Powell, M.J. *Analytical Chemistry*, **59**, 2810 (1987).

326 Noey, K.C., Liedle, S.D., Hickey, C.R. and Doane, R.W. Proceedings Symposium on Waste Management 615 (1989).

327 D'Silva, A.P., Bajie, S.J. and Zamzow, D. *Analytical Chemistry*, **65**, 3174 (1993).

328 Mukhtar, O.M., Grods, A. and Khangi, F.A. *Radiochimica Acta*, **54**, 201 (1991).

329 Tsukada, M., Yamamoto, D., Endo, R. and Nakahara, H. *Journal of Radioanalytical and Nuclear Chemistry*, **151**, 121 (1991).

330 Wisbrun, R., Schechter, I., Niessner, R., Schroeder, H. and Kompa, K.L. *Analytical Chemistry*, **66**, 2964 (1994).

331 Jerrow, M., Marr, I. and Cresser, M. *Analytical Proceedings* (London), **29**, 45 (1992).

332 Opydo, J. *Water Air and Soil Pollution*, **45**, 43 (1989).

333 Ure, A.M. *Fresenius Zeitschrift für Analytische Chemie*, **337**, 577 (1990).

334 Spevackova, U. and Kucera, J. *International Journal of Environmental Analytical Chemistry*, **35**, 241 (1989).
335 Sagar, M. *Technical Instruments Analytical Chemistry*, 12 (Hazard Met Environ 133) (1992).
336 Knizhnik, E.I., Prokopenco, U.S., Stolyarov, S.V. and Tokarevskii, V.V. *Atomic Energy*, **76**, 113 (1994).
337 Korun, M., Martincic, R. and Pucelj, B. *Nuclear Instrument Methods Phys Res, Section A*, **A300**, 611 (1991).
338 Wu, D. and Landsberger, S. *Journal of Radioanalytical and Nuclear Chemistry*, **179**, 155 (1994).
339 Del Castilho, P. and Rix, I. *International Journal of Environmental Analytical Chemistry*, **51**, 59 (1993).
340 Haynes, R.J. and Swift, R.G. *Geoderma*, **49**, 319 (1991).
341 Crosland, A.R., McGrath, S.P. and Lane, P.W. *International Journal of Environmental Analytical Chemistry*, **51**, 153 (1993).
342 Ivanova, E., Stoimenova, M. and Gentcheva, G. *Fresenius Journal of Analytical Chemistry*, **348**, 317 (1994).
343 Donaldson, E.M. *Talanta*, **26**, 543 (1989).
344 Li, M., Barban, R., Zucchi, B. and Martinotti, W. *Water Air and Pollution*, **57–58**, 495 (1991).
345 Hewitt, A.D. and Reynolds, C.M. *Atomic Spectroscopy*, **11**, 187 (1990).
346 Paudyn, A.M. and Smith, R.G. *Canadian Journal of Applied Spectroscopy*, **37**, 94 (1992).
347 Hewitt, A.D. and Reynolds, C.M. Report CRR EL–SP–90–19, CETHA–TS–CR–90052. Order No AD–A226 367 (Avail NTIS) (1990).
348 Kratochvil, B. and Mamba, S. *Canadian Journal of Chemistry*, **68**, 360 (1990).
349 Kammin, W.R. and Brandt, M. *Journal of Spectroscopy*, **4**, 49, 52 (1989).
350 Millward, C.G. and Kluckner, P.D. *Journal of Analytical Atomic Spectroscopy*, **4**, 709 (1989).
351 Lo, C.K. and Fung, Y.S. *International Journal of Environmental Analytical Chemistry*, **46**, 277 (1992).
352 Reynolds, A.R. In *Engineering Aspects of Metal – Waste Management*, Iskander, I.K. and Selin, H.M. eds, Lewis, Boca Raton, Florida pp 49–61 (1992).
353 Kingston, H.M. and Walter, P.J. *Spectroscopy*, **7**, 20 (1992).
354 Krishnamurti, G.S.R., Huang, P.M., Van Rees, K.C.J., Kozak, L.M. and Rostad, H.P.W. *Communications Soil Science and Plant Analysis*, **25**, 615 (1994).
355 Real, C., Barreiro, R. and Carballeira, A. *Science of the Total Environment*, **152**, 135 (1994).
356 Van den Akker, A.H. and Van den Heuvel, H. *Atomic Spectroscopy*, **13**, 72 (1992).
357 Sanchez, J., Garcia, R. and Millan, E. *Analusis*, **22**, 222 (1994).
358 Pickering, W.F. *CRC Critical Review Analytical Chemistry*, **12**, 233 (1981).
359 Salomons, W. and Forstner, Y. *Environmental Technology Letters*, **1**, 506 (1980).
360 Welte, B., Bles, N. and Montiel, A. *Environmental Technology Letters*, **4**, 79 (1983).
361 Forstner, U. *Fresenius Zeitschrift für Analytisch Chemie*, **17**, 604 (1983).
362 Lavkulich, L.M. and Wiens, J.H. *Soil Science of America Proceedings*, **34**, 755 (1970).
363 Chao, T.T. *Soil Science of America Proceedings*, **36**, 764 (1972).
364 Tessier, A., Campbell, P.G.C. and Bisson, M. *Analytical Chemistry*, **51**, 844 (1979).
365 Harrison, R.M., Laxen, D.P.H. and Wilson, S.J. *Journal of Environmental Science and Technology*, **15**, 1378 (1981).
366 Shuman, L.M. *Soil Science of America Journal*, **46**, 1099 (1982).
367 Chao, T.T. and Zhou, L. *Soil Science of America Journal*, **47**, 225 (1983).

368 Shuman, L.M. *Soil Science of America Journal*, **47**, 656 (1983).
369 Maher, W.A. *Bulletin of Environmental Contamination and Toxicology*, **32**, 339 (1984).
370 Miller, W.P., Martens, D.C. and Zelazny, L.W. *Soil Science of America Journal*, **50**, 598 (1986).
371 Singh, J.P., Karwasra, S.P.S. and Singh, M. *Soil Science*, **146**, 359 (1988).
372 Chao, T.T. and Sanzolone, R.F. *Soil Science of America Journal*, **53**, 385 (1989).
373 Sharpley, A.N. *Soil Science of America Journal*, **53**, 1023 (1989).
374 Griffin, T.M., Rabenhorst, M.C. and Fanning, D.S. *Soil Science of America Journal*, **53**, 1010 (1989).
375 Kodama, H. and Wang, C. *Soil Science of America Journal*, **53**, 526 (1989).
376 Guy, R.D., Chakrabarti, C.L. and Mebain, D.C. *Water Research*, **12**, 21 (1978).
377 Tipping, E., Hetherington, N.B., Hilton, J., Thompson, D.W., Bowles, E. and Hamilton-Taylor, J. *Analytical Chemistry*, **57**, 1944 (1985).
378 Tessier, A. and Campbell, P.G.C. *Analytical Chemistry*, **60**, 1475 (1988).
379 Lopez-Sanchez, R.J.F., Rubio, R. and Rauret, G. *International Journal of Environmental Analytical Chemistry*, **51**, 113 (1993).
380 Shan, V. and Chen, B. *Analytical Chemistry*, **65**, 802 (1993).
381 Desaules, A., Lischer, P., Dahinden, R. and Bachmann, H.J. *Communications in Soil Science and Plant Analysis*, **23**, 363 (1992).
382 Thompson, M., Pahlavanpaur, B., Walton, S.J. and Kirkbright, G.F. *Analyst* (London), **103**, 705 (1978).
383 Barei-Funel, G., Dalmasso, J., Ardisson, G. *Journal of Radioanalytical and Nuclear Chemistry*, **156**, 83 (1992).

Chapter 3

Determination of Metals in Non-Saline Sediments

3.1 Aluminium

3.1.1 Atomic absorption spectrometry

The determination of aluminium in non-saline sediments is discussed under multi-metal analysis in sections 3.67.2.3 and 3.67.2.4.

3.1.2 Inductively coupled plasma atomic emission spectrometry

The determination of aluminium is discussed under multi-metal analysis in section 3.67.3.1.

3.1.3 Neutron activation analysis

The determination of aluminium is discussed under multi-metal analysis in section 3.67.5.1.

3.1.4 Plasma emission spectrometry

The determination of aluminium is discussed under multi-metal analysis in section 3.67.6.1.

3.2 Americium

3.2.1 α spectrometry

See section 2.1.1.

3.3 Antimony

3.3.1 Spectrophotometric method

Abu-Hilal and Riley [1] have investigated a spectrophotometric method for the determination of antimony in sediments and clays.

In this procedure 0.5 g of the finely ground sample is weighed into a polytetrafluoroethylene beaker. This is moistened with a few drops of water and then cautiously 10 ml of 40% (w/v) hydrofluoric acid added. The beaker is covered with a PTFE lid and heated overnight on a boiling water bath. The solution is evaporated to dryness, 10 ml of redistilled nitric acid added and the solution evaporated to dryness. A further 5 ml of nitric acid is added and the solution evaporated to dryness on a hot plate at low temperature, taking care to avoid baking the residue. The latter is dissolved in 3 ml of 6 M hydrochloric acid, diluted with water and transferred quantitatively to a 1 L Erlenmeyer flask. The solution is diluted to 1 L. If necessary a preconcentration was carried out on this solution to lower the detection limits of the method. Preconcentration was achieved by a method involving coprecipitation of the antimony with hydrous zirconium oxide in which the digest is stirred with 150 mg zirconyl chloride and the pH adjusted to 5 with ammonia to coprecipitate antimony and hydrous zirconium oxide. The isolated precipitate is dissolved in 7M hydrochloric acid and 30% sulphuric acid. Antimony is then converted to the pentavalent state by successive treatment with titanium III chloride and sodium nitrite and excess nitrite destroyed by urea.

Antimony is then determined by a spectrophotometric method utilising crystal violet in which the extract is treated with this chromogenic agent and the coloured complex extracted with benzene. The benzene extract is evaluated spectrophotometrically at 610nm.

3.3.2 Inductively coupled plasma atomic emission spectrometry and inductively coupled plasma mass spectrometry

Arrowsmith [2] has discussed a laser ablation inductively coupled plasma atomic emission spectrometric method for the determination of down to 0.2 µg g^{-1} of antimony in sediments. No interference was exhibited in this method by 10 mg of acetate, arsenate, bromide, nitrate, cyanide, fluoride, iodide, nitrate, oxalate, phosphate, sulphate, thiocyanate, thiosulphate or tartrate or 0.5 mg of mercury II or 50 µg of thallium. Gold at concentrations above 0.5 µg g^{-1} unfortunately interfered.

In this method a clay sample spiked with antimony III gave recoveries in the range of 98.5 to 101.5% with an average of 100.5%. The application of these techniques to the determination of antimony is also discussed under multi-metal analysis in section 3.67.3.2 (inductively coupled plasma atomic emission spectrometry) and 3.67.4.1 (inductively coupled plasma mass spectrometry).

3.3.3 Neutron activation analysis

The determination of antimony is discussed under multi-metal analysis in section 3.67.5.1.

3.3.4 Gas chromatography

Cutter *et al.* [3] have described a selective hydride generation–gas chromatographic procedure using a photoionisation detector for the determination of down to a 3.3 p mole L $^{-1}$ of antimony III and antimony V (also arsenic III and arsenic V) in sediments.

The determination of antimony in sediments is also discussed under multi-metal analysis in section 3.67.11.1.

3.3.5 Miscellaneous

Brannon and Patrick [4] give details of studies on the distribution and mobility of antimony in sediments from several sites in rivers, waterways and coastal waters throughout the USA. Most of the naturally-occurring and added antimony in the sediments was associated with relatively immobile iron and aluminium compounds. In sediments containing added antimony, the concentrations of this metal in the interstitial water and of the exchangeable-phase antimony were high. In leaching experiments under anaerobic conditions, the greatest release of antimony occurred early for most of the sediments, suggesting that leaching of antimony from contaminated sediments was most likely to occur during the first few months of interaction between sediment and water. Under aerobic leaching conditions, the antimony moved into a less available sediment phase, thus reducing the possibility of further release. When the sediments were incubated anaerobically, there was evidence of evolution of volatile sulphur compounds.

3.4 Arsenic

3.4.1 Spectrophotometric methods

A direct spectrophotometric procedure has been described for the determination of parts per billion of hydrochloric acid releasable arsenic in river sediments [5]. In this method the arsenic in the sediment sample is reduced by stannous chloride and zinc to arsine which is then swept from the generator into silver diethyldithiocarbamate chromogenic reagent to be evaluated spectrophotometrically at 535nm. Only inorganic arsenic is included in this determination. Organically bound arsenic is not determined unless the sample is oxidised. No significant change in the recovery of 6 µg of arsenic(V) added to 5.0 g of sediment was observed when the samples were also spiked with either 200 µg (40 mg kg $^{-1}$) of chromium(VI), 300 µg (60 mg kg $^{-1}$) of copper(II), 300 µg (60 mg kg $^{-1}$) of nickel(II), 15 µg (3 mg kg $^{-1}$) of mercury(II) or 2 µg (0.4 mg kg $^{-1}$) of antimony(III). Interference by combinations of various metal ions was also studied. It appears that the recovery of arsenic generally is not affected up to a combined metal ion concentration of 300 µg per 5 g of soil

Table 3.1 Native arsenic and arsenic recovered from river sediment samples

| | Arsenic μg 15 g | | | | | |
	Native	Added	Total	Recovered	Recovery %	Standard Deviation %
Sediment (North Edisto River)	1.02	2.0	3.05	2.03	101.5	4.9
Sediment (South Edisto River)	1.27	2.0	3.18	1.91	95.5	4.2
Sediment (Caw Caw Swamp)	3.16	2.0	5.13	1.97	98.5	4.3

Source: Royal Society of Chemistry [5]

that had also been spiked with 6.0 μg of arsenic(V). Whereas higher concentrations of antimony(III) and mercury(II) enhance the apparent recovery of arsenic, the other metal ions at levels greater than the concentrations listed above decrease the recovery of arsenic. In Table 3.1 are shown the native arsenic levels found in three river sediments and the results of arsenic spiking experiments. Between 95.5 and 101.5% of the added arsenic was recovered with a standard deviation between 4.2 and 4.9%.

3.4.2 Atomic absorption spectrometry

The determination of arsenic is discussed under multi-metal analysis in section 3.67.2.6.

3.4.3 Inductively coupled plasma atomic emission spectrometry

Goulden et al. [6] have described a continuous flow semi-automated system for the determination of arsenic (and selenium) in river sediments. By use of a four-fold preconcentration step, detection limits of 0.02 μg L $^{-1}$ were achieved for arsenic. Sediments were brought into solution by fusion with sodium hydroxide in a zirconium crucible. Arsenic is determined using inductively coupled argon plasma excitation (ICAP) using a plasma power of 1400 W. The manifold used for the sodium borohydride automated hydride generation is shown in Fig. 3.1(a) while Fig. 3.1(b) shows details of the gas separation employed.

Sediment samples are pretreated in the following way. 2 g of solid sodium hydroxide and 0.2 g of the dried ground sediment were placed into a zirconium crucible. The crucible is placed in a cold furnace which is then heated to 350 °C for 2 h. The crucible is cooled and the contents are

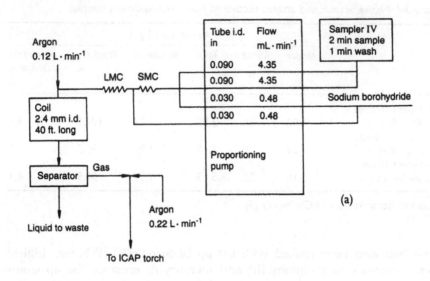

Tube i.d. in	Flow mL·min⁻¹
0.090	4.35
0.090	4.35
0.030	0.48
0.030	0.48

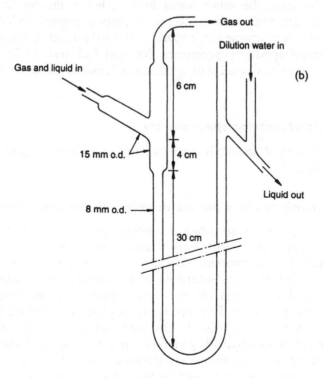

Fig. 3.1 Determination of arsenic in river sediments: (a) manifold; (b) gas separator
Source: American Chemical Society [6]

dissolved in 40 ml of 0.2N hydrochloric acid. Then 20 ml hydrochloric acid is added and the solution heated to about 90 °C for 1 h. The precipitated silicic acid is allowed to settle and the solution decanted into the sampler tubes. A calibration curve is obtained by using standard solutions of arsenic in 4N hydrochloric acid. The results obtained by this method in arsenic determinations on reference soils are shown below.

Sample	Nominal	Found
SO–1	1.9	1.9
SO–2	1.2	1.3
SO–3	2.6	2.6
SO–4	7.1	6.5
IAEA soils	93.5 ± 7,5	95.1
NBS (SRM1645)	66.0	66.4

Brzezinska-Paudyn et al. [7] compared results obtained in determinations of arsenic by conventional atomic emission spectrometry, flow injection/hydride generation inductively coupled plasma atomic emission spectrometry, graphite furnace atomic absorption spectrometry, combined furnace flame atomic absorption and neutron activation analysis. Results obtained show that all these methods can be used for the determination of down to 5 µg g^{-1} of arsenic in certified sediments. Brzezinska-Paudyn et al. [7] used an ARL 34000 inductively coupled plasma emission spectrometer with flow injection hydride generation. The 189.04nm line (3 nd order) was used for arsenic measurement. The flow injection block and Buckler peristaltic pump, as described by Liversage [8] were also used for the determination of arsenic by hydride generation. Detailed operating conditions for the inductively coupled plasma emission spectrometer have been described by Brzezinska-Paudyn et al. [9]. Details of the conventional inductively coupled plasma atomic emission spectrometer technique and the hydride generation inductively coupled plasma atomic emission spectrophotometric techniques used by Brzezinska-Paudyn et al. [7] are summarised below.

For the determination of arsenic by conventional inductively coupled plasma atomic emission spectrometry, the samples were digested in closed Teflon vessels, similar to the technique described by Brzezinska-Paudyn et al. [9].

About 0.1–0.2g of wet sediment was placed into a Teflon vessel and 3 mL of concentrated nitric acid, 0.5 mL concentrated perchloric acid and 4 mL concentrated hydrofluoric acid were added. The closed vessels were kept at room temperature for 1 h. The samples were then placed in a pressure cooker and heated for 1 h on a hot plate at a temperature of 300°C. After cooling, the vessels were uncapped and the samples evaporated to 2 mL on a hot plate at 250 °C. After cooling, 3 mL concentrated nitric acid was added. To complex the fluorides, 1 g boric acid was added to each sample. The solutions were transferred to 100 mL volumetric flasks and adjusted to volume with deionised water. Inorganic arsenic standards, having the same acid content as the samples, were used for calibration.

Inductively coupled plasma atomic emission spectrometric analysis with flow injection/hydride generation

Samples for inductively coupled plasma atomic emission spectrometric analysis with flow injection/hydride generation were digested as follows: powdered samples were dried at 80 °C for 2 h and cooled in a desiccator. About 0.5g of sample was accurately weighed into a 100 mL Teflon bleaker and 2 mL concentrated perchloric acid and 10 mL concentrated nitric acid were added. The beakers were covered with Pyrex watch glasses and heated on a hot plate at 120 °C for 2–3 h. The nitric acid was slowly evaporated until white fumes of perchloric acid were visible. After cooling, the digests were transferred to Teflon dishes. The beakers were then rinsed several times with 3–4 mL water each, which was also added to the digests. The final volume at this stage was approximately 25 mL. After addition of 5 mL concentrated hydrofluoric acid, the dishes were warmed to about 80 °C for 15 min to decompose and volatilise any siliceous residue. After cooling, the solutions were transferred to 50 mL plastic volumetric flasks, 7 mL concentrated hydrochloric acid and 1g sodium iodide were added, and the volume was adjusted with deionised water. The reductant was 0.6% sodium borohydride in alkaline solution. Inorganic arsenic standards containing the same amount of hydrochloric acid and sodium iodide as in the samples were used for calibration.

These methods were used to determine arsenic in certified sediments (Table 3.2). Conventional inductively coupled plasma atomic emission spectrometry is satisfactory for all types of samples, but its usefulness was limited to concentrations of arsenic greater than 5 µg g^{-1} dry weight. Better detection limits were achieved using the flow injection/hydride generation inductively coupled plasma technique in which a coefficient of variation of about 2% for concentrations of 10 µg g^{-1} were achieved.

Hydride generation inductively coupled plasma atomic emission spectrometry has been used to determine arsenic in non-saline sediments. This technique was found to greatly reduce sample preparation time [201]. The determination of arsenic by inductively coupled plasma atomic emission spectrometry is also discussed under multi-metal analysis in section 3.67.3.2.

Table 3.2 Comparison of As determination in sediments by different analytical methods (µg/g dry weight)

Method	NRC BESS–1 Marine sediment	NRC MESS–1 Marine sediment	NBS 1645 River sediment
Direct ICPAES	11.7 ± 2	8.6 ± 1.6	65 ± 18
ICPAES with flow injection/ hydride generation	11.9 ± 0.3	10.6 ± 0.2	65 ± 1.2
Graphite furnace	11.5 ± 0.8	10.7 ± 0.7	64 ± 8
Direct introduction of solids into furnace/flame system	12.0 ± 1.6	11.2 ± 1.2	68 ± 8
Neutron activation	12.2 ± 1.1	11.4 ± 1.2	68 ± 10
Certified values	11.1 ± 1.4	10.6 ± 1.2	66

Source: Brzezinska and Paudyn [7], Perkin–Elmer Beaconsfield [7]

3.4.4 Inductively coupled plasma mass spectrometry

Lasztity *et al.* [10] have reported an inductively coupled plasma mass spectrometric method for the determination of total arsenic in non-saline sediments. The application of this technique to the determination of arsenic is also discussed under multi-metal analysis in section 3.67.4.1.

3.4.5 Neutron activation analysis

The determination of arsenic is discussed under multi-metal analysis in section 3.67.5.1.

3.4.6 X-ray fluorescence spectroscopy

The determination of arsenic is discussed under multi-metal analysis in section 3.67.8.1.

3.4.7 X-ray spectrometry

The determination of arsenic is discussed under multi-metal analysis in section 3.67.9.1.

3.4.8 Gas chromatography

The determination of arsenic is discussed under multi-metal analysis in section 3.67.11.1.

3.4.9 Miscellaneous

Cheam and Chan [12] used certified Great Lakes performance sediments for the determination of arsenic. Brannon and Patrick [4, 11] reported on the transformation and fixation of arsenic(V) in anaerobic sediment, the long-term release of natural and added arsenic, and sediment properties which affected the mobilisation of arsenic(V), arsenic(III) and organic arsenic. Arsenic in sediments was determined by extraction with various solvents according to conventional methods. Added arsenic was associated with iron and aluminium compounds. Addition of arsenic(V) prior to anaerobic incubation resulted in accumulation of arsenic(III) and organic arsenic in the interstitial water and the exchangeable phases of the anaerobic sediments. Mobilisation of arsenic occurred over both short term and long term. Short term leaching of arsenic(III) was correlated with arsenic(III) concentrations in interstitial water and exchangeable phases, whereas long term release was related to total iron, extractable iron or calcium carbonate equivalent concentration. Arsenic leaching was most likely to be toxic during anaerobic conditions when arsenic(III) was released.

Cutter [13] has described a selective hydride generation technique as the basis for the differential determination of total arsenic in oxidatively digested river sediments. Elwaer and Belzile [70] have compared the use of a closed vessel microwave dissolution method and conventional hotplate digestion for the determination of arsenic in lake sediments. A mixture of hydrochloric acid, nitric acid and hydrofluoric acid with microwave digestion gave best recoveries. Hot plate digestion gave a poor recovery.

3.5 Barium

3.5.1 Inductively coupled plasma atomic emission spectrometry

The determination of barium is discussed under multi-metal analysis in section 3.67.3.1.

3.5.2 Neutron activation analysis

The determination of barium is discussed under multi-metal analysis in section 3.67.5.1.

3.6 Beryllium

3.6.1 Atomic absorption spectrometry

The determination of beryllium is discussed under multi-metal analysis in section 3.67.2.4.

3.7 Bismuth

3.7.1 Atomic absorption spectrometry

Ebdon *et al.* [14] and Zhe-Ming *et al.* [15] have discussed the application of this technique. Zhe-Ming *et al.* [15] determined bismuth in amounts down to 1 mg kg $^{-1}$ in river sediments by electrothermal atomic absorption spectrometry with low temperature atomisation of argon/hydrogen (90:10 L $^{-1}$). Absorption was maximal at 850–950 °C. Interference effects from the matrix were reduced, and the sensitivity increased, by using trisodium phosphate as matrix modifier. The relative standard deviation was 3.5% for replicate determinations of 2.4 mg bismuth per kg river sediment.

3.7.2 Inductively coupled plasma atomic emission spectrometry

The application of this technique is discussed under multi-metal analysis in section 3.67.3.2.

3.8 Cadmium

3.8.1 Flow injection analysis

The determination of cadmium is discussed under multi-metal analysis in section 3.67.1.1.

3.8.2 Atomic absorption spectrometry

Lum and Edgar [16] used a polarised Zeeman flame atomic absorption spectrometer to determine traces of cadmium (and silver, see later 3.49.1) in chemical extracts of river sediments. The detection limit was 0.1 mg kg $^{-1}$ based on a 0.2g sample size and a 25 ml final solution volume.

In this method 0.2g samples of the sediment were muffle ashed at 500°C for 3 h. A separate 0.2g subsample was dried in a forced air oven at 110 °C for 24 h to determine moisture content. The ashed samples were carefully wetted with doubly distilled water, transferred into PTFE beakers and 20 ml of freshly prepared aqua regia were added. The digests were reduced nearly to dryness, 15 ml of hydrofluoric acid were added and heating was continued until the samples were dry, care being taken to avoid baking. Hydrochloric acid (15 ml) and 15 ml of doubly distilled water were then added and the solutions were heated for 1 h to reduce the volume to about 15 ml. After cooling, the volumes were made up to 25 ml in a calibrated flask. Cadmium was evaluated in this solution at 228.8nm using an air acetylene flame. In Table 3.3 the results obtained in applying this procedure to reference river sediment samples are shown. The lower than expected cadmium recovery is in part due to its organic content

Table 3.3 Accuracy of determination of cadmium by Zeeman atomic absorption spectrometry

Sample NBS SRM 1645	Expected concentration µg L⁻¹	Measured concentration µg L⁻¹	Recovery %
River sediment National Institute for Environmental Studies (Japan)	185	148 ± 0.2	80
Reference pond sediment NBS SRM 1633 organic free	6.1	5.7 – 5.8	93 – 95
coal flyash	6.2	63 ± 0.8	102

Source: Royal Society of Chemistry [16]

Table 3.4 Sequential chemical extraction procedure carried out on 1 g dry weight of lake sediment

	(Total cadmium content 3.5–8.0 mg kg⁻¹)	
Extractant	In extract	% of total cadmium
0.75 mol L⁻¹ LiCl–0.25 mol L⁻¹ CsCl –60% MeOH (10 min @ 20 °C)	Readily exchangeable ions	17 ± 6
1 mol L⁻¹ CH₃COONa, pH 5.0 (5 h @ 20 °C)	Carbonate bound surface – oxide bound ions	31 ± 10
1 mol L⁻¹ NH₂OH–HCl – 25% CH₃COOH (2 h @ 20 °C)	Ions bound to Fe–Mn oxides	34 ± 7
H₂O₂, pH 2, (5 h @ 90 °C) 1.2 mol L⁻¹ CH₃COONH₄ – 20% HNO₃	Organically and sulphide bound ions	12 ± 5
Aqua Regia – HF – HCl – H₂O₂	Ions bound to the residual phase	6 ± 3

Source: Royal Society of Chemistry [16]

(1.7% from extractable oil and grease, 10.7% weight loss on ignition at 800°C). Higher recoveries were obtained with organic-free samples. Lum and Edgar [16] carried out a five-part sequential extraction procedure on 1g dry weight samples of sediment (Table 3.4). Such extractions simulate, to a certain extent, various environmental conditions to which sediments may be subjected. Although such schemes are not perfectly selective, they can provide valuable information on the mobility and availability of elements in sediments. Table 3.5 presents the results obtained in the determination of the various species of cadmium in a sediment taken from Lake Ontario.

Table 3.5 Distribution of chemical forms of cadmium suspended sediments taken from Lake Ontario

	%
Exchangeable forms	17 ± 6
Carbonate and surface oxide bound	31 ± 10
Bound to iron and manganese oxides	31 ± 7
Bound to organic matter	12 ± 5
Residual forms	6 ± 3
Total cadmium concentration	3.5 – 8.0 mg kg^{-1}

Source: Royal Society of Chemistry [16]

Table 3.6 Cadmium in river sediments by open beaker and Teflon-lined bomb digestions

Digestion conditions Vessel	Carbon content %	Teflon lined bomb	Teflon open beaker
Reagent		HF (10 ml), HNO$_3$ (4 ml)	HF (10 ml), HNO$_3$ (10 ml)
		HClO$_4$ (1 ml)	HClO$_4$ (5 ml)
Time (h)		1	12
Temperature °C		140	150
		mg kg^{-1}	mg kg^{-1}
NBS SRM 1645 river sediment 10.2 ± 1.5 mg kg^{-1} nominal cadmium		9.55 ± 0.22	9.30 ± 0.10
River Sediment 1	0.32	0.08 ± 0.01	0.08 ± 0.01
2	1.29	0.23 ± 0.01	0.21 ± 0.04
3	2.84	0.17 ± 0.01	0.18 ± 0.01
4	7.20	1.22 ± 0.12	1.22 ± 0.10

Source: Elsevier Science Ltd, Kidlington, UK [17]

Sakata and Shimoda [17] have described a simple and rapid method in which 0.5g sediment is digested for 1 h at 140 °C with a mixture of 10 ml hydrofluoric acid, 4 ml nitric acid and 1 ml perchloric acid in a Teflon-lined bomb prior to measurement by graphite furnace atomic absorption spectrometry at 228.8nm using an automatic sampling device. After digestion, 5g of boric acid is added to the solution to dissolve precipitated metal fluorides. Sodium and potassium interference is overcome by the addition of 1% ammonium sulphate matrix modifier. Table 3.6 shows cadmium contents obtained by this procedure compared to those obtained in a more lengthy open beaker digestion for a range of river

Table 3.7 Effect of matrix modifier on the determination of cadmium

Element	mg kg $^{-1}$ of sediment	Cadmium recovery %	
		Without matrix modifier	With matrix modifier
Silicon	1000	90	102
Aluminium	500	75	108
Iron	500	93	104
Sodium	200	55	95
Potassium	200	91	98
Calcium	200	81	92
Titanium	100	103	99
Manganese	100	88	98

Source: Elsevier Science Ltd, Kidlington, UK [17]

sediments containing increasing amounts of carbon, and for an NBS reference sample. The two methods have practically the same precision, there being no significant difference at the 95% confidence level in the results between the two methods in samples containing between 0.32–7.2% of carbon. A good recovery (94%) was obtained for the NBS reference sediment.

Table 3.7 shows the beneficial effect on cadmium recovery of incorporating an ammonium sulphate matrix modifier in the final solution to overcome interference effects. In the absence of the matrix modifier, the average recovery of cadmium from a river sediment was 44.5%, increasing to 100% when the modifier was used.

The application of atomic absorption spectrometry to the determination of cadmium in sediments is discussed under multi-metal analysis in sections 3.67.2.1, 3.67.2.4 and 3.67.2.5.

3.8.3 Inductively coupled plasma atomic emission spectrometry

The determination of cadmium is discussed under multi-metal analysis in section 3.67.3.1.

3.8.4 Inductively coupled plasma mass spectrometry

The determination of cadmium is discussed under multi-metal analysis in section 3.67.4.1.

3.8.5 X-ray fluorescence spectroscopy

The application of this technique is discussed under multi-metal analysis in Section 3.67.8.1.

3.8.6 X-ray spectrometry

The application of this technique is discussed under multi-metal analysis in section 3.67.9.1.

3.9 Calcium

3.9.1 Atomic absorption spectrometry

The application of this technique is discussed under multi-metal analysis in section 3.67.2.4.

3.9.2 Inductively coupled plasma atomic emission spectrometry

The application of this technique is discussed under multi-metal analysis in section 3.67.3.1.

3.9.3 Neutron activation analysis

The application of this technique is discussed under multi-metal analysis in section 3.67.5.1.

3.10 Caesium

3.10.1 Neutron activation analysis

The application of this technique is discussed under multi-metal analysis in section 3.67.5.1.

3.11 Cerium

3.11.1 Neutron activation analysis

The application of this technique is discussed under multi-metal analysis in section 3.67.5.1.

3.11.2 Inductively coupled plasma mass spectrometry

The determination of cerium is discussed under multi-metal analysis in section 3.67.4.2.

3.11.3 γ-ray spectrometry

The application of this technique is discussed under multi-metal analysis in section 3.67.10.1.

3.12 Chromium

3.12.1 Atomic absorption spectrometry

Pankow et al. [18] determined total chromium in river sediments. The samples were dried on a hot plate, approximately 1g of the dry material was weighed out and acid washed with 25 ml of 1 mol L $^{-1}$ nitric acid (to effect a removal of surface-bound chromium) and the acid wash filtered through (previously acid-washed) medium speed filter paper. The filtrate (acid wash) was collected, and analyzed by atomic absorption spectrometry using the method of standard adaition. The acid-washed sediment was then redried, weighed, and placed in a 50 ml capacity Teflon beaker. Then 15 ml concentrated hydrofluoric acid and 15 ml concentrated nitric acid were added to each sample, and then volatilised under the infrared lamps. This addition and subsequent volatilisation of acids was performed for a total of four times to ensure the complete destruction of the sediments. After the fourth cycle, the brown residue which remained was dissolved in enough 1 mol L $^{-1}$ nitric acid to total 25 ml and then filtered through acid-washed, medium speed filter paper. The average recovery achieved in the analysis of spiked sediment samples indicated an average recovery of 97% with a precision of ±4% relative. Chromium contents down to 0.5 mg kg $^{-1}$ could be determined by river sediments by this method.

Chakraborty et al. [20] determined chromium in non-saline sediments by microwave assisted sample digestion followed by atomic absorption spectrometry without the use of any chemical modifier. The determination of chromium in sediments by atomic absorption spectrometry is also discussed under multi-metal analysis in sections 3.67.2.1 and 3.67.2.4.

3.12.2 Inductively coupled plasma atomic emission spectrometry

Chromium has been determined by this technique as discussed under multi-metal analysis in section 3.67.3.1.

3.12.3 Neutron activation analysis

The determination of chromium is discussed under multi-metal analysis in section 3.67.5.1.

3.12.4 Plasma emission spectrometry

The determination of chromium is discussed under multi-metal analysis in section 3.67.6.1.

3.12.5 X-ray spectrometry

The determination of chromium is discussed under multi-metal analysis in section 3.67.9.1.

3.12.6 Miscellaneous

Scott [19] has discussed the cause and control of chromium losses during nitric acid–perchloric acid oxidation of river sediments. A three-step sequential extraction scheme has been proposed for extracting chromium from sediments. This scheme employs (1) acetic acid, (2) hydroxylamine hydrochloride and (3) ammonium acetate as extracting agents [21]. From the results obtained it is recommended to measure chromium content in steps 1 and 2 by electrothermal atomic absorption spectrometry and the chromium content of step 3 by flame atomic absorption spectrometry. Interfering effects when measuring chromium were circumvented by the use of 1% 8-hydroxyquinoline as a suppressor agent [22].

3.13 Cobalt

3.13.1 Atomic absorption spectrometry

The determination of cobalt is discussed under multi-metal analysis in sections 3.67.2.1 and 3.67.2.4.

3.13.2 Neutron activation analysis

The determination of cobalt is discussed under multi-metal analysis in section 3.67.5.1.

3.13.3 X-ray spectrometry

The determination of cobalt is discussed under multi-metal analysis in section 3.67.9.1.

3.14 Copper

3.14.1 Flow injection analysis

The application of this technique to the determination of copper is discussed under multi-metal analysis in section 3.67.1.1.

3.14.2 Atomic absorption spectrometry

The determination of copper is discussed under multi-metal analysis in sections 3.67.2.1, 3.67.2.3 and 3.67.2.4.

3.14.3 Plasma emission spectrometry

The determination of copper is discussed under multi-metal analysis in section 3.67.6.1.

3.14.4 Potentiometry

The determination of copper is discussed under multi-metal analysis in section 3.67.7.1.

3.14.5 X-ray spectrometry

The determination of copper is discussed under multi-metal analysis in section 3.67.9.1.

3.14.6 Miscellaneous

Kratchvil and Mamba [140] showed that all the zinc and copper were released from non-saline sediments within 7 min using a commercial microwave oven.

3.15 Dysprosium

3.15.1 Neutron activation analysis

The determination of dysprosium is discussed under multi-metal analysis in section 3.67.5.1.

3.15.2 Inductively coupled plasma mass spectrometry

The determination of dysprosium is discussed under multi-metal analysis in section 3.67.4.2.

3.15.3 γ-ray spectrometry

The determination of dysprosium is discussed under multi-metal analysis in section 3.67.10.1.

3.16 Europium

3.16.1 Neutron activation analysis

The determination of europium is discussed under multi-metal analysis in section 3.67.5.1.

3.16.2 Inductively coupled plasma mass spectrometry

The determination of europium is discussed under multi-metal analysis in section 3.67.4.2.

3.16.3 γ-ray spectrometry

The determination of europium is discussed under multi-metal analysis in section 3.67.10.1.

3.17 Gadolinium

3.17.1 Neutron activation analysis

The determination of gadolinium is discussed under multi-metal analysis in section 3.67.5.1.

3.17.2 Inductively coupled plasma mass spectrometry

The determination of gadolinium is discussed under multi-metal analysis in section 3.67.4.2.

3.17.3 γ-ray spectrometry

The determination of gadolinium is discussed under multi-metal analysis in section 3.67.10.1.

3.18 Gallium

3.18.1 Atomic absorption spectrometry

Xiao-quan *et al.* [23] used graphite furnace atomic absorption spectrometry with a nickel matrix modifier to determine μg kg^{-1} levels of gallium in perchloric acid digests of sediments.

3.19 Gold

3.19.1 Neutron activation analysis

The determination of gold is discussed under multi-metal analysis in section 3.67.5.1.

3.19.2 Miscellaneous

Xu and Schramel [165] have reviewed methods for the determination of gold in non-saline sediments.

3.20 Hafnium

3.20.1 Neutron activation analysis

The determination of hafnium is discussed under multi-metal analysis in section 3.67.5.1.

3.21 Indium

3.21.1 Inductively coupled plasma atomic emission spectrometry

The determination of indium is discussed under multi-metal analysis in section 3.67.3.3.

3.22 Iridium

3.22.1 Neutron activation analysis

The determination of iridium is discussed under multi-metal analysis in section 3.66.5.1.

3.23 Iron

3.23.1 Atomic absorption spectrometry

The determination of iron is discussed under multi-metal analysis in sections 3.67.2.1 to 3.67.2.4.

3.23.2 Inductively coupled plasma atomic emission spectrometry

The determination of iron is discussed under multi-metal analysis in section 3.67.3.1.

3.23.3 Neutron activation analysis

The determination of iron is discussed under multi-metal analysis in section 3.67.5.1.

3.23.4 Plasma emission spectrometry

The determination of iron is discussed under multi-metal analysis in section 3.67.6.1.

3.23.5 X-ray spectrometry

The determination of iron is discussed under multi-metal analysis in section 3.67.9.1.

3.24 Lanthanum

3.24.1 Neutron activation analysis

The determination of lanthanum is discussed under multi-metal analysis in section 3.67.5.1.

3.24.2 Inductively coupled plasma mass spectrometry

The determination of lanthanum is discussed under multi-metal analysis in section 3.67.4.2.

3.24.3 γ-ray spectrometry

The determination of lanthanum is discussed under multi-metal analysis in section 3.67.10.1.

3.25 Lead

3.25.1 Spectrophotometric method

Savvin et al. [166] have discussed a spectrophotometric method for the determination of lead in non-saline sediments.

3.25.2 Flow injection analysis

The determination of lead is discussed under multi-metal analysis in section 3.67.1.1.

3.25.3 Atomic absorption spectrometry

Various workers [24–27, 167–170] have studied the application of this technique to the determination of lead in non-saline sediments. Hinds et al. [167] investigated the application of low pressure electrothermal atomic absorption spectrometry to the determination of lead in non-saline sediments. Hinds et al. [169] also investigated the application of slurry electrothermal atomic absorption spectrometry to the determination of lead in non-saline sediments. Hinds and Jackson [170] also investigated the application of vortex mixing slurry graphite furnace atomic absorption spectrometry to the determination of lead in non-saline sediments. The determination of lead by atomic absorption spectrometry is also discussed under multi-metal analysis in sections 3.67.2.1, 3.67.2.4 and 3.67.2.5.

3.25.4 Inductively coupled plasma mass spectrometry

The determination of lead is discussed under multi-metal analysis in section 3.67.4.1.

3.25.5 Neutron activation analysis

The determination of lead is discussed under multi-metal analysis in section 3.67.5.1.

3.25.6 Anodic stripping voltammetry

Differential pulse anodic scanning voltammetry has been applied to the determination of lead in non-saline sediments [171].

3.25.7 X-ray fluorescence spectroscopy

Wegrzynek and Holynska [172] have developed a method for the determination of lead in arsenic containing sediments by energy dispersive X-ray fluorescence spectroscopy. Correction for arsenic interference is based on the use of an arsenic-free reference sample.

Koplitz et al. [173] have shown that in their X-ray fluorescence method for determining lead in sediments, by using known masses of a sediment matrix made from all of the samples, there is no need for standard additions to each separate sample.

3.25.8 Plasma emission spectrometry

The determination of lead is discussed under multi-metal analysis in section 3.67.6.1.

3.25.9 Potentiometry

The determination of lead is discussed under multi-metal analysis in section 3.67.7.1.

3.25.10 X-ray spectrometry

The determination of lead is discussed under multi-metal analysis in section 3.67.9.1.

3.25.11 Miscellaneous

Lead has been determined in sediments by using a slurry sampling technique with lead nitrate and magnesium nitrate as a chemical modifier. Results were in good agreement with known concentrations of a standard reference material [174].

3.26 Lithium

3.26.1 Atomic absorption spectrometry

The determination of lithium is discussed under multi-metal analysis in section 3.67.2.4.

3.27 Lutecium

3.27.1 Neutron activation analysis

The determination of lutecium is discussed under multi-metal analysis in section 3.67.5.1.

3.27.2 Inductively coupled plasma mass spectrometry

The determination of lutecium is discussed under multi-metal analysis in section 3.67.4.2.

3.27.3 γ-ray spectrometry

The determination of lutecium is discussed under multi-metal analysis in section 3.67.10.1.

3.28 Magnesium

3.28.1 Atomic absorption spectrometry

The determination of magnesium is discussed under multi-metal analysis in section 3.67.2.4.

3.28.2 Inductively coupled plasma atomic emission spectrometry

The determination of magnesium is discussed under multi-metal analysis in section 3.67.3.1.

3.28.3 Neutron activation analysis

The determination of magnesium is discussed under multi-metal analysis in section 3.66.5.1.

3.29 Manganese

3.29.1 Atomic absorption spectrometry

The determination of manganese is discussed under multi-metal analysis in sections 3.66.2.1, 3.66.2.3 and 3.66.2.4.

3.29.2 Inductively coupled plasma atomic emission spectrometry

The determination of manganese is discussed under multi-metal analysis in section 3.66.3.1.

3.29.3 Neutron activation analysis

The determination of manganese is discussed under multi-metal analysis in section 3.66.5.1.

3.29.4 Plasma emission spectrometry

The determination of manganese is discussed under multi-metal analysis in section 3.66.6.1.

3.29.5 X-ray spectrometry

The determination of manganese is discussed under multi-metal analysis in section 3.67.9.1.

3.30 Mercury

3.30.1 Sample digestion

The following procedure can be used to release bound mercury in solid sediment samples prior to analysis by suitable procedures such as atomic absorption spectrometry.

Pillay et al. [28] used a wet-ashing procedure with sulphuric acid and perchloric acid to digest samples. The released mercury was precipitated as the sulphide. The precipitate was then redigested using aqua regia.

Bretthaur et al. [29] described a method in which samples were ignited in a high pressure oxygen-filled bomb. After ignition, the mercury was absorbed in a nitric acid solution.

Feldman [30] digested solid samples with potassium dichromate, nitric acid, perchloric acid and sulphuric acid. Bishop et al. [31] used aqua regia and potassium permanganate for digestion. Jacobs and Keeney oxidised sediment samples using aqua regia, potassium permanganate and potassium persulphate [32]. The approved US Environmental Protection Agency digestion procedure requires aqua regia and potassium permanganate as oxidants [33].

These digestion procedures are slow and often hazardous because of the combination of strong oxidising agents and high temperatures. In some of the methods, mercuric sulphide is not adequately recovered. The oxidising reagents, especially the potassium permanganate, are commonly contaminated with mercury, which prevents accurate results at low concentrations.

Various workers have used nitric sulphuric acid [34–36] and then potassium persulphate oxidation and stannous chloride reduction [37] to digest sediments prior to the determination of mercury.

Jurka and Carter [37] claim a relative standard deviation of 6% at the 20–30 mg kg $^{-1}$ level for an automated cold vapour atomic absorption method while Agemian and Chau [36] claim 14% at the 0.1 mg kg $^{-1}$ level and 2% at the 2 mg kg $^{-1}$ level.

Horvat et al. [175] compared distillation with alkaline digestion methods for the determination of mercury in non-saline sediments by isothermal gas chromatography – cold vapour atomic fluorescence. The distillation approach produced results similar to those obtained by conventional digestion procedures, but with fewer matrix effects.

3.30.2 Spectrophotometric method

Kamburova [176] has reported a spectrophotometric method based on the formation of the mercury-triphenyltetrazolium chloride complex for the determination of mercury in non-saline sediments.

3.30.3 Atomic absorption spectrometry

Various workers have studied the application of this technique to the determination of mercury in sediments [37–41, 177–179]. Earlier work on the determination of total mercury in river sediments also includes that of Iskander et al. [35] and Craig and Morton [39]. Iskander et al. applied flameless atomic absorption to a sulphuric acid–nitric acid digest of the sample following reduction with potassium permanganate, potassium persulphate and stannous chloride. A detection limit of one part in 10^9 is claimed for this somewhat laborious method. Craig and Morton found a 2.2 µg g $^{-1}$ mean total mercury level in 136 samples of bottom deposits from the Mersey Estuary.

Jurka and Carter [37] have described an automated determination of down to 0.1 µg L $^{-1}$ mercury in river sediment samples. This method is based on the automated procedure of El-Awady et al. [40] for the determination of total mercury in waters and waste waters, in which potassium persulphate and sulphuric acid were used to digest samples for analysis by the cold vapour techniques. These workers proved that the use of potassium permanganate as an additional oxidising agent was unnecessary.

There was no significant interference due to sulphide in the solutions containing 10 mg sulphide L $^{-1}$. However, a negative interference was observed for both organic and inorganic standards containing 100 mg sulphide L $^{-1}$ which is equivalent to 25,000 mg sulphide kg $^{-1}$ in the sediment. This interference was overcome by ensuring that an excess of dichromate was present during the automated analysis.

This automated procedure was estimated to have a precision of 0.13–0.21 mg Hg kg $^{-1}$ at the 1 mg Hg kg $^{-1}$ level with standard decisions varying from 0.011 to 0.02 mg Hg kg $^{-1}$, ie relative standard deviations of 8.4–12% at the 17.2–32.3 mg Hg kg $^{-1}$ level in sediments. Recoveries in methyl mercuric chloride spiking studies were between 85 and 125%.

Abo-Rady [41] has described a method for the determination of total inorganic plus organic mercury in nanogram quantities in sediments. This method is based on the decomposition of organic and inorganic mercury compounds with acid permanganate, removal of excess permanganate with hydroxylamine hydrochloride, reduction to metallic mercury with tin and hydrochloric acid, and transfer of the liberated mercury in a stream of air to the spectrometer. Mercury was determined by using a closed recirculating air stream. Sensitivity and reproducibility of the 'closed system' were better, it is claimed, than those of the 'open system'. The coefficient of variation was 5.6% for sediment samples.

Bandyopadhyay and Das [177] extracted mercury from non-saline sediments with the liquid anion exchanger Aliquot-336 prior to determination by cold vapour atomic absorption spectrometry. A gold coated graphite furnace atomic absorption spectrometer has been used to determine mercury in non-saline sediments [218].

Azzaria and Aftabi [179] showed that stepwise compared to continuous heating of non-saline sediment samples before the determination by flameless atomic absorption spectrometry gives an increased resolution of the different phases of mercury.

3.30.4 Inductively coupled plasma atomic emission spectrometry

Smith [42] determined mercury in 1:1 hydrochloric acid:nitric acid digests of sediments in amounts down to 0.2ng L $^{-1}$ (using 200 ml sample) using isotope dilution inductively coupled plasma atomic emission spectrometry with a 201 Hg enriched spike.

Walker *et al.* [43] have recently compared inductively coupled plasma mass spectrometry with inductively coupled plasma atomic emission spectrometry (ICPAES) as methods for the determination of mercury in Great Barrier Reef sediments. Typical instrument variabilities were 1 in 10^9 for inductively coupled plasma atomic emission spectrometry and 1 in 10^{12} for inductively coupled plasma mass spectrometry.

3.30.5 Inductively coupled plasma mass spectrometry

Using individual isotope measurements, Hintelmann *et al.* measured mercury methylation rates in non-saline sediments using inductively coupled plasma mass spectrometry [49].

3.30.6 Neutron activation analysis

Pillay *et al.* [28] applied neutron activation analysis to the determination of 1.9–6.1 mg kg $^{-1}$ mercury in Lake Erie. The errors of this procedure were less than 15% at the 0.01 mg kg $^{-1}$ level and less than 5% at the 2 mg kg $^{-1}$ level. Standard deviations at these two levels were 17% and 7%. No losses of mercury occurred during the freeze drying of specimens. Losses of 10–20% did however occur during low temperature oven drying which must be avoided to ensure accurate results.

These workers studied problems associated with mercury losses throughout sampling and analysis. After neutron activation analysis, the samples were wet ashed with a mercury carrier. A preliminary precipitation is followed by further purification, and electrodeposition or precipitation were used to isolate mercury. The radioactivities ^{197}Hg and ^{197m}Hg were measured by scintillation spectrometry using a thin sodium iodide detector. Levels of mercury between 1.95 and 6.79 ppm dry weight were found in samples of sediment/silt taken in Lake Erie between 1971 and 1972. Robinson *et al.* [180] have used this technique to determine mercury in non-saline sediments. The determination of mercury in sediments is also discussed under multi-metal analysis in section 3.67.5.1.

3.30.7 Anodic stripping voltammetry

Mercury has been determined in acid-digested river sediment samples by differential pulse anodic stripping voltammetry [44]. Four types of working electrode (glassy carbon and gold rotating disk electrodes, and two types of gold film electrode, AuFe performed or in situ) were used and the analytical parameters of the procedures compared. The lowest limit of detection, 0.02 µg L $^{-1}$, was obtained with the gold rotating disc. This technique using 0.1 L $^{-1}$ perchloric acid containing a trace of hydrochloric acid as supporting electrolyte was one of two optimal procedures. The other involved determination with the gold film electrode prepared in situ in the sample extract. The latter method was more strongly affected by interferents such as iron and residual organic matter, but was quicker, and independent of film damage risk, new films being formed for each determination. Interference from iron was prevented by adding fluoride or pyrophosphate during sample pretreatment.

Tykarstra [45] has studied the application of voltammetric methods to the determination of mercury in sediments. During their study these workers found that the leachability of mercury from compost was very low.

3.30.8 X-ray spectrometry

The determination of mercury is discussed under multi-metal analysis in section 3.67.9.1.

3.30.9 High performance liquid chromatography

Hintelmann and Wilken [181] used high performance liquid chromatography to separate organomercury compounds. These were then converted to elemental mercury in a continuous flow system and detected using atomic fluorescence.

3.30.10 Gas chromatography

Empteborg et al. [182] have reported a method for determining mercury in non-saline sediments that employed supercritical fluid extraction and gas chromatography coupled to microwave induced plasma atomic emission spectroscopy. Butyl magnesium chloride was used to derivatise mercury to butyl methylmercury for gas chromatographic determination.

3.30.11 Miscellaneous

In lakes and streams, mercury can collect in the bottom sediments, where it may remain for long periods of time. It is difficult to release the mercury from these matrices for analysis. Several investigators have liberated mercury from soil and sediment samples by the application of heat to the samples and the collection of the released mercury on gold surfaces. The mercury was then released from the gold by application of heat or by absorption in a solution containing oxidising agents [46–48]. Mudrock and Kokitich [48] determined mercury in lake sediments from the St Clair Lake using a gold film mercury analyser. The mercury was extracted from the sediment by extraction with a mixture of nitric and hydrochloric acids (9:1 v/v). An accuracy of 0.02 mg kg $^{-1}$ was achieved.

Hintelmann et al. [49] have measured mercury methylation rates in sediments by individual isotope measurements using inductively coupled plasma mass spectrometry.

Hammer et al. [50] have pointed out that the Qu'Appelle river, Saskatoon is contaminated with mercury. Since the lake is eutrotropic, with reduced oxygen in the hypolimnion during stratification, a study was carried out on the effects of low oxygen concentrations on release of mercury from lake sediments and subsequent bio-accumulation by aquatic plants (Ceratophyllum demersum) and clams (Anodonta grandis). The mercury concentrations in plants and clams in experimental units with reduced oxygen concentrations (1.8 mg per litre) were significantly higher than those in oxygenated units (6.7–7.2 mg oxygen per litre) and in the controls. Possible reasons for this are discussed. It is suggested that inorganic mercury in sediments was more available for methylation under low oxygen conditions.

Lexa and Stulik [183] employed a gold film electrode modified by a film of tri-n-octylphosphine oxide to determine mercury in non-saline sediments. Concentrations of mercury as low as 0.02 mg L^{-1} were determined.

Wilken and Hintelmann [184] has reviewed methods for the determination of mercury species in non-saline sediments.

Cela *et al.* [185] attempted to correlate the behaviour of inorganic mercury and methylmercury in sediments close to waste water treatment systems and outfalls.

3.31 Molybdenum

3.31.1 Neutron activation analysis

The determination of molybdenum is discussed under multi-metal analysis in section 3.67.5.1.

3.32 Neodymium

3.32.1 Neutron activation analysis

The determination of neodymium is discussed under multi-metal analysis in section 3.67.5.1.

3.32.2 Inductively coupled plasma mass spectrometry

The determination of neodymium is discussed under multi-metal analysis in section 3.67.4.2.

3.32.3 γ-ray spectrometry

The determination of neodymium is discussed under multi-metal analysis in section 3.67.10.1.

3.33 Neptunium

3.33.1 Miscellaneous

Kim et al. [186] have demonstrated good agreement between methods for determining neptunium-237 in non-saline sediments, based on inductively coupled plasma mass spectrometry, neutron activation analysis and α-spectrometry.

3.34 Nickel

3.34.1 Atomic absorption spectrometry

The determination of nickel is discussed under multi-metal analysis in sections 3.67.2.1 and 3.67.2.4.

3.34.2 Neutron activation analysis

The determination of nickel is discussed under multi-metal analysis in section 3.67.5.1.

3.34.3 X-ray spectrometry

The determination of nickel is discussed under multi-metal analysis in section 3.67.9.1.

3.34.4 Plasma emission spectrometry

The determination of nickel is discussed under multi-metal analysis in section 3.67.5.1.

3.35 Osmium

3.35.1 Neutron activation analysis

The determination of osmium is discussed under multi-metal analysis in section 3.67.5.1.

3.36 Palladium

3.36.1 Neutron activation analysis

The determination of palladium is discussed under multi-metal analysis in section 3.67.5.1.

3.37 Platinum

3.37.1 Inductively coupled plasma atomic emission spectrometry

The determination of platinum is discussed under multi-metal analysis in section 3.67.3.3.

3.37.2 Neutron activation analysis

The determination of platinum is discussed under multi-metal analysis in section 3.67.5.1.

3.38 Plutonium

3.38.1 Inductively coupled plasma mass spectrometry

Kershaw *et al.* [51] applied inductively coupled plasma mass spectrometry to the determination of the isotopic composition of plutonium in non-saline sediments. These workers obtained good agreement of the measured isotope ratios by two mass spectrometric methods.

Kim *et al.* [187] determined the plutonium-240 to plutonium-239 ratio in non-saline sediments using the fission track method and inductively coupled plasma mass spectrometry.

3.38.2 Miscellaneous

Packed column chromatography has been used to determine various plutonium isotopes in non-saline sediments [188]. Various workers have discussed mass spectrometric and other methods for the determination of plutonium in non-saline sediments [189–191].

3.39 Polonium

3.39.1 Miscellaneous

Methods have been reviewed for the determination of polonium-210 in non-saline sediments [192, 193].

3.40 Potassium

3.40.1 Inductively coupled plasma atomic emission spectrometry

Potassium has been determined as discussed under multi-metal analysis in section 3.67.3.1.

3.40.2 Neutron activation analysis

Potassium has been determined as discussed under multi-metal analysis in section 3.67.5.1.

3.41 Radium

3.41.1 Thermal ionisation mass spectrometry

Cohen and O'Nions [194] have applied thermal ionisation mass spectrometry to the determination of 226 radium in sediments and rocks in amounts down to 4×10^{-5} pg. The chemical separation techniques employed involved initial digestion of the sample, followed by co-precipitation of radium onto strontium (as $SrRaSO_4$), spiking of the sample with 228 radium, digestion with hot concentrated sulphuric acid and then centrifuging. The washed precipitate is then counted. The abundance of 226 radium achieved by this method can be measured with 10^3 times the sensitivity achieved by conventional radioactive counting methods.

3.41.2 Miscellaneous

Various workers have reviewed the determination of radium 226 in non-saline sediments.

3.42 Rhenium

3.42.1 Inductively coupled plasma atomic emission spectrometry

Rhenium has been determined as discussed under multi-metal analysis in section 3.67.3.3.

3.43 Rubidium

3.43.1 X-ray spectrometry

Rubidium has been determined as discussed under multi-metal analysis in section 3.67.9.1.

3.44 Ruthenium

3.44.1 Neutron activation analysis

Ruthenium has been determined as discussed under multi-metal analysis in section 3.67.5.1.

3.45 Samerium

3.45.1 Neutron activation analysis

Samerium has been determined as discussed under multi-metal analysis in section 3.67.5.1.

3.45.2 Inductively coupled plasma mass spectrometry

Samerium has been determined as discussed under multi-metal analysis in section 3.67.4.2.

3.45.3 γ-ray spectrometry

Samerium has been determined as discussed under multi-metal analysis in section 3.67.10.1.

3.46 Scandium

3.46.1 Neutron activation analysis

Scandium has been determined as discussed under multi-metal analysis in section 3.67.5.1.

3.47 Selenium

The fate of selenium in natural environments such as soil and sediments is affected by a variety of physical, chemical and biological factors which are associated with changes in its oxidation state. Selenium can exist in four different oxidation states (–II, 0, IV and VI) and as a variety of organic compounds. The different chemical forms of selenium can control selenium solubility and availability to organisms. Selenate [(Se(VI)] is the most oxidised form of selenium, is highly soluble in water, and is generally considered to be the most toxic form. Selenite [Se(IV)] occurs in oxic to suboxic environments and is less available to organisms because of its affinity to sorption sites of sediment and soil constituents. Under anoxic conditions, elemental selenium and selenide(–II) are the thermodynamically stable forms. Elemental selenium is relatively insoluble, and selenide(–II) precipitates as metal selenides(–II) of very low solubility. Organic selenium (–II) compounds such as selesomethionine and selenocystine can accumulate in soil and sediments or mineralise to inorganic selenium. Therefore, Se(VI), Se(IV) and organic selenium(–II) are the most important soluble forms of selenium in natural environments.

3.47.1 Spectrofluorimetric methods

Wiersma and Lee [52] determined selenium in lake sediments. The sample is digested with 4:1 concentrated nitric acid:6% perchloric acid and the residue treated with 6M hydrochloric acid and then reduced with H_3PO_2. The fluorescence agent used was 2,3 diaminonaphthalene.

3.47.2 Atomic absorption spectrometry

Hydride generation atomic absorption spectrometry is widely used to determine the speciation of selenium in natural water, and soil–sediment extracts because of its low detection limits. Speciation of selenium is determined by subdividing sample solutions into selective treatments. Selenite is determined by directly analysing aliquots of samples without any treatments or by analysing samples acidified to pH 2 with concentrated hydrochloric acid or samples in 4–7N hydrochloric acid solutions. Selenate plus Se(IV) are determined after reduction of Se(VI) to Se(IV) in 4–7N hydrochloric acid at high temperatures (80–100 °C) and analysis for selenium to obtain Se(VI+IV) concentrations. Selenate is determined by the difference between a determination of Se(VI+IV) and a determination of Se(IV) in another subsample. Total selenium is determined by oxidising all selenium species ([organic Se(–II) and Se(IV)] to Se(VI) with hydrogen peroxide or persulphate, then reducing Se VI to Se IV with 4–7N hydrochloric acid at a high temperature (80–100 °C) and analysing for total selenium in the samples. Determination of organic

Se(–II) is calculated as the difference between the Se(VI+IV) and total selenium analyses. To separate organic Se(–II) from inorganic selenium, a technique was developed by passing an acidified sample (pH 1.6–2.2) through an XAD–8 resin column to remove hydrophobic and neutral organic Se(–II) compounds before selenium species analysis. These methods have provided valuable information about selenium speciation in natural water and soil–sediment extracts.

Some drawbacks for the speciation of selenium using hydride generation atomic absorption spectrometry have been found by some researchers. Thus Se (VI) is recovered poorly from many samples after a reduction with 6N hydrochloric acid at 100 °C. The addition of ammonium persulphate increased the recovery of Se(VI). However, part of the organic Se(–II) was included in the value reported for Se(VI) due to the oxidation of organic Se(–II) by persulphate. The reduction of Se(VI) to Se(IV) in soil extracts with 6N hydrochloric acid oxidised organic Se(–II) present in the sample resulting in an overestimation of Se(VI) concentration. XAD-resin has been used to separate hydrophobic and neutral organic Se(–II) compounds. However, hydrophillic organic Se(–II) compounds in solution, such as selenomethionine which are found in soil extracts will be detected as part of Se(VI). Also a considerable fraction of Se(IV) is removed due to a complexion of Se(IV) with humic substances when an acidified sample was passed through an XAD-8 column, thus resulting in overestimation of organic Se(–II) or Se(VI) concentrations in the samples. The studies have shown that determination of selenium speciation using hydride generation graphite furnace atomic absorption spectrometry may be possible only in solutions with little or no organic selenium(II) and humic substances, but this situation is rarely found in natural environments.

Zhang et al. [197] developed a new method to determine organic selenium(–II) in soils and sediments.

In this method persulphate is used to oxidise organic selenium(–II) and manganese oxide is used as an indicator for oxidation completion. This method was used to determine selenium speciation in soil-sediments and agricultural drainage water samples collected from the western United States. Results showed that organic selenium (–II) can be quantitatively oxidised to selenite without changing the selenate concentration in the soil–sediment extract and agricultural drainage water and then quantified by hydride generation atomic absorption spectrometry. Recoveries of spiked organic selenium(–II) and selenite were 96–105% in the soil–sediment extracts and 96–103% in the agricultural drainage water. Concentrations of soluble selenium in the soil–sediment extracts were 0.0534–2.45 μg g $^{-1}$ of which organic selenium accounted for 4.5–59.1%. Selenate is the dominant form of selenium in agricultural drainage water, accounting for about 90% of the total selenium. In contrast, organic selenium(–II) was an important form of selenium in the wetlands. These results showed that wetland sediments are more active in reducing selenate compared to evaporation pond sediments.

Itoh et al. [53] and Cutter [13] and also other workers [7, 162, 163, 164, 195, 196, 201] used hydrogen generation atomic absorption spectrometry techniques to determine selenium in non-saline sediments. The determination of selenium in sediments is also discussed under multi-metal analysis in section 3.67.2.6.

3.47.3 Inductively coupled plasma atomic emission spectrometry

This technique can determine selenium down to 0.03 µg g $^{-1}$ in the solution obtained following fusion of the sediment with solid sodium hydroxide in a zirconium crucible. A reference sample with a nominal selenium content of 0.4 mg kg $^{-1}$ gave a value of 0.49 mg kg $^{-1}$ by this method [6]. The determination of selenium by this technique is also discussed under multi-metal analysis in section 3.67.3.2.

3.47.4 Inductively coupled plasma mass spectrometry

Hydride generation inductively coupled plasma mass spectrometry has been used to determine selenium in sediments [198].

3.47.5 Neutron activation analysis

This technique has been used to determine selenium as discussed under multi-metal analysis in section 3.67.5.1.

Nadkarni and Morrison [54] estimated 47 elements in lake sediments and found 0.3–1.0 µg selenium per gram using neutron activation analysis.

3.47.6 Cathodic stripping voltammetry

Square wave cathodic stripping voltammetry has been used to determine selenium in sediments [199].

3.47.7 High performance liquid chromatography

Selenium(IV) reacts selectively with various thiols in accord with the following equation to form selenotrisulphides involving S–Se–S linkage which is generally unstable:

$$4RSH + H_2SeO_3 \rightarrow RSSeSR + RSSR + 3H_2O$$

Nakagawa et al. [200] have reported selenotrisulphide formed from penicillamine (Pen) was exceptionally stable and that the reduction proceeded in acid solution where most metal ions do not form chelates with penicillamine. These findings prompted Nakagawa et al. to apply the reaction to the selective determination of selenium (IV). Although UV absorption of penicillamine selenotrisulphide (PenSTS) enabled a parts-per-million level of selenium to be determined, the limit of detection could be lowered to a parts-per-billion level by conversion of PenSTS to a fluorophore. Thus, to develop a new method that allows assay of a low level of selenium with easy operation, Nakagawa et al. set out to find the optimum reaction conditions for the quantitative formation of PenSTS and subsequent derivatisation of PenSTS to a fluorophore using 7-fluoro-4-nitrobenz-2,1,3-oxadiazole (NBD–F), a labelling reagent for amino groups. The HPLC conditions for separation and detection of the fluorophore were also investigated. The applicability of this method is demonstrated by comparing the results for contents of selenium in standard samples with the certified values and those obtained by other established methods. The recovery of a known amount of selenium (IV) spiked to digested solution of standard sample was obtained to investigate the interferences by coexisting substances in the sample.

As a result of this work, a high performance liquid chromatography–fluorimetric method was developed for the selective determination of selenium(IV). The method involves precolumn reaction of selenium(IV) with penicillamine (Pen) to produce stable selenotrisulphide (Pen–SSeS–Pen) and subsequent derivatisation to a fluorophore by reaction with 7-fluoro-4-nitrobenz-2,1,3-oxadiazole. The fluorophore was separated by reversed-phase high performance liquid chromatography and selenium content was determined by fluorometric detection. The calibration plots showed a linear relationship in the range of 10–2000 ppb of selenium(IV) with a detection limit of 5 ppb (signal to noise ratio (S/N) >2). The method could determine total content of selenium in environmental samples after digestion of the samples and reduction of selenium(IV) to selenium(V). The results from standard samples indicated satisfactory agreement with those obtained by other established methods and certified values with good reproducibility. This method is as sensitive as, but simpler in operation than, conventional fluorometry using diaminonaphthalene.

3.47.8 Gas chromatography

de Oliveira [55] digested sediments with a mixture of nitric, perchloric acids and sulphuric acids prior to the determination of selenium by a procedure involving reaction of selenium with 4-nitro-o-phenylene-diamine to produce a volatile product which was determined in amounts down to 100 µg kg^{-1} by electron capture gas chromatography [56].

3.47.9 Miscellaneous

Cheam and Chan [12] have developed a Great Lakes reference sediment for selenium.

Cutter [13] was able to distinguish between selenite, selenate, total selenium and organic selenium in sediments. Selenium has been directly determined in sediments by PIXE [201]. Haygarth *et al.* [202] have compared the use of fluorometry, hydride generation atomic absorption spectrometry, hydride generation atomic emission spectrometry, hydride generation inductively coupled plasma mass spectrometry and radiochemical neutron activation analysis for the determination of selenium in sediments. For low concentration samples, hydride generation inductively coupled plasma mass spectrometry performed best.

Elwaer and Belzile [70] have compared the use of a closed vessel microwave assisted dissolution method and conventional hotplate digestion for the determination of selenium in lake sediments. A mixture of hydrochloric acid, nitric acid and hydrofluoric acid with microwave digestion resulted in the best recoveries of selenium. Poor recoveries were obtained with hotplate digestion.

3.48 Silicon

3.48.1 Inductively coupled plasma atomic emission spectrometry

This technique has been applied to the determination of silicon as discussed under multi-metal analysis in section 3.67.3.1.

3.49 Silver

3.49.1 Atomic absorption spectrometry

The polarised Zeeman flame atomic absorption method described under cadmium in section 3.8.2 has also been applied to the determination of silver in river sediments. The detection limit was 0.1 mg kg $^{-1}$ based on a 0.2g sample size and a 25 ml final solution volume. Silver was evaluated using the 328.1 nm line and air–acetylene flame. A certified reference sediment which had been shown by isotope dilution mass spectrometry to contain 6.4 µg L $^{-1}$ silver in the digest was shown by this method to contain 5.6 µg L $^{-1}$ silver, ie 88% recovery.

Lum and Edgar [16] carried out a five part sequential extraction procedure on 1g dry weight Moira Lake sediment (Table 3.8), showing the distribution of different forms of silver in the sediment. Regardless of core depth, most of the silver is organically or sulphide bound or bound to the residual phase. The distribution of silver found in sediment cores by these workers [16] is shown in Table 3.9.

The determination of silver by atomic absorption spectrometry is also discussed under multi-metal analysis in section 3.67.2.2.

Table 3.8 Sequential chemical extraction procedure carried out on 1 g dry weight of lake sediment

Extractant	In extract	% of total silver
(a) Sediment depth	0–1 cm	(total silver content 5.53 mg kg^{-1})
0.75 mol L^{-1} LiCl – 0.25 mol L^{-1} CsCl –60% MeOH (10 min @ 20 °C)	Readily extractable	9.4
1 mol L^{-1} CH$_3$COONa, pH 5.0 (5 h @ 20 °C)	Carbonate bound surface – oxide bound ions	0.5
1 mol L^{-1} NH$_2$OH HCl – 25% (2 h @ 20 °C)	Ions bound to Fe–Mn oxides	0.5
H$_2$O$_2$, pH 2 (5 h @ 90 °C)	Organically and sulphide bound ions	41.8
Aqua Regia – HF – HCl – H$_2$O$_2$	Ions bound to the residual phase	49
(b) Sediment depth	32–33 cm	(total silver content) 1.00 mg kg^{-1})
0.75 mol L^{-1} LiCl – 0.25 mol L^{-1} CsCl –60% MeOH (10 min @ 20 °C)	Readily extractable ions	
1 mol L^{-1} CH$_3$COONa, pH 5.0 (5 h @ 20 °C)	Carbonate bound surface – oxide bound ions	3
1 mol L^{-1} NH$_2$OHHCl – 25% CH$_3$COOH (2 h @ 20 °C)	Ions bound to Fe–Mn oxides	3
H$_2$O$_2$, pH 2 (5 h @ 90 °C)	Organically and sulphide bound ions	16
Aqua Regia – HF – HCl – H$_2$O$_2$	Ions bound to the residual phase	84

Source: Royal Society of Chemistry [16]

Table 3.9 Distribution of the chemical forms of silver in a sediment core (from [17]) (Moira Lake, Ontario)

Sediment depth cm	Total concen- tration µg g^{-1}	Exchangeable forms µg g^{-1}	Carbonate surface oxide bound µg g^{-1}	Fe, Mn oxide bound µg g^{-1}	Organic sulphide bound µg g^{-1}	Residual forms µg g^{-1}
0–1	5.53	0.52	0.03	0.03	2.31	2.70
9–10	8.05	0.63	0.03	0.03	2.02	5.40
14–15	6.60	0.90	0.03	0.03	1.30	4.40
18–19	3.66	0.12	0.03	0.03	0.24	3.30
21–22	4.30	0.10	0.03	0.03	0.36	3.84
32–33	1.00	0.03	0.03	0.03	0.16	0.84

Source: Royal Society of Chemistry [16]

3.49.2 Neutron activation analysis

This technique has been applied to the determination of silver as discussed under multi-metal analysis in section 3.67.5.1.

3.50 Sodium

3.50.1 Inductively coupled plasma atomic emission spectrometry

The determination of sodium is discussed under multi-metal analysis in section 3.67.3.1.

3.50.2 Neutron activation analysis

The determination of sodium is discussed under multi-metal analysis in section 3.67.5.1.

3.51 Strontium

3.51.1 Inductively coupled plasma atomic emission spectrometry

The determination of strontium is discussed under multi-metal analysis in section 3.67.3.1.

3.51.2 Neutron activation analysis

The determination of strontium is discussed under multi-metal analysis in section 3.67.5.1.

3.51.3 X-ray spectrometry

The determination of strontium is discussed under multi-metal analysis in section 3.67.9.1.

3.51.4 Miscellaneous

Stella et al. [203] and Ryabukhin et al. [204] have reviewed methods for the determination of radiostrontium in non-saline sediments.

3.52 Tamerium

3.52.1 Neutron activation analysis

The determination of tamerium is discussed under multi-metal analysis in section 3.67.5.1.

3.52.2 Inductively coupled plasma mass spectrometry

The determination of tamerium is discussed under multi-metal analysis in section 3.67.4.2.

3.52.3 γ-ray spectrometry

The determination of tamerium is discussed under multi-metal analysis in section 3.67.10.1.

3.53 Tantalum

3.53.1 Neutron activation analysis

The determination of tantalum is discussed under multi-metal analysis in section 3.67.5.1.

3.54 Technetium

3.54.1 Miscellaneous

Morita *et al.* [205] and Harvey *et al.* [206] have discussed the determination of technetium in non-saline sediments.

3.55 Terbium

3.55.1 Neutron activation analysis

The determination of terbium is discussed under multi-metal analysis in section 3.67.5.1.

3.55.2 Inductively coupled plasma mass spectrometry

The determination of terbium is discussed under multi-metal analysis in section 3.67.4.2.

3.55.3 γ-ray spectrometry

The determination of terbium is discussed under multi-metal analysis in section 3.67.10.1.

3.56 Thallium

3.56.1 Atomic absorption spectrometry

The determination of thallium is discussed under multi-metal analysis in section 3.67.2.5.

3.56.2 Inductively coupled plasma mass spectrometry

The determination of thallium is discussed under multi-metal analysis in section 3.67.4.1.

3.56.3 Miscellaneous

Lukaszewski and Zembrzuski [207] and Sagar [208] have reviewed methods for the determination of thallium in non-saline sediments.

3.57 Thorium

3.57.1 Spectrofluorimetric method

Mukhtar *et al.* [209] have described a laser fluorometric method for the determination of thorium (and uranium) in non-saline sediments.

3.57.2 Inductively coupled plasma mass spectrometry

Toole *et al.* [210] and Shaw and Francois [211] determined thorium (and uranium) in non-saline sediments by inductively coupled plasma mass spectrometry.

3.57.3 Neutron activation analysis

The determination of thorium is discussed under multi-metal analysis in section 3.67.5.1.

3.57.4 γ-ray spectrometry

The determination of thorium is discussed under multi-metal analysis in section 3.67.10.1.

3.57.5 Miscellaneous

Various workers [212–214] have reviewed methods for the determination of thorium in non-saline sediments. To determine thorium (and uranium) in non-saline sediments X-rays were measured by the use of a germanium plasma detector and chemometric techniques. No sample preparation was required in this method [215].

Parsa [216] has described a sequential radiochemical method for the determination of thorium (and uranium) in non-saline sediments.

3.58 Tin

3.58.1 Atomic absorption spectrometry

Dogan and Haerdi [57] applied their flameless atomic absorption method to the determination of down to 0.5 µg kg $^{-1}$ tin in humus rich lake sediments. Sample digestion was carried out using lumaton, a quaternary ammonium hydroxide, dissolved in isopropanol (available from H Kurner D-6451 Neuberg, Germany).

Long-Zhu [58] has described a graphite furnace atomic absorption spectrometric method for the determination of down to 2.5 mg kg $^{-1}$ of tin in river sediments.

Samples are decomposed in a Teflon-lined pressure vessel using perchloric, nitric and hydrofluoric acids. A mixture of ascorbic acid and iron is used as a matrix modifier. In the presence of the matrix modifier, the char temperature for tin can be raised to 1100 °C and the interferences caused by perchloric acid and sample matrices are greatly reduced. A good recovery is obtained. Between 90 and 104% of the tin is recovered.

In this procedure, 0.1g of sediment is transferred to a Teflon beaker and digested with 0,5 ml 0.01 mol L $^{-1}$ nitric acid to moisten the sample then 1 ml concentrated nitric acid. After leaving for 12 h, 1 ml 72% perchloric acid and 3 ml concentrated hydrofluoric acid are added and the container transferred to a steel bomb and heated for 6 h at 190 °C. The contents of the beaker are then heated to near dryness at 140 °C and then boiled with sufficient 0.2% oxalic acid solution. Successively 20 µL of sample solution prepared as above and 10 µL of 400 mg L $^{-1}$ iron solution are injected, it is then dried at 110 °C for 40 s, the residue charred at 1000 °C for 30 s, then atomised at 2400 °C for 5 s at 'maximum power' and 'argon flow interrupted'. Tin absorbance is measured at the resonance line of 286.3nm. Finally, the tube is cleaned at 270 °C for 3 s.

A reference sediment NBS SRM 1645 (nominal 313 ± 9 mg kg $^{-1}$ tin) gave a value of 363 mg kg $^{-1}$ by this method.

Legret and Divet [59] have described a method for the determination of tin in sedimentary hydride generation atomic absorption spectrometry. Hydride generation was carried out in a nitric acid–tartaric acid solution of the sediment. The effect of various acids and the matrix effects and interferences from other trace elements were studied. Several procedures for decomposing the samples were compared; the preferred procedure involved reflux with a mixture of nitric acid and hydrochloric acid.

3.58.2 Inductively coupled plasma atomic emission spectrometry

Brzenzinska-Paudyn and Van Loon [60] used inductively coupled plasma atomic emission spectrometry–mass spectrometry to determine tin in digested river sediments and compared results with those obtained by graphite furnace atomic absorption spectrometry with a palladium/

hydroxylamine matrix modifier. The inductively coupled plasma technique was more sensitive, achieving a detection limit of less than 1 pg of tin in the sample aliquot analysed.

3.59 Titanium

3.59.1 Inductively coupled plasma atomic emission spectrometry

This technique has been applied to the determination of titanium as discussed under multi-metal analysis in section 3.67.3.1.

3.59.2 Neutron activation analysis

This procedure has been applied to the determination of titanium as discussed under multi-metal analysis in section 3.67.5.1.

3.60 Tungsten

3.60.1 Spectrophotometric method

Quin and Brookes [61] determined tungsten in sediments by fusing the sample with potassium hydrogen sulphate and leaching the melt in 10M hydrochloric acid. A clear portion of the acid extract is heated with a solution of stannous chloride in 10M hydrochloric acid. A solution of dithiol in isoamyl acetate is added and the mixture is heated under precisely defined conditions so that a globule containing the tungsten dithiol complex is formed in >6 h. The globule is dissolved in light petroleum (boiling range 80° to 100°) and the extinction of the solution is measured at 630nm. Down to 1 ppm of tungsten can be determined and Beer's law is obeyed for up to 300 ppm. Modifications to conditions can improve the sensitivity to 0.2 ppm. Low concentrations of molybdenum do not interfere.

3.60.2 Neutron activation analysis

This technique has been applied to the determination of tungsten as discussed under multi-metal analysis in section 3.67.5.1.

3.61 Uranium

3.61.1 Spectrofluorimetric method

Mukhtar et al. [209] have described a laser fluorometric method for the determination of uranium (and thorium) in non-saline sediments.

3.61.2 Inductively coupled plasma mass spectrometry

Toole *et al.* [210] and Shaw and Francois [211] determined uranium (and thorium) in non-saline sediments by inductively coupled plasma mass spectrometry.

3.61.3 Neutron activation analysis

Uranium has been determined as discussed under multi-metal analysis in section 3.67.5.1.

3.61.4 X-ray fluorescence spectroscopy

To determine uranium (and thorium) in non-saline sediments, fluorescent X-rays were measured by the use of a germanium plasma detector and chemometric techniques [215]. No sample preparation was required in this method.

3.61.5 γ-ray spectrometry

Uranium has been determined as discussed under multi-metal analysis in section 3.67.10.1.

3.61.6 Miscellaneous

Methods have been reviewed for the determination of uranium (and thorium) in non-saline sediments [212, 217, 232].

3.62 Vanadium

3.62.1 Spectrophotometric method

Miura [218] has described a method for the determination of vanadium using 2-(8-quinolylazo)-5-(dimethylamino) phenol by reversed phase liquid chromatography spectrophotometry. This method was applied to the determination of vanadium in coal fly ash but may well have application to the analysis of sediments. In this reversed-phase high-performance liquid chromatography method for neutral and cationic metal chelates with azo dyes, tetraalkylammonium salts are added to an aqueous organic mobile phase. The tetraalkylammonium salts are dynamically coated on the reversed stationary support. As a result of the addition of tetraalkylammonium salts, the retention of the chelates is remarkably reduced. Tetrabutylammonium bromide permits rapid separation and sensitive spectrophotometric detection of the vanadium (V) chelate with 2-(8-quinolylazo)-5-(dimethylamino) phenol making it possible to determine trace vanadium(V). When a 100mm³ aqueous sample was injected, sensitivity and precision were as follows: peak height calibration curves of vanadium(V) were linear up to 800 pg at 0.005 absorbance unit full scale (AUFS) and up to 160 pg at 0.001 AUFS; the relative standard deviation for 10 determinations at 0.005 AUFS was 2.3% at a level of 320 pg of vanadium(V); the detection limit was 2.6 pg at 0.001 AUFS. Many cations including iron(III) and aluminium(III) do not interfere with the determination. Vanadium in coal fly ash can be successfully determined without pre-separation and pre-concentration of vanadium.

3.62.2 Inductively coupled plasma atomic emission spectrometry

Vanadium has been determined as discussed under multi-metal analysis in section 3.67.3.1.

3.62.3 Neutron activation analysis

Vanadium has been determined as discussed under multi-metal analysis in section 3.67.5.1.

3.62.4 Differential pulse polarography

Hasebe *et al.* [62] determined traces of vanadium in digests of pond sediments using differential pulse polarography of the vanadium IV-pyrocatecheol complex.

3.63 Ytterbium

3.63.1 Neutron activation analysis

Ytterbium has been determined as discussed under multi-metal analysis in section 3.67.5.1.

3.63.2 Inductively coupled plasma mass spectrometry

Ytterbium has been determined as discussed under multi-metal analysis in section 3.67.4.2.

3.63.3 γ-ray spectrometry

Ytterbium has been determined as discussed under multi-metal analysis in section 3.67.10.1.

3.64 Zinc

3.64.1 Atomic absorption spectrometry

Zinc has been determined as discussed under multi-metal analysis in sections 3.67.2.1, 3.67.2.3 and 3.67.2.4.

3.64.2 Inductively coupled plasma atomic emission spectrometry

Zinc has been determined as discussed under multi-metal analysis in section 3.67.3.1.

3.64.3 Inductively coupled plasma mass spectrometry

This technique has been applied to the determination of zinc as discussed under multi-metal analysis in section 3.67.4.1.

3.64.4 Neutron activation analysis

Zinc has been determined as discussed under multi-metal analysis in section 3.67.5.1.

3.64.5 Plasma emission spectrometry

Zinc has been determined as discussed under multi-metal analysis in section 3.67.6.1.

3.64.6 X-ray spectrometry

Zinc has been determined as discussed under multi-metal analysis in section 3.67.9.1.

3.64.7 Miscellaneous

Kratchvil and Mamba [140] showed all the zinc and copper were released from non-saline sediments within seven minutes using a commercial microwave oven.

3.65 Zirconium

3.65.1 Neutron activation analysis

Zirconium has been determined as discussed under multi-metal analysis in section 3.67.5.1.

Table 3.10 Relevances to rare earths

Element	Neutron activation analysis	γ-ray spectrometry	Inductively coupled plasma mass spectrometry
Cerium	3.10.1	3.11.2	3.67.4.2
	3.67.5.1	3.67.10.1	
Dysoprosium	3.1.5.1	3.1.5.2	3.67.4.2
	3.67.5.1		
Europium	3.16.1	3.16.2	3.67.4.2
	3.67.5.1	3.67.10.1	
Gadolinium	3.17.1	3.17.2	3.67.4.2
	3.67.5.1	3.67.10.1	
Lanthanum	3.24.1	3.24.2	3.67.4.2
	3.67.5.1	3.,67.10.1	
Lutecium	3.27.1	3.27.2	
	3.67.5.1	3.67.10.1	3.67.4.2
Neodynium	3.32.1	3.32.2	
	3.67.5.1	3.67.10.1	
Samerium	3.45.1	3.45.2	3.67.4.2
	3.67.5.1	3.67.10.1	
Tamerium	3.52.1	3.52.2	3.67.4.2
	3.67.5.1	3.67.10.1	
Terbium	3.55.1	3.55.2	3.67.4.2
	3.67.5.1	3.67.10.1	
Ytterbium	3.63.1	3.63.2	3.67.4.2
	3.67.5.1	3.67.10.1	

Source: Own files

3.66 Rare earths

Three techniques have been employed for the determination of rare earths in non-saline sediments, namely neutron activation analysis, γ-ray spectrometry and inductively coupled plasma mass spectrometry. The location of these techniques in the text is shown in Table 3.10. The four rare earths praseodymium, thulium (promethium), holmium and erbium are not specifically mentioned in these references.

3.67 Multi-metal analysis

3.67.1 Flow injection analysis

3.67.1.1 Heavy metals (cadmium, copper and lead)

Ma *et al.* [156] used a flow injection on-line sorbent extraction system for the determination of cadmium, copper and lead in river sediments. Detection limits were 10 times greater for lead compared to cadmium and copper.

Table 3.11 Cadmium in river sediments by open beaker and Teflon-lined bomb digestions

Digestion conditions Vessel	Carbon content %	Teflon lined bomb	Teflon open beaker
Reagent		HF (10 ml), HNO$_3$ (4 ml)	HF (10 ml), HNO$_3$ (10 ml)
		HClO$_4$ (1 ml)	HClO$_4$ (5 ml)
Time (h)		1	12
Temperature °C		140	150
		mg kg^{-1}	mg kg^{-1}
NBS SRM 1645 river sediment 10.2 ± 1.5 mg kg^{-1} nominal cadmium		9.55 ± 0.22	9.30 ± 0.10
River Sediment 1	0.32	9.55 ± 0.22	9.30 ± 0.10
2	1.29	0.23 ± 0.01	0.21 ± 0.04
3	2.84	0.17 ± 0.01	0.18 ± 0.01
4	7.20	1.22 ± 0.12	1.22 ± 0.10

Source: Elsevier Science Ltd, Kidlington, UK [17]

Table 3.12 Effect of matrix modifier on the determination of cadmium

Element	mg kg^{-1} of sediment	Cadmium recovery %	
		Without matrix modifier	With matrix modifier
Silicon	1000	90	102
Aluminium	500	75	108
Iron	500	93	104
Sodium	200	55	95
Potassium	200	91	98
Calcium	200	81	92
Titanium	100	103	99
Manganese	100	88	98

Source: Elsevier Science Ltd, Kidlington, UK [17]

3.67.2 Atomic absorption spectrometry

3.67.2.1 Heavy metals (chromium, manganese, nickel, copper, zinc, lead, iron, cobalt and cadmium)

Zink-Nielsen [63] compared direct digestion for 4 h with 1:1 nitric acid autoclaving at 120 °C with concentrated nitric acid followed by atomic absorption spectrometry as a means of determining six elements in river sediments. The results of an intercalibration exercise involving nine laboratories show that, in most cases, smaller coefficients of variation were obtained by autoclaving at 120 °C with concentrated nitric acid compared to 4 h digestion with 1:1 nitric acid. The procedure was applied to determinations of lead, iron, manganese, zinc, copper and chromium in the 0.07 to 18 mg kg $^{-1}$ concentration range.

Sinex et al. [64] have also investigated the use of nitric–hydrochloric acid in the extraction of total elements from NBS SMS1645 standard river sediments prior to analysis by atomic absorption spectrometry. They claim a recovery of 95% for chromium, manganese, nickel, copper, zinc and lead and 75% recovery for iron, cobalt and cadmium.

Sakata and Shimoda [17] have described a simple and rapid method in which 0.5g sediment is digested for 1 h at 140 °C with a mixture of 10 ml hydrofluoric acid, 4 ml nitric acid and 1 ml perchloric acid in a Teflon-lined bomb prior to measurement by graphite furnace atomic absorption spectrometry at the 228.8 nm cadmium line using an automatic sampling device. After digestion, 5g of boric acid is added to the solution to dissolve precipitated metal fluorides. Sodium and potassium interference is overcome by the addition of 1% ammonium sulphate matrix modifier. Table 3.11 shows cadmium contents obtained by this procedure compared to those obtained in a more lengthy open beaker digestion for a range of river sediments containing increasing amounts of carbon, and for an NBS reference sample. The two methods have practically the same precision, there being no significant difference at the 95% confidence level in the results between the two methods in samples containing between 0.32–7.2% of carbon. A good recovery (94%) was obtained for the NBS reference sediment.

Table 3.12 shows the beneficial effect on cadmium recovery of incorporating an ammonium sulphate matrix modifier in the final solution to overcome interference effects. In the absence of the matrix modifier, the average recovery of cadmium from a river sediment was 44.5%, increasing to 100% when the modifier was used.

Table 3.13 shows lead contents obtained by a Teflon bomb digestion procedure compared to those obtained in a more lengthy open beaker digestion for a range of river sediments containing increasing amounts of carbon, and for an NBS reference sample. The two methods have practically the same precision, there being no significant difference at the 95% confidence level in the results between the two methods in samples containing 0.32–7.2% carbon. A good recovery (96–101%) was obtained for the NBS reference sample.

Table 3.14 shows the beneficial effect on lead recovery of incorporating ammonium phosphate matrix modifier in the sample digest to overcome interference effects. In the absence of the matrix modifier, the average recovery of lead from a river sediment was 54–82%, increasing to about 100% when the modifier was used. Down to 0.1 mg kg $^{-1}$ of lead and cadmium can be determined in sediments by this procedure.

Table 3.13 Lead in river sediments by open beaker and Teflon-linked bomb digestions

Digestion conditions Vessel	Carbon content %	Teflon lined bomb	Teflon open beaker
Reagent		HF (10 ml), HNO₃ (4 ml) HClO₄ (1 ml)	HF (10 ml), HNO₃ (10 ml) HClO₄ (5 ml)
Time (h)		1	12
Temperature °C		140	150
		mg kg⁻¹	mg kg⁻¹
NBS SRM 1645 river sediment			
714 ± 28 nominal lead		685 ± 10	724 ± 43
River Sediment 1	0.32	17 ± 1	18 ± 1
2	1.29	35 ± 2	36 ± 2
3	2.84	32 ± 2	32 ± 1
4	7.20	59 ± 3	58 ± 2

Source: Elsevier Science Ltd, Kidlington, UK [17]

Table 3.14 Effect of matrix modifier on the determination of lead

Element	mg kg⁻¹ of sediment	Cadmium recovery % Without matrix modifier	With matrix modifier
Silicon	1000	99	98
Aluminium	500	76	107
Iron	50	65	99
Sodium	200	37	98
Potassium	200	46	96
Magnesium	200	95	101
Calcium	200	85	96
Titanium	100	94	99
Manganese	100	82	98

Source: Elsevier Science Ltd, Kidlington, UK [17]

Legret et al. [65] have discussed interference effects by major elements in the determination of lead, copper, cadmium, chromium and nickel in stream sediments by flameless atomic absorption spectrometry. Lead, cadmium and chromium were subject to most interferences.

In further work Legret et al. [66] demonstrated severe matrix interference in the determination of heavy metals in sediments by electrothermal atomic absorption spectrometry, particularly with regard to lead, cadmium and nickel.

Table 3.15 Accuracy of determination of cadmium by Zeeman atomic absorption spectrometry

Sample NBS SRM 1645	Expected concentration µg L^{-1}	Measured concentration µg L^{-1}	Recovery %
River sediment	185	148 ± 0.2	80
National Institute for Environmental Studies (Japan) Reference pond sediment	6.1	5.7 – 5.8	93 – 95
NBS SRM 1633 organic free coal flyash	6.2	63 ± 0.8	102

Source: Royal Society of Chemistry [16]

Hydrofluoric acid had to be eliminated by evaporation and perchloric acid was a serious interferent. A technique for the reduction of chemical interference in lead and nickel determinations was recommended, which consisted of matrix modification by ammonium dihydrogen phosphate and ascorbic acid. Rapid heating was recommended for cadmium determinations.

3.67.2.2 Cadmium and silver

Lum and Edgar [16] used a polarised Zeeman flame atomic absorption spectrometer to determine traces of cadmium and silver in chemical extracts of river sediments. The detection limit was 0.1 mg kg^{-1} based on a 0.2g sample size and a 25 ml final solution volume.

In this method 0.2g samples of the sediment were muffle ashed at 500°C for 3 h. A separate 0.2g sub-sample was dried in a forced air oven at 110°C for 24 h to determine moisture content. The ashed samples were carefully wetted with doubly distilled water, transferred into PTFE beakers and 20 ml of freshly prepared aqua regia were added. The digests were reduced nearly to dryness, 15 ml of hydrofluoric acid were added and heating was continued until the samples were dry, care being taken to avoid baking. Hydrochloric acid (15 ml) and 15 ml of doubly distilled water were then added and the solutions were heated for 1 h to reduce the volume to about 15 ml. After cooling, the volumes were made up to 25 ml in a calibrated flask. Cadmium was evaluated in this solution at 228.8nm using an air acetylene flame. In Table 3.15 the results obtained in applying this procedure to reference river sediment samples are shown. The lower than expected cadmium recovery is in part due to its organic content (1.7% from extractable oil and grease, 10.7% weight loss on ignition at 800°C). Higher recoveries were obtained with organic-free samples. Lum and Edgar [16] carried out a five part sequential extraction procedure on

Table 3.16 Sequential chemical extraction procedure carried out on 1 g dry weight of lake sediment

| Extractant | (Total cadmium content 3.5–8.0 mg kg $^{-1}$) | |
	In extract	% of total cadmium
0.75 mol L $^{-1}$ LiCl–0.25 mol L $^{-1}$ CsCl –60% MeOH (10 min @ 20 °C)	Readily exchangeable ions	17 ± 6
1 mol L $^{-1}$ CH$_3$ COONa, pH 5.0 (5 h @ 20 °C)	Carbonate bound surface – oxide bound ions	31 ± 10
1 mol L $^{-1}$ NH$_2$OH–HCl – 25% CH$_3$COOH (2 h @ 20 °C)	Ions bound to Fe–Mn oxides	34 ± 7
H$_2$O$_2$, pH 2, (5 h @ 90 °C) 1.2 mol L $^{-1}$ CH$_3$COONH$_4$ – 20% HNO$_3$	Organically and sulphide bound ions	12 ± 5
Aqua Regia – HF – HCl – H$_2$O$_2$	Ions bound to the residual phase	6 ± 3

Source: Royal Society of Chemistry [16]

1g dry weight samples of sediment (Table 3.16). Such extractions simulate, to a certain extent, various environmental conditions to which sediments may be subjected. Although such schemes are not perfectly selective, they can provide valuable information on the mobility and availability of elements in sediments.

3.67.2.3 Heavy metals (iron, manganese, copper and zinc) and aluminium

The Nordforsk Working Group on Water Analysis [67] has discussed the determination of these elements in sediments by atomic absorption spectrometry.

3.67.2.4 Heavy metals (iron, manganese, cadmium, cobalt, chromium, copper, nickel, lead, zinc) aluminium, calcium, magnesium, beryllium and lithium

Agemian and Chau [68] have studied the relative simultaneous extraction of these metals from aquatic sediments in order to obtain a rapid simple technique for measuring non-residual metals. The non-residual phase includes the exchangeable metal carbonate, organic and sulphide phases, as well as oxides and hydroxides of manganese and iron. These workers state that the extraction efficiency of the various methods studied for the 10 metals listed above is in the decreasing order hydrofluoric acid – perchloric acid – nitric acid mixture, boiling perchloric nitric acid mixture, boiling aqua regia, boiling nitric acid solution, cold 0.5M hydrochloric acid, cold 1N hydroxyammonium chloride plus 25%, acetic acid solution and 0.5N ethylene draminetetraacetic acid solution. This work is discussed in more detail under extraction of metals from sediments in section 3.67.13.

3.67.2.5 Heavy metals (lead and cadmium) and thallium

To determine lead, cadmium and thallium in non-saline sediments Garcia-Lopez *et al.* [157] suspended the sample in water containing 5% v/v hydrofluoric acid before injection into an electrothermal atomic absorption spectrometer. No modifier was needed other than the hydrofluoric acid.

3.67.2.6 Arsenic and selenium

Cutter [13] used a selective hydride generation atomic absorption spectrometer procedure as a basis for the differential determination of arsenic and selenium species in sediments.

3.67.3 Inductively coupled plasma atomic emission spectrometry

3.67.3.1 Heavy metals (cadmium, chromium, iron, manganese, zinc) aluminium, barium, calcium, potassium, magnesium, sodium, silicon, strontium, titanium and vanadium

Que-Hee and Boyle [69] studied the application of inductively coupled plasma mass spectrometry to the determination in hydrofluoric-nitric-perchloric acid digests of sediments of aluminium, barium, calcium, cadmium, chromium, iron, potassium, magnesium, manganese, sodium, phosphorus, silicon, strontium, titanium, vanadium and zinc. Electro-thermal vaporisation has been used with inductively coupled plasma atomic emission spectrometry for the direct determination of arsenic, cadmium, lead and zinc in sediments [26]. By the addition of toluene vapour to the internal furnace gas and sodium selenite to both standard solutions and solid samples, the transport losses of analytes were considerable reduced.

3.67.3.2 Arsenic, antimony, bismuth and selenium

Goldberg and Arrhenius [105] have described a semi-automated system for the determination of arsenic and selenium by hydride generation inductively coupled plasma atomic emission spectrometry. Sediments are brought into a solution by fusion with sodium hydroxide. In this method the sediment is pretreated with 2g solid sodium hydroxide in a zirconium crucible which is then heated to 350 °C for 2 h. The fused mass is dissolved

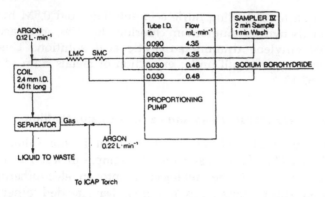

Fig. 3.2 Manifold for hydride generation
Source: Elsevier Science Ltd, Kidlington, UK [105]

Table 3.17 Results on reference sediments

Sample	As content, µg g^{-1}		Se content, µg g^{-1}	
	Recommended level	Ref [105] method	Recommended level	Ref [105] method
SO–1	1.9	1.9	0.1	0.09
SO–2	1.2	1.3	0.3	0.31
SO–3	2.6	2.6	0.05	0.08
SO–4	7.1	6.5	0.4	0.49

Source: Elsevier Science Ltd, Kidlington, UK [105]

in 40 ml 0.2N hydrochloric acid then further treated with 20 ml of hydrochloric acid to precipitate silica. The clear phase is decanted and is now ready for analysis. The manifold for hydride generation is shown in Fig. 3.2. The operating conditions are as follows: forward power 1400W, reflected power less than 10W, cooling gas flow 12 Lm^{-1}, plasma gas flow 0.12L mL^{-1}, injector flow 0.34 Lm^{-1}. The standard deviation of this procedure was 0.02 µg L^{-1} arsenic and the detection limit 0.1 µg L^{-1}. Results obtained on a selection of standard reference sediment samples are quoted in Table 3.17.

Elwaer and Belzile compared use of a closed-vessel microwave dissolution method to conventional hot plate digestion for the determination of arsenic and selenium in lake sediments [70]. A mixture of hydrochloric acid–nitric acid and hydrofluoric acid (microwave digestion) resulted in the best recoveries in samples where both elements were poorly recovered by using the hot plate technique. The volatile hydride generation inductively coupled plasma atomic emission spectrometer method [71] for the determination of arsenic, antimony and bismuth in soils, discussed in section 2.60.2.4 has also been applied to the analysis of sediments. Down to 0.1µg g^{-1} of these trace elements can be determined.

3.67.3.3 Rhenium, platinum and indium

Colodner *et al.* [72] determined rhenium, platinum and indium in sediments by flow injection isotope dilution inductively coupled plasma mass spectrometry. The minimum detectable amounts of these elements were, respectively, 3, 14 and 6 fg. Sediment digestion was achieved either by treating in a PTFE bomb with nitric acid in a microwave oven or by open tube digestion with a mixture of hydrochloric, nitric and hydrofluoric acids. The following amounts of the three elements were found in sediments: rhenium 33–281 ng g^{-1}, platinum 55 ng g^{-1}, and indium 1.5–2.2 mg g^{-1}.

3.67.3.4 Miscellaneous

Kanda and Taira [73] presented results from a study on the use of a computer-controlled rapid-scanning echelle ICPAES monochromator to determine major, minor and trace elements in sediments and soils. The concentrations of 17 elements in five standard reference materials were determined by using a single set of analytical lines without any corrections for line-overlap interferences. Lowest determinable concentrations and relative sensitivities for 13 metals are included.

3.67.4 Inductively coupled plasma mass spectrometry

3.67.4.1 Heavy metals (cadmium, lead and zinc) arsenic and antimony

Laser ablation followed by inductively coupled plasma mass spectrometry [2] has been used to determine down to 0.2 µg g^{-1} of arsenic and antimony in solid sediments.

Zaray and Kantor [26] used electrothermal vaporisation with inductively coupled plasma mass spectrometry for the determination of arsenic, cadmium, lead and zinc in non-saline sediments. By the addition of toluene vapour to internal furnace gas and sodium selenite to both standard solutions and sediment samples, the transport losses of analytes were considerably reduced.

3.67.4.2 Rare earths

Shabani and Musuda [158] have described a method of sample introduction by on-line two stage solvent extraction and back extraction to eliminate matrix interference and to enhance sensitivity in the determination of rare earth elements in sediments and rocks, by inductively coupled plasma mass spectrometry.

Two steps of extraction and back-extraction are linked together by multichannel pumping with the final back-extracts in aqueous solution being introduced to the inductively coupled plasma mass spectrometer. A mixture of 65% bis(2-ethylhexyl) hydrogen phosphate (HDEHP) and 35% 2-ethylhexyl dihydrogen phosphate (H_2MEHP) in heptane is used as the extracting agent, and octyl alcohol and nitric acid are used for the back-extraction. Parameters affecting extraction and back-extraction of rare earths in the system were examined. These include extraction coil length, pH of sample solution, concentration of complexing agent, back-extraction coil length, amount of octyl alcohol, and concentration of nitric acid for back extraction. Potential problems associated with matrix elements and deposition on the sampling cone were overcome. Analytical characteristics represent significant improvement in ICPMS sensitivity and an improvement in the accuracy and precision in the measurement of natural samples with a significant matrix. The entire process of extraction and back-extraction prior to introduction of the sample to the ICPMS can be carried out within 4 min, and a preconcentration factor of up to 1 order of magnitude for synthetic seawater has been achieved. By application of the proposed method, geological samples with final dilution factors of 5 could be introduced to the nebuliser of the ICPMS with negligible matrix problems. This represents a 100-fold improvement in sample sensitivity as compared to direct sample introduction with a dilution factor of 100. The method has been successfully applied in determination of rare earths in synthetic seawater, alkali-fused JG_{1a} standard rock, and acid-decomposed samples of JB_1.

3.67.4.3 Miscellaneous

Henshaw *et al.* [161] applied inductively coupled plasma mass spectrometry to the determination of 49 elements in lake samples in Eastern US lakes. The method was checked against an NBS SRM 1643b sediment sample and so, is obviously applicable to sediment samples.

Standard calibrations were used for 21 elements, and surrogate standards were used for 28 elements. The system detection limits, evaluated by using field blanks carried through the entire sampling and pre-treatment process, were less than $02\mu gL$ for most elements. Contamination during sampling and pretreatment was often the limiting factor. The accuracy of the determinations, as determined from the analysis of NBS SRM 1643b samples and by recoveries for spiked water samples, was typically better than ±10% for the elements determined by using standard calibration and better than ±25% for the elements determined by using surrogate standards. The long-term (12 months) precision was generally better than ±10%, expressed as relative standard deviation, for both methods of determination.

Two degrees of quantification were employed in this study. The concentrations of 21 elements were quantitatively determined by the use of calibration standards. Semi-quantitative elements of the concentrations of 28 elements were obtained by using surrogates selected from the first 21 elements. The practicality of estimating element concentrations by the use of surrogates is examined in detail.

3.67.5 Neutron activation analysis

3.67.5.1 Heavy metals (cobalt, chromium, iron, zinc, manganese, nickel, lead) arsenic, antimony, calcium, barium, strontium, magnesium, sodium, potassium, caesium, lanthanum, neodymium, samarium, gadolinium, terbium, dysprosium, tamerium, ytterbium, lutecium, europium, palladium, osmium, platinum, iridium, indium, ruthenium, gold, cerium, molybdenum, scandium, selenium, uranium, mercury, silver, tungsten, aluminium, titanium, hafnium, tantalum, thorium, vanadium and zirconium

Several workers have applied neutron activation analysis to the analysis of lake sediments [12, 54, 74, 75]. Elements determined include arsenic, selenium and mercury [12], aluminium, arsenic, barium, bromine, calcium, cerium, cobalt, chromium, caesium, dysprosium, europium, iron, gadolinium, hafnium, potassium, lanthanum, manganese, sodium, neodymium, nickel, rubidium, antimony, scandium, samarium, tantalum, terbium, thorium, uranium, vanadium, tungsten, ytterbium and zinc [12], and aluminium, vanadium, copper, titanium, bromine, magnesium, molybdenum, rubidium, gallium, samarium, thorium, uranium, iodine, chlorine, cadmium, barium, indium, neodymium, dysprosium, nickel, manganese, europium, potassium, zinc, iridium, platinum, osmium, ruthenium, gold and strontium [75]. Table 3.18 lists some elements found in sediments taken from American lakes.

This non-destructive analytical technique has also been applied by various workers [76–81] to the determination of total trace elements in river sediments. Typically [80], the sediment is wrapped in aluminium foil and sealed in a polyethylene container which is irradiated with thermal neutrons for 1–3 days, then allowed to cool for one day prior to gamma spectrometry.

The elements that have been determined are antimony, gold, barium, bromine, calcium, cerium, cobalt, chromium, europium, iron, potassium, lanthanum, molybdenum, scandium, selenium, sodium, uranium and zinc [78, 79], antimony, arsenic, barium, cobalt, chromium, iron, manganese, mercury, selenium, silver, uranium and zinc [79], antimony, arsenic, barium, bromine, chromium, cobalt, europium, iron, lanthanum,

Table 3.18 Multi-element analysis of sediments in American lakes

Element	Range µg kg⁻¹			Mean
Majors (%)				
Al	2.62	–	6.38	4.38
Ti	0.08	–	0.38	0.28
Mg	0.59	–	1.68	1.10
Na	0.30	–	0.92	0.61
K	0.50	–	2.29	1.13
Ca	1.23	–	4.00	2.69
Fe	1.47	–	3.06	2.24
Rare earths (ppm)				
La	28	–	73	37
Ce	53	–	160	85
Nd	15	–	137	52
Sm	7.86	–	28	13
Eu	0.77	–	1.94	1.28
Gd	6.37	–	22	11
Tb	0.95	–	2.39	1.52
Dy	5.26	–	15	8.76
Tm	0.19	–	0.74	0.52
Yb	2.34	–	9.34	4.46
Lu	0.52	–	1.20	0.72
Noble metals (ppm unless specified otherwise)				
Ag	<0.1	–	1.04	0.42
Au	0.25	–	19	3.55
Ru	45	–	500	160
Pd	<20	–	180	70
Os	<1	–	4.49	–
Ir	0.52	–	48	13
Pt	0.30	–	8.11	2.45
Trace elements (ppm unless specified otherwise)				
As	1.86	–	26	13
Ba	163	–	375	287
Br	23	–	96	44
Cl	<20	–	609	249
Co	3.91	–	16	7.31
Cr	16	–	50	27
Cs	0.56	–	14	2.93
Hf	1.67	–	12	7.05
In	5.3	–	19	12
Mn	214	–	4500	684
Ni	<1	–	218	38
Rb	19	–	49	35
Sb	<0.01	–	2.9	1.56
Sc	3.30	–	9.16	5.70
Se	0.03	–	1.01	0.39
Sr	<10	–	242	90
Ta	0.41	–	1.44	0.87

Table 3.18 continued

Element	Range µg kg $^{-1}$			Mean
Th	4.02	–	9.38	6.39
U	0.78	–	4.35	2.25
V	28	–	68	46
Zn	<10	–	450	278
Zr	54	–	488	263

Source: Elsevier Science Publishers, Netherlands [75]

Table 3.19 Detection limits, neutron activation analyses

Element	Isotope	γ Energy	Irradiation time 100 min mg kg $^{-1}$	400 min mg kg $^{-1}$
Ag	100 m–Ag	657.7		0.7
Ag	100 m–Ag	937.5		2
As	76–As	559.1	0.3	
Ba	131–Ba	496.3		70
Cd	115 m–in	336.3	10	
Co	60–Co	1173.2		0.08
Co	60–Co	1332.5		0.06
Cr	51–Cr	320.0		1.5
Cu	64–Cu	1345.8	10,000	
Fe	59–Fe	1099.5		70
Fe	59–Fe	1291.6		60
Hg	203–Hg	279.2		0.5
Mn	56–Mn	846.6	0.2 ... 2	
Mo	99–Mo	140.5	0.5	
Ni	58–Co	810.6		50
Sb	122–Sb	564.1	0.2	
Sb	124–Sb	1691.0		0.1
Se	75–Se	264.5		0.8
Sn	113–Sn	391.7		70
U	132–Te 239–Np	228.1	0.6	
U	140–La	1596.2		0.5
Zn	69 m–An	438.9	120 ... 250	
Zn	65–Zn	1115.5		4

Source: Bundesanstrat für Gewassarkunde Koblen [79]

manganese, potassium, scandium, sodium, tungsten and uranium [80, 81]. Detection limits achieved by Ackermann [79] ranged from 0.06 mg kg $^{-1}$ (cobalt) to 70 mg kg $^{-1}$ (barium, iron, tin) (Table 3.19).

Bart and Von Gunten [80] used neutron activation analysis to study the distribution of elements between water and sedimentary solids and found distinct differences between elements, eg iron 9 mg kg^{-1} in sediment corresponds to 15 µg L^{-1} in solution and cobalt 40 mg kg^{-1} in sediment corresponds to 0.15 µg L^{-1} in solution.

3.67.6 Plasma emission spectrometry

3.67.6.1 Heavy metals, aluminium

Welte *et al.* [82] compared two extraction methods in the speciation of heavy metals in sediments. The first method was based on extraction with ammonium acetate in nitric acid, followed by treatment with hydroxylamine chlorhydrate in acetic acid. The other method involves extraction with 0.01 mol L^{-1} nitric acid and treatment with sodium dithionite and sodium citrate. The direct residues in both cases were digested with aqua regia. The metals in the various supernatants were determined by plasma emission spectrometry for iron, manganese, zinc, copper, chromium, nickel, lead and aluminium and by atomic absorption for cadmium, cobalt and arsenic. A portion of the sediment was freeze-dried and dissolved to determine total metal content.

3.67.7 Potentiometric stripping analysis

3.67.7.1 Copper and lead

Potentiometric stripping analysis has been used to determine down to 3 mg kg^{-1} of copper [83] and lead in river and estuarine sediments with a precision of 3.9–4.5%.

This system has three electrodes – a glassy carbon cathode, a saturated calomel reference electrode, and a platinum counter electrode. The first step in the determination of metal ions in a sample solution is the electrochemical formation of a mercury film on the glassy carbon electrode. Subsequently, the metal ions are reduced and amalgamated in the mercury during the electrolysis step (plating). When the plating is terminated, the metals are stripped from the mercury film back into the solution by chemical oxidation. During this step, the potential of the carbon electrode (against SCE) vs time is recorded. The metals are identified by their stripping potentials and are quantified by measuring the stripping time for each metal.

In this procedure, approximately 0.5g of dried (105 °C) sediment was decomposed by boiling for 2 g in a mixture of 3 ml of concentrated hydrochloric acid and 1 ml of concentrated nitric acid. After the dissolution, the sample was diluted with 10 ml of redistilled water, filtered (Whatman No 41) and further diluted with redistilled water to a total volume of 100 ml. To an aliquot (25 ml) were added 1 ml of 10%

(w/v)ascorbic acid solution (antioxidant and reductant for iron (III)) and 1 g of sodium chloride (to ensure reproducible ionic activity) together with mercury solution (usually 100 μL of 1000 ppm Hg solution, but depending on metal concentration) and an internal standard (cadmium) to correct for variations in oxidation rate.

For mercury plating, 10 precoating-stripping cycles were employed. Simultaneously, the sample was deaerated by purging with helium. After precoating and deaeration of the sample, an appropriate plating time and plating potential were applied (1–32 min and –0.95 V vs SCE, respectively).

Copper was quantified by using at least two standard additions. This was done in a cyclic mode and standards were added immediately after the stripping curve had been recorded. The plating time was 4 min at –0.95V vs SCE; 2.5 μg of copper was added at each standard addition and the natural lead content of the sample was used as the internal standard.

Copper determinations carried out by this procedure were compared to expected values for a series of reference sediments. The values found were all within the certified limits. For example, NBS 5RM 1645 reference sediment (reference value 18 ± 3 mg k $^{-1}$ copper) gave a value of 16.8 mg kg $^{-1}$ and by this procedure.

This procedure has been applied to the determination of lead in river sediments in amounts down to 7 mg kg $^{-1}$ with a precision of 2.5–4.1%. Lead determinations carried out by this procedure were compared to expected values for a series of reference sediments. All values are within certified limits, for example NBS SRM 1645 sediment (reference value 714 ± 2.8 mg kg $^{-1}$ lead) gave a value of 722 ± 18 mg kg $^{-1}$.

3.67.8 X-ray fluorescence spectroscopy

3.67.8.1 Miscellaneous

Various workers have applied this technique to the determination of metals in sediments [84–86]. Prange et al. [86] determined up to 25 trace elements using total reflection X-ray fluorescence spectrometry.

3.67.9 X-ray spectrometry

3.67.9.1 Heavy metals (chromium, manganese, cobalt, nickel, copper, zinc, cadmium, iron and lead) arsenic, rubidium, strontium and mercury

Lichtfuss and Brummer [87] determined traces of chromium, manganese, cobalt, nickel, copper, zinc, arsenic, rubidium, strontium, cadmium, mercury and lead in river sediments without prior fusion of the samples. Tablets were prepared for analysis with a synthetic wax. Linear relationships were obtained between element concentration and radiation intensity. Results obtained by this method were compared with those obtained by atomic absorption spectrometry in determinations of manganese, zinc and copper in river sediments containing a range of concentrations of organic carbon, calcium carbonate and iron. At concentration levels above about 100 mg kg $^{-1}$, the X-ray method tends to give higher results. As regards the other elements, the following concentration ranges were determined: lead 24–260 mg kg $^{-1}$, chromium 39–185 mg kg $^{-1}$, arsenic 79–161 mg kg $^{-1}$, strontium 118–189 mg kg $^{-1}$, rubidium 58–105 mg kg $^{-1}$, nickel 12.7–67 mg kg $^{-1}$, cobalt 4.1–24 mg kg $^{-1}$, mercury 1.8–12.6 mg kg $^{-1}$, and cadmium 3.1–10.7 mg kg $^{-1}$.

Schneider and Weiler [88] give details of a procedure for the rapid separation of sediment particles less than 10 µm in size prior to determination of metals by totally reflecting X-ray spectrometry.

O'Day et al. [159] used X-ray absorption spectrometry (XAS) and X-ray absorption fine structure (XAFS) spectra of zinc, cadmium and lead-bearing sediments to identify the local molecular co-ordination of metals in contaminated untreated stream sediments. Quantitative analysis of the XAFS spectra, supplemented by elemental distributions on particles provided by electron microprobe and secondary ion mass spectrometry (SIMS), shows that zinc and cadmium occur in small (<1µm), residual particles of the host ore, sphalerite (ZnS) in which cadmium substitutes for zinc in the mineral structure. In half of the samples studied, analyses indicate that zinc, as it weathers from sphalerite, is scavenged primarily by zinc hydroxide and/or zinc-iron oxyhydroxide phases, depending on the total amount of iron in the system. These phases probably form as amorphous or poorly crystalline coatings on mineral surfaces. There is no evidence that zinc sorption or substitution in other mineral phases is a significant mode of uptake. In contrast, there is no spectral evidence for the association of cadmium with secondary oxide or oxyhydroxide phases in both high and low iron samples. Cadmium bound in sphalerite is found in all samples; evidence for cadmium uptake into a carbonate phase (in addition to sphalerite) was found in only one sample. This result may suggest preferential partitioning of cadmium, relative to zinc, into the aqueous phase as sphalerite weathers. XAFS spectra of lead in sediments with low total iron concentrations indicate no evidence for lead bonding in galena (PbS), the host ore, and suggest lead uptake in secondary carbonate and/or oxide phases. Uptake of metal ions from solution into secondary phases is apparently governed by competition between iron oxyhydroxide and carbonate phases that can be related to total iron in the sediments and to stream pH. This work highlights the differential chemical behaviour of three divalent metal cations in a contaminated system.

3.67.10 γ-ray spectrometry

3.67.10.1 Rare earths, thorium and uranium

Labresque *et al.* [89] determined 11 rare earth elements, thorium and uranium in river sediments, employing a germanium detector for gamma ray spectrometry.

3.67.11 Gas chromatography

3.67.11.1 Arsenic and antimony

Cutter *et al.* [3] have described a method for the simultaneous determination of arsenic and antimony species in sediments. This method uses selective hydride generation with gas chromatography using a photoionisation detector. The following species could be separately determined As(III), As(V), Sb(III), Sb(V). Detection limits are 10 p mole L $^{-1}$ (arsenic) and 3.3 p mole L $^{-1}$ (antimony).

3.67.12 Miscellaneous

McKay [90] has recently reviewed the monitoring of exposed aquatic sediments for radionucleids.

Allen [91] analysed bottom sediments, taken at the Great Stone Lake, for heavy metal content. The concentrations of metals found in the sediments were related to bedrock composition.

Atomic absorption spectrometry has been used to determine 20 elements in lake sediments [95, 96].

In addition to flame and flameless atomic absorption spectrometry, the graphite furnace technique [97] and Zeeman atomic absorption spectrometry [98] has been applied to the determination of metals in sediment extracts.

Bando *et al.* [94] discussed analytical errors associated with iron, manganese, copper, chromium and zinc determinations in particulate sedimentary matter by atomic absorption spectrometry. They investigate ashing and sonic extraction methods, and concluded that the latter was preferable as it can be applied to a wider range of elements.

Sagar [131] and Raurat *et al.* [132] have reviewed the chemical speciation and environmental mobility of metals in non-saline river sediments.

Total reflection X-ray spectrometry has been applied to studies of the speciation of metals in sediments [133].

Various workers [134–136] have carried out studies on the development of methods for the determination of trace metals in sediment standard reference materials.

Wen *et al.* [137] used surface complexation models (SCMs) to describe metal ion adsorption on pure mineral materials. However, such models have rarely been applied to model adsorption on natural materials. In this study, the surface complexation model approach was used to describe the surface properties and adsorption behaviour of natural aquatic sediment. Three typical versions of the surface complexation model were used: constant capacitance model (CCM), diffuse layer model (DLM) and triple layer model (TLM). All the model parameters were determined on the basis of the experimental data of the potentiometric titration and the metal adsorption isotherm using LeAn River (China) sediment. The experimental data of the adsorption edges were used to verify the performance of the models. This work indicated that all three models can simulate the experimental results very well. In predicting the adsorption behaviour of the sediment sample, the relative errors of these three models were quite close. The results illustrate that SCMs can be used to successfully model natural materials.

Al-Jundi et al. [138] compared neutron activation analysis and PIXE for the determination of heavy metals in sediments.

Beverage et al. [160] used ion exchange resins to evaluate the removal of 'labile' metals in contaminated non-saline sediments. Exchange materials in the H+ form extracted high proportions of copper, lead, zinc, and cadmium and recoveries depended on the strength of the acid functional groups.

Mahan et al. [92] used a microwave digestion technique in the sequential extraction of calcium, iron, chromium, manganese, lead and zinc in non-saline sediments.

The sequential extraction scheme of Tessier [150] partitions metals in sediments into exchangeable, carbonate bound, iron-manganese oxide bound, organic bound and residual binding fractions. Extraction rate experiments using conventional and microwave heating showed that microwave heating produces results comparable to the conventional procedure. Sequential microwave extraction procedures were established from the results of the extraction rate experiments. Recoveries of total metals from NBS SRM 1645 standard sediment ranged from 76% to 120% for the conventional procedure and 62% to 120% for the microwave procedure. Recoveries of total metals using the microwave and conventional techniques were reasonably comparable except for iron (62% by microwave vs 76% by conventional). Substitution of an aqua regia/hydrofluoric acid extraction for total/residual metals results in essentially complete recovery of metals. Precision obtained from 31 replicate samples of the California Gulch, Colorado, sediment yielded about an average 11% relative standard deviation excluding the exchangeable fraction which was more variable.

3.67.13 Extraction of metals from non-saline sediments

3.67.13.1 Aqueous extraction agents

In order to determine metals in sediments by means of atomic absorption spectrometry or other techniques, it is first necessary to bring them into solution. Extraction methods have been well documented and involve fusion or acid dissolution; the latter type of technique has several advantages. Mineral acids can be obtained in a sufficiently pure form that their use does not introduce any appreciable impurities and acid decomposition methods, unlike fusion techniques, do not allow large amounts of salts to be introduced into the solution: a high salt content can cause instability and lead to high instrument background readings. In addition, fusion techniques are restricted to the determination of the total metal content of silicates only. On the other hand, the concentrations of acids can be varied by dilution and therefore selective dissolution of several components of sediments can be affected.

Five mineral acids, namely hydrochloric, nitric, sulphuric, perchloric and hydrofluoric acids, have been very widely used. For the simultaneous extraction of a large number of metals, sulphuric acid has the one notable property of dissolving silica. Thus, it has been used in conjunction with nitric, hydrochloric or perchloric acid in the total decomposition of silicates. Nitric acid has been used separately or with either hydrochloric or perchloric acid. Such methods provide a high degree of metal extraction but do not dissolve silicates completely; they destroy organic matter, dissolve all precipitated and adsorbed metals, and leach out a certain amount of the metals from the silicate lattice.

Much weaker extracting agents have also been used to extract metals of a non–residual origin only. Methods involving the use of 0.5 N hydrochloric acid and 0.05N ethylenediamintetraacetic acid dissolve complexed, adsorbed and precipitated metals in sediments with minimum attack on the silicate. A mixture of 1N hydroxylammonium chloride and 25% acetic acid has been used to dissolve adsorbed trace elements in sediments and is similar to the above two methods.

Trace metal studies of sediments often include a chemical leach of the solid phase with selected reagents. Some of the leaching systems that have been employed are listed in Table 3.20 which illustrates the wide variety of reagents that have been studied. Some of these systems are discussed below in further detail.

Chester and Hughes [99] found a solution of 1N hydroxylamine-hydrochloride and 25 vol % acetic acid suitable, since this mixture would not attack lattice structures of clay minerals and reduced manganese oxide phases. Table 3.21 shows that, for most metals, 0.5N hydrochloric acid gives higher extraction than 1N hydrochloric acid. However, the levels for manganese were comparable for these two methods showing that even though 1N hydroxylamine hydrochloride plus 25% acetic acid is a weak extractant, its reducing character makes it suitable for manganese extraction. The pH of this solution was found to be 1.5, considerably lower than that for 0.5N hydrochloric acid. The reducing nature of the solution may not liberate organically bound trace elements. A comparison of the data obtained with 0.05N EDTA with that of the

Table 3.20 Selected reagents used for leaching trace metals from sediment and soil

Reagent	Ref
EDTA	
1N Ammonium chloride pH 7	58, 105
1N Magnesium chloride pH 7	106, 107
0.1 mol L $^{-1}$ Potassium pyrophosphate, pH 10	108, 109
Ammonium oxalate, pH 3	109, 110
Citrate sodium dithionite, pH 3, 4.7, 7	11, 111, 112
30% Hydrogen peroxide	113
Sodium hypochlorite, pH 8.5	107
Acetic acid–sodium acetate, pH 5.2	114
25% Acetic acid	115
25% Acetic acid – 1 mol L $^{-1}$ hydroxylamine hydrochloride	11, 99
0.1–0.5N Hydrochloric acid	58, 11, 116
Nitric-perchloric acid (1/1)	68
Aqua regia	68
Nitric acid–hydrochloric acid	58, 64, 106, 125
Ashing at 500 °C, hydrochloric-hydrofluoric acid digestion	112, 118, 119
Phthalate buffers pH 2.2 to 6.0	120
Lithium metaborate fusion	121
Ammonium acetate-nitric acid, then hydroxylamine-acetic acid	82
Nitric acid–hydrochloric acid–nitric acid	125

Source: Own files

Table 3.21 Mean background concentration of heavy metals in the sediments of the Rideau River using different extraction methods

Metal	mg kg $^{-1}$ Total[a]	4.0N HNO$_3$ +0.7N HCl	0.5N HCl	1N NH$_2$OH–HCl +25% acetic acid	0.05N EDTA
Cd	–	–	0.3	0.4	0.2
Cu	42	13	11	2	8
Pb	84	24	17	16	16
Zn	94	67	36	20	16
Ni	72	23	8	3	3
Cr	108	37	7	2	1
Co	57	17	6	2	4
Mn	582	451	305	304	275
Fe	31,000	30,000	6,700	3,000	2,600
Al	46,200	15,000	3,930	1,150	620

[a] HF–HNO$_3$–HClO$_4$ with PTFE bomb decomposition

Source: Elsevier Science Co Ltd, Kidlington, UK [99]

above reducing solution shows that there is a significant difference in the amount of copper, manganese and iron extracted. The latter method extracts more manganese and iron, and less copper. This is satisfactorily explained by the higher acidity and stronger reducing powers of the latter method and the stronger complexing power of the former method. The pH of the 0.05 N EDTA solution was 4.8, much higher than 1.5 for 1N hydroxylamine hydrochloride plus 25% acetic acid. Thus, the latter method would be expected to extract more iron. For copper, 1N hydroxylamine hydrochloride plus 25% acetic acid gave lowest levels of all methods. It is well known that copper is highly correlated with organic matter [100, 101]. Thus it appears that this method is not suitable for copper determinations. Holmes *et al.* [102] used this solution satisfactorily, to extract zinc and cadmium, and Kronfeld and Navrat [103] used it to extract cadmium, chromium, lead and zinc from the adsorbed phase and the ferromanganese and carbonate mineral phases of sediments. They stated that this technique does not dissolve the authigenically formed sulphide minerals or organic complexes. From Table 3.21 this extraction technique is seen to be essentially equivalent to 0.5N EDTA except for copper and cobalt.

Chowdbury and Bose [104] have shown that copper complexed with humic compounds isolated from soils is readily liberated with dilute hydrochloric acid (pH 1.0). Therefore 0.5N hydrochloric acid liberates copper from organic matter. Thus, unlike 1N hydroxylamine hydrochloride plus 25% acetic acid, the above method is suitable for copper extraction. The mean level of copper with 0.5N hydrochloric acid to be of the same order of magnitude as that of 0.05N EDTA. It is interesting to note that the mean levels of lead with all four partial extraction methods were very similar.

Table 3.22 represents the contrast of the different metals with each method for the four major anomalies at locations B, T, W and X. The contrast for copper and lead is essentially constant for all methods with the exception of method 3 for copper and method 1 for lead. For the other four trace elements, the contrast increases as the strength of the extractant decreases from methods 1 to 4.

From the above discussion, a few conclusive points arise. The 1N hydroxylamine hydrochloride plus 25 vol % acetic acid is satisfactory for the extraction of ferromanganese and carbonate minerals and adsorbed trace elements, although it has no effect on authigenically formed sulphide minerals and organic complexes. Therefore, it is not suitable for the simultaneous extraction of a large number of the non-silicate authigenically formed elements. The 4.0N nitric acid plus 0.7N hydrochloric acid solution has a considerable effect on the aluminosilicate crystal lattice. Thus it may provide misleading environmental information if the mineralogy of the area of study is not constant. It also gives rise to lower contrast of anomalous to background samples than the

Table 3.22 Contrast (mg kg $^{-1}$ anomaly/mg kg $^{-1}$ background) for the four most anomalous sample

Site	Cd				Cu				Pb			
	Method Used*											
	1	2	3	4	1	2	3	4	1	2	3	4
B	–	4.7	–	9	9.3	9.1	7.5	8.1	12	16	16	17
T	–	45	39	78	4.4	42	1.5	5.3	6.5	8.2	7.7	8.9
W	–	6.0	5.0	9	4.2	4.1	1.5	5.3	6.5	7.9	7.4	8.7
X	–	3.3	4.0	6	2.8	3.0	1.5	3.6	13	17	16	17

Site	Zn				Ni				Cr			
	Method Used*											
	1	2	3	4	1	2	3	4	1	2	3	4
B	10	18	32	34	–	–	–	2	–	2.1	3.6	4.4
T	8.5	16	28	34	5.9	13	18	32	3.7	6.6	15	18
W	5.1	7.3	13	15	–	–	2.7	–	–	–	3.7	6.4
X	4.9	7.8	14	15	–	–	–	–	–	–	–	–

* 1 = 4.0N HNO$_3$ + 0.7N HCL; 2 = 0.05N HCl; 3 = 1N NH$_2$OH – HCl + 25% acetic acid; 4 = 0.05N EDTA

Source: Geochemical Exploration [104]

other methods and thus is not as preferable. The other two methods are most informative and advantageous because they provide high contrast of anomalous to background samples, and simultaneously extract a large number of metals from sediments without attacking the aluminosilicate minerals. The 0.5N hydrochloric acid method is the preferred method for measuring authigenically formed metals in sediments since, due to some natural processes, the adsorbed metals in some sediments are bonded more strongly, making the 0.05N EDTA method incapable of extracting them completely.

Agemian and Chau [68] studied the relative simultaneous extraction of a large number of metals from aquatic sediments in order to obtain a rapid simple technique for measuring non-residual metal. The non-residual metal phase includes the exchangeable metal carbonate, organic and sulphide phases, as well as oxides and hydroxides of manganese and iron.

Leaching conditions
The size of sediment particles strongly influences the extractable metal content of the samples. The < 80-mesh portion of an air dried (at 20 °C)

sediment provides the greatest contrast between anomalous and background samples. To facilitate the dissolution necessary for determining the total metal, a sub-sample from the < 80-mesh (0.177 mm) portion was ground to about 200-mesh.

Cold-extractable metal content

A 5g sample of sediment was shaken overnight at room temperature with 100 ml solutions of 0.05 N ethylenediaminetetraacetic acid at pH 4.8, 1N hydroxylammonium chloride plus 25% acetic acid, and 0.5N hydrochloric acid.

Acid extractable metal content

A 1g amount of sediment was digested with 25 ml of nitric acid, boiled to dryness twice, with 25 ml of aqua regia and boiled to dryness twice and with 25 ml of nitric–perchloric (1 + 1) acids and boiled to dryness twice. The residue was dissolved in dilute hydrochloric acid in each instance.

Total metal content

The < 80–mesh sample was crushed to about 200-mesh and 100 mg of this powder was digested with 6 ml of hydrofluoric acid and 1 ml of perchloric acid in a PTFE bomb. Table 3.23 shows the degree of extraction of several metals by use of the methods under consideration. The methods used were three of the four types of extraction techniques, namely those which extract total (last column), acid-extractable (columns 5–7) and cold-extractable (columns 2–4) metal. The acid leaching techniques (columns 5–7) show varying degrees of attack on the crystal lattice and thus give an intermediate value between cold-extractable and total metal extractions. The results (Table 3.23) reflect this postulation.

It is apparent from Table 3.23 that, for the sample studied, perchloric acid does not liberate all of the metal from the silicate matrix. The amount of metal extracted by perchloric acid depends on the type of sample (both type of mineral and organic matter content). For many types of sample, this acid is suitable for total metal extraction.

The nitric acid used in this method (Table 3.23, column 7) serves only as a safety measure if large amounts of organic matter are present. The use of perchloric acid alone for the sample in Table 3.23 gave results identical with those obtained using nitric-perchloric acids.

There was one unsatisfactory recovery with perchloric acid, this being that of chromium. The low boiling point of chromyl chloride (CrO_2Cl_2), 116 °C, compared with about 200 °C for perchloric acid, probably results in volatilisation losses. With nitric acid or aqua regia, these losses do not occur because the boiling points of nitric and hydrochloric acids are lower.

Aqua regia and nitric acid are weaker extracting agents than perchloric acid. Aqua regia (Table 3.23) is a stronger oxidising and extracting agent

Table 3.23 Comparison of the extraction of metals from a sediment containing 2.8% of organic carbon and 0.17% of organic nitrogen using different extraction systems

Metal	0.5N EDTA	1N NH$_2$OH$_2$HCl + 25% CH$_3$COOH	0.5N HCl	HNO$_3$ (boiling)	Aqua regia (boiling)	HNO$_3$–HClO$_4$ (1 + 1) (boiling)	HF–HNO$_3$–HClO$_4$ (6 + 4 + 1) PTFE bomb
				All results are in mg kg^{-1}			
Al	400	970	4,000	15,500	25,300	38,500	43,000
Ca	21,400	22,700	23,100	27,000	27,000	30,000	30,000
Fe	4,800	6,800	12,500	32,000	32,000	34,000	42,000
Mg	3,800	4,800	6,900	12,000	13,000	12,000	16,000
Mn	550	620	620	750	800	750	4,500
Ba	20	80	100	1,100	1,600	2,600	2,700
Cd	2.2	2.2	2.0	4.0	4.0	6.0	40
Co	6	6	8	31	31	36	200
Cr	5.5	15	22	49	48	15	110
Cu	23	16	33	44	40	50	50
Li	0.4	1.0	5.4	33	38	44	50
Ni	16	20	28	32	28	33	200
Pb	53	52	56	70	70	70	100
Zn	97	122	149	218	206	229	290

Source: Royal Society of Chemistry [68]

than nitric acid as a result of the presence of free or nascent chlorine. Nitric acid, aqua regia and perchloric acid have their strongest leaching effect when they are boiling. Perchloric acid, especially, is a strong leaching dehydrating and oxidising agent only when it is hot and concentrated.

Cold extraction methods are the weakest as they do not attack the silicate lattice appreciably. With such methods it is usually desirable to extract the non-residual metal from the sediments. The three methods of this type studied are compared in Table 3.24, together with a total extraction method for manganese, iron and aluminium, for 13 sediments of different types. The extraction efficiency of the three methods is in the increasing order 0.05N ethylenediaminetetraacetic acid, 1N hydroxyl-ammonium chloride plus 25% acetic acid, and 0.5N hydrochloric acid. This trend correlates with the decreasing pHs of 4.8, 1.5 and 0.3 respectively, for the above reagents. From their chemical properties, it would be expected that these methods would extract the adsorbed, precipitated and complexed metals.

The results given in Table 3.24 indicate that the cold-extraction methods extract only a very small fraction of the total aluminium from the sediments (compare methods a, b and c with method d). On the other hand, a considerable amount of the manganese is extracted. The fraction of total iron extracted by use of these methods is intermediate between that of manganese and aluminium. Both iron and aluminium are mainly found in the residual phase. About 14 ± 10, 17 ± 10 and 32 ± 6% of the total iron was extracted by methods a, b and c respectively (Table 3.24). For aluminium, about 2 ± 15.4, 4 ± 4 and 10 ± 4 % of the total was extracted, respectively, for methods a, b and c. This shows that these methods do not affect the crystalline structure appreciably. To substantiate further the above contention, the amount of silicon extracted by the total metal and 0.5N hydrochloric acid extraction methods was measured. Table 3.24 gives the total amount of silicon in each of the samples. It was found that the mean amount of silicon extracted by use of the 0.5N hydrochloric acid method was about 1% of the total.

Table 3.25 shows the results obtained for the same samples, with the three partial extraction methods, for the seven trace elements studied. The order of efficiency of extraction of the methods is again increasing from 0.05N ethylenediamine tetraacetic acid to 0.5N hydrochloric acid, except for copper, where 1N hydroxylammonium chloride plus 25% acetic acid gives the lowest extraction. Hydrochloric acid (pH 1.0) liberates copper complexed with humic compounds isolated from soils. However, 1N hydroxylammonium chloride plus 25% acetic acid mixture is not a strong enough complexing agent to compete in complex equilibria, nor acidic enough to cause dissociation of the natural complexes. The result of this effect is seen with copper (which correlates highly with organic matter), where ethylenediaminetetraacetic acid, which is the weakest extracting

Table 3.24 Extraction of major metals from sediments of various organic carbon contents with four extraction methods

All results are in mg kg^{-1}

Sample	Organic carbon %	Silicon %	Sample description	Sample location[a]	Manganese				Iron				Aluminium			
					a	b	c	d	a	b	c	d	a	b	c	d
1	0.15	31.3	Orange sand	Lake Superior	99	120	120	227	990	1,400	1,400	11,700	970	1,400	1,700	20,300
2	0.43	24.2	Clay	Rideau River	140	220	300	655	1,800	3,770	10,940	35,000	660	1,800	6,400	54,700
3	0.90	28.0	Sand	Calgary	140	160	160	253	1,980	2,580	4,170	14,400	140	240	1,000	41,500
4	2.25	2.25	Silt	Rideau River	320	380	380	718	3,360	4,160	8,790	30,900	790	1,500	4,800	45,100
5	2.47	19.0	Clay	Rieau River	900	1,010	1,070	1,100	5,780	7,740	12,900	36,300	340	930	4,000	49,900
6	2.81	20.6	Silt	Lake Ontario	550	620	620	787	4,750	12,520	34,700	400	970	4,000	49,700	
7	2.95	20.3	Silt and fine sand	Rideau River	240	300	280	633	3,190	3,740	8,970	30,800	620	1,100	5,500	48,700
8	3.15	22.8	Silt	Ottawa River	160	200	220	600	3,780	4,540	8,770	27,300	1,400	2,600	5,400	39,500
9	3.35	20.3	Dark brown clay	Lake Huron	1,090	1,210	1,170	1,340	8,300	10,500	17,500	38,800	1,400	2,700	7,600	43,400
10	4.10	20.8	Grey clay	Cardigan Bay	100	120	140	417	6,750	7,110	9,960	27,500	7,500	9,700	15,000	30,300
11	4.85	19.9	Silt and clay	Rideau River	400	440	440	1,070	3,140	3,570	7,710	24,600	650	1,200	5,100	48,500
12	10.75	20.0	Silt and clay	Rideau River	400	440	480	–	3,380	3,770	8,120	–	1,100	1,900	6,500	–
13	12.30	20.3	Silt and clay	Rideau River	380	440	420	533	4,180	3,790	6,160	22,300	1,200	1,500	4,000	38,800

[a] All sample locations are in Canada

a = 0.05N ethylenediaminetetraacetic acid (pH 4.8); b = 1N hydroxylammonium chloride + 25% acetic acid; c = 0.5N hydrochloric acid; d = HF–HNO$_3$–HClO$_4$ (total media)

Source: Royal Society of Chemistry [68]

agent, shows higher values for copper than the hydroxylammonium solution. The samples listed in Table 3.25 have a wide range of organic matter contents. Table 3.26 presents the results for the extraction of copper by the three methods, related to results for extraction with 0.5N hydrochloric acid. It can be seen that, as the organic matter content increases, the relative extraction of copper by 1N hydroxylammonium chloride plus 25% acetic acid is reduced. This evidence confirms the inability of this last solution to extract copper from its organic complexes. However, 0.5N hydrochloric acid and 0.05N ethylenediaminetetraacetic acid liberate copper from its stable organic complexes. This property is desirable because organic matter plays a very important role in smaller lakes and rivers.

The 1N hydroxylammonium chloride plus 25% acetic acid is a special solution for the dissolution of ferromanganese minerals, the hydroxylammonium chloride dissolving the manganese oxide phase. The role of the 25% acetic acid is to dissolve the iron oxide phases. Thus, the mixture of the above reagents satisfactorily dissolved ferromanganese minerals and is suitable for other metals that do not form strong complexes with organic matter. Because of this property, it is not suitable for the simultaneous extraction of a large number of metals.

In conclusion Agemian and Chau [68] state that the extracting efficiency of the methods for the 10 metals studied is in the decreasing order hydrofluoric-perchloric-nitric acid mixture, boiling perchloric-nitric acid mixture, boiling aqua regia, boiling nitric acid solution, cold 0.5N hydrochloric acid solution, cold 1N hydroxylammonium chloride plus 25% acetic acid solution, and 0.5N ethylenediamenetetraacetic acid solution. The two exceptions to this order are boiling perchloric–nitric acid solution if used for chromium and cold 1N hydroxylammonium chloride plus 25% acetic acid solution if used for copper. With the former, chromium is lost by volatilisation and, with the latter, copper in the form of organic complexes is not extracted.

Although boiling perchloric-nitric acid may give total extraction of some metals from some non-resistant samples, complete destruction of the silica matrix by hydrofluoric acid is necessary for total extraction of all metals of interest.

Of the cold-extractable metal extraction techniques, 1N hydroxyl-ammonium chloride plus 25% acetic acid is inadequate for the simultaneous extraction of a large number of metals from a variety of sample types because it does not extract copper that has been complexed by organic matter. Both 0.5N hydrochloric acid and 0.05N ethylenediaminetetraacetic acid are suitable for the simultaneous extraction of ten elements from the adsorbed organic and precipitated phases of aquatic sediments. The 0.5N hydrochloric acid method is preferred because, owing to some natural processes, the adsorbed metals in some sediments are bonded more strongly, rendering the 0.05N

Table 3.25 Extraction of trace elements from sediments of various organic carbon contents with three extraction systems

All results are in mg kg^{-1}

Sample	Organic carbon %	Cadmium			Chromium			Copper			Cobalt			Nickel			Lead			Zinc		
		a	b	c	a	b	c	a	b	c	a	b	c	a	b	c	a	b	c	a	b	c
1	0.15	0.4	0.6	0.6	0.8	1.8	2.4	8	8	9	0	2	2	4	0	4	6	6	6	6	7	8
2	0.43	0.4	0.4	0.4	1.2	4.4	17	12	16	24	4	6	8	4	4	18	8	10	10	3	9	32
3	0.90	0.6	0.8	0.8	1.4	2.0	2.8	8	4	9	4	4	4	4	8	8	14	10	12	14	17	25
4	2.25	0.0	0.0	0.0	1.4	2.8	9.6	8	2	12	4	4	8	2	8	10	20	18	16	14	17	36
5	2.47	2.0	2.0	2.0	4.2	14	20	22	16	28	6	6	8	18	20	26	54	50	56	92	127	175
6	2.81	2.2	2.2	2.2	5.5	15	22	25	16	33	6	6	8	16	20	28	55	52	56	97	122	149
7	2.95	0.6	0.4	0.6	1.8	2.8	8.7	10	2	14	4	4	8	4	4	10	16	16	17	13	18	39
8	3.15	1.0	0.8	1.0	2.6	4.7	10	13	3	19	2	4	4	4	4	10	22	22	26	39	57	84
9	3.35	1.2	1.2	1.2	1.2	4.8	16	31	20	39	6	8	10	20	24	32	40	36	40	30	43	76
10	4.10	1.0	0.8	1.0	4.6	10	12	59	23	74	457	434	478	10	10	16	22	22	22	39	39	54
11	4.85	0.2	0.2	0.2	1.6	2.4	8.4	11	1	13	4	2	6	2	4	10	14	14	16	15	18	47
12	10.75	0.2	0.2	0.2	1.0	2.8	9.1	15	1	15	6	4	8	4	4	10	28	24	28	27	32	61
13	12.30	2.0	1.2	2.0	3.2	4.4	6.2	19	2	20	4	2	6	8	6	12	54	50	54	114	126	141

a = 0.05N ethylenediaminetetraacetic acid (pH 4.8); b = 1N hydroxylammonium chloride + 25% acetic acid; c = 0.5N hydrochloric acid

Source: Royal Society of Chemistry [68]

Table 3.26 Effect of organic matter on the extraction of copper by three methods

Sample	Organic matter %	Copper/mg kg $^{-1}$ by method a, b or c Copper/mg kg $^{-1}$ with 0.5N HCl extraction		
		a	b	c
1	0.15	0.90	0.90	1.00
2	0.43	0.50	0.67	1.00
3	0.90	0.90	0.44	1.00
4	2.25	0.67	0.17	1.00
5	2.47	0.79	0.57	1.00
6	2.81	0.76	0.48	1.00
7	2.95	0.71	0.14	1.00
8	3.15	0.68	0.16	1.00
8	3.15	0.68	0.16	1.00
9	3.35	0.79	0.51	1.00
10	4.10	0.80	0.32	1.00
11	4.85	0.85	0.08	1.00
12	10.75	1.00	0.07	1.00
13	12.30	0.95	0.10	1.00

a = 0.05N ethylenediaminetetraacetic acid; b = 1N hydroxylammonium chloride + 25% acetic acid; c = 0.5N hydrochloric acid
Source: Royal Society of Chemistry [68]

ethylenediaminetetraacetic acid extraction method incapable of extracting them.

Malo [112] evaluated four procedures employing 0.3N hydrochloric acid, citrate-dithionite pH3, nitrate-dithionite pH7, acetic acid-hydroxylamine, and nitric acid-hydrogen peroxide-hydrochloric acid for their potential for removing surface metals from river sediments. Prior to application of the partial extraction procedures, the samples were treated with 30% hydrogen peroxide on a steam bath to destroy organic matter.

The following are brief descriptions of the four partial extraction procedures and the nitric acid dry ashing treatment.

Hydrochloric acid
The sediment (5–10g) water mixture, following hydrogen peroxide treatment, is diluted to 200 mL with deionised water. Then 10 ml of 6 mol L $^{-1}$ hydrochloric acid is added, and the suspension is mixed and heated to just below boiling on a hot plate. Heating at this temperature is continued for 30 min. The hot mixture is filtered through a Whatman No 42 filter paper, or equivalent, and the filtrate is collected in a 250 mL volumetric flask. The residue on the filter is washed at least three times with hot dilute 5 vol % hydrochloric acid and the filtrate is cooled and brought to volume with 5 vol % hydrochloric acid.

pH buffered citrate-dithionite

Add 40 mL of pH$_3$ citrate buffer (160.0g citric acid and 71.3g sodium citrate dissolved in deionised water to give a final volume of 1000 mL) to 5–10g of peroxide treated sample and heat with occasional swirling for 1 h at 80 °C in an oven. Add 80 mL of purified dithionite-citrate solution (95.6g sodium citrate and 36.8g citric acid dissolved in 1L of deionised water, extracted by shaking with 1g ammonium pyrollidone diethyldithiocarbamate, 100 mL ethyl propionate, and 50g sodium dithionite, and re-extracted using only 100 mL of solvent) and hold at 80°C with occasional swirling for 3 h. If the solution temperature exceeds 80°C, insoluble sulphides will form. Remove from oven and flocculate the sediment with saturated sodium nitrate solution, centrifuge, and decant into a 250 ml volumetric flask. Wash the sediment twice with 10 mL citrate buffer, combine washings and extract, and adjust to volume with deionised water.

pH7 buffered citrate-dithionite

Add 50 ml of 0.3 mol L $^{-1}$ sodium citrate (88g Na$_3$C$_6$H$_5$O$_7$2H$_2$O L $^{-1}$) and 5 ml 1 mol L $^{-1}$ sodium bicarbonate (84g NaHCO$_3$ $^{-1}$) to the peroxide treated sample (5–10g). Heat to 80 °C in an oven, remove and quickly add 10 mL of sodium dithionite solution (100g Na$_2$S$_2$O$_4$L $^{-1}$). Heat to 80 °C in an oven and continue heating for 15 min. If the solution temperature exceeds 80 °C, insoluble sulphides will form. Cool and flocculate the sediment with saturated sodium chloride solution, centrifuge, and decant the clear supernate into a 250 mL volumetric flask. Wash residue twice with 10 mL portions of 0.3 mol L $^{-1}$ sodium citrate, and combine washings and extract. Adjust to volume with deionised water.

1 mol L $^{-1}$ hydroxylamine hydrochloride in 25 vol % acetic acid

Add 50 mL of mixed acid-reducing reagent (mix 150 mL of 25% (w/v) hydroxylamine hydrochloride and 350 mL of 35 vol % acetic acid) to the peroxide treated sample (5–10g) in a 250 mL conical flask. Place on a mechanical shaker and shake at room temperature for 4 h. Filter through a No 42 Whatman paper into a 250 mL volumetric flask. Wash the residue on the paper several times with deionised water, combine washings and filtrate, and adjust to volume with deionised water.

Nitric acid dry ashing extraction

Add 35 mL concentrated nitric acid and 5 mL 30% hydrogen peroxide to 5–10g of sample in a beaker and evaporate to dryness on a hot plate. Ash at 400–425 °C for 1 h in a muffle furnace and cool. Add 25 mL of acid mixture (200 mL concentrated nitric acid, 50 mL concentrated hydrochloric acid and 750 mL deionised water), 20 mL 10% ammonium chloride and 1 mL Ca(NO$_3$)$_2$4H$_2$O (11.8g/100mL). Heat gently for 15 min and cool for at least 5 min. Separate the residue by filtration or centrifugation, wash the residue twice with deionised water, and combine washings and extract in a 250 mL volumetric flask and adjust to volume with demineralised water.

Table 3.27 Mean concentration of major constitutients (all values in mg kg $^{-1}$ unless noted)

Determination	Method	Sample	
		1	2
SiO$_2$ %	(a)	50.3	84.1
	(b)	–	–
	(c)	1.05	0.14
	(d)	0.84	0.18
	(e)	0.24	0.045
	(f)	0.36	0.024
Al	(a)	106,000	9,690
	(b)	31,900	4,600
	(c)	19,200	1,280
	(d)	21,500	1,440
	(e)	10,700	313
	(f)	13,400	223
Fe	(a)	61,700	13,700
	(b)	60,200	10,000
	(c)	36,900	3,130
	(d)	53,000	9,000
	(e)	25,000	2,890
	(f)	12,700	876
Mn	(a)	701	203
	(b)	519	155
	(c)	693	139
	(d)	783	166
	(e)	524	17.8
	(f)	478	88.2

(a) = fusion; (b) = ashing HNO$_3$–H$_2$O$_2$–HCl; (c) = 0.3 mol L $^{-1}$ HCl; (d) = citrate-dithionite pH 3; (e) = citrate-dithionate pH 7; (f) = acetic acid hydroxylamine

Source: American Chemical Society [112]

The mean concentration of four replicate analyses for the major constituents by each of the six analytical procedures is presented in Table 3.27. Similar data for minor constituents are in Table 3.28. The ashing procedure was used as a reference point to determine the relative recoveries of trace metals because trace metal concentrations recovered by the ashing procedure approach their total concentration in the sediments. Tables 3.29 and 3.30 present data showing the mean recovery and standard deviation from the mean of major and minor sediment constituents relative to the total amount present and to the concentrations as determined by the nitric acid dry ashing digestion, respectively. Note that the pH7 citrate-dithionite and the 25% acetic acid extractions recovered significantly less of all constituents than the 0.3 mol L $^{-1}$ hydrochloric acid and pH 3 citrate-dithionite procedures.

Table 3.28 Mean concentration of minor constituents (all concentrations in mg kg $^{-1}$)

Determination	Method	Sample	
		1	2
Cu	(b)	1.62	169
	(c)	0.67	172
	(d)	0.87	159
	(e)	0.12	4.77
	(f)	0.11	126
Ni	(b)	7.07	238
	(c)	5.83	67.6
	(d)	5.69	111
	(e)	1.06	35.5
	(f)	1.18	25.5
Pb	(b)	50.9	91.9
	(c)	5.64	28.4
	(d)	20.3	45.4
	(e)	8.82	23.6
	(f)	2.49	17.7
Zn	(b)	12.9	123
	(c)	7.03	73.6
	(d)	5.27	82.5
	(e)	0.00	39.2
	(f)	0.01	35.2
Cd	(b)	0.06	0.88
	(c)	0.11	0.88
	(d)	0.64	1.22
	(e)	0.21	0.75
	(f)	0.04	0.62
Cr	(b)	2.90	18
	(c)	1.35	58.7
	(d)	1.64	184
	(e)	0.92	45.4
	(f)	0.02	17.3
Co	(b)	3.44	53.3
	(c)	2.55	48.9
	(d)	3.99	49.1
	(e)	1.64	30.5
	(f)	0.74	30.1

(b) = ashing HNO_3–H_2O_2–HCl; (c) = 0.3 mol L $^{-1}$ HCl; (d) = citrate dithionite pH 3; (e) = citrate-dithionate pH 7; (f) acetic acid–hydroxylamine

Source: American Chemical Society [112]

Table 3.29 Recovery of major constituents from all samples relative to total amount present (all values in %) = mg kg^{-1} method under test × 100

$$\frac{\text{mg kg}^{-1} \text{ method under test} \times 100}{\text{mg kg}^{-1} \text{ nitric acid dry ashing}}$$ (see Table 3.27)

		Ashing	0.3 mol L^{-1} HcL	pH 3 citrate-dithionite	pH 7 citrate-dithionite	Hydroxylamine acetic acid
SiO$_2$	X (mean)	a	1.5	0.89	0.27	0.32
	R (range)	a	0.13–5.14	0.11–2.23	0.03–0.73	0.01–0.93
	σ (std dev)	a	0.50	0.24	0.09	0.12
Al	X	30.9	10.0	8.2	2.7	3.0
	R	6.2–49.0	2.1–18.1	1.0–20.3	0.2–10.1	0.1–12.6
	σ	4.5	1.7	1.9	1.0	1.2
Fe	X	71.5	35.8	50.9	18.0	9.3
	R	29.6–97.7	13.1–71.9	16.4–85.9	4.1–40.5	2.6–20.6
	σ	7.8	6.4	7.3	3.6	2.1
Mn	X	62.0	73.2	68.1	36.3	49.8
	R	27.7–88.6	32.9–106	16.8–112	7.4–82.6	11.4–103
	σ	7.1	9.7	10.9	10.8	9.3

a SiO$_2$ is not brought back into solution following dry ashing of samples. No data are therefore available on SiO$_2$ concentration extracted.

Source: American Chemical Society [112]

Table 3.30 Mean recovery of metals from all samples relative to ashing procedure (all values in %) = mg kg^{-1} method under test × 100

$$\frac{\text{mg kg}^{-1} \text{ method under test} \times 100}{\text{mg kg}^{-1} \text{ nitric acid dry ashing}}$$ (see Table 3.28)

	0.3 mol L^{-1} HCl		pH 3 citrate-dithionite		pH 7 citrate-dithionite		Hydroxylamine acetic acid	
Al	34.1 ±	4.2	26.7 ±	5.2	8.8 ±	3.1	10.2 ±	4.0
Fe	50.0 ±	7.3	73.2 ±	8.9	26.0 ±	4.4	14.7 ±	4.5
Mn	117 ±	20.3	96.7 ±	15.9	51.7 ±	15.1	69.9 ±	14.0
Cu	93.0 ±	13.8	73.4 ±	9.7	3.2 ±	1.1	52.8 ±	10.5
Ni	60.1 ±	5.1	71.1 ±	11.9	23.2 ±	5.2	32.0 ±	9.5
Pb	48.9 ±	7.3	85.2 ±	4.5	23.5 ±	3.1	36.2 ±	6.1
Zn	75.9 ±	6.1	62.4 ±	6.7	34.3 ±	10.1	37.8 ±	7.8
Cd	110 ±	12.6	146 ±	22.0	70.9 ±	13.7	80.8 ±	15.6
Cr	59 8 ±	5.8	80.1 ±	11.3	46.5 ±	6.6	30.6 ±	6.9
Co	81.4 ±	5.9	104 ±	14.2	39.0 ±	3.9	47.2 ±	5.6

Source: American Chemical Society [112]

Table 3.31 Concentrations of metals in samples of varying particle size distribution (all values in μg/g)

	HCl	pH 3 citrate-dithionite
Sample No 4 (98.9% sand; ;1.1% silt and clay)		
Al	522	256
Fe	1,600	2,130
Mn	52.9	53.9
Cu	0.67	0.87
Ni	5.83	5.69
Pb	5.64	20.3
Zn	7.03	5.27
Cd	0.11	0.64
Cr	1.35	1.64
Co	2.55	3.99
Sample No 1 (68.8% sand; 31.2% silt and clay)		
Al	3,930	2,310
Fe	7,890	12,300
Mn	109	119
Cu	73.8	67.4
Ni	12.7	13.1
Pb	93.4	103
Zn	99.3	65.8
Cd	1.38	2.02
Cr	22.2	25.5
Co	5.11	9.42
Sample No 3 (11.0% sand; 89% silt and clay)		
Al	19,200	21,500
Fe	36,900	53,000
Mn	693	783
Cu	57.2	41.9
Ni	48.8	45.4
Pb	56.6	65.9
Zn	254	241
Cd	1.32	1.57
Cr	14.9	22.7
Co	32.9	33.8

Source: American Chemical Society [112]

Examination of the data in Table 3.30 indicates that 0.3 mol L^{-1} hydrochloric acid recovers more zinc and less cadmium, chromium and cobalt than the pH 3 extract. There is no difference in the recovery of copper, nickel and lead by either procedure. Table 3.31 indicates that this relationship is generally valid for samples of widely varying particle size distribution.

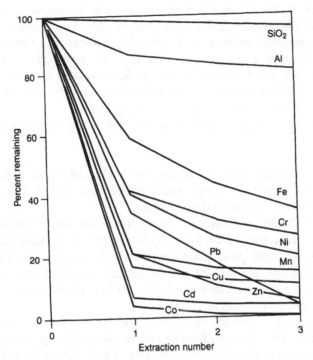

Fig 3.3 Mean recovery of metals by successive extractions with 0.3 mol L −I HCl solution
Source: American Chemical Society [112]

In addition to the extraction efficiency, another criterion applied to the selection of a procedure is the limitation of structural degradation. The concentration of silica and aluminium in the extracts is assumed to reflect the extent of degradation since the principal mineral structure in the samples consists of alumino-silicates. Data in Table 3.29 shows that both of the more efficient procedures recover approximately 1% of the total silica and 10% of the total aluminium. The lower end of the range in each case was for sample #4, a predominantly sand sample from the Mississippi River. Sample #8, a sediment of lateritic origin having highly weathered silica, is at the upper end of the silica range.

To test the significance with regard to structural degradation of extraction of approximately 1% silica and 10% aluminium, five samples were selected and subjected to three successive extractions with 0.3 mol L⁻¹ hydrochloric acid. The data for each sample are presented in Fig. 3.3. From these data, the recovery of trace metals more closely approximate the recovery of iron and manganese than silica and aluminium for the series of extractions. This ensures neither that all of the minor elements present in the surface coatings are recovered nor that there is no structural degradation. It does indicate that, for a wide range of physical and chemical sample characteristics, a significant part of the surface coating is being removed while solution of structural components is minimised. The extent to which a given sample will only be stripped of its surface coatings or will be degraded upon application of a specific chemical treatment is determined by mineral type, degree of weathering, composition of the coating matrix, and particle size.

Based on the results of this study, the following conclusions can be made:

- The acetic acid hydroxylamine hydrochloride and pH7 citrate-dithionite extractions are unacceptable because of low recovery efficiency and operational problems.
- The pH 3 citrate-dithionite procedure, although possibly more efficient than the 0.3 mol L^{-1} hydrochloric acid extraction, is not suitable for high production large-scale laboratory use principally because of the number of manipulations involved and difficulties associated with analysing the high solids concentration extracts by atomic absorption spectrometry.
- The 0.3 mol L^{-1} hydrochloric acid extraction requires fewer manipulations than the citrate-dithionite procedure, yields approximately the same recovery efficiency with minimum structural degradation, and results in an easily analysed solution.
- Serial extraction with 0.3 mol L^{-1} hydrochloric acid indicates that the solution of trace metals more closely follows the solution of iron and manganese than the structural components silicon and aluminium.
- The precision data obtained for the combined extraction analysis for metals by the three procedures compare favourably with published precision data for direct metal analysis by atomic absorption spectrophotometric methods.

Agemian and Chau [119] also carried out more detailed studies of some of these extraction techniques (aqua regia, 0.5N hydrochloric acid, 1N hydroxylamine hydrochloride plus 25% acetic acid, and 0.05N EDTA at pH 4.8 – see Table 3.23) for the determination of heavy metals in sediments from the River Rideau, Canada. Their aim was to find the metal extraction procedure which gave the greatest contrast between background sediment samples and anomalous sediment samples with enhanced metal levels due to mineral deposits occurring in the area from which the samples were taken. Sediment extracts were analysed by atomic absorption spectrometry.

In this study, the analysis was performed on the 80-mesh portion of the sediments. As this portion of a sediment provides the greatest contrast between anomalous and background samples [122], Oliver [123] showed that the particle size of the sediment strongly influenced the metal

Table 3.32 Mean background concentration of heavy metals in the sediments of the Rideau River using different extraction methods

| Metal | mg kg^{-1} | | | | |
	Total[a]	4.0N HNO$_3$ +0.7N HCl	0.5N HCl	1N NH$_2$OH–HCl +25% acetic acid	0.05N EDTA
Cd	–	–	0.3	0.4	0.2
Cu	42	13	11	2	8
Pb	84	24	17	16	16
Zn	94	67	36	20	16
Ni	72	23	8	3	3
Cr	108	37	7	2	1
Co	57	17	6	2	4
Mn	582	451	305	304	275
Fe	31,000	30,000	6,700	3,000	2,600
Al	46,200	15,000	3,930	1,150	620

[a] HF–HNO$_3$–HClO$_4$ with PTFE bomb decomposition

Source: Springer Verlag Life Science Journals, New York [119]

content. The metal content increased rapidly for surface areas up to about 10 m²/g and then levelled off. Thus when samples with different surface areas are compared, the 80–mesh portion of the sediment with surface area greater than 10 m² g^{-1} must be used to normalise all samples. The samples were air-dried and sieved to 80-mesh. Agemian and Chau [119] treated a representative sub-sample of the 80-mesh fraction of each sediment in each of the five following ways:

(a) 100 mg of sample was accurately weighed into a Parr 4745 acid digestion bomb and digested with a mixture of hydrofluoric nitric and perchloric acids, as described by Agemian and Chau [119]. This provides a measure of total metal in the sediment.
(b) 1g of sample was accurately weighed and digested in 100 ml of 4.0N nitric–0.7N hydrochloric acid solution for 2 h at 70–90°C.
(c) 5g of sample was digested with shaking in 100 ml of 0.5N hydrochloric acid solution overnight at room temperature.
(d) As in (b) but digested in 100 ml of 1N hydroxylamine hydrochloride plus 25% acetic acid solution.
(e) As in (b) but digested in 100 ml of 0.05N EDTA solution of pH 4.8.

Fig. 3.4a–g presents the levels of cadmium, copper, lead, zinc, nickel, chromium and cobalt along the river, using the four partial extraction methods, together with a total extraction method involving Parr bomb digestion. For cadmium, the total and 4.0N nitric acid plus 0.7N

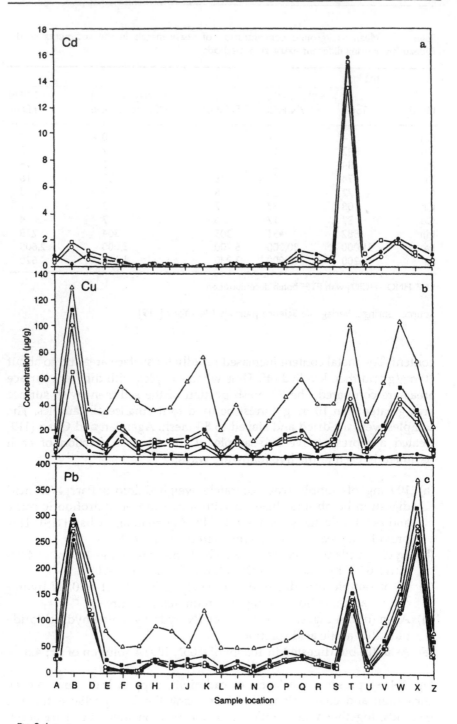

Fig. 3.4 a–c

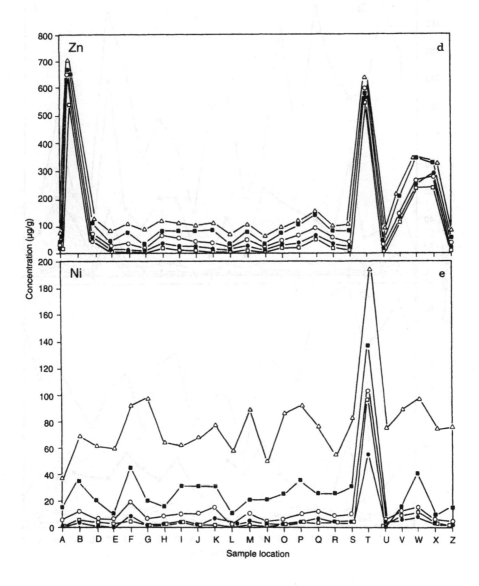

Fig 3.4 d–e

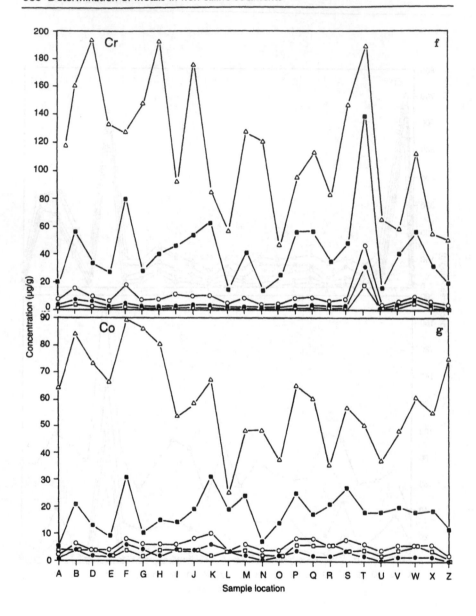

Fig. 3.4 a–g: The concentration (μg/g) of Cd, Cu, Pb, Zn, Ni, Cr and Co as a function of sample location in the bottom sediments of the Rideau River, extracted by the following methods: Δ = HF–HNO$_3$–HClO$_4$ (6 + 4 + 1) in Teflon bomb; ■ = 4.0N HNO$_3$ + 0.7N HCl; 0 = 0.05N HCl; • = 1N NH$_2$OH–HCl + 25% CH$_3$COOH; $\square$ = 0.05N EDTA.

Source: Springer Verlag Life Science Journals, New York [119]

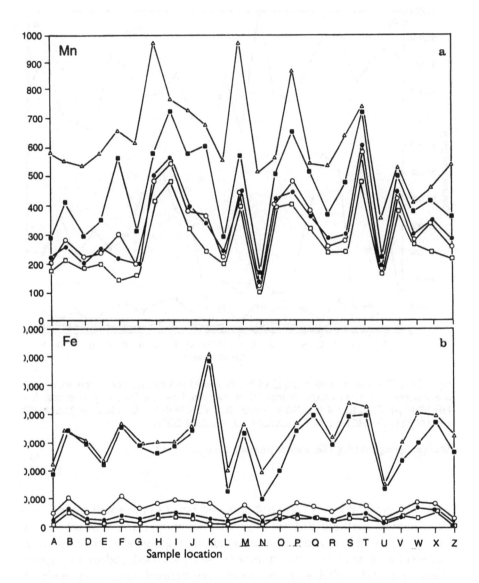

Fig. 3.5 a–b

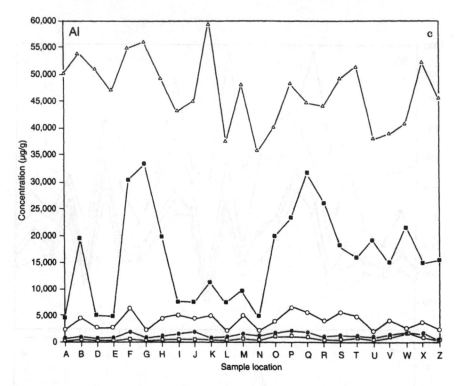

Fig. 3.5 a–c The concentration (µg/g) of Mn, Fe and Al as a function of sample location in the bottom sediments of the Rideau River, extracted by the following methods: Δ = HF–HNO$_3$–HClO$_4$ (6 + 4 + 1) Teflon bomb; ■ = 4.0N HNO$_3$ + 0.7 HCl; 0 = 0.05N HCl; • = 1N NH$_2$OH–HCl + 25% CH$_3$COOH; □ = 0.05N EDTA.

Source: Springer-Verlag Life Science Journals, New York [119]

hydrochloric acid levels were not obtained because the small sample size used in these methods did not provide quite enough metal in solution to give a detectable signal. The anomalous regions are visually quite apparent in spite of the small number of samples used, cadmium, copper, lead, zinc and nickel showing very well-defined anomalies with all methods. For chromium this is less clear, while no significant anomalies are seen for cobalt. Fig. 3.5a–c provide similar data for iron, manganese and aluminium. These three are major elements and they have no significant anomalies. All methods, except the total and 4.0N nitric acid plus 0.7N hydrochloric acid, provide the same type of anomalies and the same distribution of metal levels along the river. This means that the three weaker extraction methods (0.5N hydrochloric acid, 1N NH$_2$OH HCl plus 25% acetic acid and 0.5 N EDTA) attacked essentially the same part of the sediment. The other two methods showed the same type of distribution as the three weaker methods only in the cases of copper, lead, zinc, nickel and manganese. The absolute amount of metal extracted decreases as the extracting solution becomes weaker. Of the partial extraction methods, the 4.0N nitric–hydrochloric acid solution is highest followed by the 0.5N hydrochloric acid solution. The amount of metal extracted by the other two weaker solutions, relative to each other, depends on the metal involved. Table 3.32 shows the mean background levels of each metal in the Rideau River sediments calculated by excluding the anomalous samples.

The extracting efficiency of the methods will depend on the acid strength and the oxidising or reducing powers of the solutions. The extraction mixture of 4.0N nitric acid plus 0.7N hydrochloric acid at 90°C is certainly the strongest acidic and oxidising mixture other than the total extraction method. Therefore it is expected that all adsorbed metals, precipitated salts, oxides and hydroxides of iron and manganese and most organically complexed metals would be extracted. Weak hydrochloric acid will remove loosely bonded and adsorbed metals, precipitated salts, and possibly attack some of the less resistant silicates such as layered silicates. Both 4.0N nitric acid plus 0.7N hydrochloric acid and 0.5N hydrochloric acid may attack some silicates. This, in addition to the more effective extractions, has resulted in the higher means values shown in Table 3.32.

Of the three major elements studied, aluminium and iron are mainly found in the aluminosilicate crystal lattice. Only a small portion of manganese is found in this phase. This is confirmed in Fig. 3.5a where the distribution of manganese along the river is essentially the same with all five extraction methods. This means that they are all attacking the same phase of manganese in the sediment. Fig. 3.5b and Table 3.32, however, show that only the total and 4.0N nitric acid plus 0.7N hydrochloric acid methods extract the bulk of iron and aluminium respectively.

The three weaker methods show only a very small percentage of the total iron and aluminium extracted, with the same uniform distribution along the river. This indicates that they are attacking only the non-detrital part of the sediment. With iron, 4.0N nitric acid plus 0.7N hydrochloric acid provides results nearly identical to the total extraction. For aluminium a different distribution is obtained with considerable attack on the silicate lattice.

As for the trace elements, the total and 4.0N nitric acid plus 0.7N hydrochloric acid extraction methods give different distributions to the weak extraction methods, especially for chromium and cobalt. They also show anomalous metal levels at locations where the weaker methods show only background levels. This is an indication of variations in the rock matrix of the area and attack of the detrital phase of the sediment by both of these methods.

The above discussion showed that the three weaker extraction methods satisfy the essential requirement of minimal attack on the silicate detrital

Table 3.33 Trace metal concentration in a River Arno sediment after extraction with inverse aqua regia with and without hydrofluoric acid addition (mg kg $^{-1}$)

	Hg	Cd	Pb	Cu	Ni	Mn	Zn
4.5 ml HNO$_3$ 1.5 ml HCl	0.99	1.01	64.2	64.6	63.0	573	259
4.5 ml HNO$_3$ 1.5 ml HCl 2.0 ml HF	0.91	1.03	60.7	59.5	60.0	558	264

Source: Springer Verlag Chemie GmbH, Germany [125]

lattice. The remaining essential requirements are liberation of the metals in question from organic matter, satisfactory simultaneous extraction of the metals in question from the non-detrital phases in the sediment, and a high contrast of anomalous to background samples.

Van Valin and Morse [124] showed that metal removal from sediments by a number of different leaching methods can be related to the pH of the leaching solution. As the pH decreases, metal release to solution increases. Many of the published techniques are pH related.

Breder [125] used different mixtures of nitric, hydrochloric and hydro-fluoric acids in his determinations of total metals in river sediments. In the absence of hydrofluoric acids, 1g of River Arno sediment yielded 600 mg of silicaceous residue when digested for 3 h with 4.5 ml nitric acid and 1.5 ml hydrochloric acid at 150°C under pressure. The inclusion of 2 ml of hydrofluoric acid reduced the residue to 200 mg but further increases in the hydrofluoric acid addition up to 6 ml did not further reduce the weight of residue. He also observed that practically the same metal analyses were obtained with and without hydrofluoric acid (Table 3.33).

Breder [125] also points out that the use of hydrofluoric acid can lead to analytical errors. Arsenic and selenium fluorides, for example, are very volatile. Also, the compensation system for unspecified light losses in the atomic absorption spectrometry can fail to operate at high fluoride concentration. Breder [125] claims that Agemian and Chau [68] did not appreciate this and consequently obtained high results in nitric-perchloric-hydrofluoric acid digestions (Table 3.33).

Table 3.34 shows results obtained in 3 h pressure digestions of 0.2–1g of a River Rhine reference sediment (ICEI) with 4–8 ml of various concentrated acids, *viz* 3:1 nitric acid–hydrochloric acid in PTFE vessels or 5:1 nitric-perchloric acid or 3:1 hydrochloric-nitric acid (aqua regia) or nitric acid alone.

Only pressurised digestion with nitric acid provided good precision. The mean values for cobalt and nickel agree well with the reference values. The mercury data are somewhat lower. The data for other elements tend to be somewhat higher than the ICEI mean values. Only a small part of the silver compounds are brought into solution by nitric acid. An addition of hydrochloric acid, however, achieves dissolution and stabilisation by the formation of silver chlorocomplexes. Distinctly lower concentrations were found for silver, cobalt, nickel, lead and zinc after the open digestion with aqua regia for 1 h in a 100 ml quartz Erlenmeyer flask, covered with quartz watch glasses. After 3 h heating in 250 ml quartz flasks with reflux condenser, the concentrations of metals approach those of the pressurised digestion except for nickel. The precision of the analytical data from the digestion experiments with aqua regia and also with nitric–perchloric acid (5:1) is only satisfactory if large amounts of sediment and acids are used. Nitric-perchloric acid incompletely digests silver and nickel. Microwave oxygen plasma digestion gives excellent precision but cannot be applied in the case of mercury, selenium and arsenic.

Table 3.35 shows the excellent agreement for total element contents in various reference sediments obtained by pressurised digestion with 3:1 nitric-hydrochloric acid.

Welte *et al.* [82] compared two extraction methods in the speciation of heavy metals in sediments. The first method was based on extraction with ammonium acetate in nitric acid, followed by treatment with hydroxylamine chlorhydrate in acetic acid. The other method involves extraction with 0.01 mol L^{-1} nitric acid and treatment with sodium dithionite and sodium citrate. The direct residues in both cases were digested with aqua regia. The metals in the various supernatants were determined by plasma emission spectrometry for iron, manganese, zinc, copper, chromium, nickel, lead and aluminium and by atomic absorption for cadmium, cobalt and arsenic. A portion of the sediment was freeze-dried and dissolved to determine total metal content.

Trefrey and Metz [120] studied the effect of pH on metal release/uptake processes in sediments. In these experiments they contacted 0.4g of sediment with 20 mL phthalate buffers ranging in pH from 2.2 to 6.0 for periods up to 24 h at 20°C, then centrifuged off the aqueous phase prior to analysis for cadmium, copper, iron, manganese, lead and zinc by flame or flameless atomic absorption spectrometry.

Total dissolution of the samples was also carried out to establish the fraction of metal removed at each pH. Digestion of 0.400g samples was performed in Teflon beakers by using concentrated nitric acid, hydrofluoric acid and perchloric acid in a three-step process. First 1 mL of perchloric acid, 1 mL of nitric acid and 3 mL of hydrofluoric acid were added, a Teflon watch cover was put in place, and the sample was heated at 50°C until a moist paste was obtained. The mixture was heated for 3 h at 80°C with an additional 2 mL of nitric acid and 3 mL of hydrofluoric acid and then brought to dryness. Finally, 1 mL of nitric acid and 30 mL of deionised distilled water were added to the sample and heated to

Table 3.34 Precision of the data after digestion procedures on River Rhine Reference Sediment (ICE I)

	Ag	Cd	Co	Cr	Cu	Hg	Mn	Ni	Pb	Zn
	9.32	9.65	21.9	450	244	4.4	704	79	170	950
	±1.6	±1.2	±2	±27	±14	±0.33	±51	±6.6	±9.7	±4.2
0.5 g Sediment, 5 ml HNO₃ 3 h pressurized digestion, n = 5										
x (µg g⁻¹)	–	10.3	23.0	483	265	4.1	763	77.1	180	1.028
2 $s_{\bar{x}}$ (µg g⁻¹)	–	0.33	0.998	15.0	4.0	0.12	18.9	4.8	6.8	31.5
V (%)	–	3.2	4.3	3.1	1.5	2.9	2.5	6.2	3.8	3.1
0.4 g Sediment 3 ml HCl/l ml HNO₃ 3 h pressurized digestion, n = 5										
x (µg g⁻¹)	8.1	9.5	16.0	449	256	4.4	687	59.9	154	875
2 $s_{\bar{x}}$ (µg g⁻¹)	0.2	0.24	0.26	5,3	3.3	0.24	9.3	1.1	2.7	12.2
V (%)	2.5	2.5	1.6	1.2	1.3	2.9	1.4	1.8	1.8	1.4
0.3 g Sediment, 3 mL HNO₃/l mL HCl 1h open digestion, n = 3										
x (µg g⁻¹)	8.1	9.5	16.0	449	256	4.4	687	59.9	154	875
2 $s_{\bar{x}}$ (µg g⁻¹)	0.2	0.24	0.26	5.3	3.3	0.24	9.3	1.1	2.7	12.2
V (%)	2.5	2.5	1.6	1.2	1.3	5.5	1.4	1.8	1.8	1.4
1 g Sediment 15 mL HCl/5 mL HNO₃ 3h øpen digestion with reflux condenser, n = 3										
x (µg g⁻¹)	8.7	9.4	18.0	403	244	4.3	778	58.7	162	912
2 $s_{\bar{x}}$ (µg g⁻¹)	0.84	0.24	1.04	20.4	7.2	0.52	23.7	3.5	5.8	10.7
V (%)	9.7	2.6	4.8	5.1	3.0	12.1	3.0	6.0	3.6	1.2

Table 3.34 continued

	Ag 9.32 ±1.6	Cd 9.65 ±1.2	Co 21.9 ±2	Cr 450 ±27	Cu 244 ±14	Hg 4.4 ±0.33	Mn 704 ±51	Ni 79 ±6.6	Pb 170 ±9.7	Zn 950 ±4.2
1 g Sediment, 15 mL HNO$_3$/3mL HClO$_4$ 3 h open digestion with reflux condenser, $n = 5$										
x (µg g^{-1})	–	10.2	22.3	451	264	4.7	773	61.2	157	976
2 $s_{\bar{x}}$ (µg g^{-1})	–	0.32	2.8	20.2	15.9	0.16	8.8	3.6	6.7	32.8
V (%)	–	3.1	12.6	4.5	6.0	3.4	1.1	5.9	4.3	3.4
0.3 g Sediment, 4 h digestion in oxygen plasma[a] $n = 3$										
x (µg g^{-1})	7.0	9.2	14.3	495	237	2.9	657	45.1	152	853
2 $s_{\bar{x}}$ (µg g^{-1})	0.2	0.26	0.24	6.6	3.3	0.24	22.4	0.18	2.3	14.5
V (%)	2.9	2.8	1.7	1.3	1.4	8.3	3.4	0.4	1.5	1.7
0.4 g Sediment, 3 mL HNO$_3$/1 mL HCl 3 h pressurized digestion, $n = 17$										
x (µg g^{-1})	8.5	9.8	21.6	471	262	4.6	762	75.7	173	972
2 $s_{\bar{x}}$ (µg g^{-1})	0.36	0.16	1.1	9.7	4.7	0.12	12	2.6	3.7	12.9
V (%)	4.2	1.6	5.1	2.1	1.8	2.6	1.6	3.4	2.1	1.3

[a] The digestions were carried out in quartz dishes in anoxygen plasma produced by microwave (International Plasma Corporation 4000). The pressure was 0.1–5 mbar, the oxygen flow 300 ml min^{-1}, the electrical power 100–150 W. The residue was shaken with 1 ml of tridistilled water and 1 mL HCl.
[b] ±2 standard deviation

Source: Springer Verlag Chemie GmbH, Germany [125]

Table 3.35 Trace metal determinations in Standard Reference Sediments (values in mg kg^{-1}) dry weight

Standard Reference Material	Cd	Pb	Cu	Mn	Cr	Ni	Zn
NBS 1645 (River Sediment)							
certified value	10.2 ± 1.5	714 ± 28	109 ± 19	785 ± 97	—*	45.8 ± 2.9	1720 ± 16.9
own value	9.6	706	105	793	—	45.7	1642
IAEA SL–1 (Lake Sediment)							
certified value	0.26 ± 0.5	37.7 ± 7.4	30 ± 5	—*	104 ± 9	44.9 ± 8.0	223 ± 10
own value	0.30	36.8	28	—	94	42.0	211
NRC Canada, MESS–1 (Marine Sediment)							
certified value	0.59 ± 0.10	34.0 ± 6.,1	25.1 ± 3.8	513 ± 25	71 ± 11	29.5 ± 2.7	191 ± 17
own value	0.66	31.6	21.0	493	68	29.2	191
NRC Canada, BCSS–1 (Marine Sediment)							
certified value	0.25 ± 0.04	22.7 ± 3.4	18.5 ± 2.7	229 ± 15	123 ± 14	55.3 ± 3.6	119 ± 12
own value	0.25	19.2	20.0	245	116	54.8	103

0.2 g Sediment pressure digested with 3 ml nitric and 1 mL hydrochloric acid
* No trace amount

Source: Springer Verlag Chemie GmbH [125]

dissolve perchlorate salts and reduce the volume. The completely dissolved and clear samples were then diluted to 25 mL with deionised distilled water. Analysis by flame or flameless atomic absorption spectrometry followed.

Plots of metal leached vs pH for Mississippi River suspended particulates provide a good representation of the overall trends observed. Copper, iron and lead concentrations follow similar patterns of sharp increases in concentration below pH 4. At pH values >4, < 1 µg of copper and lead (g $^{-1}$) is leached. For both copper and lead, the same percentage (20%) of total metal is removed at pH 2.3, whereas only 2% of the total iron is removed at the lowest pH.

In contrast to the iron group, copper and manganese show a more continuous increase in the amount leached with decreasing pH. At pH 2.3, > 80% of the total cadmium and > 50% of the total manganese are removed from this sample. At higher pH values (4–6), 10–40% of the total cadmium and manganese is removed. Zinc removal follows the gradual increase with decreasing pH found for the manganese group, but with a small percentage of total metal leached at each pH, a characteristic of the iron group.

These results show that a higher percentage of the total cadmium and manganese relative to copper, iron, lead and zinc will be released from suspended particulates with decreasing pH. In general, metal leaching is related to the final pH of the leaching solution.

A pH approach to leaching may be less practical in cases where samples have a large percentage of carbonates, metal oxides or other acid-consuming components. In carbonate-rich (30 –> 95% carbonate) samples, Trefrey and Metz [120] approached the acid consumption problem by slowly titrating samples with 0.01N hydrochloric acid (or 1N hydrochloric acid for large samples or carbonates) until the carbonate reaction is completed. At this point, the pH may be adjusted by adding acid or base. Such increased handling is less desirable, but use of a greater buffer capacity solution in such samples will only increase the matrix problems already introduced by high calcium concentrations.

Del Castilho and Rix [139] have reviewed the suitability of the ammonium acetate extraction method for the prediction of heavy metal availability in sediments.

Voutsinan-Taliadouri [129] has reported an acid extraction method for surface sediments that can give the 'anthropogenic fingerprint' of the sample. The basis of this claim is that the cold dilute hydrochloric extractant should affect only the non–residual part of the metals.

3.67.13.2 Microwave extraction

In addition to the bomb digestion techniques, the technique of digesting sediments in PTFE lined bombs in a microwave oven has also been

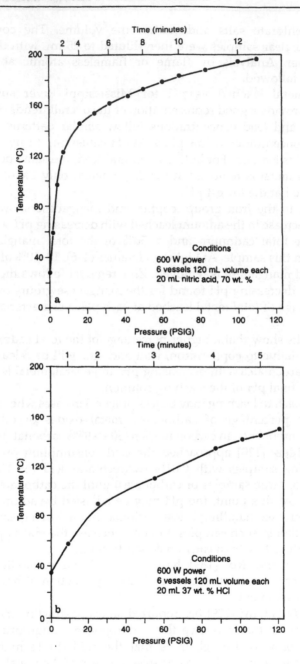

Fig. 3.6 a,b CEM microwave digestion system Model MDS–81D; temperature-pressure relationship with (a) concentrated nitric acid; (b) concentrated hydrochloric acid
Source: Pittsberg Conference and Exposition on Analytical Chemistry and Applied Spectroscopy, American Chemical Society [93]

discussed [92, 93]. It has been demonstrated [93] that closed vessel microwave sediment digestions can reduce sample dissolution times, be applicable to a wide variety of sample matrices, allow for dissolution flexibility and produce analytical data comparable to those obtained by conventional digestion procedures. Digestion at 200°C with 70% nitric acid in a 120 mL closed vessel produces an internal pressure of 120 psig, whilst digestion with 37% hydrochloric acid at 153°C produces an internal pressure of 100 psig.

Fig. 3.6a is a temperature–pressure plot obtained for the simultaneous heating of six closed vessels, each containing 20 mL of concentrated nitric acid, at 600 W power. As can be seen, the temperature and pressure rise smoothly to 193°C and 100 psig in 12 min.

Fig. 3.6b is a temperature-pressure plot obtained for the heating of six closed vessels, each containing 20 mL of concentrated hydrochloric acid, at 600 W power. For this acid, the curve also increases smoothly to 140°C and 101 psig in about 4.5 min.

At such elevated temperatures, these and other acids become more corrosive. Materials that digest slowly, or will not digest at the atmospheric boiling points of the acids, become more soluble, so dissolution times are greatly reduced.

The closed vessel system used and shown in Fig. 3.7 consists of a vessel body, safety pressure relief valve, vessel cap, venting nut and tubing, all of which are constructed of Teflon PFA material. Teflon PFA is transparent to microwave energy, allowing microwaves to pass through the vessel and couple directly with a digesting acid, thus making it possible to obtain elevated temperatures very rapidly. This vessel system is designed to operate at internal pressures up to 120 psig. Above 120 psig, the safety valve will open, allowing the system to vent into a collection container, thus lowering the pressure inside the vessel. The valve then reseals, allowing pressure to increase again. This automatic venting is a safety feature of the closed vessel system to ensure that vessel rupture, due to excessive pressure, will not occur.

Figs. 3.8a,b shows the temperature-pressure development plots obtained for 1g of NBS SRM 1645 standard river sediment during the first 30 min of digestion, respectively with the stated quantities of 50% nitric acid and nitric acid–hydrogen peroxide. Excellent agreement with reference values were obtained using either method of digestion (Table 3.36).

Various workers have applied microwave extraction techniques to the digestion and extraction of metals in non-saline sediments [20, 70, 92, 140–148, 233].

Kingston and Walter [146] compared microwave digestion with conventional dissolution methods for the determination of metals in sediments.

Mahan et al. [92] used a microwave digestion technique in the sequential extraction of calcium, iron, chromium, manganese, lead and zinc in non-saline sediments (see section 2.67.12).

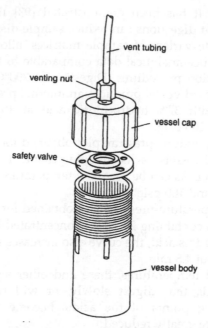

vent tubing

venting nut

vessel cap

safety valve

vessel body

Fig. 3.7 Digestion vessel assembly
Source: Pittsberg Conference and Exposition on Analytical Chemistry and Applied
Spectroscopy, American Chemical Society [93]

Table 3.36 NBS SRM 1645 sediment, microwave digestion

Element	1 : 1 HNO₃ : H₂O digestion^a Amount Recovered %	HNO₃–H₂O₂ digestion^b Amount Recovered %	Certified Value %
As	0,0060, 0.0060	0,0075, 0.0070	0.0066
Cd	0.0012, 0.00012	0.0011, 0.0012	0.0012 ± 0.00015
Cr	3.00, 2.98	3.04, 2.96	2.96 ± 0.28
Cu	0.0122, 0.0113	0.0118, 0.0119	0.0109 ± 0.0019
Mg	0.72, 0.72	0.70, 0.70	0.74 ± 0.02
Mn	0.0790, 0.0780	0.0720, 0.0725	0.0785 ± 0.0097
Ni	0.0050, 0.0050	0.0044, 0.0044	0.00458 ± 0.00029
Pb	0.0736, 0.0737	0.0736, 0.0733	0.0714 ± 0.0028
Se	0.0001, 0.0001	0.0001, 0.0001	0.00015
Zn	0.170, 0.168	0.160, 0.160	0.172 ± 0.017

^a 1 g sample digested with 20 mL 1 : 1 HNO₃ : H₂O for 10 min, then microwave power reduced to
maintain temperature at 180°C and pressure at 1000 psig for further 50 min
^b 1 g sample digested with 1 : 1 HNO₃ : H₂O in open vessel for 10 min at 180 W, cooled and 5 mL
conc HNO₃ and 3 mL 30% H₂O₂ added. Vessel sealed and power applied for 15 min at 180 W
(115°C) then H₂ 15 min at 300 W (152°C) at 38 psig

Source: American Chemical Society Pittsberg Conference and Exposition on Analytical
Chemistry and Applied Spectroscopy, American Chemical Society [93]

Millward and Kluckner [142] have shown metals showing the poorest precision in conventional digestion methods showed the greatest improvements when microwave digestion was used to study a standard reference sediment.

Kammin and Brandt [141] showed that microwave digestion had considerable promise as a high speed alternative to the Environmental Protection Agency digestion method 305.0 for the determination of trace metals in non-saline sediments.

Kratchvil and Mamba [140] showed that all the zinc and copper were released from non-saline sediments within 7 min using a commercial microwave oven.

Nieuwenholze et al. [147] analysed six reference sediments after microwave aqua regia extraction. The results obtained showed close agreement with the reference values, and microwave extraction gave the same or slightly higher results than those obtained by conventional reflux extraction methods for seven metals tested in 30 samples.

Elwaer and Belzile [70] have compared the use of a closed vessel microwave dissolution method and conventional hotplate digestion for the determination of arsenic in lake sediments. A mixture of hydrochloric acid, nitric acid and hydrofluoric acid with microwave digestion gave best recoveries. Hotplate digestion gave a poor recovery.

Elawer and Belzile [70] have compared the use of a closed vessel microwave assisted dissolution method and conventional hotplate digestion for the determination of selenium in lake sediments. A mixture of hydrochloric acid, nitric acid and hydrofluoric acid with microwave digestion gave best recoveries.

Chakrabarty et al. [20] determined chromium in non-saline sediments by microwave-assisted sample digestion followed by atomic absorption spectrometry without the use of any chemical modifier.

Sturgeon et al. [148] have demonstrated that a continuous flow microwave-assisted digestion of sediment samples gave an average recovery of trace elements of 90% with good recovery.

3.67.13.3 Sequential extraction

The terms *dissolved, suspended* and *sediment* fraction are frequently referred to in the study of natural water systems whether it be in a stream, lake or marine environment. In addition to a knowledge of the overall composition of these three primary fractions, there is interest in the exchange of chemical constituents among these primary fractions and the potential effect upon biota and human health. Relatively little is known about the way that heavy metals are bound to sediments or the ease with which they may be released, despite the fact that sediments are considered to be the ultimate sink for metals released into the environment. For example, if one is interested in the mobility, availability, uptake and transport of lead in aquatic systems, little information is gained from the measurement of *total* lead in the sediment other than an estimate of the total lead burden in the sink. Knowledge of how the lead is bound in the sediment is necessary to accurately assess potential modes of transport and future impact on the environment. To address this problem, a number of sequential extraction schemes have been developed that are designed to mimic environmental conditions that might be encountered in nature. These are procedures that sequentially partition metals bound in sediments or other particulates into fractions that are likely to be affected by differing environmental conditions. The scheme developed principally by Tessier *et al.* [150] partitions metals into five fractions (see section 2.60.12.7 for further detail).

Mahan *et al.* [92] carried out sequential extractions of calcium, iron, chromium, manganese, lead and zinc and obtained comparable results to conventional techniques in less time.

Kheboian and Bauer [126] have studied the accuracy of selective extraction procedures for metal speciation in model aquatic sediments. These workers have pointed out that aquatic sediments consist of several different geochemical phases that act as reservoirs of trace metals in the environment. These phases include carbonates, sulphides, organic matter, iron and manganese oxides, and clays, all of which may occur in a variety of structural forms. Historically, analytical determination of the distribution of metals among these phases has been approached by phase-selective extractions involving single or multiple 'selective' extracting reagents. Although such procedures are one of the few ways of exploring an important aspect of environmental trace element chemistry, their status as useful analytical tool is controversial. The controversy embraces both philosophical and practical matters. Philosophically, selective extraction requires existence of discrete phases that may be dissolved independently. Many sediments in fact may not contain chemically or physically distinct phases; thus element 'distribution' patterns produced by extraction methods may be artifactual. In addition, even when phases are discrete, there are technical difficulties associated with achieving complete and selective dissolution and recovery of trace metals from those phases.

Despite these problems, interest is growing rapidly in development and application of these methods. More than 10 different sequential extraction procedures have been developed. In particular the approach of Tessier *et al.* [150] (section 2.60.12.7) has been applied widely and has become a benchmark against which more recent developments have been compared. Because of this status, Kheboian and Bauer [126] focused on this method. They recognised two major experimental problems with sequential procedures: non-selectivity of extractants, and trace element redistribution among phases during extraction. Generally it is difficult to associate a given extractant reagent with a particular physicochemical phase. Instead the specificity of an extractant is operationally defined according to what it extracts, eg acetate buffer extracts the 'acetate-buffer-extractable' phase. Although it is often believed that this particular phase includes carbonate minerals, these minerals may not be *uniquely* attacked by this reagent. This is the non-selectivity problem. For the other problem – redistribution – trace elements liberated by one extractant have an opportunity to reassociate with remaining undissolved sediment components before recovery of the extract. Both processes scramble the true phase association of the trace elements thus complicating interpretations.

Most validation studies have focused on selectivity and extraction completeness rather than redistribution, and predominantly from the standpoint of major species rather than trace elements. For example, Tessier *et al.* [150] measured aluminium, silicon, calcium, sulphur and inorganic carbon, and organic carbon in leachates and residues of river sediments to establish removal of major phases. A significant limitation of these studies is that one must assume that selective removal of a major phase means also selective removal of its associated trace elements. There has been only a minimal amount of work on selectivity with respect to trace elements and this has been limited to extractions of single phases with single reagents.

These experiments overlook the non-trivial problem of redistribution, which appears only when extractions are performed in the presence of mixed substrates.

Kheboian and Bauer [126] prepared the major sedimentary geochemical phases and characterised them by X-ray diffraction and scanning electron microscopy. A specific trace metal was doped into each phase by adsorption or coprecipitation, content being verified by atomic absorption spectrometry. The phases and metals were calcite (Pb), iron sulphides (Zn), humic acid (Cu) and iron oxyhydroxides (Cu or Ni). Three different model sediments were prepared by combining phases and diluting with silica or illite. Measurements of extracts by atomic absorption spectrometry and inductively coupled plasma indicated that trace metals were not recovered in the appropriate fractions. This implies that 'selective extraction' procedures as currently practiced may not be suitable for distinguishing the phase association of metals in real sediments.

Jones [151] has suggested that these experiments of Kheboian and Bauer [126] suffer from a number of severe flaws related to (i) the representativity of their model sediments, (ii) the validity of the approach itself, and (iii) experimental problems. In the opinion of Kheboian and Bauer [126], these flaws sufficiently weakened their argument so as to make their general conclusion unwarranted.

Representativity of the model sediments
In preparing their model sediments, Kheboian and Bauer [126] combined in equal proportions large quantities of both freshly prepared iron oxides

(amorphous iron oxides or goethite of poor crystallinity) and sulphides (poorly crystalline mackinawite or greigite) ie phases that would not coexist in natural sediments. These two types of compounds are indeed mutually exclusive in a given sediment stratum; recently formed iron oxides (represented in their model sediment by amorphous iron oxides or goethite) are present in the upper oxic layer of natural sediments whereas recently formed sulphides (represented in their model sediment by mackinawite or greigite) exist in the lower anoxic strata. In oxic sediments, only resistant sulphides (eg pyrites) should be present whereas in the FeS-bearing strata, reactive forms of iron(III) should have disappeared. Kheboian and Bauer [126] justify the representativity of their mixture of phases by stating that (i) biological activity can carry oxygen to the anoxic zone of the sediments, thus oxidising some iron, and (ii) collection of a sediment layer (oxic or anoxic) may inadvertently include material from another layer. Biological activity (bioturbation) would only lead to trace quantities of freshly formed iron oxides in the anoxic strata, not to the large quantities initially present in their model sediments. As Kheboian and Bauer [126] suggest, (careless) sampling methods might lead to roughly equivalent proportions of both types of solid phases (fresh iron oxides and sulphides), as used in their model sediments; however, in such a case, possible problems in determining metal partitioning should be attributed to inadequate sampling practices rather than to the limitations of the extraction methods. This lack of representativity of the model sediments precludes extrapolation of their results to natural sediments.

Validity of the approach
The model sediments, as prepared by mixing a phase doped with a trace metal with other phases not doped with that metal, differ markedly from natural sediments where the metal would have already been distributed among the various sediment phases and where the driving force for redistribution during extraction would correspondingly be much lower; in other words, the model system is biased toward trace metal redistribution. Based on work on competitive adsorption, one would indeed expect that a trace metal added initially in a given phase would tend to redistribute among the various phases of the model sediments (from the doped phase to the uncontaminated ones), even in the absence of any chemical extractants. The final distribution will depend on the relative binding strengths of each component for the trace metal and the number of available binding sites of each component. It can thus be expected that application of a sequential extraction procedure to such an inherently unstable system will lead to incomplete recovery of the added trace metal in its initial form, ie in the fraction corresponding to the phase in which it was added to the model sediments. It is thus in the view of Jones [151] incorrect to attribute this incomplete recovery solely to

artifacts in the extraction procedures. To be valid for the study of the extent of postextraction readsorption, an approach using model sediments should involve solid phases that are all initially equilibrated with the metals under study; this would eliminate the 'normal re-equilibration' effect that is present in the experiments of Kheboian and Bauer [126]. By comparison of the known initial partitioning with that obtained with sequential extractions, it would then be possible to determine without ambiguity the degree of postextraction readsorption.

A more appropriate approach to test the degree of readsorption is a 'standard addition' procedure, in which trace metals are added to the extracting solution during the extraction of natural sediments and the recovery of the added metals is verified.

Experimental inconsistencies
The experiments described by Kheboian and Bauer [126] also suffer from several experimental inconsistencies that make their results difficult to interpret.

Reagent concentrations
The sequential extraction procedure was used without due consideration of the unusual nature of the model sediments; reagent concentrations should have been adjusted to take into account the fact that the concentrations of certain phases in their model sediments were much higher than those for which the procedure was developed. For example, their model sediments contain up to 18% iron oxides, which is much higher than the concentrations usually found in natural sediments (typically less than 4%). As they neglected to increase the concentration of hydroxylamine hydrochloride prescribed in the procedure (0.04M), no excess of this reagent was available to reduce all the amorphous Fe(III) oxyhydroxides present in some of their model sediments (eg final concentration 0.8M Fe(III) for 1g of sediment P in 20 mL of extraction). In the experience of Jones [151], an excess of hydroxylamine hydrochloride is necessary to achieve reduction of Fe(III).

Similarly, because they used relatively high concentrations of solid carbonates in their model sediments (up to 17% calcite), Kheboian and Bauer [126] should also have increased the concentration of acetate buffer that was suggested in the procedure for sediments of low carbonate content. Unfortunately Kheboian and Bauer [126] did not determine the efficiency of the acetate buffer in dissolving the carbonate phase in their complete model sediments.

Xiao-Quan and Biu Chen [127] evaluated sequential extraction methods for the extraction of a range of elements from humic sediments. The elements discussed included calcium, strontium, vanadium, iron, aluminium, copper, manganese, nickel, zinc, cobalt, lead and chromium. The three-stage sequential extraction procedure for speciation of heavy metals, proposed by the Commission of the European Communities Bureau of Reference (BCR), was found to be repeatable and reproducible with some modifications [152].

Thomas et al. [153] examined the Commission of the European Communities Bureau of Reference (BCR) sequential extraction procedure for river sediments and concluded that the method worked well.

On the other hand, Whalley and Grant [154] have questioned the reproducibility of the Commission of the European Communities Bureau of Reference (BCR) three stage extraction procedure for the speciation of metals in sediments.

In a sequential extraction study of chromium from contaminated aquifer sediments Asikainen and Nikolaidis [155] found that 65% of the chromium was extractable. Of this amount 25% was exchangeable, 11% was bound to organic matter and 30% was bound to iron and manganese oxide surfaces.

In a study of sequential extraction of heavy metals in anoxic sediment slurries, Kong and Shiu [130] found that the amendment of different amounts of metals into the sediment slurries resulted in a different percentage distribution of metals in the various fractions extracted.

Sahuquillo et al. [21] studied the sequential extraction of chromium from sediments. These workers proposed a three-step sequential extraction scheme using acetic acid, hydroxylamine hydrochloride and ammonium acetate as extracting agents. From the results obtained, it is recommended that chromium is measured in the acetic acid and hydroxylamine extraction steps by electrothermal atomic absorption spectrometry and chromium in the ammonium acetate extraction step by flame atomic absorption spectrometry.

Ure [128] has recently reviewed single and sequential extraction schemes for trace metal speciation in sediments. A weak acid extraction method for surface sediments has been reported that can give the 'anthropogenic fingerprint' of the sample. The basis of this claim is that the cold dilute hydrochloric acid extractant should affect only the non-residual part of the metals [129].

In a study of sequential extraction of heavy metals in anoxic sediment slurries, Kong found that the amendment of different amounts of metals into the sediment slurries resulted in a different percentage distribution of metals in the various fractions extracted [130].

3.67.13.4 Miscellaneous

Zaggia et al. [149] has studied the extraction of anthropogenic heavy metals from sediments and found it is possible to over- or under-estimate the anthropogenic concentration, considering the mineralogical composition of the sediment.

3.68 Analysis of sediment pore waters

3.68.1 Heavy metals

3.68.1.1 Atomic absorption spectrometry

Revera-Duarte and Flegal [219] used liquid–liquid preconcentration with organic chelators, coupled with graphite furnace atomic absorption spectrometry to determine silver, cadmium, cobalt, copper, iron, nickel and lead in sediment pore waters.

3.68.1.2 Voltammetric method

Brendel and Luther [220] have developed a solid state voltammetric gold amalgam microelectrode to measure iron, manganese and sulphur(II) in pore water. They used the electrode to measure depth profiles of these species at millimetre resolution.

3.68.2 Rhenium, platinum and iridium

3.68.2.1 Isotope dilution inductively coupled plasma mass spectrometry

Analytical methods for the determination of rhenium, iridium and platinum in sediment pore waters include graphite furnace atomic absorption spectrometry (GFAAS) for all three elements [221, 222]. Resonance ion mass spectrometry (RIMS) for rhenium [223], secondary ion mass spectrometry (SIMS) for rhenium [224], negative thermal ionisation mass spectrometry (NTIMS) for rhenium and iridium [225], instrumental and radiochemical neutron activation analysis (INAA and RNAA) for rhenium and iridium [226, 227] and absorptive cathodic stripping voltammetry (ACSV) for platinum [228]. While the detection limits of these techniques (after preconcentration) are sufficient for analysis of sediments (Table 3.37), none of them allow simultaneous determination of all three elements. Additionally GFAAS, RIMS, SIMS and NTIMS require extensive purification of the elements in order to remove interferences and reduce matrix effects. The relative freedom of the isotope dilution inductively coupled plasma mass spectrometric method described by Colodner *et al.* [72] from interferences in variable matrices makes it possible to simply these preconcentration procedures significantly.

Colodner *et al.* [72] have described the application of flow injection isotope dilution inductively coupled plasma mass spectrometry to the determination of pbenium, platinum and iridium in marine sediment pore waters. A stable isotope-enriched spike is added to the sample before processing. Sediments are dissolved in all-Teflon digestion vessels using a modified standard microwave oven. Anion exchange of the chloro

Table 3.37 Comparison of instrumental detection limits

| method[e] | ref | detection limit, pg | | |
		Re	Ir	Pt
GFAAS	6, 7	750[a]	30[a]	20[a]
SIMS	9	<40[b]		
RIMS	8	200[c]		
NTIMS	10	100[c]	<40[b]	
INAA	11		<0.02[b]	
ACSV	13			0.1[d]
FI–ID–ICPMA	f	5[a]	6[a]	14[a]

[a] Three times background. [b] Detection limit not stated; smallest sample mesaured. [c] Detection limited by filament background, quoted as 3 times background. [d] $3\sigma_{sample}$. [e] GFAAS, graphite furnace atomic absorption spectcrophotometry; SIMS, secondary ion mass spectrometry; RIMS, resonance ion mass spectrometry; NTIMS, negative thermal ionization mass spectrometry; INAA, coincidence/anticoincidence instrumental neutron activation analysis; ACSV, adsorptive cathodic stripping voltamettry; FI–ID–ICPMS, flow injection isotope dilution inductively coupled plasma mass spectrometry.

Source: American Chemical Society [72]

complexes of iridium and platinum and of the perrhenate ion (ReO_4) is used to preconcentrate the elements and to separate them from concomitants which produce molecular ions in the argon plasma resulting in isobaric interferences. Samples are then introduced into the inductively coupled plasma mass spectrometer in a small volume (300–600 µL) using flow injection. Overall recoveries were 90 ± 10% for all three elements, although the effects of variable recovery efficiency were minimised by the isotope dilution technique. The method has detection limits (three times background) of approximately 5 pg of Re, 6 pg of Ir and 14 pg of Pt, ie less than 100 µg L $^{-1}$.

3.68.3 Formate

3.68.3.1 High performance liquid chromatography

There is considerable interest in the role of formic acid and other volatile fatty acids in the early diagenesis of organic matter in lacustrine and marine sediments. Formic acid is an important fermentation product or substrate for many aerobic and anaerobic bacteria and for some yeasts.

Despite its potential importance, formic acid has proven difficult to quantify at submicromolar levels in natural water samples. Formidable analytical difficulties are associated with its detection in highly saline samples. Ion exclusion, anion exchange, and reversed-phase high-performance liquid chromatography (RP–HPLC) techniques based on the direct detection of formic acid in aqueous samples are prone to interferences (especially from inorganic salts) that ultimately limit the sensitivity of these methods.

A potentially more sensitive and selective approach involves reaction of formic acid with a reagent to form a chromophore or fluorophore, followed by chromatographic analysis. A wide variety of alkylating and silylating reagents have been used for this purpose. Two serious drawbacks to this approach are that inorganic salts and/or water interfere with the derivitisation reaction, and these reactions are generally non-specific for formic acid or other carboxylic acids. These techniques are prone to errors from adsorption losses, contamination and decomposition of the components of interest. Enzymatic techniques, in contrast, are ideal for the analysis of natural pore water samples, since they are compatible with aqueous media and involve little or no chemical or physical alterations of the sample (eg pH, temperature).

Kieber et al. [229] have described a method for the determination of submicromolar quantities of formate in saline sediment pore waters by a coupled enzymatic–high performance liquid chromatographic technique. The method is based on the oxidation of formate by formate dehydrogenase with corresponding reduction of β-nicotin–amide adenine dinucleotide (βNAD+) to reduced β-NAD+ (β-NADH); β-NADH is quantified by reversed-phase high-performance liquid chromatography with fluorometric detection. An important feature of this method is that the enzymatic reaction occurs directly in aqueous media, even seawater, and does not require sample pretreatment other than sample filtration. The reaction proceeds at room temperature at a slightly alkaline pH (7.5–8.5) and is specific for formate with a detection limit of 0.5 µM (S/N = 4) for a 200 µL injection. The precision of the method was 4.6% relative standard deviation (n = 6) for a 0.6 µM standard addition of formate to Sargasso seawater. Average recoveries of 2 µM additions of formate to seawater, pore water or rain were 103, 103 and 87% respectively. Intercalibration with a Dionex ion chromatographic system showed an excellent agreement of 98%. Concentrations of formate present in Biscayne Bay sediment porewater were in the range 0.9–8.4 µM.

3.68.4 Miscellaneous

Bufflap and Allen [168] compared four pore water sampling techniques including centrifugation (the recommended method), squeezing, vacuum filtration and dialysis.

Teasdale et al. [230] have used in-situ dialysis samples to collect sediment pore waters.

Zhang et al. [231] used polyacrylamide gels as in-situ probes for the determination of trace metals in sediment pore waters.

References

1 Abu-Hilal, A.H. and Riley, J.P. *Analytica Chimica Acta*, **131**, 175 (1981).
2 Arrowsmith, P. *Analytical Chemistry*, **59**, 1437 (1987).
3 Cutter, L.S., Cutter, G.A. and San Siego McGlove, M.L.C. *Analytical Chemistry*, **63**, 1138 (1991).
4 Brannon, J.M. and Patrick, W.H. *Environmental Pollution* (Series B), **9**, 107, (1985).
5 Sandhu, A.S. *Analyst (London)*, **106**, 311 (1981).
6 Goulden, P.D., Anthony, D.H.J. and Austen, K.D. *Analytical Chemistry*, **53**, 2027 (1981).
7 Brzezinska-Paudyn, A., Van Loon, J.C. and Hancock, R. *Atomic Spectroscopy*, **7**, 72 (1986).
8 Liversage, B.R., Van Loon, J.C. and de Andrade, J.C. *Analytica Chimica Acta*, **161**, 275 (1984).
9 Brzezinska-Paudyn, A., Balicka, A. and Van Loon, J.C. *Water Air and Soil Pollution*, **323**, (1983).
10 Lasztity, A., Krushevska, A., Kotrebai, M., Barnes, R.M. and Amarasiriwardena, D.J. *Analytical Atomic Spectroscopy*, **10**, 505 (1995).
11 Brannon, J.M. and Patrick, W.H. *Science and Technology*, **21**, 450 (1987).
12 Cheam, V. and Chau, A.S.Y. *Analyst (London)*, **109**, 775 (1984).
13 Cutter, G.A., Electric Power Research Institute, Palo Alto, California, Report EPRI–EA4641, Vol 1. Speciation of selenium and arsenic in natural waters and sediments, arsenic speciation (1986).
14 Ebdon, L., Hutton, R.C. and Ottaway, J.M. *Analytica Chimica Acta*, **95**, 117 (1977).
15 Zhe-Ming, N., Xiao-Chun, L. and Heng-Bin, H. *Analytica Chimica Acta*, **186**, 147 (1986).
16 Lum, K.R. and Edgar, D.G. *Analyst (London)*, **108**, 918 (1983).
17 Sakata, M. and Shimoda, O. *Water Research*, **16**, 231 (1982).
18 Pankow, J.F., Leta, D.P., Lin, J.W., Ohl, S.E., Shum, W.P. and Janner, G.E. *Science of the Total Environment*, **7**, 17 (1977).
19 Scott, K. *Analyst (London)*, **103**, 754 (1978).
20 Chakraborty, R., Das, A.K., Cervera, M.L. and De La Guardia, M. *Journal Analytical Atomic Spectroscopy*, **10**, 3536 (1995).
21 Sahuquillo, A., Lopez–Sanchez, J.F., Rubio, R., Rauret, G. and Hatjie, V. *Fresenius Journal of Analytical Chemistry*, **351**, 197 (1995).
22 Sahuquillo, A., Lopez–Sanchez, J.F., Rubio, R. and Rauret, G. *Mikrochimica Acta*, **119**, 251 (1995).
23 Xiao-Quan, S., Zhi-Neng, Y. and Zhe-Ming, N. *Analytical Chemistry*, **57**, 857 (1985).
24 Chen, T.C. and Hong, A. *Journal of Hazardous Materials*, **41**, 147 (1995).
25 Lopez-Garcia, L., Sanchez-Merlos, M. and Hernandez-Cordoba, M. *Analytica Chimica Acta*, **328**, 19 (1996).
26 Zaray, G. and Kantor, T. *Spectrochimica Acta Part B*, **50B**, 489 (1995).
27 Manceau, A., Boisset, M.C., Sarret, G., Hazemann, J.L., Mench, M., Cambier, P. and Prost, R. *Environmental Science and Technology*, **30**, 1540 (1996).
28 Pillay, K.J.S., Thomas, C.C., Sondel, J.A. and Hyche, C.M. *Analytical Chemistry*, **43**, 1419 (1971).
29 Bretthaur, E.W., Moghissi, A.A., Snyder, S.S. and Matthews, N.W. *Analytical Chemistry*, **46**, 445 (1974).
30 Feldman, C. *Analytical Chemistry*, **46**, 1606 (1974).
31 Bishop, J.N., Taylor, L.A. and Neary, B.P. in *The Determination of Mercury in Environmental Samples*, Ministry of the Environment, Canada (1973).

32 Jacobs, L.W. and Keeney, D.R. *Environmental Science and Technology*, **8**, 267 (1976).
33 Environmental Protection Agency, Methods for the Analysis of Water and Wastes, Cincinnati, Ohio, p 134 (1974).
34 Agemian, H. and Chau, A.S.Y. *Analytica Chimica Acta*, **75**, 297 (1974).
35 Iskander, K., Syers, J.K., Jakobs, L., Feeney, D. and Gilmour, J.T. *Analyst (London)*, **97**, 388 (1972).
36 Agemian, H. and Chau, A.S.Y. *Analyst (London)*, **101**, 91 (1976).
37 Jurka, A.M. and Carter, M.J. *Analytical Chemistry*, **50**, 91 (1978).
38 Kozuchowski, J. *Analytica Chimica Acta*, **99**, 293 (1978).
39 Craig, P.J. and Morton, S.F. *Nature (London)*, **261**, 126 (1976).
40 El-Awady, A.A., Miller, R.B. and Carter, M.J. *Analytical Chemistry*, **48**, 110 (1976).
41 Abo-Rady, M.D.K. *Fresenius Zeitschrift für Analitische Chemie*, **299**, 187 (1979).
42 Smith, R.G. *Analytical Chemistry*, **65**, 2485 (1993).
43 Walker, G.S., Ridd, M.J. and Brunskill, G.J. *Rapid Commun Mass Spectrom*, **10**, 96 (1996).
44 Hatle, M., Golimowsk, J. and Orzechowska, A. *Tolanta*, **34**, 1001 (1987).
45 Tykarstra, A. *Fresenius Journal of Analytical Spectroscopy*, **351**, 656 (1995).
46 Leong, P.C. and Ong, H.P. *Analytical Chemistry*, **43**, 940 (1971).
47 Anderson, D.H., Evans, J.H., Murphy, J.J. and White, W.W. *Analytical Chemistry*, **43**, 1511 (1971).
48 Mudrock, A. and Kokitich, E. *Analyst (London)*, **112**, 709 (1987).
49 Hintelmann, H., Evans, R.D. and Villeneuve, J.Y. *Analytical Atomic Spectrometry*, **10**, 619 (1995).
50 Hammer, U.T., Merkowsky, A.J. and Huang, P.M. *Archives of Environmental Contamination and Toxicology*, **17**, 257 (1988).
51 Kershaw, P.J., Sampson, K.E., McCarthy, W. and Scott, D.R. *Journal Radioanalytical and Nuclear Chemistry*, **198**, 113 (1995).
52 Wiersma, J.H. and Lee, G.F. *Environmental Science and Technology*, **5**, 1203 (1971).
53 Itoh, K., Chikuma, M. and Tanaka, H. *Fresenius Zeitschrift für Analytische Chemie*, **330**, 600 (1988).
54 Nadkarni, R.A. and Morrison, G.H. *Analytica Chimica Acta*, **99**, 133 (1978).
55 de Oliveira, E., McLaren, J.W. and Berman, S.S. *Analytical Chemistry*, **55**, 2047 (1983).
56 Siu, K.W.M. and Berman, S.S. *Analytical Chemistry*, **55**, 1603 (1983).
57 Dogan, S. and Haerdi, W. *International Journal of Environmental Analytical Chemistry*, **8**, 249 (1980).
58 Long-Zhu, J. *Atomic Spectroscopy*, **5**, 91 (1984).
59 Legret, M. and Divet, L. *Analytica Chimica Acta*, **189**, 313 (1986).
60 Brzenzinska-Paudyn, A. and Van Loon, J.C. *Fresenius Zeitschrift für Analytische Chemie*, **331**, 707 (1988).
61 Quin, B.F. and Brooks, R.R. *Analytica Chimica Acta*, **58**, 301 (1972).
62 Hasche, K., Kakizaki, T. and Tochida, H. *Fresenius Zeitschrift für Analytische Chemie*, **322**, 486 (1985).
63 Zink-Nielsen, I. *Vatten*, **1**, 14 (1977).
64 Sinex, S.A., Cantillo, A.Y. and Helz, G.R. *Analytical Chemistry*, **52**, 2342 (1980).
65 Legret, M., Demare, D., Marchandise, P. and Robbe, D. *Analytica Chimica Acta*, **149**, 107 (1983).
66 Legret, M., Divet, L. and Demare, D. *Analytica Chimica Acta*, **175**, 203 (1985).
67 Zink-Nielsen, I.: Nordforsk Working Group on Water Analysis, Water Quality Institute, 11 Agern Alle, DK 2970 Horsholm, Denmark, private communication.

68 Agemian, H. and Chau, A.S.Y., *Analyst (London)*, **101**, 761 (1976).
69 Que-Hee, S.G. and Boyle, J.R. *Analytical Chemistry*, **60**, 1033 (1988).
70 Elwaer,N. and Belzile, N. *International Journal of Environmental Analytical Chemistry*, **61**, 189 (1995).
71 Pahlavanpour, B., Thompson, M. and Thorne, L. *Analyst (London)*, **105**, 756 (1980).
72 Colodner, D.C., Boyle, E.A. and Edmond, J.M. *Analytical Chemistry*, **65**, 1419 (1993).
73 Kanda, Y. and Taira, M. *Analytica Chimica Acta*, **207**, 269 (1988).
74 Madoro, M. and Moanuro, A. *Journal of Radioanalytical and Nuclear Chemistry*, **90**, 129 (1985).
75 Nadkarni, R.A. and Morrison, G.H. *Analytica Chimica Acta*, **99**, 133 (1975).
76 Slavic, I., Draskovic, R., Tasovac, T. *Radosavljevic*, **9**, 87 (1973).
77 Anders, U.W. *Analytical Chemistry*, **44**, 1930 (1972).
78 Lieser, K.H., Lalmano, W., Heuss, H. and Neitzert, V. *Journal of Radioanalytical and Nuclear Chemistry*, **77**, 717 (1977).
79 Ackermann, E. *Deutsche Gewässer Kundliche Mittelungen*, **21**, 53 (1977).
80 Bart, G. and Von Gunten, H.R. *International Journal of Environmental Analytical Chemistry*, **6**, 25 (1979).
81 Bonifort, R., Madaro, M. and Moauro, A. *Journal of Radioanalytical and Nuclear Chemistry*, **84**, 441 (1984).
82 Welte, B., Bleo, N. and Montiel, A. *Environmental Technology Letters*, **4**, 223 (1983).
83 Madsen, P.P., Drabach, I. and Sorenson, J. *Analytica Chimica Acta*, **151**, 479 (1983).
84 Hellman, H.Z. *Fresnius Zeitschrift für Analytische Chemie*, **263**, 14 (1973).
85 Hellman, H.Z. *Fresnius Zeitschrift für Analytische Chemie*, **254**, 192 (1971).
86 Prange, A., Knoth, J., Stossel, R.P., Baddaber, H. and Kramer, K. *Analytica Chimica Acta*, **195**, 275 (1987).
87 Lichtfuss, R. and Brummer, G. *Chemical Geology*, **21**, 51 (1978).
88 Schneider, B. and Weiler, K. *Environmental Technology Letters*, **5**, 245 (1984).
89 Labresque, J.J., Rosales, P.A. and Meijas, G. *Applied Spectroscopy*, **40**, 1232 (1986).
90 McKay, W.A. *Journal of Radiological Protection*, **15**, 159 (1995).
91 Allen, R.J. *Environmental Geology*, **3**, 49 (1979).
92 Mahan, K.I., Fuderaro, T.A., Ganza, T.I., Martines, R.M., Maroney, M.R., Trivisanne, M.R. and Willging, E.M. *Analytical Chemistry*, **59**, 938 (1987).
93 Reve, A.R. and Hasty, E. Recovery study using an elevated pressure temperature microwave dissolution technique presented at 1987 Pittsburgh Conference and Exposition on Analytical Chemistry and Applied Spectroscopy March (1987).
94 Bando, R., Galanti, G. and Varini, P.G. *Analyst (London)*, **108**, 722 (1983).
95 Helinke, P.A., Schomberg, P.J. and Iskander, I.K. *Environmental Science and Technology*, **11**, 984 (1977).
96 Agemian, H. and Chau, A.S.Y. *Analytica Chimica Acta*, **80**, 61 (1975).
97 Bettinelli, M., Pastorelli, N. and Borani, U. *Analytica Chimica Acta*, **185**, 109 (1986).
98 Van San, M. and Muntan, H. *Fresenius Zeitschrift für Analytische Chemie*, **328**, 390 (1987).
99 Chester, R. and Hughes, M.I. *Chemical Geology*, **2**, 249 (1967).
100 Schnitzer, M. and Skinner, S.I.M. *Soil Science*, **102**, 361 (1966).
101 Rashid, M.A. *Chemical Geology*, **13**, 115 (1974).
102 Holmes, C.W., Slade, E.A. and McLerran, C.J. *Environmental Science and Technology*, **8**, 255 (1974).
103 Kronfeld, K. and Navrat, J. *Environmental Pollution*, **6**, 281 (1974).

104 Chowdbury, A.N. and Bose, B.B. *Geochemical Exploration CIM Special*, **11**, 410 (1971).
105 Goldberg, E.I.D. and Arrhenius, G.D.S. *Geochimica Cosmochimica Acta*, **13**, 153 (1958).
106 Carrol, D. and Starkey, H. *Clays Clay Mineral* Proceedings 7th National Conference pp 80–101 (1960).
107 Gibbs, R.J. *Geological Society of America Bulletin*, **88**, 829 (1977).
108 Bascomb, C.L.J. *Soil Science*, **19**, 251 (1968).
109 Arshad, M.A., St Arnaud, R.J. and Huang, P.M. *Canadian Journal of Soil Science*, **52**, 19 (1972).
110 McKeague, J.A. and Day, J.A. *Canadian Journal of Soil Science*, **43**, 7 (1963).
111 Caffin, D.E. *Canadian Journal of Soil Science*, **43**, 7 (1963).
112 Malo, B.A. *Environmental Science and Technology*, **11**, 277 (1977).
113 Presley, B.J., Kolodny, Y., Nissenbaum, A. and Kaplan, I.R. *Geochimica Cosmochimica Acta*, **36**, 1073 (1973).
114 Rosholt, J.N., Emiliani, C., Geiss, J., Kiczy, F.F. and Wangersky, P.J. *Geology*, **69**, 162 (1961).
115 Hirst, D.M. and Nicholls, G.D. *Sediment Petrology*, **28**, 461 (1958).
116 Sorenson, R.C., Oelsligle, D.D. and Knudsen, D. *Soil Science*, **111**, 352 (1971).
117 Kitano, Y., Sakata, M. and Matsumoto, E. *Geochimica Cosmochimica Acta*, **44**, 1279 (1980).
118 Marchandise, F., Ohl, J.L., Robbe, D. and Legret, M. *Environmental Technology Letters*, **3**, 157 (1982).
119 Agemian, H. and Chau, A.S.Y. *Archives of Environmental Contamination and Toxicology*, **6**, 69 (1977).
120 Trefrey, J.H. and Metz, S. *Analytical Chemistry*, **56**, 745 (1984).
121 Cantillo, A.Y., Sinex, S.A. and Helz, G.R. *Analytical Chemistry*, **56**, 33 (1984).
122 Hawkes, H.E. and Webb, I.S. in *Geochemistry in Mineral Exploration*, Harper & Row, New York, NY (1962).
123 Oliver, B.G. *Environmental Science and Technology*, **7**, 135 (1973).
124 Van Valin, R. and Morse, J. *Marine Chemistry*, **11**, 535 (1982).
125 Breder, R. *Fresenius Zeitschrift für Analytische Chemie*, **313**, 395 (1982).
126 Kheboian, C. and Bauer, C.F. *Analytical Chemistry*, **59**, 1417 (1987).
127 Xiao-Quan, S. and Biu Chen. *Analytical Chemistry*, **65**, 802 (1993).
128 Ure, A.M., Davidson, C.M. and Thomas, R.P. *Technical Instrumentation in Analytical Chemistry*, **17**, 505 (1995).
129 Voutsinou-Taliadouri, F. *Mikrochim Acta*, **119**, 243 (1995).
130 Kong, I., Shiu-Mei, C. *Ecotoxicological Environmental Safety*, **32**, 34 (1995).
131 Sagar, M. *Technical Instruments, Analytical Chemistry*, **12**, (Hazard Met Envir) 133 (1992).
132 Rauret, G., Rubio, R., Lopez–Sanchez, J.F. and Casassas, E. *International Journal of Environmental Analytical Chemistry*, **35**, 89 (1989).
133 Battiston, G.A., Gerbasi, R., Degetto, S. and Sbrignadello, G. *Spectrochimica Acta Part B*, **48B**, 217 (1993).
134 Krumgalz, B.S. and Fainshtein, G. *Analytica Chimica Acta*, **218**, 335 (1989).
135 Cheam, V., Aspila, I. and Chau, A.S.Y. *Science of the Total Environment*, **87–88**, 517 (1989).
136 Epstein, M.S., Diamondstone, B.I. and Gills, T.E. *Talanta*, **36**, 141 (1989).
137 Wen, X., Du, Q. and Tang, H. *Environmental Science and Technology*, **32**, 870 (1998).
138 Al-Jundi, J., Mamas, C., Earwaker, L.G., Randle, K. and West, J. *Analytical Proceedings (London)*, **30**, 153 (1993).
139 Del Castilho, P. and Rix, I. *International Journal of Environmental Analytical Chemistry*, **51**, 59 (1993).
140 Kratchvil, B. and Mamba, S. *Canadian Journal of Chemistry*, **68**, 360 (1990).

141 Kammin, W.R. and Brandt, M. *Journal of Spectroscopy*, **4**, 49, 52 (1989).
142 Millward, C.G. and Kluckner, P.D. *Journal of Analytical Atomic Spectroscopy*, **4**, 708 (1989).
143 Li, M., Barban, R., Zucchi, B. and Martinotti, W. *Water Air and Soil Pollution*, **57–58**, 495 (1991).
144 Paudyn, A.M. and Smith, R.G. *Canadian Journal of Applied Spectroscopy*, **37**, 94 (1992).
145 Hewitt, A.D. and Reynolds, C.M., Report CRREL–SP–90–19 CETHA–TS–CR–90052 Order No AD–A–226367 (Avail NTIS) (1990).
146 Kingston, H.M. and Walter, P.J. *Spectroscopy*, **7**, 20 (1992).
147 Nieuwenholze, J., Poley-Vos, C.H., Van den Akker, A.H. and Van Delft, W. *Analyst (London)*, **116**, 347 (1991).
148 Sturgeon, R., Willie, S.N., Methven, B.A., Lam, J.W.H. and Mutusiewicz, H. *Journal of Atomic Spectroscopy*, **10**, 981 (1995).
149 Zaggia, L., Argese, E. and Zonta, R. *Toxiological Environmental Chemistry*, **54**, 11 (1996).
150 Tessier, A., Campbell, P.G.C. and Bissan, M. *Analytical Chemistry*, **51**, 844 (1979).
151 Lex, I. and Stulik, K. *Analytical Chemistry*, **60**, 1475 (1988).
152 Davidson, C.M., Thomas, R.P., McVey, S.E., Derala, R., Littlejohn, D. and Ure, A.M. *Analytica Chimica Acta*, **291**, 277 (1994).
153 Thomas, R.P., Ure, A.M., Davidson, C.M., Littlejohn, D., Rauret, G., Rubio, R. and Lopez-Sanchez, J.F. *Analytica Chimica Acta*, **286**, 423 (1994).
154 Whalley, C. and Grant, A. *Analytica Chimica Acta*, **291**, 287 (1994).
155 Asikainen, J.M. and Nickolads, N.P. *Ground Water Monitoring*, **14**, 185 (1994).
156 Ma, R., Mol, V.W. and Adams, F. *Analytica Chimica Acta*, **285**, 33 (1994).
157 Garcia-Lopez, I., Sanchez-Merlos, M. and Hernandez-Cordoba, M. *Analytica Chimica Acta*, **328**, 19 (1996).
158 Shabani, M.B. and Masuda, A. *Analytical Chemistry*, **63**, 2099 (1991).
159 O'Day, P.A., Carroll, S.A. and Waychunas, G.A. *Environmental Science and Technology*, **32**, 943 (1998).
160 Beverage, A., Waller, P. and Pickering, W.F. *Talanta*, **36**, 535 (1989).
161 Henshaw, J.M., Heithmar, E.M. and Hinners, T.A. *Analytical Chemistry*, **61**, 335 (1989).
162 Joshi, S.R. *Applied Radiation and Isotopes*, **40**, 691 (1989).
163 Livens, F.R. and Singleton, D.L. *Analyst (London)*, **114**, 1097 (1989).
164 Hwang, J.D., Huxley, H.P., Diomiguardi, J.P. and Vaughn, W. *Journal of Applied Spectroscopy*, **44**, 491 (1990).
165 Xu, L. and Schramel, P. *Fresenius Journal of Analytical Chemistry*, **342**, 179 (1992).
166 Savvin, S.B., Petrova, T.U., Dzherayan, T.G. and Reikhshat, M.M. *Fresenius Journal of Analytical Chemistry*, **340**, 217 (1991).
167 Hinds, M.W., Latimer, K.E. and Jackson, J.W. *Journal of Analytical Atomic Spectroscopy*, **7**, 171 (1992).
168 Bufflap, S.E. and Allen, H.E. *Water Research*, **29**, 2051 (1995).
169 Hinds, M.W., Latimer, K.E. and Jackson, K.W. *Journal of Analytical Atomic Spectroscopy*, **6**, 473 (1991).
170 Hinds, M.W. and Jackson, K.W. *Atomic Spectroscopy*, **12**, 109 (1991).
171 Fernando, A.R. and Plambeck, J.A. *Analyst (London)*, **117**, 39 (1992).
172 Wegrzynek, D. and Holynska, B. *Applied Radiation Isot*, **44**, 1101 (1993).
173 Koplitz, L.V., Urbanik, J., Harris, S. and Mills, D. *Environmental Science and Technology*, **28**, 538 (1994).
174 Bermejo-Barrera, P., Barciel-Alonso, C., Aboal-Somaza, M. and Bermejo-Barrera, A. *Journal of Atomic Spectroscopy*, **9**, 469 (1994).

175 Horvat, M., Bloom, N.S. and Liang, L. *Analytica Chimica Acta*, **281**, 135 (1993).
176 Kamburova, M. *Talanta*, **40**, 719 (1993).
177 Bandyopadhyay, S. and Das, A.K. *Journal of Indian Chemical Society*, **66**, 427 (1989).
178 Lee, H.S., Jung, K.H. and Lee, D.S. *Talanta*, **36**, 999 (1989).
179 Azzaria, L.M. and Aftabi, A. *Water Air and Soil Pollution*, **56**, 203 (1991).
180 Robinson, L., Dyer, F.F., Combs, D.W., Wade, W., Teasley, N.A. and Carlton, J.E. *Journal of Radioanalytical and Nuclear Chemistry*, **179**, 305 (1994).
181 Hintelmann, H. and Wilken, R.D. *Applied Organometallic Chemistry*, **7**, 173 (1993).
182 Emteborg, H., Bjoerklund, E., Oedman, F., Karlsson, L., Mathiasson, L., Fresh, W. and Baxter, P.C. *Analyst (London)*, **121**, 19 (1996).
183 Lexa, J. and Stulik, K. *Talanta*, **36**, 843 (1989).
184 Wilken, R.D. and Hintelmann, H., NATO ASI Series G 1990–23 (Met Speciation Environ) 339 (1990).
185 Cela, R., Lorenzo, R.A., Rubl, E., Botana, A., Valino, M. and Casais, C. *Journal of Radioanalytical and Nuclear Chemistry*, **130**, 443 (1989).
186 Kim, C.K., Takaku, A., Yamamoto, M., Kawamura, H., Shiraiski, K., Igarashi, Y., Igarashi, S., Takayama, H. and Ikeda, N. *Journal of Radioanalytical and Nuclear Chemistry*, **132**, 131 (1989).
187 Kim, C.K., Oura, Y., Takaku, N.H., Igarashi, V. and Ikeda, N. *Journal of Radioanalytical and Nuclear Chemistry*, **136**, 353 (1989).
188 Jia, G., Testa, C., Desideri, D. and Mell, M.A. Analytica Chimica Acta, **220**, 103 (1989).
189 Green, L.W., Miller, F.C., Sparling, J.A. and Joshi, S.R. *Journal of American Society of Mass Spectrometry*, **2**, 240 (1991).
190 Holgye, Z. *Journal of Radioanalytical Nuclear Chemistry*, **149**, 275 (1991).
191 Barci-Funel, G., Dalmasso, J. and Ardisson, G. *Journal of Radioanalytical and Nuclear Chemistry*, **156**, 83 (1992).
192 Nakanishi, T., Satoh, M., Takei, M., Ishikawa, A., Murata, M., Dairyoh, M. and Higuchi, S. *Journal of Radioanalytical and Nuclear Chemistry*, **138**, 321 (1990).
193 El-Daoushy, F. and Garcia-Tenorio, R. *Journal of Radioanalytical and Nuclear Chemistry*, **138**, 5 (1990).
194 Cohen, A.S. and O'Nions, R.K. *Analytical Chemistry*, **63**, 2705 (1991).
195 Williams, L.R., Leggett, R.W., Espegren, M.L. and Little, C.A. *Environmental Monitoring Assessment*, **12**, 83 (1989).
196 Hafez, A.F., Moharram, B.M., El-Khatib, A.M. and Abel-Naby, A. *Isotopenpraxis*, **27**, 185 (1991).
197 Zhang, Y., Moore, J.N. and Frankenberger, W.T. *Environmental Science and Technology*, **33**, 1652 (1999).
198 McCurdy, E.J., Lange, J.D. and Haygarth, P.M. *Science of the Total Environment*, **135**, 131 (1993).
199 Rojas, C.L., de Maroto, S.B. and Valanta, D. *Fresenius Journal of Analytical Chemistry*, **348**, 775 (1994).
200 Nakagawa, T., Aoyama, E., Hasegawa, N., Kobayashi, N. and Tanaka, H. *Analytical Chemistry*, **61**, 233 (1989).
201 Cruvinel, P.E. and Flocchini, R.G. *Nuclear Instrumental Methods Physics Research Section B*, **B75**, 415 (1993).
202 Haygarth, P.M., Rowland, A.P., Sturup, S. and Jones, K.C. *Analyst (London)*, **118**, 1303 (1993).
203 Stella, R., Ganzerli, U.M.T. and Maggi, L. *Journal of Radioanalytical and Nuclear Chemistry*, **161**, 413 (1992).
204 Ryabukhin, V.A., Volynets, M.P., Myasoedoo, B.F., Radionova, I.M. and Tuzova, A.M. *Fresenius Journal of Analytical Chemistry*, **341**, 636 (1991).

205 Morita, S., Kim, C.K., Takaku, Y., Seki, R. and Ikeda, N. *Applied Radiation and Isotopes*, **42**, 531 (1991).
206 Harvey, B.R., Williams, K.J., Lovett, M.B. and Ibbett, R.D. *Journal of Radioanalytical and Nuclear Chemistry*, **158**, 417 (1992).
207 Lukazewski, Z. and Zembrzuski, W. *Talanta*, **39**, 221 (1992).
208 Sagar, M. *Mikrochimica Acta*, **106**, 241 (1992).
209 Mukhtar, O.M., Ghods, A. and Khangi, F.A. *Radiochimica Acta*, **54**, 201 (1991).
210 Toole, J., McKay, K. and Baxter, M. *Analytica Chimica Acta*, **245**, 83 (1990).
211 Shaw, T.J. and Francois, R. *Geochimica and Cosmochimica Acta*, **55**, 2075 (1991).
212 Shuktomova, I.I. and Kochan, I.G. *Journal of Radioanalytical Chemistry*, **129**, 245 (1989).
213 Lazo, E.N. Report 1988 DOE (Department of Environment) /OR/0033–T424 Order No DE89010612 Avail NTIS 350 pp (1988).
214 Noey, K.C., Liedle, S.D., Hickey, C.R. and Doane, R.W. Proceedings on Symposium on Waste Management 615 (1989).
215 Lazo, E.N., Doessier, G.S. and Bervan, B.A. *Health Physics*, **6**, 231 (1991).
216 Parsa, B. *Journal of Radioanalytical and Nuclear Chemistry*, **157**, 65 (1992).
217 Lazo, E.N. Report 1988 DOE (Department of Environment) /OR/0033–T424 Order No DE89010612 Avail NTIS 350 pp (1988).
218 Miura, J. *Analytical Chemistry*, **62**, 1424 (1990).
219 Revera-Duarte, I. and Flegal, A.R. *Analytica Chimica*, **328**, 13 (1996).
220 Brendel, P.J. and Luther, G.W. *Environmental Science and Technology*, **29**, 751 (1995).
221 Hodge, V.F., Stallard, M., Koide, M. and Goldberg, E. *Analytical Chemistry*, **58**, 616 (1986).
222 Koide, M., Hodge, V., Yang, J.S. and Goldberg, E.D. *Analytical Chemistry*, **59**, 1802 (1987).
223 Walker, R.J. *Analytical Chemistry*, **60**, 1231 (1988).
224 Luck, J.M. Geochemie du Rhenium-Osmium Methode et applications Docteur es Sciences, University of Paris VII (1982).
225 Creaser, R.A., Papanastassiou, D.A. and Wasserburg, G.J. *Geochimica Cosmochimica Acta*, **55**, 397 (1991).
226 Murali, A.V., Parekh, P.P. and Cummings, J.B. *Geochimico Cosmochimica Acta*, **54**, 889 (1990).
227 Keays, R.R., Ganapathy, R., Laul, J.C., Krakenbuhl, U.R.S. and Morgan, J.W. *Analytica Chimica Acta*, **72**, 1 (1974).
228 Van den Berg, C.M.G. and Jacinto, G.S. *Analytica Chimica Acta*, **211**, 129 (1988).
229 Kieber, D.J., Vaughan, G.M. and Mopper, K. *Analytical Chemistry*, **60**, 1654 (1988).
230 Teasdale, P.R., Batley, G.E., Apte, S.C. and Webster, I.T. *Trends Analytical Chemistry*, **14**, 250 (1995).
231 Zhang, H., Davison, W. and Grime, G.W. *ASTM Special Technical Publication STP*, 1293, 170 (1995).
232 Noey, K.C., Liedle, S.D., Hickey, C.R. and Doane, R.W. Proceedings on Symposium on Waste Management 615 (1989).
233 Hewitt, A.D., Reynolds, C.M. *Atomic Spectroscopy*, **66**, 187 (1990).

Chapter 4

Determination of metals in marine sediments

4.1 Aluminium

4.1.1 Inductively coupled plasma atomic emission spectrometry

The determination of aluminium in marine sediments is discussed under multi-metal analysis in sections 4.44.3.1 and 4.44.3.2.

4.1.2 X-ray secondary emission spectrometry

The determination of aluminium is discussed under multi-metal analysis in section 4.44.8.1.

4.2 Antimony

4.2.1 Inductively coupled plasma atomic emission spectrometry and inductively coupled plasma mass spectrometry

The determination of antimony is discussed under multi-metal analysis in section 4.44.3.3 (inductively coupled plasma atomic emission spectrometry) and section 4.44.4.1 (inductively coupled plasma mass spectrometry).

4.2.2 Neutron activation analysis

The determination of antimony is discussed under multi-metal analysis in section 4.44.5.1.

4.2.3 Photon activation analysis

The determination of antimony is discussed under multi-metal analysis in section 4.44.7.1.

4.3 Arsenic

4.3.1 Spectrophotometric method

Maher [1] has described a procedure for the determination of total arsenic in marine sediments. The sample is first digested with a mixture of nitric, sulphuric and perchloric acids. Then arsenic is converted into arsine using a zinc reductor column, the evolved arsine is trapped in a potassium iodide-iodine solution, and the arsenic determined spectrophotometrically at 866nm as the arseno-molybdenum blue complex. The detection limit is 0.05 mg kg $^{-1}$ dry sediment and the coefficient of variation 5.1% at this level. The method is free from interferences by other elements at levels normally found in marine sediments. Values of 9.7 ± 0.3 and 13.2 ± 0.4 mg kg $^{-1}$ obtained for NBS reference waters SRM 1S71 and SRM 1566 were in good agreement, respectively, with the nominal values of 10.2 and 13.4 mg kg $^{-1}$.

In this method sediments were freeze-dried and ground (to less than 200μm) before analysis.

A weighed sample (less than 0.5g) was placed in a 30mL Pyrex centrifuge tube, 5 mL of concentrated nitric acid were added and the mixture was allowed to stand for at least 12 h at room temperature to ensure complete dissolution (this avoids foaming on heating). The tube was then placed in an aluminium heating block and refluxed until the evolution of brown fumes ceased. After cooling, 5 mL of a nitric-sulphuric-perchloric acid mixture ($5 + 1 + 3$ V/V) were added and heating continued until dense fumes of sulphur trioxide appeared. The digest was diluted with 5 mL of 1.5M hydrochloric acid, 1 mL of stannous chloride-potassium iodide reducing solution was added, and the solution allowed to stand for 40 min to reduce all of the inorganic arsenic to the trivalent form. The solution was then made up to 25 mL in a calibrated flask with 1.5M hydrochloric acid. The apparatus was assembled as in Fig. 4.1 with 1.5 mL of the potassium iodide-iodine solution in the centrifuge tube. The nitrogen gas flow rate was adjusted to 150 ml min $^{-1}$, 1 mL of sample was injected onto the zinc column and the evolved arsine collected for 2 min in the potassium iodide-iodine solution. Between each sample, 1 mL of 1.5M hydrochloric acid was injected to prevent any accumulation of potentially interfering material on the column. After the arsine had been trapped, the centrifuge tube was removed and 0.5 mL of the mixed spectrophotometric molybdate reagent added. The solution was mixed by means of a vortex mixer and allowed to stand for 30 min to develop the arsenomolybdenum blue. The absorbance was measured at 866nm. Calibration graphs of absorbance *versus* amount of arsenic (0.1 and 0.5 μg) were prepared by using arsenic standards carried through the entire analytical procedure.

Arsenic recoveries from the zinc column in the range 0.1–5 μg mL $^{-1}$ arsenic exceeded 97%. The concentrations at which certain elements

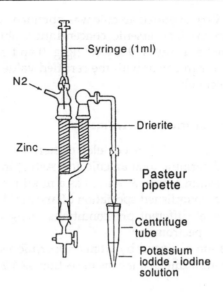

Fig. 4.1 Reduction apparatus for determination of arsenic in marine sediments
Source: Royal Society of Chemistry [1]

Table 4.1 Effect of inorganic ions on the generation and trapping of arsine

All tests used 0.1 µg of As(III) in 1 ml of 1.5m hydrochloric acid. Optimised hydride generation and trapping conditions were used

Species	Cu(II)	Hg(II)	Mo(VI)	Sb(III)	Se(IV)	Si(IV)	V(V)
Interference level/µg	50	2.5	400	25	0.1	50	250

Source: Royal Society of Chemistry [1]

interfere are shown in Table 4.1. Various other elements (Al(III), B(III), Ca(II), Cd(II), Co(II), Cr(VI), Fe(III), K(I), Li(I), Mg(II), Mn(II), Na(I), Ni(II), Pb(II), S(VI), Sn(II) and Zn(II)) showed no significant interference at the 500µg level. Only low selenium concentrations in extracts can be tolerated. However, few environmental samples contain appreciable amounts of selenium. As selenium is not reduced to hydrogen selenide on the column, selenium will not interfere in the final determination step, but probably suppresses either arsenic reduction or arsine formation. Selenium appears to suppress arsine generation at high arsenic concentrations but causes a slight enhancement at low arsenic concentrations (around 0.1 µg) which could not be traced to arsenic impurities in the selenium standard used.

Complete recovery of added arsenic was obtained within experimental error for a sediment. The arsenic concentration obtained by replicate analysis of the orchard leaves (9.7 + 0.3 µg g $^{-1}$) and oyster tissue (13.2 + 0.4 µg g $^{-1}$) were in agreement with the certified values of 10 ± 2 and 13.4 ± 1.9 µg g $^{-1}$ respectively.

4.3.2 Atomic absorption spectrometry

Bermejo-Barrera *et al.* [2] used electrothermal atomic absorption spectrometry to determine total arsenic and As(III) in marine sediments. Palladium-magnesium nitrate was used as a modifier for total arsenic.

Soto *et al.* [86] accomplished speciation of arsenic(III) and arsenic(V) in marine sediments at different pH conditions using hydride generation atomic absorption spectrometry.

The determination of arsenic by atomic absorption spectrometry is also discussed under multi-metal analysis in section 4.44.2.3.

4.3.3 Inductively coupled plasma atomic emission spectrometry

The determination of arsenic is discussed under multi-metal analysis in section 4.44.3.3.

4.3.4 Photon activation analysis

The determination of arsenic is discussed under multi-metal analysis in section 4.44.7.1.

4.3.5 Gas chromatography

Siu *et al.* [3] determined arsenic in marine sediments down to 1 µg kg $^{-1}$ by digestion with concentrated acid and derivatisation with 2.3 mercapto-propanol and electron capture gas chromatography.

4.3.6 Ion exchange chromatography

Maher [4] has discussed the application of this technique to the determination of inorganic arsenic (fraction 1), also monomethyl arsonic and dimethylarsinic acids (fraction 2) in estuarine sediments. The organic arsenic species are isolated by extraction with toluene, separated by ion-exchange chromatography and selectively determined by arsine generation. Recoveries of spikes of 5 and 10 µg arsenic taken through the whole procedure were 92 to 96%.

The arsenic in fractions 1 and 2 was determined by reduction to the corresponding arsine in the zinc reductor column, decomposition of the arsine evolved in a heated carbon tube furnace, and by measurement of atomic absorption of the arsenic at 193.7nm. The principal advantages of using the zinc reductor column are its rapidity and freedom from interferences by other elements.

Initial reduction of inorganic arsenic(V), monomethylarsonic acid and dimethyl arsinic acid to the trivalent oxidation state was required for quantitative reduction to the corresponding arsine on the column. Conversion of the first of the above eluates was achieved by the addition of 1.5 ml of concentrated hydrochloric acid and 0.5 ml of stannous chloride-potassium iodide reducing agent. The other eluates were evaporated to dryness after the addition of 2 ml of concentrated nitric acid to the monomethylarsonic fractions. The residues were dissolved in 2 ml of concentrated hydrochloric acid and 1 ml of reducing agent added. After 30 min (to allow complete reduction) the solutions were diluted to 10 ml.

Before injection of 0.5–1 ml of solution into the zinc column, the inert gas flow rate was adjusted to 0.7 L min⁻¹, and the furnace (1700 °C) and recorder were turned on and allowed to establish a stable baseline (approximately 10 s). The solution was injected as quickly as possible using a syringe and the furnace turned off when the recorder signal had returned to the previously established baseline (approximately 20 s).

Blanks for the entire procedure were typically less than 4 ng mL⁻¹ and derived mainly from the hydrochloric acid.

Fig. 4.2 shows the separation achieved of arsenic(V) and organic arsenic compounds.

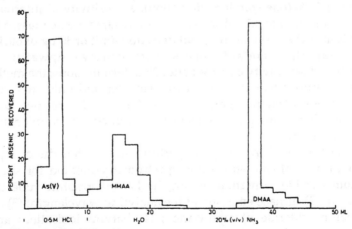

MMAA = Monomethyl arsonic acid
DMAA = dimethyl arsinic acid

Fig. 4.2 Elution of arsenic compounds from Dowex 50AG–X8 (100–200 mesh) resin (50 µg As for each species)
Source: Elsevier Science Publishers BV, Netherlands [4]

4.3.7 Miscellaneous

Brzezinska-Paudyn *et al.* [5] compared five methods for the determination of arsenic in marine and river sediments. The results obtained by graphite furnace atomic absorption spectrometry, neutron activation analysis, combined furnace flame atomic absorption inductively coupled plasma atomic emission spectrometry and flow injection/hydride generation inductively coupled plasma atomic emission spectrometry showed that all methods were appropriate for arsenic determinations higher than 5 mg kg $^{-1}$ in sediments. Graphite furnace atomic absorption spectrometry had a detection limit of 0.5–1 mg kg $^{-1}$.

4.4 Barium

4.4.1 Atomic absorption spectrometry

Bishop [6] has shown that his method for the determination of barium in sea water is also applicable to the determination of barium in sediment pore waters in amounts down to 100 pg L $^{-1}$ of sample.

Bishop [6] pointed out that the precise determination of barium in saline sediments and seawater by isotope dilution mass spectrometry is a demanding analytical problem as this technique requires extensive sample manipulation (addition of isotope spike, purification by ion exchange, evaporation, etc) and is limited to the determination of six samples per day.

Direct injection graphite furnace atomic absorption spectrometry was studied as a means of shortening analysis time. This method gave a precision of 13% (one standard deviation), a sensitivity at internal flow rate of 100 mL/min of 0.16 absorbance s/ng (characteristic mass, $M_0 = 28$ pg/0.0044 absorbance s) barium, and detection limit of 30 pg of barium in a 50–µL sample; standard curves were concave upward; severe degradation of sensitivity occurred after 25 determinations; the method of standard additions had to be used for seawater and only two samples could be determined per graphite tube. Sample throughput was not significantly better than that achieved by inductively coupled plasma mass spectrometry.

Many factors explain the poor performance of the direct injection graphite furnace atomic absorption spectrometry method for seawater: (1) barium carbide formation within the heated graphite tube during atomisation; (2) furnace emission at the analytical wavelength; (3) smoke from sodium chloride at atomisation; (4) barium ionisation; and (5) chloride vapour phase interferences. Items 3–5 could be overcome with appropriate modifiers or were unimportant, but carbide formation, short tube life and furnace emission appeared to remain unsolved. The method below described by Bishop [6] appears to overcome these remaining problems allowing barium to be routinely determined in seawater by direct injection graphite furnace atomic absorption spectrometry to a precision of better than 2% with a better than four–fold improvement in sensitivity. This technique has the potential to replace inductively coupled plasma mass spectrometric methods which achieve comparable precision yet by comparison are slow and cumbersome.

In the graphite furnace method, the V_2O_5/SI modifier added to undiluted samples promotes injection, sample drying, graphite tube life and the elimination of most seawater components in a slow char at 1150–1200 °C. Atomisation is at 2600 °C. Detection is at 553.6nm and calibration is by peak area. Sensitivity is 0.8 absorbance s/ng (M_0 = 5.6 pg/0.0044 absorbance s) at an internal argon flow of 60 mL/min. Detection limit is 2.5 pg (one standard deviation) barium in a 25–μL sample. Precision is 1–2% and accuracy is 2–3% for natural seawater (5.6–2.8 μg L $^{-1}$). The method works well in sediment pore waters.

A Perkin–Elmer Zeeman 5000 atomic absorption spectrometry equipped with a HGA–500 graphite furnace, AS–40 autosampler and RS–232C communications interface were used. Argon is used as purge gas. Ringsdorff and Schunk and Ebe pyrolytically coated graphite tubes were used. Furnace cooling was maintained with a Neslab CFT–75 refrigerated recirculator. Peak response (peak area, peak height and associated error statistics) was computed by MS–DOS (or MAC) computer after acquisition of high speed (60 Hz) raw digital data directly from the PE–5000. Instrumental settings were 553.6nm, slit 0.4nm, read 15 s, and lamp current 50 mA (energy 58–60), peak area, Zeeman background correction was used.

4.4.2 Photon activation analysis

This technique has been applied to the determination of barium as discussed under multi-metal analysis in section 4.44.7.1.

4.5 Beryllium

4.5.1 Atomic absorption spectrometry

This technique has been applied to the determination of beryllium as discussed under multi-metal analysis in section 4.44.2.2.

4.5.2 Inductively coupled plasma atomic emission spectrometry

This technique has been applied to the determination of beryllium as discussed under multi-metal analysis in section 4.44.3.1.

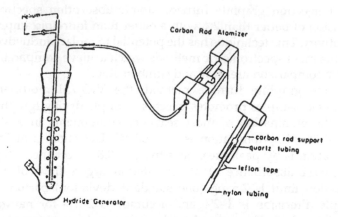

Fig. 4.3 Modified carbon rod atomiser for determination of bismuth in marine sediments
Source: American Chemical Society [7]

4.6 Bismuth

4.6.1 Atomic absorption spectrometry

Hydride generation and atomic absorption detection provide a very sensitive method for metalloid analyses. Collection of the hydride in a liquid nitrogen trap and then subsequent rapid introduction into an atomiser improved the sensitivities of such elements as arsenic and tin by an order of magnitude compared to the continuous method. However, such a collection method is only applicable to the elements whose hydrides are stable enough to be handled at ambient temperature. Bismuth is unfortunately too unstable for this technique. Upon warming bismuthine from liquid nitrogen temperatures only 5–15% of the trapped bismuthine was trapped, the remainder decomposed to overcome this problem. Lee [7] used a heated graphite tube to collect the generated bismuthine. Some 72% of the generated bismuthine was collected in the tube reproducibly. In this method the bismuth is reduced in solution by sodium borohydride to bismuthine, stripped with helium gas and collected in situ in a modified carbon rod atomiser (Fig. 4.3). The collected bismuth is subsequently atomised by increasing the atomiser temperature and detected by an atomic absorption spectrophotometer. The absolute detection limit is 3 pg of bismuth. The precision of the method is 2.2% for 150 pg and 6.7% for 25 pg of bismuth. Down to 6 μg kg^{-1} bismuth could be determined in sediments. High concentrations of cobalt, copper, gold, molybdenum, nickel, palladium, platinum, selenium, silver and tellurium interfere in this procedure. Varying amounts of bismuth were found in seawater 0.08–0.63 ng L^{-1} in Narragansett Bay and 0.05 ng L^{-1} in the

North Pacific. Bismuth contents on sediments taken in Narragansett Bay ranged from 0.27 to 6.4 mg kg $^{-1}$ and from the North Pacific Ocean from 0.10 to 0.12 mg kg $^{-1}$ indicating very high concentration factors.

4.6.2 Inductively coupled plasma atomic emission spectrometry

This technique has been applied to the determination of bismuth as discussed under multi-metal analysis in section 4.44.3.3.

4.7 Cadmium

4.7.1 Atomic absorption spectrometry

This technique has been applied to the determination of cadmium as discussed under multi-metal analysis in sections 4.44.2.1 and 4.44.2.2.

4.7.2 Inductively coupled plasma atomic emission spectrometry and inductively coupled plasma mass spectrometry

Wavelength modulation inductively coupled plasma echelle spectrometry has been used to determine cadmium [8]. This method was applied to two standard marine sediments, one (US Geological Survey MAG–1) from a relatively unpolluted marine sediment, and the other (Environment Canada WQB–2) from a freshwater harbour of an industrial city. A correction was necessary to overcome interference by the 214.445nm and 226.505nm lines of iron which interfere at both the cadmium 214.438nm and the cadmium 226.502 nm lines. Cadmium determinations were in good agreement with nominal values for the two standard sediments; WQW–2, expected 1.5 ± 0.2 mg kg $^{-1}$ found 2.0 ± 0.5 mg kg $^{-1}$, and MAG–1, expected 0.17 ± 0.1 mg kg $^{-1}$ found less than 0.2 mg kg $^{-1}$.

The determination of cadmium by inductively coupled plasma mass spectrometry is discussed under multi-metal analysis in section 4.44.4.1.

4.8 Caesium

4.8.1 Neutron activation analysis

This technique has been applied to the determination of caesium as discussed under multi-metal analysis in section 4.44.5.1.

4.9 Calcium

4.9.1 Atomic absorption spectrometry

This technique has been applied to the determination of calcium as discussed under multi-metal analysis in section 4.44.2.1.

4.9.2 Inductively coupled plasma atomic emission spectrometry

This technique has been applied to the determination of calcium as discussed under multi-metal analysis in section 4.44.3.1.

4.9.3 Photon activation analysis

This technique has been applied to the determination of calcium as discussed under multi-metal analysis in section 4.44.7.1.

4.9.4 X-ray secondary emission spectrometry

This technique has been applied to the determination of calcium as discussed under multi-metal analysis in section 4.44.8.1.

4.10 Cerium

4.10.1 Inductively coupled plasma atomic emission spectrometry

This technique has been applied to the determination of cerium as discussed under multi-metal analysis in section 4.44.3.2.

4.11 Chromium

4.11.1 Atomic absorption spectrometry

This technique has been applied to the determination of chromium as discussed under multi-metal analysis in sections 4.44.2.1 and 4.44.2.2.

4.11.2 Inductively coupled plasma atomic emission spectrometry and inductively coupled plasma mass spectrometry

This technique has been applied to the determination of chromium as discussed under multi-metal analysis in section 4.44.3.2 (inductively coupled plasma atomic emission spectrometry) and section 4.44.4.1 (inductively coupled plasma mass spectrometry)

4.11.3 Neutron activation analysis

This technique has been applied to the determination of chromium as discussed under multi-metal analysis in section 4.44.5.1.

4.11.4 Photon activation analysis

This technique has been applied to the determination of chromium as discussed under multi-metal analysis in section 4.44.7.1.

4.11.5 X-ray secondary emission spectrometry

This technique has been applied to the determination of chromium as discussed under multi-metal analysis in section 4.44.8.1.

4.12 Cobalt

4.12.1 Spectrophotometric method

This technique has been applied to the determination of cobalt as discussed under multi-metal analysis in section 4.44.1.1.

4.12.2 Atomic absorption spectrometry

This technique has been applied to the determination of cobalt as discussed under multi-metal analysis in sections 4.44.2.1 and 4.44.2.2.

4.12.3 Inductively coupled plasma atomic emission spectrometry

This technique has been applied to the determination of cobalt as discussed under multi-metal analysis in sections 4.44.3.1 and 4.44.3.2.

4.12.4 Neutron activation analysis

This technique has been applied to the determination of cobalt as discussed under multi-metal analysis in section 4.44.5.1.

4.12.5 Photon activation analysis

This technique has been applied to the determination of cobalt as discussed under multi-metal analysis in section 4.44.7.1.

4.13 Copper

4.13.1 Spectrophotometric method

This technique has been applied to the determination of copper as discussed under multi-metal analysis in section 4.44.1.1.

4.13.2 Atomic absorption spectrometry

This technique has been applied to the determination of copper as discussed under multi-metal analysis in sections 4.44.2.1 and 4.44.2.2.

4.13.3 Inductively coupled plasma atomic emission spectrometry

This technique has been applied to the determination of copper as discussed under multi-metal analysis in sections 4.44.3.1 and 4.44.3.2.

4.13.4 Differential pulse anodic stripping voltammetry

Differential pulse anodic stripping voltammetry using in situ plating techniques has been used to determine down to 0.1 µg kg $^{-1}$ copper in acid extracts of marine sediments [9]. This technique is also discussed under multi-metal analysis in section 4.44.6.1.

4.13.5 X-ray secondary emission spectrometry

This technique has been applied to the determination of copper as discussed under multi-metal analysis in section 4.44.8.1.

4.13.6 Electron probe microanalysis

This technique has been applied to the examination of copper distribution in marine sediments as discussed under multi-metal analysis in section 4.44.9.1.

4.13.7 Miscellaneous

Burzminskii and Fernando [10] studied methods for the extraction of copper as its oxalato complex from deep sea ferromanganese nodules.

Gerriuga et al. [88] have reviewed methods for the determination of copper in marine sediments.

4.14 Dysprosium

4.14.1 Inductively coupled plasma atomic emission spectrometry

This technique has been applied to the determination of dysprosium as discussed under multi-metal analysis in section 4.44.3.2.

4.15 Gallium

4.15.1 Inductively coupled plasma atomic emission spectrometry

This technique has been applied to the determination of gallium as discussed under multi-metal analysis in section 4.44.3.2.

4.16 Iridium

4.16.1 Atomic absorption spectrometry

Hodge et al. [11] determined picogram amounts of iridium in marine sediments by isolation of anionic forms on a single ion exchange bead followed by graphite furnace atomic absorption spectrometry. Yields varied between 35 and 90%.

4.17 Iron

4.17.1 Atomic absorption spectrometry

This technique has been applied to the determination of iron as discussed under multi-metal analysis in section 4.44.2.1.

4.17.2 Inductively coupled plasma atomic emission spectrometry

This technique has been applied to the determination of iron as discussed under multi-metal analysis in sections 4.44.3.1 and 4.44.3.2.

4.17.3 Neutron activation analysis

This technique has been applied to the determination of iron as discussed under multi-metal analysis in section 4.44.5.1.

4.17.4 Photon activation analysis

This technique has been applied to the determination of iron as discussed under multi-metal analysis in section 4.44.7.1.

4.17.5 X-ray secondary emission spectrometry

This technique has been applied to the determination of iron as discussed under multi-metal analysis in section 4.44.8.1.

4.17.6 Miscellaneous

Lucotte and D'Anglejan [12] compared five different methods for the extraction and determination of iron hydroxides and/or associated phosphorus in estuarine particulate matter from the St Lawrence and Eastmain Rivers estuaries. The extraction reagents examined were citrate-dithionite-bicarbonate, ammonium oxalate-oxalic acid, hydroxylamine hydrochloride, calcium nitrilotriacetic acid, and acetate-tartrate. The procedure using the citrate-dithionite bicarbonate extraction reagent was generally more specific and reproducible than the others. This extraction procedure was included in an extraction sequence which differentiated the exchangeable phosphorus and iron fraction, the iron hydroxides and associated phosphorus and the organic phosphorus and lithogenous iron.

4.18 Lead

4.18.1 Atomic absorption spectrometry

This technique has been applied to the determination of lead as discussed under multi-metal analysis in sections 4.44.2.1 and 4.44.2.2.

4.18.2 Inductively coupled plasma atomic emission spectrometry and inductively coupled plasma mass spectrometry

The wavelength modulation inductively coupled plasma echelle spectrometric technique [8] has also been applied to the determination of lead in marine sediments. The 220.353nm lead line was used. Only aluminium at concentrations higher than that usually encountered in marine sediments is expected to interfere in the determination of lead at 220.353nm. Lead determinations in standard reference samples were in good agreement with expected values.

Reference sediment	Expected	Lead mg kg^{-1} determined
WQB–2	79 ± 4	75 ± 7
MAG–1	25 ± 1	25 ± 1

The application of these techniques to the determination of lead in sediments is also discussed under multi-metal analysis in section 4.44.3.1 (inductively coupled plasma atomic emission spectrometry) and section 4.44.4.1 (inductively coupled plasma mass spectrometry).

4.18.3 Potentiometric stripping analysis

The application of this technique to the determination of lead is discussed under multi-metal analysis in section 4.44.6.1.

4.18.4 Photon activation analysis

The application of this technique to the determination of lead is discussed under multi-metal analysis in section 4.44.7.1.

4.19 Lanthanum

4.19.1 Inductively coupled plasma atomic emission spectrometry

The application of this technique to the determination of lanthanum is discussed under multi-metal analysis in section 4.44.3.2.

4.20 Magnesium

4.20.1 Inductively coupled plasma atomic emission spectrometry

The application of this technique to the determination of magnesium is discussed under multi-metal analysis in section 4.44.3.1.

4.20.2 Photon activation analysis

The application of this technique to the determination of magnesium is discussed under multi-metal analysis in section 4.44.7.1.

4.21 Manganese

4.21.1 Atomic absorption spectrometry

The application of this technique to the determination of manganese is discussed under multi-metal analysis in section 4.44.2.1.

4.21.2 Inductively coupled plasma atomic emission spectrometry

The application of this technique to the determination of manganese is discussed under multi-metal analysis in sections 4.44.3.1 and 4.44.3.2.

4.21.3 X-ray secondary emission spectrometry

The application of this technique to the determination of manganese is discussed under multi-metal analysis in section 4.44.8.1.

4.22 Mercury

4.22.1 Fluorescence spectroscopy

Hutton and Preston [13] have described a non-dispersive atomic fluorescence method using cold vapour generation for the determination of down to 0.04 mg kg^{-1} mercury in marine sediments. In this procedure, mercury in the sediment was reduced to its elemental form with acidic stannous chloride solution, and swept with argon into the fluorimeter. Marine and estuarine sediments examined by this procedure were found to contain between 1 and 16 µg mercury kg^{-1}.

4.22.2 Atomic absorption spectrometry

Abo-Rady [14] has described a method for the determination of total inorganic plus organic mercury in nanogram quantities in sediments. This method is based on the decomposition of organic and inorganic mercury compounds with acid permanganate, removal of excess permanganate with hydroxylamine hydrochloride, reduction to metallic mercury with tin and hydrochloric acid, and transfer of the liberated mercury in a stream of air to the spectrometer. Mercury was determined by using a closed recirculating air stream. Sensitivity and reproducibility of the 'closed system' were better, it is claimed, than those of the 'open system'. The coefficient of variation was 5.6% for sediment samples.

4.22.3 Inductively coupled plasma mass spectrometry

Walker *et al.* [89] compared inductively coupled plasma mass spectrometric and inductively coupled plasma atomic emission spectrometric methods for the determination of mercury in Great Barrier Reef sediments. Typical instrument variabilities were 1 in 10^9 for inductively coupled plasma atomic emission spectrometry and 1 in 10^{12} for inductively coupled plasma mass spectrometry.

4.22.4 Miscellaneous

The microbial methylation of inorganic mercury to methyl mercury observed in lakewater sediments has also been observed in marine sediments. Mercury has been found in some sediments but at very low concentrations, mainly from areas of known mercury pollution. It represents usually less than 1% of the total mercury in the sediment, and frequently less than 0.1% [15–18]. Micro-organisms within the sediments are considered to be responsible for the methylation [15, 19] and it has been suggested that methylmercury may be released by the sediments to the sea water, either in dissolved form or attached to particulate material and thereafter rapidly taken up by organisms [20–22].

Fresnet-Robin and Ottmann [23] studied the distribution of mercury between water and sediment in samples taken from the Loire Estuary between 1972 and 1975, and observed a decrease in mercury pollution during this period. The ratio of mercury in sediment/mercury in water increased with increasing salinity of the estuary water. Mercury fixes most on suspensions in the surface water, as these suspensions comprise smaller particles.

Bothner *et al.* [24] have studied mercury loss from contaminated estuarine sediments. A systematic decrease of mercury with an apparent half-life of about 1.3 years was observed in oxidising sediments sampled. In situ measurements of the mercury flux from sediments could not account for this decrease. It is suggested that removal is associated with sediment particles transport. The decrease was modelled using a steady-state mixing model. Mercury concentration in anoxic interstitial waters was 126 times higher than that in overlying sea water.

4.23 Molybdenum

4.23.1 Spectrophotometric method

Pavlova and Yatsimirskii [25] employed a spectrophotometric method based on the catalytic acceleration by molybdenum of the oxidation of dithiooxamine by hydrogen peroxide for the determination of down to 2 ng of molybdenum in seawater sediments.

4.23.2 Inductively coupled plasma mass spectrometry

The application of this technique is discussed under multi-metal analysis in sections 4.44.4.1.

4.24 Nickel

4.24.1 Atomic absorption spectrometry

This technique has been applied to the determination of nickel as discussed under multi-metal analysis in sections 4.44.2.1 and 4.44.2.2.

4.24.2 Inductively coupled plasma atomic emission spectrometry and inductively coupled plasma mass spectrometry

These techniques have been applied to the determination of nickel as discussed under multi-metal analysis in sections 4.44.3.1 and 4.44.3.2 (inductively coupled plasma atomic emission spectrometry) and 4.44.4.1 (inductively coupled plasma mass spectrometry).

4.24.3 Photon activation analysis

This technique has been applied to the determination of nickel as discussed under multi-metal analysis in section 4.44.7.1.

4.24.4 X-ray secondary emission spectrometry

This technique has been applied to the determination of nickel as discussed under multi-metal analysis in section 4.44.8.1.

4.24.5 Miscellaneous

Burzminskii and Fernando [10] studied methods for the extraction of nickel as its oxalato complexes from deep sea ferromanganese nodules.

4.25 Platinum

4.25.1 Atomic absorption spectrometry

Hodge *et al.* [11] determined picogram quantities in marine sediments by isolation of anionic forms on a single ion exchange bead followed by graphite furnace atomic absorption spectrometry.

4.26 Potassium

4.26.1 Photon activation analysis

This technique has been applied to the determination of potassium as discussed under multi-metal analysis in section 4.44.7.1.

4.26.2 X-ray secondary emission spectrometry

This technique has been applied to the determination of potassium as discussed under multi-metal analysis in section 4.44.8.1.

4.27 Rhenium

4.27.1 Atomic absorption spectrometry

Rhenium is one of the last stable elements discovered, one of the least abundant metals in the earth's crust, and one of the most important sentinels of reducing aqueous environments through its abundance in sediments. Although its chemistry is fairly well circumscribed, its marine chemistry is as yet poorly developed. In addition, the understanding of rhenium's marine chemistry will provide an entry to the understanding of the marine chemistry of technetium, an element which is just above rhenium in group VIIA (group 7 in 1985 notation) of the periodic table. Technetium has only unstable isotopes whose origins are primarily in nuclear weapon detonations and in nuclear reactor wastes. These two elements have remarkably similar chemistries.

A graphite furnace atomic absorption method described by Koide et al. [87] for the determination of rhenium at parts-per-billion levels in marine sediments is based upon the isolation of heptavalent rhenium species upon anion exchange resins. All steps are followed with [186]Re as a yield tracer. A crucial part of the procedure is the separation of a rhenium from molybdenum, which significantly interferes with the graphite furnace detection when the Mo/Re ratio is two or greater. The separation is accomplished through an extraction of tetra-phenylarsonium perrhenate into chloroform in which the molybdenum remains in the aqueous phase.

4.28 Rubidium

4.28.1 Neutron activation analysis

This technique has been applied to the determination of rubidium as discussed under multi-metal analysis in section 4.44.4.1.

4.29 Scandium

4.29.1 Spectrophotometric method

To determine scandium in marine sediments, Shimizu [26] digested the sample with hydrochloric acid–hydrofluoric acid and then separated scandium from other elements by successive cation and anion exchange chromatography prior to coprecipitation of scandium with ferric hydroxide. Finally, scandium is determined spectrophotometrically at 610nm with bromopyrogallol red.

4.29.2 Neutron activation analysis

This technique has been applied to the determination of scandium as discussed under multi-metal analysis in section 4.44.5.1.

4.30 Selenium

4.30.1 Spectrophotometric method

Terada et al. [27] determined 0.3–1 ppm selenium in marine sediments after converting the element to selenium bromide which was distilled off and assayed colorimetrically as piazselenol.

4.30.2 Atomic absorption spectrometry

Willie et al. [28] applied hydride generation atomic absorption spectrometry with in situ concentration in a graphite furnace to the determination of selenium in marine sediments. A custom-made Pyrex cell was used to generate selenium hydride which was carried to a quartz tube and then to a preheated furnace. All the selenium was identified when the furnace was operated at 400 °C. Pyrolitic graphite-coated tubes could be used and the life of the tubes exceeded 1200 firings. A 10% solution of the dimethyldichlorosilane was used to deactivate internal surfaces of the generation cell which could be used for 130 determinations before requiring resilylation. No interferences were found in this study:

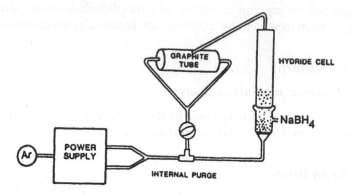

Fig. 4.4 Assembly of hydride generation apparatus for determination of selenium in marine sediments
Source: American Chemical Society [28]

1000 µg of iron, 6 µg. of copper, 15 µg of nickel and 2.5 µg of arsenic had no effect on the signal from 2 ng selenium in 5 ml of 0.05M hydrochloric acid solution. An absolute detection limit of 70 pg selenium equivalent to 30 ng g $^{-1}$ in sediment was achieved.

In this method a 0.5g portion of sediment was decomposed by acid digestion in a PTFE bomb according to the procedure described by Siu and Berman [29, 30] and diluted to 50 mL in 1M hydrochloric acid. Total selenium was determined by using 500 µL aliquots delivered into the hydride cell containing 5 mL of 0.5 M hydrochloric acid. Alternatively 0.5g samples were dry ashed using magnesium nitrate as an ashing aid [30] to prevent volatilisation losses of selenium. The solution was diluted to 50 mL. Total selenium was determined by using 50 µL aliquots diluted to 5 mL with 0.5M hydrochloric acid. Reagent blanks were processed through identical steps for decomposition of sediments.

In the method described by Willie *et al.* [28] atomic absorption measurements were made with a Perkin–Elmer 5000 spectrometer fitted with a Model HGA 500 graphite furnace and Zeeman effect background correction system. Peak absorbance signals were recorded with a Perkin–Elmer PRS–10 printer-sequencer. A selenium electrodeless discharge lamp (Perkin–Elmer Corp) operated at 6W was used as the source. Absorption was measured at the 196.0nm line. The spectral band pass was 0.7nm. Standard Perkin–Elmer pyrolytic graphite-coated tubes were used in all studies.

A custom-made Pyrex cell [31] was used to generate selenium hydride. The internal purge gas supply line to the furnace was routed through a stopcock made of Teflon that permitted the operator to select gas flow into either the bottom of the hydride cell or into the furnace, as illustrated in Fig. 4.4.

Table 4.2 Analytical results[a]

Sample	Determined	Accepted value
NASS–I, ng/g	0.024 ± 0.002	(0.025 ± 0.001)[b]
BCSS–I, µg/g	0.46 ± 0.04	(0.43 ± 0.06)[c]
TORT–I, µg/g	6.74 ± 0.17	6.88 ± 0.47
NBS 1566, µg/g	2.13 ± 0.10	2.1 ± 0.5

[a] Precision expression as standard deviation based on five determinations. [b] Uncertified value. [c] Uncertified value

Source: American Chemical Society [28]

In this manner an argon flow could be used to strip the generated hydride from solution and carry it out the top of the cell where it was directed, via a 1 mm i.d. × 1.5 mm o.d. quartz tube, into the sample introduction port of a preheated furnace.

Prior to use, the cell and transfer line were silylated to deactivate the internal surfaces [32]. A 10% solution of dimethyl-dichlorosilane in toluene was used to rinse the cell, followed by successive rinses in toluene and methanol. The surfaces were then dried at room temperature in a stream of nitrogen.

Sodium borohydride solution was pumped into the cell using a rack-mounted Ismatec peristaltic pump. The sequence of operations describing selenium hydride generation, collection and atomisation is given below. During collection the stopcock was closed to direct internal purge gas through the hydride cell, and the furnace was preheated for 10 s at 600 °C. The sodium borohydride solution was pumped into the cell at a rate of 4 mL/min for 30 s for 5 mL samples (60 s for 50 mL seawater samples), during which time the selenium hydride was swept via the generated hydrogen gas stream (≈ 50 mL/min under these conditions) into the furnace where it was trapped. Internal purge gas flow was automatically initiated at the end of sodium borohydride addition, and the cell purged for 120 s at a flow rate of 100 mL/min for 5 mL sample volumes (190–210 s for 50 mL seawater samples). At the end of this cycle, thermal programming of the furnace was terminated, this quartz transfer line removed from the sample introduction port, and the stopcock opened to permit internal purge gas to flow through the furnace. The sample was then atomised at 2600 °C using maximum power heating and internal gas stop, followed by a cleaning cycle at 2700 °C with 300 mL/min internal purge gas flow. The furnace programme is shown in Table 4.3.

Internal purge gas was again diverted through the reaction cell, which was emptied and rinsed with doubly distilled water. The sodium borohydride solution was withdrawn from the injector tip by reversing

Table 4.3 Furnace program

	Program	Step	Temp °C	Time s	Internal purge, mL/min
Generation–collection	1	1	600	10	0
		2	600	30	0
		3	600	120	100
Atomisation	2	1	2600	4	0
		2	2700	2	300

Source: American Chemical Society [28]

the direction of the peristaltic pump. The next sample aliquot was then added to the cell and the measurement process repeated. Replicate measurements could be made every 3–4 min.

Application of this method to standard reference sediment samples gave results very close to accepted values as shown in Table 4.2.

Itoh *et al.* [33] have developed an automated flow system to eliminate interferences from transition metal ions during the hydride generation/atomic absorption spectrometric determination of selenium in estuarine sediments. Experimental conditions were adjusted to minimise interferences. Low concentrations of tetrahydroborate in 6 mol L^{-1} hydrochloric acid were preferred. Iron(III), added as chloride at a concentration of 2 mmol L^{-1}, removed the depression of selenium signals by ions such as copper(II) and bismuth(III).

The application of atomic absorption spectrometry to the determination of selenium in marine sediments is also discussed under multi-metal analysis in section 4.44.2.3.

4.30.3 Inductively coupled plasma atomic emission spectrometry

de Oliveira *et al.* [34] have described a method for the determination of selenium in potassium hydroxide fusion or nitric-perchloric-hydrofluoric acid digests of marine sediments.

The application of inductively coupled plasma atomic emission spectrometry to the determination of selenium is also discussed under multi-metal analysis in section 4.44.3.3.

4.30.4 Gas chromatography

Siu and Berman [29] determined down to a 0.2 pg selenium in marine sediments with a precision of 7%. The procedure involved acid digestion in a PTFE bomb, conversion to 5–nitropiazselenol, and electron capture gas chromatography of the toluene extract. This method is based on the fact that 1,2 diaminolienzene (o–phenylene diamine) and its derivatives react selectively and quantitatively with selenium(IV) (average recovery from 94 ± 5%) from piazselenols that are both volatile and stable. Piazoselenols can be determined by electron capture gas chromatography. The sediments were digested as follows:

A 0.5g sample was placed in a poly(tetrafluoroethylene) pressure decomposition vessel. A concentrated acid mixture comprising 3 mL of nitric, 3 mL of hydrochloric and 1 mL of perchloric was added. The closed vessel was immersed in boiling water for 2h. The vessel was opened after cooling, and the contents transferred to a PTFE beaker with doubly distilled water. The solution was taken to dryness overnight on a hot plate at about 90 °C. The residue was dissolved in about 20 mL of 1M hydrochloric acid and the solution was again taken to dryness at about 90°C. This ensured complete reduction of Se(VI) to Se(IV). The residue was again dissolved and diluted to 50 mL with 1M hydrochloric acid. This solution was washed twice with about 20 mL of toluene.

One millilitre of sediment solution was allowed to stand with 0.1 mL of the 1% 4-nitro-o-phenylenediamine hydrochloride solution in a 10 mL glass vial for 2 h. One millilitre of toluene was added. The vial was capped and shaken vigorously for 2 min. One microlitre of the toluene layer was injected into the gas chromatograph.

A Varian Aerograph Model 1200 gas chromatograph equipped with a Tracor ^{63}Ni electron capture detector (ECD) was used. The column was a 2 m borosilicate tube packed with 3% OV 225 on Chromosorb W 80/100 mesh. It was normally kept at 200 °C. Nitrogen (<10 ppm oxygen), further purified by passage through molecular sieve 5A and a heated oxygen scavenger (Supelco) was used as carrier and detector make-up gas. The flow rates were usually about 25 mL/min^{-1}. The ECD was heated to 320°C and operated in constant voltage mode at an optimal voltage of usually –13V.

4.31 Silicon

4.31.1 Inductively coupled plasma atomic emission spectrometry

This technique has been applied to the determination of silicon as discussed under multi-metal analysis in section 4.44.3.2.

4.31.2 X-ray secondary emission spectrometry

The determination of silicon is discussed under multi-metal analysis in section 4.44.8.1.

4.32 Silver

4.32.1 Atomic absorption spectrometry

Bloom [35] has used Zeeman corrected graphite furnace atomic absorption spectrometry to determine silver in marine sediments.

4.33 Sodium

4.33.1 Inductively coupled plasma atomic emission spectrometry

This technique has been applied to the determination of sodium as discussed under multi-metal analysis in section 4.44.3.1.

4.33.2 Photon activation analysis

This technique has been applied to the determination of sodium as discussed under multi-metal analysis in section 4.44.7.1.

4.34 Strontium

4.34.1 Neutron activation analysis

This technique has been applied to the determination of strontium as discussed under multi-metal analysis in section 4.44.5.1.

4.34.2 Photon activation analysis

This technique has been applied to the determination of strontium as discussed under multi-metal analysis in section 4.44.7.1.

4.34.3 Inductively coupled plasma mass spectrometry

The determination of strontium as discussed under multi-metal analysis in section 4.44.4.1.

4.35 Thallium

4.35.1 Atomic absorption spectrometry and anodic stripping voltammetry

Riley and Siddique [36] described procedure for the determination of thallium in deep sea sediments. It involves preliminary concentration by adsorption on a strongly basic anion-exchange resin as the tetrachloro-thallate ion, followed by elution with sulphur dioxide, and evaporation, before determination by graphite furnace atomic absorption spectrometry or differential pulse anodic stripping voltammetry. There was good agreement between the results obtained by the two methods.

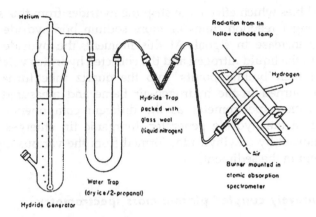

Fig. 4.5 Apparatus for the determination of stannae and the organotin hydrides
Source: American Chemical Society [37]

4.35.2 Inductively coupled plasma mass spectrometry

The determination of thallium as discussed under multi-metal analysis in section 4.44.4.1.

4.36 Tin

4.36.1 Atomic absorption spectrometry

Hodge *et al.* [37] have described a hydride generation atomic absorption spectrometric method for the measurement of nanogram or subnanogram amounts of inorganic tin(IV) and various alkyl and aryl tin halides in sediments and algal residues from Narragansett Bay, California. In this method, approximately 1g samples of oven dried (110°C) material was completely digested with nitric acid–perchloric acid and hydrofluoric acid– hydrochloric acid. Ultimately the algae residues are dissolved in 50 mL of 2N nitric acid and the sediments end up in 50 mL of an equal volume of 6N hydrochloric acid and 2N nitric acid – resulting from the total destruction procedure.

The hydride generator is changed with 0.1 ml or less of the sediment solution delivered with an Eppendorf tip pipette into the hydride generator, followed by 1 ml 2M acetic acid, diluted to 100 ml with double distilled water and reacted with sodium borohydride. The solution is acidified with 1 mL of 2N acetic acid and purged with the helium carrier gas for a few minutes, during which time the hydride trap is cooled in liquid nitrogen (Fig. 4.5). One mL of the purified 4% sodium borohydride is injected over a period of 25 s into the hydride generator through a Teflon-backed silicone rubber septum, producing a smooth evolution of hydrogen bubbles which effectively strip the hydrides from this solution in 5 min. Longer reaction times or more sodium borohydride do not produce any increase in signal. At this moment, the hydride trap is removed from the liquid nitrogen and the collected hydrides volatilise, in the order of their boiling points, into the quartz tube furnace. The hydrides are burned in the hydrogen-air flame and measured by the atomic absorption spectrometer as the different components emerge separately from the hydride generator. Inorganic tin emerges first as stannane followed by alkylhydrides formed from the various organotin labides present in the sediment.

4.36.2 Inductively coupled plasma mass spectrometry

The determination of tin is discussed under multi-metal analysis in section 4.44.4.1.

4.37 Titanium

4.37.1 Inductively coupled plasma atomic emission spectrometry

This technique has been applied to the determination of titanium as discussed under multi-metal analysis in section 4.44.3.2.

4.37.2 Photo activation analysis

This technique has been applied to the determination of titanium as discussed under multi-metal analysis in section 4.44.7.1.

4.37.3 X-ray secondary emission spectrometry

This technique has been applied to the determination of titanium as discussed under multi-metal analysis in section 4.44.8.1.

4.38 Uranium

4.38.1 Photon activation analysis

This technique has been applied to the determination of uranium as discussed under multi-metal analysis in section 4.44.7.1.

4.38.2 Inductively coupled plasma mass spectrometry

The determination of uranium is discussed under multi-metal analysis in section 4.44.4.1.

4.39 Vanadium

4.39.1 Inductively coupled plasma atomic emission spectrometry

This technique has been applied to the determination of vanadium as discussed under multi-metal analysis in sections 4.44.3.1 and 4.44.3.2.

4.40 Ytterbium

4.40.1 Inductively coupled plasma atomic emission spectrometry

This technique has been applied to the determination of ytterbium as discussed under multi-metal analysis in section 4.44.3.2.

4.41 Yttrium

4.41.1 Inductively coupled plasma atomic emission spectrometry

This technique has been applied to the determination of yttrium as discussed under multi-metal analysis in section 4.44.3.2.

4.42 Zinc

4.42.1 Atomic absorption spectrometry

This technique has been applied to the determination of zinc as discussed under multi-metal analysis in section 4.44.2.1.

4.42.2 Inductively coupled plasma atomic emission spectrometry and inductively coupled plasma mass spectrometry

These techniques have been applied to the determination of zinc as discussed under multi-metal analysis in sections 4.44.3.1 and 4.44.3.2 (inductively coupled plasma atomic emission spectrometry) and section 4.44.4.1 (inductively coupled plasma mass spectrometry).

4.42.3 Photon activation analysis

This technique has been applied to the determination of zinc as discussed under multi-metal analysis in section 4.44.7.1.

4.42.4 X-ray secondary emission spectrometry

This technique has been applied to the determination of zinc as discussed under multi-metal analysis in section 4.44.8.1.

4.42.5 Electron probe microanalyses

This technique has been applied to the determination of zinc as discussed under multi-metal analysis in section 4.44.9.1.

4.43 Zirconium

4.43.1 Spectrophotometric method

Sastry *et al.* [38] separated zirconium from marine sediments by digestion with hydrochloric acid, then coprecipitated zirconium with cerium iodide prior to spectrophotometric determination as the Alizarin R–Mordant Red-S chromophore.

4.43.2 Inductively coupled plasma atomic emission spectrometry

This technique has been applied to the determination of zirconium as discussed under multi-metal analysis in section 4.44.3.2.

4.43.3 Photon activation analysis

This technique has been applied to the determination of zirconium as discussed under multi-metal analysis in section 4.44.7.1

4.44 Multi-metal analysis

4.44.1 Spectrophotometric method

4.44.1.1 Cobalt and copper

Spectrophotometric methods, involving the use respectively of dimethylglyoxime and 2-nitroso-1 naphthol, have been used to determine nickel and cobalt in marine sediments [39].

4.44.2 Atomic absorption spectrometry

4.44.2.1 Heavy metals (chromium, manganese, iron, cobalt, nickel, copper, zinc, cadmium and lead) and calcium

Pilkington and Warren [40] have described a method for studying the distribution of heavy metals (lead, cadmium and zinc) in marine sediments in which the various sediment components, including organic matter, are separated into distinct bands by centrifuging in tetrabromoethane. Total elements are then determined using direct flame and graphite furnace atomic absorption spectrometry. Table 4.4 shows the mineral components of the density subfractions and the heavy metal contents of these. It may be seen from Table 4.4 that the mass balances

Table 4.4 Mineral components of the density subfractions

Density subfraction	Sample ES 26 (+10 μm)	Sample ES 35 (+10 μm)	Mass balance μg		
			Pb	Cd	Zn
<2.40	Organics (9 wt %); mostly organic debris; a few soft white conglomerates of magnesian calcite and quartz	Clay conglomerates (8 wt %); soft golden brown conglomerates of magnesian calcite, mica, kaolin, quartz and goethite; a few black vitreous particles	19.74	0.985	22.01
2.40–2.55	Conglomerates (6 wt %); soft white conglomerates of magnesian calcite, quartz, a little feldspar	Conglomerates (40 wt %); soft brown and off-white conglomerates of magnesian calcite, mica, some kaolin quartz, traces goethite, feldspar	5.74	0.325	7.13
2.55–2.66	Quartz + calcite shells (46 wt %); quartz and calcitic coralline stems, hexagonal rods, microshells	Quartz + calcite shells (42 wt %); mostly quartz; some calcitic shell fragments, a few hexagonal rods	11.04	0.470	18.12
2.66–2.75	Magnesian calcite shells (20 wt %); opaque white shell fragments and whole microshells of magnesian calcite; some coralline stems and tubes	Magnesian calcite shells (4 wt %); as for ES 26 but including some black particles of irregular shape	6.64	0.585	9.23
2.75–2.95	Aragonite shells (19 wt %); thin shiny shell fragments of aragonite; some transparent platelets	Aragonite shells (6 wt %); as for ES 26	2.33	0.193	3.58
>2.95	Heavy minerals (0.1 wt %) as for ES 35	Heavy minerals (0.2 wt %); translucent brown and green particles of tourmaline ilmenite, opaque black particles of magnetite, opaque golden rounded particles	Weight too small to analyse		
Total			45.5	2.56	60.0
Unfractionated sample			45.5	2.36	72.4

Source: American Chemical Society [40]

obtained for lead and cadmium were satisfactory, but that there was apparently some loss of zinc during fractionation. The concentrations of lead, cadmium and zinc were found to vary between the density bands in each sample and between similar bands in different samples. Zinc concentrations were highest in the organic fraction. Lead and cadmium were also preferentially concentrated in the organics subfraction. The technique avoids the problem of non-selective dissolution encountered with chemical methods of determining metal distributions, and provided that the sediments contain only a small proportion of discrete ultrafine particles, accurate and reproducible data can be obtained. The concentration and distribution patterns revealed are useful in evaluating the environmental significance of contaminated sediments.

Sinex et al. [41] have evaluated the accuracy and precision of a method for the determination of chromium, manganese, iron, cobalt, nickel, copper, zinc, cadmium, lead and calcium in estuarine sediments. This method involved a 4 h hot digestion of 5g of dried (at 105 °C) sample with 100 ml 9:1 v/v nitric acid-hydrochloric acid followed by atomic absorption spectrometry of the 25 ml of concentrated extract. High recoveries and good reproducibility were obtained with coefficients of variation being 10% or less. They compared results obtained by this procedure with those obtained by a lithium metaborate fusion-d.c. plasma emission spectroscopic method [42, 43] on a NBS SRM–1645 river sediment and on an estuarine sediment (US Geological Survey Marine mud sample MAG–1). In the lithium metaborate fusion method, 100 mg of the dried sediments were fused with 0.5 g of lithium metaborate in a graphite crucible at 900 °C for approximately 15 min. The molten lithium borate glass bead was poured into a 150 ml Teflon FEP beaker containing 100 ml of 4% nitric acid. The beaker was placed on a magnetic stirrer, and the contents were stirred for approximately 10 min. The solution was then transferred to a 125 ml polyethylene bottle. Considerably higher and more reproducible recoveries were obtained by fusion than by acid digestion. When the acid extraction method was applied to estuarine acids (MAG–1), the resulting trace metal concentrations, although reproducible, were usually substantially lower than total concentrations determined by lithium metaborate fusion d.c. plasma emission spectroscopy. Excellent recoveries were obtained when the test was applied to the NBS SRM 1645 reference river sediment; chromium, manganese, nickel, copper, zinc and lead recovery was complete, while for iron, cobalt and cadmium, modest under recovery (13–25%) was observed.

4.44.2.2 Heavy metals (cadmium, copper, lead, nickel, cobalt and chromium) and beryllium

Sturgeon et al. [44] demonstrated quantitative recovery of cadmium, lead, copper, nickel, cobalt, beryllium and chromium from estuarine sediments using a digestion-graphite furnace atomic absorption spectrometric with a L'vov platform method.

In general, it is agreed that there is a lack of well characterised and representative standard reference estuarine sediments for the evaluation of methods of analysis. The National Research Council of Canada Marine Analytical Standards Programme has prepared two near–shore marine sediment reference materials, MESS–1 and BCSS–1, for which reliable results are available for 12 major and minor constituents and 13 trace metals. Sturgeon *et al.* [44] used these standards to evaluate their method.

Digestions on dried (105 °C) sediments (0.5 g) were performed in PTFE beakers using hydrochloric acid (5 mL), nitric acid (2 mL), hydrofluoric acid (5 mL), and water (4 mL) for 2 h at 90 °C. After gentle evaporation to dryness, a further 5 mL of concentrated nitric acid and 2 mL perchloric acid were added and the solutions again taken to dryness. The residue was dissolved in 20 mL L mol^{-1} hydrochloric acid, filtered and made up to 50 mL with water.

It is shown in Table 4.5 that determined results are in good agreement with reference values for all seven elements.

4.44.2.3 Arsenic and selenium

Cutter [45] used a selective hydride generation procedure as a basis for the differential determination of arsenic and selenium species in marine sediments.

4.44.3 *Inductively coupled plasma atomic emission spectrometry*

4.44.3.1 Heavy metals (cobalt, copper, iron, manganese, nickel, lead and zinc), aluminium, calcium, magnesium, sodium, vanadium and beryllium

McLaren *et al.* [46] have described a procedure which permits the simultaneous determination of six major and minor elements (aluminium, iron, calcium, magnesium, sodium and phosphorus) and eight trace elements (beryllium, cobalt, copper, manganese, nickel, lead, vanadium and zinc) in near shore marine sediments by inductively coupled plasma atomic emission spectrometry. Dissolution of the samples is achieved with a mixture of nitric, perchloric and hydrofluoric acids in sealed Teflon vessels. Accurate calibration for all elements can be achieved with simple aqueous standards, provided that proper correction for various spectroscopic interferences is made. The method was not suitable for determination of arsenic, cadmium and molybdenum in these materials because of inadequate sensitivity nor for chromium and thallium because of incomplete dissolution. Detection limits varied from 0.5 mg kg^{-1} (beryllium) to 5 mg kg^{-1} (arsenic).

Table 4.5 Comparison of graphite furnace atomic absorption spectrometry results with accepted values for MESS–1

Element	Concentration (µg g $^{-1}$)^a by different methods				
	Standard additions		L'vov platform	Accepted value	
	Direct injection	Chelation–extraction^d	Direct injection		
Cd	0.54 ± 0.06 (9)	0.58 ± 0.11 (9)	0.65 ± 0.13 (3)	0.59 ± 0.1	
Pb	26.4 ± 2.7 (9)	28.2 ± 6.5 (9)	27.3 ± 1.6 (3)	34.0 ± 6.1^b	
Cu	21/9 ± 2.3 (8)	25.4 ± 3.0 (10)	21.3 ± 1.6 (3)	25.1 ± 3.8	
Ni	28.7 ± 2.9 (10)	31.2 ± 4.1 (11)	28.6 ± 4.8 (3)	29.5 ± 2.7	
Co	11.7 ± 1.8 (14)	12.2 ± 2.7 (11)	11.7 ± 2.9 (3)	10.8 ± 1.9	
Be	1.8 ± 0.2 (8)	ND^c	1.8 ± 0.1 (3)	1.9 ± 0.2	
Cr	70.3 ± 5.0 (6)	ND	ND	71 ± 11	

Comparison of graphite furnace atomic absorption spectrometry results with accepted values for BCSS–1

Element	Concentration (µg g $^{-1}$)^a by different methods				
	Standard additions		L'vov platform	Accepted value	
	Direct injection	Chelation–extraction^d	Direct injection		
Cd	0.24 ± 0.04 (9)	0.26 ± 0.10 (3)	0.23 ± 0.03 (3)	0.25 ± 0.04	
Pb	22.0 ± 3.0 (6)	22.0 ± 10.1 (3)	20.0 ± 5.4 (3)	22.7 ± 3.4	
Cu	15.0 ± 3.4 (3)	16.2 ± 2.2 (4)	14.6 ± 2.5 (3)	18.5 ± 2.76	
Ni	54.7 ± 5.1 (7)	56.6 ± 5.0 (5)	53.0 ± 2.2 (3)	55.3 ± 3.6	
Co	12.7 ± 3.6 (3)	11.9 ± 6.9 (3)	10.7 ± 1.3 (3)	11.4 ± 2.1	
Be	1.1 ± 0.2 (8)	ND^c	1.1 ± 0.1 (3)	1.3 ± 0.3	
Cr	122 ± 7 (6)	ND	ND	123 ± 14	

a Precision expressed as the 95% confidence interval for a single result (n) = number of determinations
b Accepted value excludes results from graphite furnace atomic absorption spectrometry
c Not determined
d 8 ml of dissolved sediment diluted with 30 ml 10 mol L $^{-1}$ hydrochloric acid, iron removed by extraction with MIKB. Digest evaporated to dryness and diluted to 100 mL with water. 10 mL aliquot diluted to 50 mL with water used to prepare standard addition plot by extracting samples with ammonium pyrroalidone diethyl dithiocarbamate in MIKB. Then trace metals back extracted into 5 ml mol L $^{-1}$ nitric acid
Numbers in brackets = numbers of determinations

Source: Elsevier Science Publishers [44]

Sediment samples (0.5g) were placed in Teflon pressure decomposition vessels and to each were added 3 mL of concentrated nitric acid, 1 mL of concentrated perchloric acid and 3 mL of concentrated hydrofluoric acid. The sealed vessels were completely immersed in a boiling water bath for

Table 4.6 Trace element analysis of two marine sediments

	MESS–1[a]		BCSS–1[a]	
Element	ICA–AES	Accepted value	ICP–AES	Accepted value
Be	2.0 ± 0.2	1.9 ± 0.2	1.4 ± 0.2	1.3 ± 0.3
Co	10.5 ± 2.2	10.8 ± 1.9	10.4 ± 3.6	11.4 ± 2.1
Cu	24.8 ± 2.5	25.1 ± 3.8	17.0 ± 1.3	18.5 ± 2.7
Mn	529 ± 34	513 ± 25	238 ± 4	229 ± 15
Ni	30.2 ± 2.2	29.5 ± 2.7	55.9 ± 3.6	55.3 ± 3.6
Pb	36.7 ± 4.9	34.0 ± 6.1	24.8 ± 5.4	22.7 ± 3.4
V	74.2 ± 2.7	72.4 ± 5.3	92.0 ± 4.2	93.4 ± 4.9
Zn	183 ± 10	191 ± 17	112 ± 7	119 ± 12
Al_2O_3	11.05 ± 0.71	11.03 ± 0.38	11.89 ± 0.78	11.83 ± 0.41
Fe_2O_3	4.42 ± 0.27	4.36 ± 0.25	4.77 ± 0.20	4.70 ± 0.14
Na_2O	2.68 ± 0.20	2.50 ± 0.15	2.80 ± 0.18	2.72 ± 0.21
MgO	1.42 ± 0.,04	1.44 ± 0.09	2.25 ± 0.16	2.44 ± 0.23
CaO	0.674 ± 0.022	0.674 ± 0.064	0.763 ± 0.013	0.760 ± 0.074
P_2O_3	0.153 ± 0.004	0.146 ± 0.014	0.168 ± 0.111	0.154 ± 0.016

a Results in %; precision expressed as the 95% confidence interval for a single result

Source: American Chemical Society [46]

1–2 h. After being cooled, the contents of the vessels were transferred to Teflon beakers and evaporated to dryness on a low temperature hot plate. The residues were dissolved and diluted to 50 ml with 1 mol L $^{-1}$ hydrochloric acid. Determined values by this procedure agree well with authenticated values on reference sediments MESS–1 and BCSS–1 (Table 4.6).

McLaren et al. [47] also applied isotope dilution inductively coupled plasma mass spectrometry to the determination of metals in the two reference marine sediments referred to above. The precision obtained in replicate analyses was better than that achieved by alternative calibration strategies in inductively coupled plasma mass spectrometry.

4.44.3.2 Heavy metals (chromium, manganese, iron, cobalt, nickel, copper and zinc), aluminium, silicon, titanium, vanadium, gallium, yttrium, lanthanum, cerium, dysprosium, ytterbium and zirconium

The application of this technique to estuarine sediment analysis has been investigated by several workers [42, 48–50].

Cantillo et al. [42] used a procedure involving lithium metaborate fusion followed by measurement with a single–channel direct current plasma emission spectrometer in a large scale study of major and minor elements in estuarine sediments. The coefficients of variation of replicate analyses were in the range 2–10% which is comparable to the reproducibility in routine atomic absorption work, but worse than that in

Table 4.7 Fusion results of duplicate analyses of the NBS estaurine sediment (SRM 1646) (µg g⁻¹) except where percent)

	Analyst I (n = 6)[a]		Analyst II (n = 6)		F[c]	t[c]
Al (%)	5.3	(2)[b]	5.4	(1)	5.2*	
Si (%)	30	(3)	30	(1)	1.6	0
Ti (%)	0.36	(2)	0.37	(2)	6.4*	
V	89	(3)	91	(3)	1.4	1.4
Cr	74	(1)	70	(4)	9.0*	
Mn	280	(2)	280	(2)	1.0	0
Fe (%)	2.9	(2)	3.0	(0)		
Co	17	(18)	20	(14)	1.2	1.8
Ni	31	(5)	28	(6)	5.6*	
Cu	28	(9)	28	(6)		
Zn	130	(0)	120	(13)		
Ga	19	(8)	19	(6)	1.9	0
Y	17	(8)	15	(4)	5.0	3.2*
Zr	270	(4)	300	(4)	1.4	4.0
La	36	(2)	36	(3)	2.5	0
Ce	110	(4)	110	(7)	3.3	0
Dy	2.8	(7)	3.1	(14)	4.1	1.6
Yb	2.2	(4)	2.0	(5)	1.5	3.8*

[a] Number of samples
[b] Coefficient of variation (%) = 100 σ/mean
[c] F and t test statistics, asterisk denotes a difference between the determined and reference values at the 5% level of significance

Source: American Chemical Society [42]

the best inductively coupled plasma emission analysis. Systematic errors occurred in analysis of lanthanum, cerium and zirconium. Results for aluminium, chromium and iron were slightly low, while those for ytterbium and zirconium were substantially low. Detection limits ranged from 1 mg kg⁻¹ (ytterbium) to 60 mg kg⁻¹ (cerium). The sediment fusion was carried out as follows. Approximately 0.2 g of sample was weighed to the nearest tenth of a milligram into a 7.5 cm³ drill point graphite crucible (Ultra Carbon Crop, Bay City, k MI) containing 1.0 g of lithium metaborate. The sediment was placed in a small depression made in the borate flux to prevent incomplete fusion problems encountered when sediment is in direct contact with graphite.

The sediment-flux mixture was fused in a muffle furnace at 950 ± 50 °C for 15 min. The liquid borate bead was poured into a 150 mL Teflon FEP beaker containing 100 mL of 5% nitric acid (measured from a graduated cylinder). The beaker was then placed on a magnetic stirrer and the contents were stirred for approximately 10 min. The resulting solution was transferred to an acid-washed 125 mL linear polyethylene bottle.

The precision of this method was evaluated by performing replicate analysis of NBS SRM 1646 estuarine sediment (Table 4.7). The test result indicated that aluminium, titanium, chromium and copper have significantly different variances at the 5% level from the nominal concentrations in the sediment (ie F^c >5).

Ren and Salin [90] showed that direct analysis of solid samples is possible, by using furnace vaporisation with Freon modification and inductively coupled plasma atomic emission spectrometry. The relative standard deviation of several metals in marine reference sediments varied from 3 to 15%.

4.44.3.3 Arsenic, antimony, bismuth and selenium

The optimal reaction conditions for the generation of the hydrides can be quite different for the various elements. The type of acid and its concentration in the sample solution often have a marked effect on sensitivity. Additional complications arise because many of the hydride-forming elements exist in two oxidation states which are not equally amenable to borohydride reduction. For example, potassium iodide is often used to prereduce As(V) and Sb(V) to the 3+ oxidation state for maximum sensitivity, but this can also cause reduction of Se(IV) to elemental selenium from which no hydride is formed. For this and other reasons Thompson *et al.* [51] found it necessary to develop a separate procedure for the determination of selenium in soils and sediments although arsenic, antimony and bismuth could be determined simultaneously [52]. A method for simultaneous determination of As(III), Sb(III) and Se(IV) has been reported in which the problem of reduction of Se(IV) to Se(0) by potassium iodide was circumvented by adding the potassium iodide after the addition of sodium borohydride. Goulden *et al.* [53] have reported this simultaneous determination of arsenic, antimony, selenium, tin and bismuth but it appears that in this case the generation of arsine and stibine occurs from the 5+ oxidation state.

de Oliveira *et al.* [34] have described the application of a simple continuous hydride generation system coupled to a low power (1.4 kW) inductively coupled plasma atomic emission spectrometric technique to the determination of trace concentrations of arsenic, antimony and selenium in marine sediments. A variety of sample dissolution procedures and hydride generation reaction conditions were evaluated in an attempt to establish optimal conditions for the simultaneous determination of all three elements. In addition, the effect of the oxidation state of the elements on hydride formation in dilute hydrochloric acid solution was studied.

Four methods of dissolution were evaluated for marine sediments:

(1) The acid digestion procedure described above for biological tissues. Crock and Lichte [54] have described a similar procedure, involving

hydrofluoric as well as nitric, perchloric and sulphuric acids, for dissolution of geological materials prior to arsenic and antimony determination by atomic absorption spectrometry.

(2) Fusion with sodium hydroxide, as described by Goulden *et al.* [53], but using porcelain or nickel crucibles.

(3) Acid digestion with a mixture of nitric, perchloric and hydrofluoric acids in sealed Teflon vessels, as described by McLaren *et al.* [50].

(4) Fusion with potassium hydroxide at 500 °C.

All four dissolution procedures studied were found to be suitable for arsenic determinations in marine samples, but only one (potassium hydroxide fusion) yielded accurate results for antimony in marine sediments and only two (sodium hydroxide fusion or a nitric/perchloric/ hydrofluoric acid digestion in sealed Teflon vessels) were appropriate for determination of selenium in marine sediments. Thus, the development of a single procedure for the simultaneous determination of arsenic, antimony and selenium (and perhaps other hydride-forming elements) in marine materials by hydride generation inductivley coupled plasma atomic emission spectrometry requires careful consideration not only of the oxidation–reduction chemistry of these elements and its influence on the hydride generation process but also of the chemistry of dissolution of these elements.

The apparatus used by de Oliveira [34] *et al.* consisted of a custom ICP-echelle spectrometer. Hydride generation was accomplished in a continuous mode by using two channels of a four-channel peristaltic pump (Gilson Instrument Co., Minipuls Il) to deliver sample and borohydride reagent to a phase separator modified from that of Thompson *et al.* [55], a schematic diagram of the assembly employed is shown in Fig. 4.6. An air bubble maintained by the surface tension at the junction of the two horizontal arms of the 'T' prevents mixing of the reagent and sample until the two solutions begin to flow down the vertical arm into the phase separator. This results in a smooth and continuous generation of hydrogen which is not significantly disturbed by changeover from one sample to the next, or to the blank. The gaseous hydrides and hydrogen are swept from the phase separator and into the inductively coupled plasma by a continuous flow of argon.

This technique has been applied to the determination of 0.1 mg kg $^{-1}$ arsenic, antimony and selenium in marine sediments. The following wavelengths were used: arsenic 193.696nm, antimony 206.833nm, and selenium 196.026nm.

Digestion procedure for antimony and arsenic
The marine sediment (0.5 g) was digested with 4 g potassium hydroxide in a nickel crucible. The crucible was placed in a furnace and then heated to 500 °C for 30 min. The crucible was then cooled and the contents

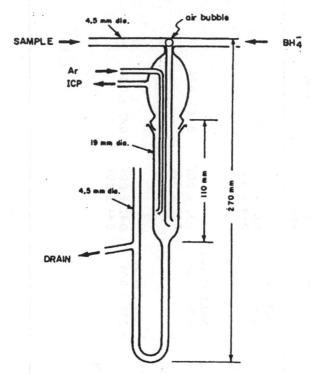

Fig. 4.6 Hydride generation phase separation system for determination of arsenic, antimony, bismuth and selenium marine sediments
Source: American Chemical Society [34]

dissolved in 50 ml of 1 mol L $^{-1}$ hydrochloric acid. The precipitated silicic acid was allowed to settle or was filtered before analysis.

Digestion procedure for selenium

As above, but using solid sodium hydroxide or fusion with nitric-perchloric-hydrofluoric acids [46]. The addition of potassium bromide or concentrated hydrochloric acid and heating to 50 °C for 50 min converts selenium to the Se(IV) state, leaving arsenic and antimony in the pentavalent state. Reduction with sodium borohydride then converts all three metalloids to their hydrides.

Results obtained by this procedure for NBS and NRCC (National Research Council of Canada) reference marine sediments are shown in Table 4.8. For arsenic, all values agreed well with the values reported by the US National Bureau of Standards or the National Research Council of Canada. For antimony, only the potassium hydroxide fusion procedure yielded accurate results. Two dissolution procedures provided consistent results for selenium – sodium hydroxide fusion and acid digestion in sealed Teflon vessels.

Table 4.8 Results for near-shore marine sediments for four different dissolution procedures

Digestion procedure	As, found, µg g^{-1}	As, certified, µg g^{-1}	Sb, found, µg g^{-1}	Sb, certified, µg g^{-1}	Se, found, µg g^{-1}	Se, certified, µg g^{-1}
				NBS 1646		
Procedure 1	12.2 ± 1.0	11.6 ± 1.3	e	(0.4)	0.61 ± 0.02	(0.6)
Procedure 2	11.0 ± 0.9		c		0.59 ± 0.06	
Procedure 3	11.0 ± 0.9		0.36 ± 0.09			
				NRCC MESS–1		
Procedure 1	11.6 ± 0.3	11.1 ± 1.4	c	0.73 ± 0.08	0.33 ± 0.05	0.34 ± 0.06
Procedure 2	12.2 ± 1.1		c		0.35 ± 0.02	
Procedure 3	11.4 ± 0.8		0.71 ± 0.03			
				NRCC BCSS–1		
Procedure 1	11.6 ± 1.8	10.6 ± 1.2	c	0.59 ± 0.06	0.42 ± 0.09	0.43 ± 0.06
Procedure 2	11.4 ± 1.1		c		0.44 ± 0.03	
Procedure 3	11.2 ± 0.5		0.61 ± 0.06			

Procedure 1: Fusion with KOH
Procedure 2: Fusion with NaOH
Procedure 3: Fusion with nitric-perchloric-hydrofluoric acids in teflon

Source: American Chemical Society [34]

Table 4.9 Operating conditions for isotopic analysis by ICP–MS

ICP	
plasma Ar	14 L min^{-1}
auxiliary Ar	2.0 L min^{-1}
nebuliser Ar	0.9 L min^{-1}
rf power	1.2 kW
Mass spectrometer	
sampler	nickel, 1.2 mm orifice
skimmer	nickel, 0.9 mm orifice
Operating pressures	
interface region	~1 torr
mass spectrometer chamber	~ 4 × 10^{-5} torr
Lens voltages	
photon stop (S2)	−7.0 V
Bessel box barrel (B)	+2.95V
einzel lenses 1 and 3 (E1)	−12.0 V
Bessel box end lenses (P)	−11.3 V
einzel lens 2	−130 V
entrance a.c. rods	0 V
exit a.c. rods	−5 V

Source: American Chemical Society [47]

4.44.4 *Inductively coupled plasma mass spectrometry*

4.44.4.1 **Heavy metals (cadmium, nickel, zinc, chromium and lead) and strontium, molybdenum, tin, antimony, thallium and uranium**

McLaren *et al.* [47] applied isotope dilution inductively coupled plasma mass spectrometry to the analysis of two marine sediment reference samples HESS–1 and BCSS–1. Isotope dilution inductively coupled plasma mass spectrometry was applied to the determination of these 11 trace elements (Cr, Ni, Zn, Sr, Mo, Cd, Sn, Sb, Ti, Pb and U) in the marine sediment reference materials MESS–1 and BCSS–1. Accuracy and especially precision are better than those that can be easily achieved by other ICP–MS calibration strategies, as long as isotopic equilibration is achieved and the isotopes used for the ratio measurement are free of isobaric interferences by molecular species. The measurement of the isotope ratios on unspiked samples provides a sensitive diagnostic of such interferences.

Measurements were carried out using an ELAN 250 instrument using the operating conditions outlined in Table 4.9.

Table 4.10 Results from analyses of a national reference sediment, NBS SRM 1645 river sediment and NBS SRM 1646 estaurine sediment

Standard	Metal content (µg g $^{-1}$)[a]			
	Lead		Copper	
	Found	Certified[b]	Found	Certified[b]
Reference sediment				
(Lillebaeit)	19.6 ± 0.8 (7)	19.0 ± 2.0	25.5 ± 1.0 (6)	24.7 ± 1.7
SRM 1645	722 ± 18 (8)	714 ± 28	111 ± 5 (5)	109 ± 19
SRM 1646	26.5 (1)	28.2 ± 1.8	16.8 (2)	18 ± 3

[a] Dry sediments were processed in all cases. Numbers in parentheses indicate the number of separate determinations used for calculation of the standard deviations shown
[b] Certified values with 95% confidence intervals

Source: Elsevier Science Publishers BV, Netherlands [58]

4.44.5 Neutron activation analysis

4.44.5.1 Heavy metals (cobalt, chromium, iron), caesium, rubidium, antimony, scandium and strontium

This technique has been applied to the determination of cobalt, chromium, caesium, iron, rubidium, antimony, scandium and strontium in suspended matter isolated from seawater [56].

4.44.5.2 Miscellaneous metals

Ellis and Chattopadhya [57] investigated the determination of 28 elements in estuarine suspended solids by neutron activation analysis and found detection limits for most of the elements ranged between 0.37 and 1.0 ng. Average precision ranged between 10 and 16% and accuracy within 10%.

4.44.6 Potentiometric stripping analysis.

4.44.6.1 Copper and lead

Madsen et al. [58] applied this technique to marine sediments for the determination of lead (7–700 µg g $^{-1}$) and copper (3–110 µg g $^{-1}$). The precision obtained was in the range 2.5–4.1% for lead and 3.9–4.5% for copper. The accuracy of this method was checked on standard reference materials (NBS SRM 1645 river sediment) and was found to be typically better than 7%.

In Table 4.10 are shown the results of repetitive analyses of a Danish marine reference sediment ('Lillebaelt' sediment) and SRM 1645 river

Table 4.11 Results obtained in the ICES sediment intercalibration. MS1 (NBS SRM 1645 river sediment), MS2 and MS3 are the three intercalibration samples

Sample	Metal content (μg g^{-1}) Lead		Copper	
	Found	x ± S^a	Found	x ± S^a
MS1	724	688 ± 59	104	108 ± 7
MS2	64	52 ± 13	35	39 ± 4
MS3	7.5	4.5 ± 0.6	2.6	1.7 ± 0.8

^{a} Mean and estimated deviation obtained in the intercalibration tests for the dry sediments. The certified results for MS1 are given in Table 4.10

Source: Elsevier Science Publishers BV, Netherlands [58]

Table 4.12 Results from analysis of intercalibration samples from the 17th Danish Intercalibration arranged by the Danish National Environmental Protection Agency

Sample	Metal content (μg g^{-1}) Lead		Copper	
	Found	x ± S^a	Found	x ± S^a
Sediment	18.6	21.5 ± 4.0	14.5	18.1 ± 1.0

^{a} Mean and estimated deviation obtained in the intercalibration tests for dried samples

Source: Elsevier Science Publishers BV, Netherlands [58]

sediment from the National Bureau of Standards (USA). For lead, the precisions found were 4.1% and 2.5% respectively, and for copper 3.9% and 4.5% respectively. The values found were in all cases within the certified limits, indicating good accuracy (Pb 3.2% and 1.1% respectively; Cu 3.2% and 1.8% respectively). Also in Table 4.10 are some results from analyses of the NBS SRM 1646 estuarine sediment.

Table 4.11 gives the results obtained during collaboration in the ICES (International Committee for Exploration of the Sea) marine sediment intercalibration [59]. The sample marked MS1 was identical with NBS SRM 1645 river sediment. Except for the lead in MS3, agreement is satisfactory. The results of analysis of samples from intercalibration runs on sediments arranged by the Danish National Environmental Protection Agency are listed in Table 4.12. There are minor discrepancies between the results and the intercalibration results for copper in the sediment.

Table 4.13 Trace elements and minor constituents of marine sediment reference material BCSS–1 as determined by instrumental photon activation analysis (IPAA)

	IPAA[a]			Certified value[b]			Detection limits
Arsenic	11.6	±	1.3	11.1	±	1.4	0.8
Barium	347	±	36	330			12
Antimony	0.6	±	0.10	0.59	±	0.06	0.2
Chromium	123	±	9	123	±	14	30
Cobalt	10.9	±	0.8	11.4	±	2.1	2.6
Lead	23.0	±	1.6	22.7	±	3.4	2.1
Magnesium	213	±	17	229	±	15	19
Nickel	54.8	±	3.2	55.3	±	3.6	0.7
Strontium	101	±	12	96			9
Uranium	2.51	±	0.12	3			0.3
Zinc	122	±	16	119	±	12	7
Zirconium	226	±	21				0.4
				Minor constituents (%)			
C	2.25	±	0.03	2.19	±	0.09	0.08
Na$_2$O	2.59	±	0.15	2.72	±	0.21	0.02
MgO	2.38	±	0.56	2.44	±	0.23	0.004
Cl	1.11	±	0.18	1.12	±	0.05	0.01
K$_2$O	2.25	±	0.27	2.17	±	0.05	0.01
CaO	0.72	±	0.09	0.760	±	0.074	0.02
TiO$_2$	0.71	±	0.03	0.734	±	0.024	0.004
Fe$_2$O$_3$	4.62	±	0.15	4.70	±	0.14	0.16

[a] Concentrations determined by using NBS 1632a coal as a calibration standard. Precision is expressed as a standard deviation
[b] Precision is expressed as 95% tolerance limits

Source: American Chemical Society [60]

4.44.7 Photon activation analysis

4.44.7.1 Heavy metals (chromium, cobalt, lead, nickel, zinc and iron), arsenic, barium, antimony, magnesium, strontium, uranium, zirconium, sodium, potassium, calcium and titanium

This technique has been used [60] to determine twelve trace elements and eight minor elements in a certified marine sediment standard reference material (BCSS–1). Photon activation analysis was carried out at the National Research Council of Canada electron linear accelerator using the Bremsstrahlung produced by the impact of a focused electron beam on a tungsten converter. An example of a gamma ray spectrum of the marine sediment following photo initiation at 25 MeV beam energy is shown in Fig. 4.7. Table 4.13 compares the analyses obtained by photon activation analysis and certified values for the BCSS–1 sediments. Detection limits

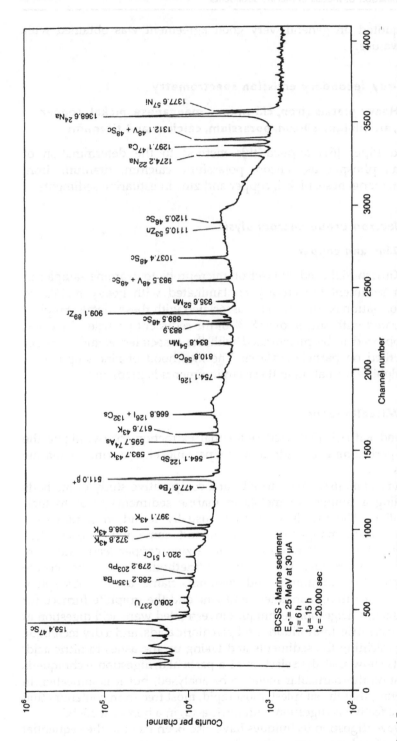

Fig. 4.7 γ-ray spectrum taken with a Ge(Li) spectrometer following bombardment of BCSS marine sediment with bremsstrahlung flux produced by 25 MeV electrons impinging upon a 6 mm thick tungsten converter. After an irradiation period of 6 h and a cooling off period of 15 h, the γ-ray spectrum was accumulated for 20,000 s. Peaks are labelled with γ-ray energy in KeV and assigned a parent radionuclide. Source: American Chemical Society [60]

are also quoted. In general, very good agreement was obtained with reference values.

4.44.8 X-ray secondary emission spectrometry

4.44.8.1 Heavy metals (iron, chromium, manganese, nickel, copper and zinc), aluminium, silicon, potassium, calcium and titanium

Baker and Piper [61] applied this method to the determination of aluminium, phosphorus, silicon, potassium, calcium, titanium, iron, chromium, manganese, nickel, copper and zinc in estuarine sediments.

4.44.9 Electron probe microanalysis

4.44.9.1 Zinc and copper

Lee and Kittrich [62] used an electron microprobe to examine samples of a harbour sediment (anaerobic), contaminated with heavy metals, to obtain information on the elements associated with these metals. Sulphur was associated with zinc in over 83% of the sediment particles. Zinc and copper appeared to be precipitated with the associated elements rather than adsorbed on particle surfaces. The likelihood of zinc sulphide or copper sulphide formation in these environments is predicted.

4.44.10 Miscellaneous

Hickley and Kittrick [63] used sequential extraction to investigate the chemical partitioning of cadmium, copper, nickel and zinc in marine sediments.

Belhomme et al. [64] have carried out a comparative study of methods for digesting a number of metals in marine sediments prior to their determination. The metals were those usually found in sediments – aluminium, iron, manganese, titanium and some which occur less frequently and at very low concentrations – zinc, copper, lead, cadmium, chromium and arsenic. The analytical methods used were visible spectrometry for aluminium and titanium, flame atomic absorption spectrometry for iron, manganese and zinc and the graphite furnace for the other metals. Digestion techniques investigated were acid digestion in hydrochloric, nitric, hydrofluoric and perchloric acids, and a dry technique involving calcining the sediments and taking up the ashes in nitric acid. The results show that the selection of a particular digestion technique is dependent on the particular metal to be analysed, but it is indicated, in general terms, that the simplest, most rapid, most free from contamination and most effective is digestion with nitric acid, in a bomb, at 150°C.

Microwave digestion techniques have also been used in the sequential extraction of calcium, iron, chromium, manganese, lead and zinc in sediments [65]. The extraction rates of metals in each of the five binding fractions of marine sediments (metal exchangeable, carbonate bound, iron-manganese oxide bound, organic bound, and residual) were determined for both conventional and microwave heating techniques. Microwave digested sediment samples (20 ml) were centrifuged at 10,000 rpm for 30 min. Metals were determined by flame atomic absorption spectrometry. To prepare material for the determination of extraction rates for one particular fraction, a quantity of material was taken through the conventional sequence of steps up to the fraction for which the rate was determined. Sequential microwave extraction procedures were then established from the results of the rate experiments. Recoveries of total metals from standard sediment samples were comparable for both conventional and microwave techniques except for iron (62% by microwave and 76% by conventional method).

Magni [66] studied the optimisation of the extraction of metal humic acid complexes from marine sediments.

Seuten and Charlier [91] have reviewed methods for the analysis of heavy metals in estuarine and coastal water sediments.

Bryan *et al.* [67] evaluated three burrowing species, the polychaete *Nereis diversicolor* and the deposit-feeding bivalves *Scrobicularia plana* and *Macoma balthica* as indicators of the biological availability of metals in estuarine sediments. Their ability to accumulate a range of 13 trace metals was determined from observations of the concentration of these elements in the tissues of these animals and in the surrounding sediments from 30 different estuaries in southwest England and South Wales. Results demonstrated that availability differed by several orders of magnitude between contaminated and uncontaminated sites. *Scobicularia* was studied in most detail, and was the best accumulator of metals, with a slow exchange rate making it a good indicator of long-term changes in chronic pollution. The polychaete *Nereis diversicolor* also proved to be a useful indicator for silver, cadmium, copper and mercury, but not for zinc, which it regulates. The results permitted a preliminary classification of the estuaries according to their severity of pollution for each metal.

Prange *et al.* [68] studied the use of total reflection X-ray fluorescence spectrometry for the simultaneous determination of up to 25 trace elements. Results obtained by this method on samples of sediments and suspended solids are summarised.

4.44.11 Separation and fractionation of sediments from seawater

Sedimentary solid matter is of two main classes, that which is settled on the sea bed or river bed and that which is present in suspension in the water column. The former can be collected by hand in shallow waters or by using a suitable solid sampler in deeper waters as discussed by

Butman [69]. Particulates present in samples of the water column can be isolated by a variety of means including filtration and centrifugation.

In the first place, it is generally agreed that the distribution of particle sizes in the oceans is continuous, from the whale to the simple single molecule [70, 71]. The size at which one calls an aggregate of molecules a particle is therefore arbitrary. In the case of sea water, the dividing line between dissolved and particulate has been chosen as 0.45 μm, largely because the first commercially available membrane filters had that as their pore size.

4.44.11.1 Separation by filtration

It should be mentioned at this point that the acceptance of 0.45 μm as the dividing line is purely nominal, since few workers in this field actually use filters with this pore size. The glass fibre filters used by many workers have pore sizes which are considerably larger, ranging from 0.7 μm for Whatman GF/F to 1.32 μm for GF/C. With these filters, all particles larger than the nominal pore size are retained, but many smaller particles are also trapped. The silver filters, and most particularly the 0.4 μm size, contain relatively large and variable amounts of carbon, which must be removed by combustion. After this combustion, the pore sizes are considerably enlarged, with the 0.45 μm filter approaching 0.8 μm in pore size. The nominal 0.8 μm pore size filter is used by many investigators because the pore size changes very little under heat treatment. Thus, although 0.45 μm has been accepted as the minimum size particulate matter by definition, the filters actually used have a somewhat larger pore size, and retain particles which are considerably smaller than the nominal cut-off size [72].

The choice of filter can determine the amount of material considered as particulate, sometimes with unexpected results. Thus, the Whatman GF/C filter with its larger pore size actually retains about three times as much particulate organic carbon as does the 0.8 μm silver filter. Presumably the difference results from the larger number of small particles retained by the glass fibre filters.

The method of calculation of the blank can also influence the determined sediment content. If surface sea water is filtered through a pad consisting of two or more filters, either glass fibre or silver, the bottom filter will often contain a small amount of sediment above the blank value. Some workers have maintained that this is due to the adsorption of dissolved organic matter on to the filter and that this value should therefore be subtracted from the weight of sediment found on the top filter [73, 74]. Other workers feel that the material caught in the second filter is largely composed of smaller particles passing through the first filter. Depending on the way in which the particulate fraction is defined, the material caught by the second filter should either be added to that collected on the first filter [6, 75, 76] or ignored. It can easily be seen that the choice of blank calculation can cause a considerable difference to the final values given for sediment content, at least in surface waters. As far as the particulate fraction is defined, not in terms of particle size, but in terms of material caught on a specific filter, it is recommended that only one filter, rather than a pad of two or more, be used, since the material caught on subsequent filters is irrelevant by definition.

When uniform methods of collection and analysis are used, the deeper layers of the oceans give remarkably consistent results. Replicate samples, taken with a Niskin rosette sampler rigged to close six 5 L bottles simultaneously, displayed a standard deviation of ±1.3 μg of carbon per litre [77].

Methods which collect a greater proportion of the smaller particles have also been employed. For example, a layer of fine inorganic particulate matter deposited on a filter of coarser porosity has been used to separate the particulate from the dissolved fraction. Thus, Fox et al. [78] used layers of calcium hydroxide and magnesium hydroxide, while Ostapenya and Kovalevskaya [79] used powdered glass. These filters suffered from three disadvantages – they were troublesome to construct, the nominal pore size was irreproducible, and adsorption of truly dissolved material was possible. The techniques were abandoned with the advent of the first membrane filters having graduated pore size.

Many sediment collection methods used are biased towards those particles falling very slowly. If the residence time of a particle in the water column is only a few days, the probability of being caught in a 5 L Niskin bottle is small. This has been pointed out by the work of Bishop et al. [6]. These investigations used an in situ pump and filtration apparatus to filter very large quantities (5–30 m³) of sea water and caught many classes of particles never seen in Niskin bottle samples. Their results are not comparable to those obtained in other filtration methods.

4.44.11.2 Separation by centrifugation

A method for removing particles, which is not limited in volume sampled, and which suffers less from problems of overlapping classification, is continuous flow centrifugation. Separation into density classes can be achieved by choice of speed of centrifugation. Jacobs and Ewing [80] used continuous centrifugation to collect total suspended matter in the oceans, and Lammers [81] discussed the possible uses of the method. The biggest drawback seems to be that the separation is governed by particle density, rather than particle size. However, the method holds considerable promise for the collection of colloidal material as a separate fraction.

4.44.11.3 Fractionation by filtration

Some work has been done on size fractionation of particulate matter in water samples by the use of graduated filters. Since the filters in common use do not display a sharp cut-off in particle size retention, interpretation of the results is difficult. Repeated filtration of a single sample through filters of different pore size does not divide the particulate matter into definite size classes, since each filter retains particles smaller than the nominal pore size. The results of the filtration of separate aliquots through a series of filters can only be reported in terms of 'particles smaller than' the nominal pore size and are equally difficult to interpret. Although such size fractionation has been reported [82], the conclusions can only be accepted in the broadest possible sense. Particle size distributions based on filtration should be supported by Coulter counter data before any conclusions can be drawn.

4.44.11.4 Fractionation by chemical leaching

Different chemical reagents dissolve different proportions of a sediment sample and the concentration of inorganic or organic substances will differ in each fraction. Thus the following reagents dissolve the different percentages of a sediment and the stated percentage of the total cadmium content of the unfractionated samples is found in each fraction.

Extractant	Species extracted	Cadmium. % of total cadmium content of unfractionated sample found in the fraction
LiCl–CsCl MeOH @ 20°C	readily exchanged ions	17
CH₃COONa, pH 5.0 @ 20°C	carbonate bound surface oxide bound ions	31
NH₂OH–HCl–CH₃COOH @ 20°C	ions bound to Fe and Mn oxides	34
H₂O₂, pH 2.0 @ 90°C	organically and sulphide bound ions	12
aqua regia–HF–HCl–H₂O₂	ions bound to residual phase	6

4.44.11.5 Fractionation by sedimentation

A dispersion of dry sediment in sea water can be used to carry out sedimentation measurements of the particle size distribution of sediments. In this procedure the sediment to be analysed is introduced as a 5% suspension in water into the sedimentation cell by means of a peristaltic pump. A finely collimated beam of X-rays is passed through the suspended medium. Radiation is detected as pulses by a scintillation detector on the opposing side of the cell to the source. The concentration of sediment in the beam is proportional to the X-ray intensity. The sedimentation cell is continuously moved across the beam to reduce the

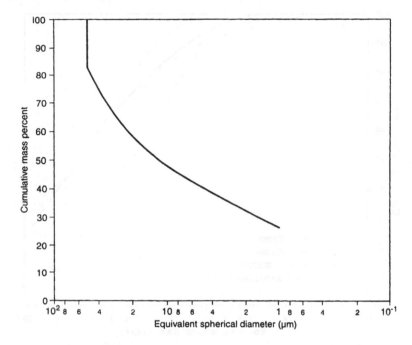

Fig. 4.8 Particle size analysis of natural silt sample using sedigraph, sieves and sedimentation balance methods
Source: Own files

analysis time. A computer solves Stokes Law and presents the result linearly as a cumulative mass per cent distribution. Fig. 4.8 depicts curves drawn from results obtained on a silt sample.

Cumulative per cent distribution curves obtained by the sedigraph and three other methods are compared in Fig. 4.9. It is clear that 'different' results are obtained using various instruments. The choice of instrument will determine the 'grain size' measured. This choice will depend largely on convenience availability, reproducibility and comparability of results to those obtained by other research workers in related fields. The sedigraph fulfils these criteria adequately and it is therefore suggested that it could be employed as standard tool for the measurement of sediment particle size distribution.

4.44.11.6 Fractionation by centrifuging

Centrifuging a dry sediment in tetraobromoethane provided several fractions based on density (Table 4.14). The mass balance of lead, cadmium and zinc in each fraction is also given. It is seen that the majority of the heavy metals occur in the organic and conglomerate fractions obtained from this method.

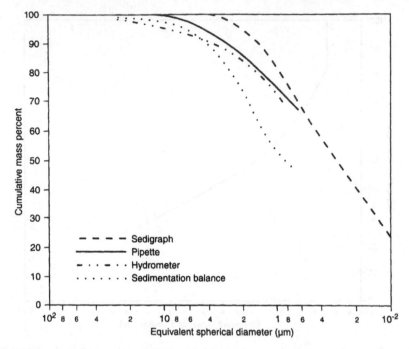

Fig. 4.9 Particle size analysis of natural A clay sample according to various methods
Source: Own files

Table 4.14 Centrifuging dry sediment in tetrabromomethane

Density	Fraction	Fractions obtained %	Mass balance			Concentration in fraction µg kg⁻¹		
			Pb	Cd	Zn	Pb	Cd	Zn
<2.4	Organics	9	20	1	22	2220	110	2460
2.4–2.55	Conglomerates	6	6	0.3	7	1000	50	1160
2.55–2.66	Quartz and Calcite	46	11	0.5	18	240	11	390
2.66–2.75	Magnesium and calcite	20	6	0.6	9	300	30	450
2.75–2.95	Aragonite	19	2	0.2	4	105	10	210
>2.95	Heavy minerals	0.1	<	<	<	<	<	<

Source: Own files

4.44.11.7 Other fractionation methods

Kiff [83] and Dines and Wharfe [85] have described procedures for sampling and particle size analysis of estuarine sediments. For the determination of particle size down to 45 µm, sieving is used; for the sub-sieve range, gravitational and centrifugal sedimentation methods and the

Coulter counter are applied. By means of the disc centrifuge, sizes down to 0.1 μm can be determined. The scheme is designed to be accurate and reproducible for long periods of operation and not to require elaborate calibration procedures.

Hakanson [84] investigated the relation between physical and chemical characteristics of sediments in marine sediments and lakes, for the purpose of determining how representative sediment data can be established for recent sediments. He developed a mathematical model for predicting the distribution of physical and chemical parameters in sediments.

References

1 Maher, W.A. *Analyst (London)*, **108**, 939 (1983).
2 Bermejo-Barrera, P., Barciela-Alonso, M., Ferrou-Novais, M. and Bermejo-Barrera, A. *Journal of Atomic Spectroscopy*, **10**, 247 (1995).
3 Siu, K.W.M., Roberts, S.Y. and Berman, S.S. *Chromatographia*, **19**, 398 (1984).
4 Maher, W.A. *Analytica Chimica Acta*, **126**, 157 (1981).
5 Brzezinska-Paudyn, A., Van Hoon, J. and Hancock, R. *Atomic Spectroscopy*, **7**, 72 (1986).
6 Bishop, J.K.B. *Analytical Chemistry*, **62**, 553 (1990).
7 Lee, D.S. *Analytical Chemistry*, **54**, 1682 (1982).
8 McLaren, J.W. and Berman, S.S. *Applied Spectroscopy*, **35**, 403 (1981).
9 Heggie, D.T. *Study of Reservoir Fluxes and Pathways in an Alaskan Fjord*, PhD dissertation University of Alaska 217 pp (1970).
10 Burzminskii, M.J. and Fernando, Q. *Analytical Chemistry*, **50**, 1177 (1978).
11 Hodge, V., Stallard, M., Kiode, M. and Goldberg, E.D. *Analytical Chemistry*, **58**, 616 (1986).
12 Lucotte, M. and D'Anglejan, B. *Chemical Geology*, **48**, 257 (1985).
13 Hutton, R.C. and Preston, B. *Analyst (London)*, **105**, 981 (1980).
14 Abo-Rady, M.D.K. *Fresenius Zeitschrift für Analytische Chemie*, **299**, 187 (1979).
15 Olsen, B.H. and Cooper, R.C. *Water Research*, **10**, 113 (1976).
16 Andren, A.W. and Harris, R.C. *Nature (London)*, **245**, 256 (1973).
17 Bartlett, P.O., Craig, P.J. and Morton, S.F. *Private communication*.
18 Windon, H., Gardner, W., Stephens, J. and Taylor, F. *East Coast Marine Station*, **4**, 579 (1976).
19 Shin, E. and Krenkel, P.A. *J Water Pollution Control Federation*, **48**, 44 (1976).
20 Gillespie, D.C. *J Fisheries Research Board, Canada*, **29**, 1035 (1972).
21 Jernelov, A. *Luminology and Oceanography*, **15**, 958 (1970).
22 Langley, D.G. *J Water Pollution Control Federation*, **48**, 473 (1973).
23 Fresnet-Rohin, M. and Ottmann, F. *Estuarine and Marine Coastal Science*, **7**, 425 (1978).
24 Bothner, M.H., Jahnke, R.A., Peterson, M.L. and Carpenter, R. *Geochimica and Cosmochimica Acta*, **44**, 273 (1980).
25 Pavlova, U.K. and Yatasimirskii, K.B. *Zhur Analit Khim*, **24**, 1347 (1969).
26 Shimizu, T. *Bulletin Chemical Society Japan*, **42**, 1561 (1969).
27 Terada, K., Ooba, T. and Kiba, T. *Talanta*, **22**, 41 (1975).
28 Willie, S.N., Sturgeon, R.E. and Berman, S.S. *Analytical Chemistry*, **58**, 1140 (1986).
29 Siu, K.W.M. and Berman, S.S. *Analytical Chemistry*, **55**, 1603 (1983).
30 Siu, K.W.M. and Berman, S.S. *Talanta*, **31**, 1010 (1984).

31 Sturgeon, R.E., Willie, S.N. and Berman, S.S. *Analytical Chemistry*, **57**, 2311 (1985).
32 Reamer, D.C., Veillon, G. and Tokousbalides, P.T. *Analytical Chemistry*, **53**, 245 (1981).
33 Itoh, K., Chikuma, M. and Tanaka, H. *Fresenius Zeitschrift für Analytische Chemie*, **330**, 600 (1988).
34 de Oliveira, E., McLaren, J.W. and Berman, S.S. *Analytical Chemistry*, **55**, 2047 (1983).
35 Bloom, N. *Atomic Spectroscopy*, **4**, 204 (1983).
36 Riley, J.P. and Siddique, S.A. *Analytica Chemica Acta*, **181**, 117 (1986).
37 Hodge, V.F., Seidel, S.L. and Goldberg, E.D. *Analytical Chemistry*, **51**, 1256 (1979).
38 Sastry, V.N., Krislmamoorthy, T.M. and Sarma, T.P. *Current Science (India)*, **38**, 279 (1969).
39 Yatsimirskii, K.B., Emel, Gamoo, E.M., Pavlova, K. and Savichenko, Ya.S. *Okeanologiya*, **10**, 1111 (1970).
40 Pilkington, E.S. and Warren, E.T. *Environmental Science and Technology*, **13**, 295 (1979).
41 Sinex, S.A., Cantillo, A.Y. and Helz, G.R. *Analytical Chemistry*, **52**, 2342 (1980).
42 Cantillo, A.Y., Sinex, S.A. and Helz, G.R. *Analytical Chemistry*, **56**, 33 (1984).
43 Suhr, N.H. and Ingasmello, C.O. *Analytical Chemistry*, **38**, 730 (1968).
44 Sturgeon, R.E., Desauliniers, J.A.H., Berman, S.S. and Russell, D.S. *Analytica Chimica Acta*, **134**, 283 (1982).
45 Cutter, G.A., Electric Power Research Institute, Palo Alto, California, *Report EPRCEA 4641* Vol 1 (100 pp) (1986).
46 McLaren, J.W., Berman, S.S., Boyko, V.J. and Russell, D.S. *Analytical Chemistry*, **58**, 1802 (1986).
47 McLaren, J.W., Beauchemin, D. and Berman, S.S. *Analytical Chemistry*, **59**, 610 (1987).
48 McQuaker, N.R., Kluckner, P.D. and Chang, G.N. *Analytical Chemistry*, **51**, 888 (1979).
49 Walsh, J.N. and Howie, R.A. *Min Management*, **43**, 967 (1980).
50 McLaren, J.W., Berman, S.S., Bayco, V.J. and Russell, D.S. *Analytical Chemistry*, **53**, 1802 (1981).
51 Thompson, M., Pahlavanpour, B. and Pullen, J.H. *Analyst (London)*, **105**, 274 (1980).
52 Pahlavanpour, B., Thompson, M. and Thorne, L. *Analyst (London)*, **105**, 756 (1980).
53 Goulden, P.D., Anthony, D.H.J. and Austen, K.D. *Analytical Chemistry*, **53**, 2027 (1981).
54 Crock, J.G. and Lichte, F.E. *Analytica Chimica Acta*, **144**, 223 (1982).
55 Thompson, M., Pahlavanpour, B. and Walton, S.J. *Analyst (London)*, **103**, 568 (1978).
56 Piper, D.T. and Goles, G.G. *Analytica Chimica Acta*, **47**, 560 (1969).
57 Ellis, K.M. and Chattopadhya, V.A. *Analytical Chemistry*, **51**, 942 (1979).
58 Madsen, P.P., Drabaek, I. and Sørensen, J. *Analytica Chimica Acta*, **151**, 479 (1983).
59 Centre National pour l'Exploration des Oceans Metaux–Traces dans les Sediments Marins (1980).
60 Lansberger, S. and Davidson, W.F. *Analytical Chemistry*, **57**, 197 (1985).
61 Baker, E.T. and Piper, D.Z. *Deep Sea Research*, **23**, 1181 (1976).
62 Lee, F.Y. and Kittrick, J.A. *Soil Science Society of America Journal*, **48**, 548 (1984).
63 Hickley, M.G. and Kittrick, J.A. *Journal of Environmental Quality*, **13**, 372 (1984).
64 Belhomme, J.M., Erb, R., Dequidt, J. and Philippo A. *Revue Français des Sciences de l'Eau*, **1**, 205 (1982).

65 Mahan, K.I., Foderano, T.A., Garza, T.L., Martinez, R.M., Maroney, G.A., Trivisonno, M.R. and Willging, E.M. *Analytical Chemistry*, **59**, 938 (1978).
66 Magni, E., Giusto, T. and Frache, R. *Analytical Proceedings*, **32**, 267 (1995).
67 Bryan, G.W., Langston, W.J. and Hummerstone, L.G., Marine Biological Association Plymouth UK Occasional Publication No 1 *The use of biological indicators of heavy metal contamination in estuaries with special reference to the biological availability of metals in estuarine sediments from South West Britain* 73 pp (1980).
68 Prange, A., Knoth, J., Stossel, R.P., Boddeker, H. and Kramer, R. *Analytica Chimica Acta*, **195**, 275 (1987).
69 Butman, A. *Journal of Marine Research*, **44**, 645 (1986).
70 Sharp, J.H. *Marine Chemistry*, **1**, 211 (1973).
71 Sharp, J.H. *Limnology and Oceanography*, **18**, 441 (1976).
72 Sheldon, R.W. and Sutcliffe, W.H. *Limnology and Oceanography*, **14**, 441 (1969).
73 Banoub, M.W. and Williams, P.J. *Journal of Marine Research*, **19**, 443 (1972).
74 Menzel, D.W. *Deep Sea Research*, **14**, 220 (1967).
75 Gordon, D.C. and Sutcliffe, W.H. *Limnology and Oceanography*, **19**, 989 (1974).
76 Sharp, J.H. *Limnology and Oceanography*, **19**, 984 (1974).
77 Wangersky, P.K. *Limnology and Oceanography*, **19**, 980 (1974).
78 Fox, D.L., Oppenheimer, C.H. and Kettridge, J.S. *Journal of Marine Research*, **12**, 223 (1953).
79 Ostapenya, J.P. and Kovalevskaya, R.Z. *Okeanologiya*, **4**, 694 (1965).
80 Jacobs, M.J.M. and Ewing, M. *Science*, **163**, 180 (1969).
81 Lammers, W.G. *Environmental Science and Technology*, **1**, 52 (1967).
82 Mullin, M.M. *Limnology and Oceanography*, **10**, 459 (1965).
83 Kiff, P.R. *Laboratory Practice*, **22**, 259 (1973).
84 Hakanson, L. *Water Resources Research*, **17**, 1625 (1981).
85 Dines, E. and Wharf, J. *Water Bulletin*, **250**, 6 (1987).
86 Soto, E.G., Rodriguez, E.A., Rodriguez, D.P., Mahia, P.L. and Lorenzo, S.M. *Science of the Total Environment*, **141**, 87 (1994).
87 Koide, M., Hodge, V., Yang, J.S. and Goldberg, E.D. *Analytical Chemistry*, **59**, 1802 (1987).
88 Gerringa, L.J.A., Van der Meer, J. and Cauvet, G. *Marine Chemistry*, **36**, 51 (1991).
89 Walker, G.S., Ridd, M.J. and Brunskill, G.J. *Rapid Communicational Mass Spectrometry*, **10**, 96 (1996).
90 Ren, J.M. and Salin, E.D. *Spectrochimica Acta Part B*, **49B**, 567 (1994).
91 Seuten, J.R. and Charlier, R.H. *International Journal of Environmental Analytical Chemistry*, **37**, 79 (1991).

Chapter 5

Determination of metals in estaurine sediments

5.1 Aluminium

5.1.1 Inductively coupled plasma atomic emission spectrometry

This technique has been applied to the determination of aluminium as discussed under multi-element analysis in section 5.26.2.1.

5.1.2 X-ray secondary emission spectrometry

This technique has been applied to the determination of aluminium as discussed under multi-element analysis in section 5.26.4.1.

5.2 Arsenic

5.2.1 Ion exchange chromatography – hydride generation atomic absorption spectrometry

Maher [1] determined inorganic and methylated arsenic species in estuarine sediments. The procedure is discussed in section 4.3.6.

5.3 Beryllium

5.3.1 Atomic absorption spectrometry

This technique has been applied to the determination of beryllium as discussed under multi-element analysis in section 5.26.1.2.

5.4 Cadmium

5.4.1 Atomic absorption spectrometry

This technique has been applied to the determination of cadmium as discussed under multi-element analysis in sections 5.26.1.1 and 5.26.1.2.

5.5 Calcium

5.5.1 X-ray secondary emission spectrometry

This technique has been applied to the determination of calcium as discussed under multi-element analysis in section 5.26.4.1.

5.6 Cerium

5.6.1 Inductively coupled plasma atomic emission spectrometry

This technique has been applied to the determination of cerium as discussed under multi-element analysis in section 5.26.2.1.

5.7 Chromium

5.7.1 Atomic absorption spectrometry

This technique has been applied to the determination of chromium as discussed under multi-element analysis in sections 5.26.1.1 and 5.26.1.2.

5.7.2 Inductively coupled plasma atomic emission spectrometry

This technique has been applied to the determination of chromium as discussed under multi-element analysis in section 5.26.2.1.

5.7.3 X-ray secondary emission spectrometry

This technique has been applied to the determination of chromium as discussed under multi-element analysis in section 5.26.4.1.

5.8 Cobalt

5.8.1 Atomic absorption spectrometry

This technique has been applied to the determination of cobalt as discussed under multi-element analysis in sections 5.26.1.1 and 5.26.1.2.

5.8.2 Inductively coupled plasma atomic emission spectrometry

This technique has been applied to the determination of cobalt as discussed under multi-element analysis in section 5.26.2.1.

5.9 Copper

5.9.1 Atomic absorption spectrometry

This technique has been applied to the determination of copper as discussed under multi-element analysis in sections 5.26.1.1 and 5.26.1.2.

5.9.2 Inductively coupled plasma atomic emission spectrometry

This technique has been applied to the determination of copper as discussed under multi-element analysis in section 5.26.2.1.

5.9.3 X-ray secondary emission spectrometry

This technique has been applied to the determination of copper as discussed under multi-element analysis in section 5.26.4.1.

5.10 Dysprosium

5.10.1 Inductively coupled plasma atomic emission spectrometry

This technique has been applied to the determination of dysprosium as discussed under multi-element analysis in section 5.26.2.1.

5.11 Iron

5.11.1 Atomic absorption spectrometry

This technique has been applied to the determination of iron as discussed under multi-element analysis in section 5.26.1.1.

5.11.2 Inductively coupled plasma atomic emission spectrometry

This technique has been applied to the determination of iron as discussed under multi-element analysis in section 5.26.2.1.

5.11.3 X-ray secondary emission spectrometry

This technique has been applied to the determination of iron as discussed under multi-element analysis in section 5.26.4.1.

5.11.4 Miscellaneous

Lucotte and D'Anglejan [2] compared five different methods for the extraction and determination of iron hydroxides and/or associated phosphorus in estuarine particulate matter from the St Lawrence and Eastmain Rivers estuaries. The extraction reagents examined were citrate-dithionite-bicarbonate, ammonium oxalate-oxalic acid, hydroxylamine hydrochloride, calcium nitrilotriacetic acid, and acetate-tartrate. The procedure using the citrate-dithionite bicarbonate extraction reagent was generally more specific and reproducible than the others. This extraction procedure was included in an extraction sequence which differentiated the exchangeable phosphorus and iron fraction, the iron hydroxides and associated phosphorus, and the organic phosphorus and lithogenous iron.

5.12 Lanthanum

5.12.1 Inductively coupled plasma atomic emission spectrometry

This technique has been applied to the determination of lanthanum as discussed under multi-element analysis in section 5.26.2.1.

5.13 Lead

5.13.1 Atomic absorption spectrometry

This technique has been applied to the determination of lead as discussed under multi-element analysis in sections 5.26.1.1 and 5.26.1.2.

5.14 Manganese

5.14.1 Atomic absorption spectrometry

This technique has been applied to the determination of manganese as discussed under multi-element analysis in section 5.26.1.1.

5.14.2 Inductively coupled plasma atomic emission spectrometry

This technique has been applied to the determination of manganese as discussed under multi-element analysis in section 5.26.2.1.

5.14.3 X-ray secondary emission spectrometry

This technique has been applied to the determination of manganese as discussed under multi-element analysis in section 5.26.4.1.

5.15 Mercury

5.15.1 Miscellaneous

Fresnet-Robin and Ottmann [3] studied the distribution of mercury between water and sediment in samples taken from the Loire estuary between 1972 and 1975, and observed a decrease in mercury pollution during this period. The ratio of mercury in sediment/mercury in water increased with increasing salinity of the estuary water. Mercury fixes most on suspensions in the surface water, as these suspensions comprise smaller particles.

Bothner et al. [4] have studied mercury loss from contaminated estuarine sediments. A systematic decrease of mercury with an apparent half-life of about 1.3 years was observed in oxidising sediments sampled. In situ measurements of the mercury flux from sediments could not account for this decrease. It was suggested that removal is associated with sediment particles transport. The decreased was modelled using a steady-state mixing model. Mercury concentration in anoxic interstitial waters was 126 times higher than that in overlying sea water.

5.16 Nickel

5.16.1 Atomic absorption spectrometry

This technique has been applied to the determination of nickel as discussed under multi-element analysis in sections 5.26.1.1 and 5.26.1.2.

5.16.2 Inductively coupled plasma atomic emission spectrometry

This technique has been applied to the determination of nickel as discussed under multi-element analysis in section 5.26.2.1.

5.16.3 X-ray secondary emission spectrometry

This technique has been applied to the determination of nickel as discussed under multi-element analysis in section 5.26.4.1.

5.17 Potassium

5.17.1 X-ray secondary emission spectrometry

This technique has been applied to the determination of potassium as discussed under multi-element analysis in section 5.26.4.1.

5.18 Selenium

5.18.1 Hydride generation atomic absorption spectrometry

Itoh et al. [5] have developed an automated flow system to eliminate interferences from transition metal ions during the hydride generation/atomic absorption spectrometric determination of selenium in estuarine sediments. Experimental conditions were adjusted to minimise interferences. Low concentrations of tetrahydroborate in 6 mol L $^{-1}$ hydrochloric acid were preferred. Iron(III), added as chloride at a concentration of 2 mmol L $^{-1}$, removed the depression of selenium signals by ions such as copper(II) and bismuth(III).

5.19 Silicon

5.19.1 Inductively coupled plasma atomic emission spectrometry

This technique has been applied to the determination of silicon as discussed under multi-element analysis in section 5.26.2.1.

5.19.2 X-ray secondary emission spectrometry

This technique has been applied to the determination of silicon as discussed under multi-element analysis in section 5.26.4.1.

5.20 Titanium

5.20.1 Inductively coupled plasma atomic emission spectrometry

This technique has been applied to the determination of titanium as discussed in section 5.26.2.1.

5.20.2 X-ray secondary emission spectrometry

This technique has been applied to the determination of titanium as discussed under multi-element analysis in section 5.26.4.1.

5.21 Vanadium

5.21.1 Inductively coupled plasma atomic emission spectrometry

This technique has been applied to the determination of vanadium as discussed under multi-element analysis in section 5.26.2.1.

5.22 Ytterbium

5.22.1 Inductively coupled plasma atomic emission spectrometry

This technique has been applied to the determination of ytterbium as discussed under multi-element analysis in section 5.26.2.1.

5.23 Yttrium

5.23.1 Inductively coupled plasma atomic emission spectrometry

This technique has been applied to the determination of yttrium as discussed under multi-element analysis in section 5.26.2.1.

5.24 Zinc

5.24.1 Atomic absorption spectrometry

This technique has been applied to the determination of zinc as discussed under multi-element analysis in section 5.26.1.1.

5.24.2 Inductively coupled plasma atomic emission spectrometry

This technique has been applied to the determination of zinc as discussed under multi-element analysis in section 5.26.2.1.

5.24.3 X-ray secondary emission spectrometry

This technique has been applied to the determination of zinc as discussed under multi-element analysis in section 5.26.4.1.

5.25 Zirconium

5.25.1 Inductively coupled plasma atomic emission spectrometry

This technique has been applied to the determination of zirconium as discussed under multi-element analysis in section 5.26.2.1.

5.26 Multi-metal analysis

5.26.1 Atomic absorption spectrometry

5.26.1.1 Heavy metals (chromium, manganese, iron, cobalt, nickel, copper, zinc, cadmium and lead).

Sinex et al. [6] have evaluated the accuracy and precision of a method for the determination of chromium, manganese, iron, cobalt, nickel, copper, zinc, cadmium and lead in estuarine sediments. This method involved a 4 h hot digestion of 5 g of dried (at 105 °C) sample with 100 mL 9:1 v/v nitric acid–hydrochloric acid followed by atomic absorption spectrometry of the 25 ml of concentrated extract. High recoveries and good reproducibility were obtained with coefficients of variation being 10% or less. They compared results obtained by this procedure with those obtained by a lithium metaborate fusion–d.c. plasma emission spectroscopic method [7, 8], NBS SRM–1645 river sediment and on an estuarine sediment (US Geological Survey Marine mud sample MAG–1). In the lithium metaborate fusion method, 100 mg of the dried sediments were fused with 0.5g of lithium metaborate in a graphite crucible at 900°C for approximately 15 min. The molten lithium borate glass bead was poured into a 150 mL Teflon FEP beaker containing 100 mL of 4% nitric acid. The beaker was placed on a magnetic stirrer, and the contents were stirred for approximately 10 min. The solution was then transferred to a 125 mL polyethylene bottle. Considerably higher and more reproducible recoveries were obtained by fusion than by acid digestion. When the acid extraction method was applied to estuarine acids (MAG–1), the resulting trace metal concentrations, although reproducible, were usually substantially lower than total concentrations determined by lithium metaborate fusion–d.c. plasma emission spectroscopy. Excellent recoveries were obtained when the test was applied to the NBS SRM 1645 reference river sediment; chromium, manganese, nickel, copper, zinc and lead recovery was complete, while for iron, cobalt and cadmium, modest under recovery (13–25%) was observed.

5.26.1.2 Heavy metals (cadmium, lead, copper, nickel, cobalt and chromium) and beryllium

Sturgeon et al. [9] demonstrated quantitative recovery of cadmium, lead, copper, nickel, cobalt, beryllium and chromium from estuarine sediments using a digestion–graphite furnace atomic absorption spectrometric with a L'vov platform method.

In general, it is agreed that there is a lack of well-characterised and representative standard reference estuarine sediments for the evaluation of methods of analysis. The National Research Council of Canada Marine Analytical Standards Programme has prepared two near–shore marine sediment reference materials, MESS–1 and BCSS–1, for which reliable results are available for 12 major and minor constituents and 13 trace metals. Sturgeon et al. [9] used these standards to evaluate their method.

Digestions on dried (105°C) sediments (0.5 g) were performed in PTFE beakers using hydrochloric acid (5 mL), nitric acid (2 mL) hydrofluoric acid (5 mL) and water (4 mL) for 2 h at 90°C. After gentle evaporation to dryness, a further 5 mL of concentrated nitric acid and 2 mL perchloric acid were added and the solutions again taken to dryness. The residue was dissolved in 20 mL 1 mol L^{-1} hydrochloric acid, filtered and made up to 50 mL with water.

It is shown in Table 5.1 that determined results are in good agreement with reference values for all seven elements.

5.26.2 Inductively coupled plasma atomic emission spectrometry

5.26.2.1 Heavy metals (chromium, manganese, iron, cobalt, nickel, copper and zinc) and aluminium, silicon, titanium, vanadium, yttrium, zirconium, lanthanum, cerium, dysprosium and ytterbium

The application of this technique to estuarine sediment analysis has been investigated by several workers [7, 10–14].

Cantillo et al. [7] used a procedure involving lithium metaborate fusion followed by measurement with a single-channel direct current plasma emission spectrometer in a large-scale study of major and minor elements in estuarine sediments. The coefficients of variation of replicate analyses were in the range 2–10%, which is comparable to the reproducibility in routine atomic absorption work, but worse than that in the best inductively coupled plasma emission analysis. Systematic errors occurred in analysis of lanthanum, cerium and zirconium. Results for aluminium, chromium and iron were slightly low, while those for ytterbium and

Table 5.1 Comparison of graphite furnace atomic absorption spectrometry results with accepted values for MESS–1

Element	Concentration (µg g $^{-1}$) ^{a} by different methods				
	Standard additions			L'vov platform	Accepted value
	Direct injection	Chelation–extraction ^{d}		Direct injection	
Cd	0.54 ± 0.06 (9)	0.58 ± 0.11 (9)		0.65 ± 0.13 (3)	0.59 ± 0.1
Pb	26.4 ± 2.7 (9)	28.2 ± 6.5 (9)		27.3 ± 1.6 (3)	34.0 ± 6.1 ^{b}
Cu	21.9 ± 2.3 (8)	25.4 ± 3.0 (10)		21.3 ± 1.6 (3)	25.1 ± 3.8
Ni	28.7 ± 2.9 (10)	31.2 ± 4.1 (11)		28.6 ± 4.8 (3)	29.5 ± 2.7
Co	11.7 ± 1.8 (14)	12.2 ± 2.7 (11)		11.7 ± 2.9 (3)	10.8 ± 1.9
Be	1.8 ± 0.2 (8)	ND ^{c}		1.8 ± 0.1 (3)	1.9 ± 0.2
Cr	70.3 ± 5.0 (6)	ND		ND	71 ± 11

Comparison of graphite furnace atomic absorption spectrometry results with accepted values for BCSS–1

Element	Concentration (µg g $^{-1}$) ^{a} by different methods				
	Standard additions			L'vov platform	Accepted value
	Direct injection	Chelation–extraction ^{d}		Direct injection	
Cd	0.24 ± 0.04 (9)	0.26 ± 0.10 (3)		0.23 ± 0.03 (3)	0.25 ± 0.04
Pb	22.0 ± 3.0 (6)	22.0 ± 10.1 (3)		20.0 ± 5.4 (3)	22.7 ± 3.4
Cu	15.0 ± 3.4 (3)	16.2 ± 2.2 (4)		14.6 ± 2.5 (3)	18.5 ± 2.76
Ni	54.7 ± 5.1 (7)	56.6 ± 5.0 (5)		53.0 ± 2.2 (3)	55.3 ± 3.6
Co	12.7 ± 3.6 (3)	11.9 ± 6.9 (3)		10.7 ± 1.3 (3)	11.4 ± 2.1
Be	1.1 ± 0.2 (8)	ND ^{c}		1.1 ± 0.1 (3)	1.3 ± 0.3
Cr	122 ± 7 (6)	ND		ND	123 ± 14

^{a} Precision expressed as the 95% confidence interval for a single result (*n*) = number of determinations
^{b} Accepted value excludes results from graphite furnace atomic absorption spectrometry
^{c} Not determined
^{d} 8 ml of dissolved sediment diluted with 30 ml 10 mol L $^{-1}$ hydrochloric acid, iron removed by extraction with MIKB. Digest evaporated to dryness and diluted to 100 mL with water. 10 mL aliquot diluted to 50 mL with water used to prepare standard addition plot by extracting samples with ammonium pyrroalidone diethyl dithiocarbamate in MIKB. Then trace metals back extracted into 5 ml mol L $^{-1}$ nitric acid

Source: Elsevier Science Publishers [9]

zirconium were substantially low. Detection limits ranged from 1 mg kg $^{-1}$ (ytterbium) to 60 mg kg $^{-1}$ (cerium). The sediment fusion was carried out as follows. Approximately 0.2 g of sample was weighed to the nearest

Table 5.2 Fusion results of duplicate analyses of the NBS estaurine sediment (SRM 1646) (μg g^{-1}) except where percent)

	Analyst I (n = 6)[a]		Analyst II (n = 6)		F[c]	t[c]
Al (%)	5.3	(2)[b]	5.4	(1)	5.2*	
Si (%)	30	(3)	30	(1)	1.6	0
Ti (%)	0.36	(2)	0.37	(2)	6.4*	
V	89	(3)	91	(3)	1.4	1.4
Cr	74	(1)	70	(4)	9.0*	
Mn	280	(2)	280	(2)	1.0	0
Fe (%)	2.9	(2)	3.0	(0)		
Co	17	(18)	20	(14)	1.2	1.8
Ni	31	(5)	37	9(6)	2.1	5.5*
Cu	28	(9)	28	(6)	5.6*	
Zn	130	(0)	120	(13)		
Ga	19	(8)	19	(6)	1.9	0
Y	17	(8)	15	(4)	5.0	3.2*
Zr	270	(4)	300	(4)	1.4	4.0
La	36	(2)	36	(3)	2.5	0
Ce	110	(4)	110	(7)	3.3	0
Dy	2.8	(7)	3.1	(14)	4.1	1.6
Yb	2.2	(4)	2.0	(5)	1.5	3.8*

[a] Number of samples
[b] Coefficient of variation (%) = 100 σ/mean
[c] F and t test statistics, asterisk denotes a difference between the determined and reference values at the 5% level of significance

Source: American Chemical Society [7]

tenth of a milligram into a 7.5 cm^3 drill point graphite crucible (Ultra Carbon Crop, Bay City, k MI) containing 1.0 g of lithium metaborate. The sediment was placed in a small depression made in the borate flux to prevent incomplete fusion problems encountered when sediment is in direct contact with graphite.

The sediment-flux mixture was fused in a muffle furnace at 950 ± 50°C for 15 min. The liquid borate bead was poured into a 150 mL Teflon FEP beaker containing 100 mL of 5% nitric acid. The beaker was then placed on a magnetic stirrer, and the contents were stirred for approximately 10 min. The resulting solution was transferred to an acid-washed 125 mL linear polyethylene bottle.

The precision of this method was evaluated by performing replicate analysis of NBS SRM 1646 estuarine sediment (Table 5.2). The test result indicated that aluminium, titanium, chromium and copper have significantly different variances, at the 5% level, from the nominal concentrations in the sediment (ie F^c > S).

5.26.3 Neutron activation analysis

5.26.3.1 Miscellaneous metals

Ellis and Chattopadhya [13] investigated the determination of 28 elements in estuarine suspended solids by neutron activation analysis and found detection limits for most of the elements ranged between 0.37 and 1.0 ng. Average precision ranged between 10 and 16% and accuracy within 10%.

5.26.4 X-ray secondary emission spectrometry

5.26.4.1 Heavy metals (iron, chromium, manganese, nickel, copper and zinc) and aluminium, silicon, potassium, calcium and titanium

Baker and Piper [14] applied this method to the determination of aluminium, phosphorus, silicon, potassium, calcium, titanium, iron, chromium, manganese, nickel, copper and zinc in estuarine sediments.

References

1 Maher, W.A. *Analytica Chimica Acta*, **126**, 157 (1981).
2 Lucotte, M. and D'Anglejan, B. *Chemical Geology*, **48**, 257 (1985).
3 Fresnet-Robin, M. and Ottman, E. *Estuarine and Marine Coatsal Science*, **7**, 425 (1978).
4 Bothner, M.H., Jahnke, R.A., Peterson, M.L. and Carpenter, R. *Geochimica and Cosmochimica Acta*, **44**, 273 (1980).
5 Itoh, K., Chikuma, M. and Tanaka, H. Fresenius *Zeitschrift für Analytische Chemie*, **330**, 600 (1988).
6 Sinex, S.A., Cantillo, A.Y. and Helz, G.R. *Analytical Chemistry*, **52**, 2342 (1980).
7 Cantillo, A.Y., Sinex, S.A. and Helz, G.R. *Analytical Chemistry*, **56**, 33 (1984).
8 Suhr, N.H. and Ingasmello, C.O. *Analytical Chemistry*, **38**, 730 (1968).
9 Sturgeon, R.E., Desauliniers, J.A.H., Berman, S.S. and Russell, D.S. *Analytica Chimica Acta*, **134**, 283 (1982).
10 McQuaker, N.R., Kluckner, P.D. and Chang, G.N. *Analytical Chemistry*, **51**, 888 (1979).
11 Walsh, J.N. and Howie, R.A. *Min Management*, **43**, 967 (1980).
12 McLaren, J.W., Berman, S.S., Bayko, V.J. and Russel, D.S. *Analytical Chemistry*, **53**, 1802 (1981).
13 Ellis, K.M. and Chattopadhya, V.A. *Analytical Chemistry*, **51**, 942 (1979).
14 Baker, E.T. and Piper, D.Z. *Deep Sea Research*, **23**, 1181 (1976).

Chapter 6

Determination of metals in sludges

6.1 Aluminium

6.1.1 Atomic absorption spectrometry

This technique has been applied to the determination of aluminium in sludges as discussed under multi-metal analysis in sections 6.53.1.3 and 6.53.1.5.

6.1.2 Neutron activation analysis

This technique has been applied to the determination of aluminium as discussed under multi-metal analysis in section 6.53.3.1.

6.2 Antimony

6.2.1 Atomic absorption spectrometry

Kunselman and Huff [15] have applied hydride generation atomic absorption spectrometry to the determination of antimony in sewage effluents. This technique has also been applied to the determination of antimony as discussed under multi-metal analysis in section 6.53.1.6.

6.2.2 Neutron activation analysis

This technique has been applied to the determination of antimony as discussed under multi-metal analysis in section 6.53.3.1.

6.2.3 Photon activation analysis

This technique has been applied to the determination of antimony as discussed under multi-metal analysis in section 6.53.4.1.

6.3 Arsenic

6.3.1 Atomic absorption spectrometry

Webster [1] has investigated the determination of arsenic in the concentration range 0.5–1.0 mg kg^{-1} dry solids in sewage sludge. The method involves the use of sodium borohydride to generate arsenic hydrides, and their introduction into a silica furnace, maintained at dull red heat by the air–acetylene flame of an atomic absorption spectrophotometer. Predigestion of sewage sludge with nitric/perchloric acid is recommended. Recoveries, using a standard addition technique, were 80–100% and precision between 16 and 18%, and some interference from heavy metals was observed. Results from a 12-month study of sludges in the Lothian region showed arsenic concentrations between 0.5 and 7 mg kg^{-1} dry solids.

Webster [1] made a close study of various parameters influencing results obtained in this procedure such as sample preparation using the standard nitric perchloric acid method [2] hydrochloric acid strength at the reduction stage, concentration and volume of sodium borohydride, volume of nitrogen for purging the reduction cell and effect of arsenic valency all have a bearing on results obtained [3].

Kunselman and Huff [15] have applied hydride generation atomic absorption spectrometry to the determination of traces of arsenic in sewage effluents. A standard UK hydride generation atomic absorption method [3] has been applied to the determination of arsenic in sewage sludges. The application of atomic absorption spectrometry in sludges is also discussed under multi-metal analysis in sections 6.53.1.3 and 6.53.1.6.

6.3.2 Inductively coupled plasma atomic emission spectrometry

Atsuya and Akatsuka [4] have described a method for determining trace amounts of arsenic. The technique, which uses capacitatively coupled microwave plasma with an arsine generation system, has been used to determined arsenic in sewage sludge.

6.3.3 Neutron activation analysis

This technique has been applied to the determination of arsenic as discussed under multi-metal analysis in section 6.53.3.1.

6.3.4 Photon activation analysis

This technique has been applied to the determination of arsenic as discussed under multi-metal analysis in section 6.53.4.1.

6.4 Barium

6.4.1 Neutron activation analysis

This technique has been applied to the determination of barium as discussed under multi-metal analysis in section 6.53.3.1.

6.4.2 Photon activation analysis

This technique has been applied to the determination of barium as discussed under multi-metal analysis in section 6.53.4.1.

6.5 Bismuth

6.5.1 Atomic absorption spectrometry

This technique has been applied to the determination of bismuth as discussed under multi-metal analysis in section 6.53.1.6.

6.5.2 Neutron activation analysis

This technique has been applied to the determination of bismuth as discussed under multi-metal analysis in section 6.53.3.1.

6.5.3 Photon activation analysis

This technique has been applied to the determination of bismuth as discussed under multi-metal analysis in section 6.53.4.1.

6.6 Cadmium

6.6.1 Fluorometric method

Pal et al. [5] give details of a direct fluorometric method for determination of very low concentrations of cadmium, using the fluorescent chelate, morin (3,5,7,2',4'–pentahydroxyflavone), which had a constant and stable fluorescent intensity over the pH range 8.0–9.4. Beer's law was obeyed for cadmium concentrations ranging from 0.5 ppb to 0.8 ppm. Results obtained on several synthetic mixtures and sewage sludges are presented.

6.6.2 Atomic absorption spectrometry

This technique has been applied to the determination of cadmium as discussed under multi-metal analysis in sections 6.53.1.1 to 6.53.1.4.

6.6.3 Inductively coupled plasma atomic emission spectrometry

This technique has been applied to the determination of cadmium as discussed under multi-metal analysis in section 6.53.2.1.

6.6.4 Neutron activation analysis

Esprit *et al.* [6] give details of equipment and procedure for determination of cadmium by activation analysis with fast neutrons, obtained by irradiation of a thick beryllium target with deuterons. The resulting cadmium-111m was separated by liquid–liquid extraction with zinc diethyldithiocarbamate in chloroform and measured with a germanium (lithium) gamma spectrometer. For low concentrations, cadmium can be precipitated as cadmium ammonium phosphate after the extraction. Results obtained on a number of reference samples (city waste incineration ash and sewage sludge) are presented and discussed. For cadmium concentrations between 3 and 500 ug per g, the relative standard deviation ranged from 5 to 3%. The application of this technique to the determination of cadmium is also discussed under multi-metal analysis in section 6.53.3.1.

6.6.5 Photon activation analysis

This technique has been applied to the determination of cadmium as discussed under multi-metal analysis in section 6.53.4.1.

6.6.6 Differential pulse polarography

This technique has been applied to the determination of cadmium as discussed under multi-metal analysis in section 6.53.6.1.

6.7 Caesium

6.7.1 Neutron activation analysis

This technique has been applied to the determination of caesium as discussed under multi-metal analysis in section 6.53.3.1.

6.7.2 Photon activation analysis

This technique has been applied to the determination of caesium as discussed under multi-metal analysis in section 6.53.4.1.

6.8 Calcium

6.8.1 Atomic absorption spectrometry

This technique has been applied to the determination of calcium as discussed under multi-metal analysis in sections 6.53.1.2 and 6.53.1.5.

6.8.2 Neutron activation analysis

This technique has been applied to the determination of calcium as discussed under multi-metal analysis in section 6.53.3.1.

6.8.3 Photon activation analysis

This technique has been applied to the determination of calcium as discussed under multi-metal analysis in section 6.53.4.1.

6.9 Cerium

6.9.1 Neutron activation analysis

This technique has been applied to the determination of cerium as discussed under multi-metal analysis in section 6.53.3.1.

6.9.2 Photon activation analysis

This technique has been applied to the determination of cerium as discussed under multi-metal analysis in section 6.53.4.1.

6.10 Chromium

6.10.1 Atomic absorption spectrometry

Thomas and Wagstaff [7] have described a simple atomic absorption spectrophotometric method utilising an air–acetylene flame for the determination of down to 0.005 µg mL $^{-1}$ of chromium in sewage sludges which utilises a concentration by evaporation technique, in which ammonium perchlorate is incorporated into the sample solution in order to minimise inter-element effects from the sample matrix. These workers found that the use of sequential background correction using the 357.3 nm non-resonance line gave satisfactory results. In this method, volumes (50 ± 0.5 mL) of the samples, standards and blanks were placed in 100 mL borosilicate glass beakers and 4 mL (± 0.1 mL) of 25% V/V hydrochloric acid, 2 mL (± 0.05 mL) of 10% m/V ammonium perchlorate solution and some aluminium oxide anti-bumping granules were added. The beakers were then placed on a hot-plate with the temperature set such that gentle simmering occurred. When the volume of the solution had decreased to

Table 6.1 Optimised instrumental operating conditions

Wavelength	357.9nm
Background correction wavelength	357.3nm (Pb)
Slit width	0.5nm
Air flow	As recommended in handbook
Acetylene flow	Flame on verge of luminosity (no yellow luminosity visible)
Distance from top of burner grid to position where the grid just intercepted the light beam (0.01 absorbance)	3.5 mm
Integration period	3 s
Wash solution	3% V/V hydrochloric acid (36% m/m)

Source: Royal Society of Chemistry [7]

Table 6.2 Recovery test results

		Chromium recovered μg mL⁻¹			
		Flame on verge of luminosity (Table 6.1)			Luminous flame with chromium added (0.04 μg mL⁻¹) and no ammonium perchlorate
Sample	Sample No.	Chromium added (0.04 μg mL⁻¹) plus ammonium perchlorate	Chromium added (0.04 μg mL⁻¹) plus ammonium perchlorate	Chromium added (0.04 μg mL⁻¹) plus no ammonium perchlorate	
Sewage final effluent	5	0.0386	0.0387	0.312	0.243
Sewage final effluent	6	0.0382	0.370	0.306	0.246

Source: Royal Society of Chemistry [7]

20 mL (± 5 ml), 0.5 mL (± 0.05 mL) of hydrogen peroxide (6% m/m) was then added. This ensured that any chromium(VI) would be converted into chromium(III). The evaporation was then continued until the final volume was about 5 mL (± 1 mL) and this took approximately 1.5 h. The solutions were allowed to cool and the contents transferred into the 10 mL calibrated borosilicate glass tubes. The beakers were washed out using three approximately 1.5 mL washes with de-ionised water. The contents of the tube were then diluted to volume and shaken and any suspended matter was allowed to settle prior to nebulisation.

The final acid concentration during the evaporation step, in conjunction with the addition of hydrogen peroxide, should ensure adequate digestion of particulate and organically bound chromium in natural waters and effluents. Avoid boiling the solution to dryness by controlling the hot-plate temperature. Instrumental operating conditions are shown in Table 6.1. Table 6.2 shows the recoveries obtained after the addition of 0.04 and 0.4 µg mL^{-1} of chromium(III) to sewage effluents and also shows some recoveries obtained in the absence of ammonium perchlorate. The natural chromium level in the samples was below the detection limit. It can be seen from Table 6.2 that the method is satisfactory for sewage final effluent analysis and that in the absence of ammonium perchlorate poor recoveries were observed, especially if the flame conditions were set for maximum response (ie a distinctly luminous flame).

The application of atomic absorption spectrometry to the determination of chromium is also discussed under multi-metal analysis in sections 6.53.1.1 and 6.53.1.2.

6.10.2 Inductively coupled plasma atomic emission spectrometry

This technique has been applied to the determination of chromium as discussed under multi-metal analysis in section 6.53.2.1.

6.10.3 Neutron activation analysis

This technique has been applied to the determination of chromium as discussed under multi-metal analysis in section 6.53.3.1.

6.10.4 Photon activation analysis

This technique has been applied to the determination of chromium as discussed under multi-metal analysis in section 6.53.4.1.

6.10.5 Differential pulse polarography

This technique has been applied to the determination of chromium as discussed under multi-metal analysis in section 6.53.6.1.

6.10.6 Miscellaneous

Chakraborty et al. [8] determined chromium in sewage sludge without the use of any chemical modifier after microwave-assisted sample digestion.

6.11 Cobalt

6.11.1 Atomic absorption spectrometry

This technique has been applied to the determination of cobalt as discussed under multi-metal analysis in sections 6.53.1.1, 6.53.1.2 and 6.53.1.7.

6.11.2 Neutron activation analysis

This technique has been applied to the determination of cobalt as discussed under multi-metal analysis in section 6.53.3.1.

6.11.3 Photon activation analysis

This technique has been applied to the determination of cobalt as discussed under multi-metal analysis in section 6.53.4.1.

6.12 Copper

6.12.1 Atomic absorption spectrometry

This technique has been applied to the determination of copper as discussed under multi-metal analysis in sections 6.53.1.1 and 6.53.1.2.

6.12.2 Inductively coupled plasma atomic emission spectrometry

This technique has been applied to the determination of copper as discussed under multi-metal analysis in section 6.53.2.1.

6.12.3 Neutron activation analysis

This technique has been applied to the determination of copper as discussed under multi-metal analysis in section 6.53.3.1.

6.12.4 Differential pulse polarography

This technique has been applied to the determination of copper as discussed under multi-metal analysis in section 6.53.6.1.

6.12.5 Potentiometric stripping analysis

This technique has been applied to the determination of copper as discussed under multi-metal analysis in section 6.53.7.1.

6.12.6 Miscellaneous

Senesi and Sposito [9] studied copper(II) anionic surfactant complexes using electron spin resonance spectrometry. The surfactants used were linear alkyl aryl sulphonate, sodium dodecyl benzene sulphonate, and sodium lauryl sulphate. Copper-ligand molar ratios ranged from 0.1 to 1.0. Electron spin resonance spectra of frozen (77K) aqueous solutions showed that all three surfactants formed inner sphere complexes with copper(II) ions held in four oxygen–ligand square planar binding sites.

The sulphonate type surfactants had a higher affinity for copper(II) than did the ester sulphate type. The copper(II)–anionic surfactant complexes were different from, and less stable than, copper(II)–fulvic acid complexes. Solution spectra yielded more structural information on the complexes than did solid state spectra. It was concluded that undegraded anionic surfactants in sewage sludge did not participate as isolated independent ligands, but may participate as co-ligands with other oxygen-containing functional groups, or as moities incorporated into the fulvic acid structure.

These workers [10] also obtained electron spin resonance spectra for both frozen aqueous solutions and solid samples of purified fulvic acid extracted from two anaerobically digested sewage sludges and from organic matter in a soil, and the characteristics of the spectra were examined. Residual divalent copper complexes were present in all the samples. The types of binding involved are discussed.

6.13 Dysprosium

6.13.1 Neutron activation analysis

This technique has been applied to the determination of dysprosium as discussed under multi-metal analysis in section 6.53.3.1.

6.14 Europium

6.14.1 Neutron activation analysis

This technique has been applied to the determination of europium as discussed under multi-metal analysis in section 6.53.3.1.

6.15 Gadolinium

6.15.1 Neutron activation analysis

This technique has been applied to the determination of gadolinium as discussed under multi-metal analysis in section 6.53.3.1.

6.16 Gold

6.16.1 Neutron activation analysis

This technique has been applied to the determination of gold as discussed under multi-metal analysis in section 6.53.3.1.

6.17 Hafnium

6.17.1 Neutron activation analysis

This technique has been applied to the determination of hafnium as discussed under multi-metal analysis in section 6.53.3.1.

6.18 Indium

6.18.1 Neutron activation analysis

This technique has been applied to the determination of indium as discussed under multi-metal analysis in section 6.53.3.1.

6.18.2 Photon activation analysis

This technique has been applied to the determination of indium as discussed under multi-metal analysis in section 6.53.4.1.

6.19 Iridium

6.19.1 Neutron activation analysis

This technique has been applied to the determination of iridium as discussed under multi-metal analysis in section 6.53.3.1.

6.20 Iron

6.20.1 Atomic absorption spectrometry

This technique has been applied to the determination of iron as discussed under multi-metal analysis in sections 6.53.1.1, 6.53.1.2 and 6.53.1.5.

6.20.2 Neutron activation analysis

This technique has been applied to the determination of iron as discussed under multi-metal analysis in section 6.53.3.1.

6.20.3 Photon activation analysis

This technique has been applied to the determination of iron as discussed under multi-metal analysis in section 6.53.4.1.

6.20.4 Differential pulse polarography

This technique has been applied to the determination of iron as discussed under multi-metal analysis in section 6.53.6.1.

6.21 Lanthanum

6.21.1 Neutron activation analysis

This technique has been applied to the determination of lanthanum as discussed under multi-metal analysis in section 6.53.3.1.

6.22 Lead

6.22.1 Atomic absorption spectrometry

This technique has been applied to the determination of lead as discussed under multi-metal analysis in sections 6.53.1.1 to 6.53.1.4.

6.22.2 Inductively coupled plasma atomic emission spectrometry

This technique has been applied to the determination of lead as discussed under multi-metal analysis in section 6.53.2.1.

6.22.3 Neutron activation analysis

This technique has been applied to the determination of lead as discussed under multi-metal analysis in section 6.53.3.1.

6.22.4 Photon activation analysis

This technique has been applied to the determination of lead as discussed under multi-metal analysis in section 6.53.4.1.

6.22.5 Differential pulse polarography

This technique has been applied to the determination of lead as discussed under multi-metal analysis in section 6.53.6.1.

6.22.6 Potentiometric stripping analysis

This technique has been applied to the determination of lead as discussed under multi-metal analysis in section 6.53.7.1.

6.22.7 Thin-layer chromatography

Jahns et al. [11] have described a thin-layer chromatographic method for the determination of lead in sewage sludge, based on the use of ammonium pyrrolidinedithiocarbamate for the extraction and enrichment of lead. Instead of the previously reported conversion and visualisation of the lead complex in the short wavelength region using dithizone, the lead carbamate was converted to lead sulphide with the aid of a 6% solution of sodium sulphide in methanol/water (3:1) on silica gel plates, after development of the plates with toluene. Details of the application of the method are given for samples of sewage sludge, together with the extraction procedure, linearity and reproducibility of results.

6.23 Lutecium

6.23.1 Neutron activation analysis

This technique has been applied to the determination of lutecium as discussed under multi-metal analysis in section 6.53.3.1.

6.24 Magnesium

6.24.1 Atomic absorption spectrometry

This technique has been applied to the determination of magnesium as discussed under multi-metal analysis in sections 6.53.1.2 and 6.53.1.5.

6.24.2 Neutron activation analysis

This technique has been applied to the determination of magnesium as discussed under multi-metal analysis in section 6.53.3.1.

6.24.3 Photon activation analysis

This technique has been applied to the determination of magnesium as discussed under multi-metal analysis in section 6.53.4.1.

6.25 Manganese

6.25.1 Atomic absorption spectrometry

This technique has been applied to the determination of manganese as discussed under multi-metal analysis in sections 6.53.1.1, 6.53.1.2 and 6.53.1.7.

6.25.2 Neutron activation analysis

This technique has been applied to the determination of manganese as discussed under multi-metal analysis in section 6.53.3.1.

6.25.3 Photon activation analysis

This technique has been applied to the determination of manganese as discussed under multi-metal analysis in section 6.53.4.1.

6.25.4 Differential pulse polarography

This technique has been applied to the determination of manganese as discussed under multi-metal analysis in section 6.53.6.1.

6.26 Mercury

6.26.1 Atomic absorption spectrometry

Magyar et al. [12] have used Zeeman atomic absorption spectrometry to determine mercury in sludge. The suitability of using this method is discussed. Samples were also analysed by the cold vapour method after wet oxidation of samples in closed teflon vessels.

Watanabe et al. [13] have described an evaporation–concentration method for the quantitative analysis of mercury prior to analysis by atomic absorption spectrometry. Chlorides (eg sodium, magnesium and ferric) and oxidising agents (chromium and vanadium) inhibited vaporisation of mercury in the procedure. Increase in recovery could be effected with nitric acid but not sulphuric acid. Recoveries were over 90% when samples of sewage were used (0.2 to 0.05 of their original volumes, with 30 ml nitric acid, 10 ml sulphuric acid and 250 mg sodium (as sodium chloride)). Determination of mercury to 0.01 ppb was possible using the evaporation–concentration method and preconcentration to 0.05 original volume.

Further applications of atomic absorption spectrometry to the determination of mercury in sludge are discussed under multi-metal analysis in sections 6.53.1.3 and 6.53.1.4.

6.26.2 Neutron activation analysis

This technique has been applied to the determination of mercury as discussed under multi-metal analysis in section 6.53.3.1.

6.26.3 Photon activation analysis

This technique has been applied to the determination of mercury as discussed under multi-metal analysis in section 6.53.4.1.

6.26.4 Miscellaneous

A standard UK method for the determination of mercury in sludges [28] supplements but does not supersede the previous standard procedure for mercury in waters, effluents and sludges. It presents a group of four additional methods of very high sensitivity for liquids or particulate solids, in which either flameless AAS or atomic fluorescence spectrometry could be used for detection of trace amounts of mercury down to 1 ng per litre for liquids and 0.01 ug per g for solids.

6.27 Molybdenum

6.27.1 Spectrophotometric method

A standard UK method [14] has been described for the determination of molybdenum in sewage sludge and soils.

6.27.2 Neutron activation analysis

This technique has been applied to the determination of molybdenum as discussed under multi-metal analysis in section 6.53.3.1.

6.27.3 Photon activation analysis

This technique has been applied to the determination of molybdenum as discussed under multi-metal analysis in section 6.53.4.1.

6.27.4 Atomic absorption spectrometry

The determination of molybdenum is discussed under multi-metal analysis in section 6.53.1.7.

6.28 Neodynium

6.28.1 Neutron activation analysis

This technique has been applied to the determination of neodynium as discussed under multi-metal analysis in section 6.53.3.1.

6.29 Nickel

6.29.1 Atomic absorption spectrometry

This technique has been applied to the determination of nickel as discussed under multi-metal analysis in sections 6.53.1.1 and 6.53.1.2.

6.29.2 Inductively coupled plasma atomic emission spectrometry

This technique has been applied to the determination of nickel as discussed under multi-metal analysis in section 6.53.2.1.

6.29.3 Neutron activation analysis

This technique has been applied to the determination of nickel as discussed under multi-metal analysis in section 6.53.3.1.

6.29.4 Photon activation analysis

This technique has been applied to the determination of nickel as discussed under multi-metal analysis in section 6.53.4.1.

6.30 Potassium

6.30.1 Atomic absorption spectrometry

This technique has been applied to the determination of potassium as discussed under multi-metal analysis in section 6.53.1.2.

6.30.2 Neutron activation analysis

This technique has been applied to the determination of potassium as discussed under multi-metal analysis in section 6.53.3.1.

6.30.3 Photon activation analysis

This technique has been applied to the determination of potassium as discussed under multi-metal analysis in section 6.53.4.1.

6.31 Rubidium

6.31.1 Neutron activation analysis

This technique has been applied to the determination of rubidium as discussed under multi-metal analysis in section 6.53.3.1.

6.31.2 Photon activation analysis

This technique has been applied to the determination of rubidium as discussed under multi-metal analysis in section 6.53.4.1.

6.32 Samarium

6.32.1 Neutron activation analysis

This technique has been applied to the determination of samarium as discussed under multi-metal analysis in section 6.53.3.1.

6.33 Scandium

6.33.1 Neutron activation analysis

This technique has been applied to the determination of scandium as discussed under multi-metal analysis in section 6.53.3.1.

6.33.2 Photon activation analysis

This technique has been applied to the determination of scandium as discussed under multi-metal analysis in section 6.53.4.1.

6.34 Selenium

6.34.1 Atomic absorption spectrometry

A procedure described by Kunselman and Huff [15] and a standard UK method [3] have been applied to the determination of selenium in sludges.

6.34.2 Neutron activation analysis

This technique has been applied to the determination of selenium as discussed under multi-metal analysis in section 6.53.3.1.

6.34.3 Photon activation analysis

This technique has been applied to the determination of selenium as discussed under multi-metal analysis in section 6.53.4.1.

6.35 Silicon

6.35.1 Spectrophotometric method

This technique has been applied to the determination of silicon in sludges [16].

6.35.2 Neutron activation analysis

This technique has been applied to the determination of silicon as discussed under multi-metal analysis in section 6.53.3.1.

6.35.3 Photon activation analysis

This technique has been applied to the determination of silicon as discussed under multi-metal analysis in section 6.53.4.1.

6.36 Silver

6.36.1 Atomic absorption spectrometry

This technique has been applied to the determination of silver as discussed under multi-metal analysis in sections 6.53.1.3 and 6.53.1.7.

6.36.2 Neutron activation analysis

This technique has been applied to the determination of silver as discussed under multi-metal analysis in section 6.53.3.1.

6.36.3 Photon activation analysis

This technique has been applied to the determination of silver as discussed under multi-metal analysis in section 6.53.4.1.

6.37 Sodium

6.37.1 Atomic absorption spectrometry

This technique has been applied to the determination of sodium as discussed under multi-metal analysis in section 6.53.1.2.

6.37.2 Neutron activation analysis

This technique has been applied to the determination of sodium as discussed under multi-metal analysis in section 6.53.3.1.

6.37.3 Photon activation analysis

This technique has been applied to the determination of sodium as discussed under multi-metal analysis in section 6.53.4.1.

6.38 Strontium

6.38.1 Neutron activation analysis

This technique has been applied to the determination of strontium as discussed under multi-metal analysis in section 6.53.3.1.

6.38.2 Photon activation analysis

This technique has been applied to the determination of strontium as discussed under multi-metal analysis in section 6.53.4.1.

6.39 Tantalum

6.39.1 Neutron activation analysis

This technique has been applied to the determination of tantalum as discussed under multi-metal analysis in section 6.53.3.1.

6.40 Terbium

6.40.1 Neutron activation analysis

This technique has been applied to the determination of terbium as discussed under multi-metal analysis in section 6.53.3.1.

6.41 Tellurium

6.41.1 Atomic absorption spectrometry

Kunselman and Huff [15] have applied hydride generation atomic absorption spectrometry to the determination of tellurium in sewage effluents. The determination of tellurium is also discussed under multi-metal analysis in section 6.53.1.6.

6.41.2 Neutron activation analysis

This technique has been applied to the determination of tellurium as discussed under multi-metal analysis in section 6.53.3.1.

6.41.3 Photon activation analysis

This technique has been applied to the determination of tellurium as discussed under multi-metal analysis in section 6.53.4.1.

6.42 Thallium

6.42.1 Atomic absorption spectrometry

This technique has been applied to the determination of thallium as discussed under multi-metal analysis in section 6.53.1.6.

6.42.2 Neutron activation analysis

This technique has been applied to the determination of thallium as discussed under multi-metal analysis in section 6.53.3.1.

6.42.3 Photon activation analysis

This technique has been applied to the determination of thallium as discussed under multi-metal analysis in section 6.53.4.1.

6.43 Thorium

6.43.1 Neutron activation analysis

This technique has been applied to the determination of thorium as discussed under multi-metal analysis in section 6.53.3.1.

6.44 Tin

6.44.1 Atomic absorption spectrometry

Legret and Divet [17] determined tin in sewage sludges by atomic absorption spectrometry with hydride generation. Results are presented from studies to assess the optimal operating conditions for the determination of tin in sediments and sludges by atomic absorption spectrometry with hydride generation in a nitric acid/tartaric acid solution. The effects of various acids and the matrix effects and interferences from other trace elements were investigated. Several procedures for decomposing the samples were compared; the preferred procedure involved reflux with a mixture of nitric acid and hydrochloric acid.

The application of atomic absorption spectrometry to the determination of tin in sludges is also discussed under multi-metal analysis in section 6.53.1.7.

6.44.2 Neutron activation analysis

This technique has been applied to the determination of tin as discussed under multi-metal analysis in section 6.53.3.1.

6.44.3 Photon activation analysis

This technique has been applied to the determination of tin as discussed under multi-metal analysis in section 6.53.4.1.

6.45 Titanium

6.45.1 Neutron activation analysis

This technique has been applied to the determination of titanium as discussed under multi-metal analysis in section 6.53.3.1.

6.45.2 Photon activation analysis

This technique has been applied to the determination of titanium as discussed under multi-metal analysis in section 6.53.4.1.

6.46 Tungsten

6.46.1 Neutron activation analysis

This technique has been applied to the determination of tungsten as discussed under multi-metal analysis in section 6.53.3.1.

6.47 Vanadium

6.47.1 Atomic absorption spectrometry

This technique has been applied to the determination of vanadium as discussed under multi-metal analysis in section 6.53.1.3 and 6.53.1.6.

6.47.2 Neutron activation analysis

This technique has been applied to the determination of vanadium as discussed under multi-metal analysis in section 6.53.3.1.

6.47.3 Photon activation analysis

This technique has been applied to the determination of vanadium as discussed under multi-metal analysis in section 6.53.4.1.

6.48 Uranium

6.48.1 Neutron activation analysis

This technique has been applied to the determination of uranium as discussed under multi-metal analysis in section 6.53.3.1.

6.49 Ytterbium

6.49.1 Neutron activation analysis

This technique has been applied to the determination of ytterbium as discussed under multi-metal analysis in section 6.53.3.1.

6.50 Yttrium

6.50.1 Neutron activation analysis

This technique has been applied to the determination of yttrium as discussed under multi-metal analysis in section 6.53.3.1.

6.50.2 Photon activation analysis

This technique has been applied to the determination of yttrium as discussed under multi-metal analysis in section 6.53.4.1.

6.5I Zinc

6.5I.I Spectrophotometric method

Kurochkina *et al.* [18] have described a photometric determination of zinc in sewage sludge utilising dithizone. In this method the sample is heated with concentrated hydrochloric acid and the resulting solution passed down an EDE–10P anion exchange column. The column is washed with 1:1 hydrochloric acid then zinc eluted with 1:3 hydrochloric acid. After neutralisation of acidity with aqueous sodium hydroxide the solution is buffered with sodium borate and ethanolic dithizone added. The coloured product is evaluated spectrophotometrically using a green filter.

6.5I.2 Atomic absorption spectrometry

This technique has been used to determine zinc as discussed under multi-metal analysis in sections 6.53.1.3 to 6.53.1.3.

6.5I.3 Inductively coupled plasma atomic emission spectrometry

This technique has been used to determine zinc as discussed under multi-metal analysis in section 6.53.2.1.

6.5I.4 Neutron activation analysis

This technique has been used to determine zinc as discussed under multi-metal analysis in section 6.53.3.1.

6.5I.5 Photon activation analysis

This technique has been used to determine zinc as discussed under multi-metal analysis in section 6.53.4.1.

6.5I.6 Differential pulse polarography

This technique has been applied to the determination of zinc as discussed under multi-metal analysis in section 6.53.6.1.

6.52 Zirconium

6.52.I Neutron activation analysis

This technique has been used to determine zirconium as discussed under multi-metal analysis in section 6.53.3.1.

Table 6.3 Conditions used for homogenisation of sludge samples

Sludge	Time (min)	Speed (rpm)
Primary	15	6000
Digested	10	8000
Surplus activated	7	10000

Source: Elsevier Science Publishers BV, Netherlands [26]

6.52.2 Photon activation analysis

This technique has been used to determine zirconium as discussed under multi-metal analysis in section 6.53.4.1.

6.53 Multi-metal analysis

Earlier methods for the determination of heavy metals in sewage sludges required preliminary acid digestion of the organic materials before the metal was determined spectrophotometrically [19]. Other earlier techniques that were applied to the determination of heavy metals included emission spectrometry [20], atomic absorption spectrometry [21–23], neutron activation analysis [24] and inverse polarography [25].

There has been much published work on this subject. This is discussed below in date order.

6.53.1 Atomic absorption spectrometry

6.53.1.1 Heavy metals (chromium, iron, nickel, copper, cobalt, cadmium, zinc, lead and manganese)

Lester *et al.* [26] have described rapid flameless atomic absorption spectrometric methods for the determination of lead, cadmium and copper [26] and chromium nickel and zinc [27] in sewage sludges. These methods are reviewed below.

(i) Lead, cadmium, copper [26]

Composite samples of sludge (800 mL) were collected and shaken vigorously in a closed polyethylene container to ensure homogeneity. A subsample (200 mL) was withdrawn and added to 1% nitric acid (800 mL) in a polyethylene bottle and then vigorously shaken, this constituted the sample. Prior to analysis a subsample (100 mL) was withdrawn and added to 1% nitric acid (900 mL) in a 2 L tall-form beaker. The resultant mixture was homogenised with the Ultra–Turrax Model T45N (Scientific Instrument Co, London) according to the conditions in Table 6.3.

Table 6.4 Conditions used for flameless atomic absorption analysis

Conditions	Metal					
	Cr	Ni	Zn	Cd	Cu	Pb
Drying temperature (°C)	99	99	99	99	99	99
Drying time (s)	30	30	30	30	30	30
Ashing time (°C)	1207	971	520	350	901	550
Ash time (s)	30	30	60	30	30	30
Atomisation temperature (°C)	2660	2627	2231	1800	2600	2040
Atomisation time (s)	10	10	5	10	10	10
Analytical line (nm)	357.9	232.0	307.6	228.8	324.7	283.3

Source: Elsevier Science Publishers, BV, Netherlands [26]

Analysis of homogenised sludge samples

An indication of the metal content of the sample was obtained by flameless atomic absorption analysis using the conditions set out in Table 6.4, by comparison with aqueous standards in 1% nitric acid. Accurate analysis was then performed after addition of varying quantities of metal standards and dilution to a standard volume according to the method of standard additions. Aliquots (25 µL) were then injected into the flameless atomiser with an Excel micropipette and analysed according to the programmes in Table 6.4.

Borosilicate glass is a known source of metallic contamination, and the stainless steel generator (stator–rotor) of the Ultra–Turrax was also expected to be a source of metals. Hence, a blank solution of 1% nitric acid in distilled water (the same acidity as acidified sludge samples) was homogenised with the Ultra–Turrax in the normal manner.

Acid digestion procedure

The sulphuric acid–nitric acid digestion procedure [19] was used. The apparatus recommended for the digestion of sewages was used. In all other respects the procedure was carried out was recommended. 'Analar' grade reagents were used throughout. All but the most difficult sludges are satisfactorily homogenised and dispersed after 5 min at 8000 rpm. Sludges which are not readily dispersed by this treatment may require preliminary treatment with a coarser homogeniser, for example T30/2G (Scientific Instrument Co, London). Even so, the combined treatments need not exceed 10 min at 8000 rpm.

Bomb digestion procedure

A 25 ml capacity bomb (Uniseal Decomposition Vessels Ltd, Haifa, Israel) was used throughout. Prior to each digestion the 'teflon' cup was leached

Table 6.5 Results of analysis of samples from a sewage treatment works

| Sample | Suspended solids | Metal (μg mL $^{-1}$) wet volume | | |
	(g/l)	Cd	Cu	Pb
Mixed primary sludge	41.40	0.30	6.90	3.90
Primary sludge	45.02	0.32	15.80	15.75
Digested sludge	21.30	1.80	16.00	25.00
Activated sludge	8.20	0.06	3.93	3.75

Source: Elsevier Science Publishers BV, Netherlands [26]

in 1% 'Aristar' nitric acid and rinsed in distilled water. To each 2 mL sludge sample (total solids 3.25%), 2.5 mL of 'Aristar' nitric acid (SG 1.42) were added. Digestion was undertaken in a 130°C oven for 4 h, after which 2 mL of the supernatant were transferred to a volumetric flask and made up to 10 mL with distilled water. This procedure is complete within 1 h.

Nitric–sulphuric digestion procedure
The sulphuric–nitric acid digestion procedure [19] was used. Rather than the 50 mL beaker recommended for sludges, which was prone to contamination, the apparatus recommended for the digestion of sewages was used, after the substitution of a 500 mL flask for the original 250 mL one. In all other respects, the procedure was carried out as recommended. 'Analar' grade reagents were used throughout. This procedure requires 6 h for completion.

All analyses were performed using a Perkin–Elmer Model 305 double beam atomic absorption spectrophotometer fitted with a deuterium background corrector. For flameless analyses, this instrument was used in conjunction with an HGA 72 flameless atomiser. Samples were shaken vigorously at all stages of preparation, which was undertaken in borosilicate glassware leached in 50% 'Analar' nitric acid and rinsed in distilled water.

Analysis was performed after the addition of varying quantities of metal standards and dilution to a standard volume according to the method of standard additions. Aliquots (25 μL) were then injected into the flameless atomiser with a micropipette and analysed according to the programmes in Table 6.4.

Some typical results from cadmium, copper and lead obtained on two sewage sludges, one with a low and one with a high input of industrial effluent are tabulated in Table 6.5.

No difficulties were encountered with any of the differing types of sludges analysed (mixed primary, primary, digested and activated). The repeatability of the analysis was high. The coefficient of variation for 12

Table 6.6 Metal blanks (µg mL^{-1})

	Cd	Cu	Pb	Cr	Ni	Zn
1% HNO$_3$ sstood in borosilicate glass vessel	–	–	–	0.001	–	0.0017
1% HNO$_3$ in borosilicate glass – Ultra Turrax (stainless steel)	–	0.009	0.002	0.275	0.135	0.020
1% HNO$_3$ in borosilicate glass – Ultra Turrax (titanium)	0.0001	0.0015	0.0008	0.0002	0.004	0.009

Source: Elsevier Science Publishers BV, Netherlands [26]

consecutive injections of a sample analysed for cadmium was 8% and other metals gave a result of the same order.

(ii) Chromium, nickel and zinc [27]

Homogenisation
Samples were collected and diluted in 1% 'Analar' nitric acid as discussed above [26]. The 50-fold diluted sludge was homogenised with an Ultra–Turrax Model T45N (Scientific Instrument Co, London). However the usual stainless steel shaft and rotor were replaced with a replica manufacture from 99.7% pure titanium. This substitution was necessary due to the excessively high blank values for both nickel and chromium obtained from the stainless steel Ultra Turrax shaft.

The blank values for the various metals due to borosilicate glass, 'Analar' nitric acid and the distilled water can be seen in Table 6.6. Only trace contamination of zinc and chromium resulted from these sources. The blank values for the titanium shaft (Table 6.6) clearly indicated the suitability of this titanium shaft for the preparation of samples for the analysis of six elements.

The results presented in Table 6.7 indicate good agreement between the mean values of all three methods. It is also apparent that in no case does the flameless procedure give the largest spread of results as indicated by the ranges.

The heavy metal content of composite sludge samples from two sewage treatment works appear in Table 6.8. These examples are fairly typical of the results obtained from these works, one of which had a low and the other a high input of industrial effluent.

Table 6.7 Comparison of the digestion/flame and flameless/homogenisation techniques of sewage sludge analysis (μg mL^{-1} wet volume)

Metal	Flame AA				Flameless AA	
	H$_2$SO$_4$/HNO$_3$		Teflon bomb		Homogenisation	
	Mean	Range	Mean	Range	Mean	Range
Ni	2.3	2.2–2.6	2.3	2.1–2.3	2.3	2.2–2.5
Cr	1.5	1.5–1.6	1.5	1.4–1.5	1.4	1.4–1.5
Zn	18	18–18	21	20–22	19	18–19

Source: Elsevier Science Publishers BV, Netherlands [26]

Table 6.8 Results of analysis of samples from a sewage treatment works

Sample	Suspended solids	Metal (μg mL^{-1}) wet volume		
	(g/l)	Cd	Cu	Pb
Mixed primary sludge[a]	39.20	1.7	1.7	19
Primary sludge[a]	41.30	6.0	3.2	120
Digested sludge[b]	23.24	1.6	1.5	24
Activated sludge[b]	8.15	0.5	0.4	34

[a] High industrial input
[b] Low industrial input

Source: Elsevier Science Publishers BV, Netherlands [26]

Mitchell et al. [29] describe a simple rapid method for the determination of cadmium, copper and lead in sewage sludge using a microsampling cup technique with a nitrous oxide acetylene flame. Detection limits, linearity and precision data are tabulated, and the results obtained by two standardisation procedures show no significant difference. The method is simple and rapid. Sewage sludge samples are thoroughly mixed by ultrasonic treatment, diluted as necessary and pipetted into microsampling cups. Samples are dried at 105°C for 20 min and injected into a nitrous oxide–acetylene flame.

Instrumentation and equipment
The microsampling cup system has been described by Kahl et al. [30]. Basically molybdenum cups are injected into a position about 1 mm below the entrance hole of a silicon carbide absorption tube. A nitrous oxide–acetylene flame, burning at the 0.38 × 57 mm slot of a specially designed 'safe' burner, heats the cup and absorption tube. The sample is volatilised into the tube and atomised, and an absorption sample is obtained by measuring the fraction of resonance radiation absorbed as it

Table 6.9 Detection limit, linearity and precision data

Metal	Wavelength (nm)	Spectral band-ass (nm)	Linear range µg L^{-1}	Detection limit (ng)[b]	s$_r$ (%)[a]
Cd	228.8	0.33	0–25	0.02	7.3
Cu	324.7	0.17	0–250	0.2	3.5
Pb	283.3	0.33	0–250	0.2	2.5

[a] Relative standard deviation for 12 replicate measurements of aqueous solutions giving a peak absorbance of ca 0.075
[b] Weight of analyte which can be determined in aqueous solution with a relative standard deviation of 50%

Source: Elsevier Science Publishers [29]

passes through the tube. Hot molybdenum cups are protected from aerial oxidation by withdrawing them from the flame into a sheath tube supplied with nitrogen. Measurements were carried out using either a Model AA5 (Varian–Techtron, Palo Alto, California) or a Model 151 (Instrumentation Laboratory, Lexington, Mass) atomic absorption spectrometer. In both cases, absorption signals were displayed on a strip chart recorder and peak absorbances were measured from the recorder training.

Eppendorf pipettes were used to measure 100 µL aliquots of sample and standard solutions.

A Model C 1000 Sonoverter (Insonator, Ultrasonic Systems Inc, Farmingdale, New York) with a 3 mm tip was used to treat the sludge samples before analysis.

Reagents
Aqueous standards were prepared daily from stock solutions (500 mgL^{-1}) of lead (as lead nitrate), cadmium (as cadmium nitrate) and copper (as copper(II) nitrate) respectively, prepared from analytical reagent–grade salts.

Sample treatment
Sludge samples were mixed by ultrasonic treatment for 60 s at 50 W to ensure a homogeneous mixture. The mixed sludge (10 g) was diluted with deionised water as necessary (usually 1 + 99). Aliquots (100 µL) were pipetted into cups, dried at 105°C for 10 min and analysed. The aqueous standards were in the range 0–250 µg L^{-1} for lead and cadmium, and 0–25 µg L^{-1} for cadmium. These samples were also analysed by the method of standard additions, with 0, 0.25, 0.5 and 0.75 ng added cadmium and 0, 2.5, 5.0 and 12.5 ng added lead and copper. Detection limits achieved by this method are tabulated in Table 6.9. The linear concentration ranges are

Table 6.10 Comparison of results obtained by the aqueous standard (Aq) and standard addition (Sa) methods for cadmium, copper and lead in sludges (Results are given as µg g⁻¹ of solids)

Sample	Solids present (%)	Cadmium		Copper		Lead	
		Sa	Aq	Sa	Aq	Sa	Aq
1	1.01	28.2	24.9	1740	1800	150	130
2	8.09	16.8	16.4	3420	3650	240	210
3	13.8	7	7	440	410	10	10
4	1.6	14.2	12.9	840	900	110	100
5	1.20	35.8	35.2	830	760	60	60
6	1.47	8	8	1420	1270	180	170
7	0.81	<8ᵃ	<8	1250	1270	150	180
8	3.20	5	6	330	300	300	290
9	1.42	6	6	290	350	410	390
10	0.92	<9ᵃ	<9	280	270	230	230
11	0.46	15	14	870	930	130	130
12	3.26	7	6	380	390	270	260
13	4.47	9.4	8.1	490	510	260	280

ᵃ Minimum reportable concentrations changed slightly from sample to sample because of variations in sample dilution and day-to-day instrumental variations

Source: Elsevier Science Publisheres [29]

for a 100 µL sample and can be readily extended a further order of magnitude by varying the sample volume from 20 to 200 µL.

The microsampling cup procedure is exceedingly rapid and yields the same results whether aqueous standards or standard additions are used for calibration. There is no evidence that total metal levels are determined, but the results are higher than those obtained by a nitric acid digestion procedure. The major application for this procedure is a rapid (and possibly semi-quantitative) method for monitoring sludge. For this purpose, it is probably not essential to obtain accurate metal levels, particularly since there may be large sampling errors. Simple analytical technique and speed of analysis are more important.

In Table 6.10, are listed results obtained on a range of sewage samples using the aqueous standard (Aq) and the standard addition calibration methods. No significant difference occurs in the results obtained by the two calibration procedures.

Rees and Hilton [31] have described a rapid method for the determination of zinc, copper, cadmium, chromium, lead and nickel in sewage sludges. Dried sludge is heated for 2 h on a water bath with concentrated hydrochloric acid and 30% hydrogen peroxide, filtered into lanthanum chloride (18% solution) and analysed by atomic absorption spectrophotometry. The results do not differ significantly from those of previously used methods, and the time is considerably reduced.

These workers showed that the assumption that organic matter must be destroyed before heavy metal determination can be carried out is invalid. Rees and Hilton [31] compared their results with those obtained in other methods in which organic matter is removed and demonstrated organic matter removal is, indeed, unnecessary.

In their method Rees and Hilton [31] dried the sample overnight at 105°C then mixed and ground the sample with a mortar and redried it for 1 h at 105°C. A suitable weight of dried sludge is heated on a water bath for 2 h with 5 mL concentrated hydrochloric acid and 1 mL 30% hydrogen peroxide (to remove sulphide). This solution is filtered into a 50 mL flask containing 5 mL 18% lanthanum chloride and made up to the mark. Elements are then determined by atomic absorption spectrometry. Results obtained by this procedure were not significantly different from those obtained by lengthy oxidative treatments with mixtures of nitric and hydrochloric acids.

Corrondo et al. [32] studied the effect of inorganic conditioning agents used the treatment of sewage sludge (viz lime, ferric chloride and aluminium chlorohydrate and of cationic polyelectrolytes) on the determination of cadmium, chromium, copper, nickel, lead and zinc in sewage sludge. These agents are often used to aid the de-watering of sewage sludges prior to disposal to agricultural land and might interfere in the electrothermal atomic absorption spectrophotometric analysis of heavy metals.

Corrondo et al. [32] analysed for heavy metals by flame and electro-thermal atomisation methods in conditioned and unconditioned samples. The organic polyelectrolytes tested did not interfere, nor did most inorganic conditioners at rates of addition consistent with normal sewage treatment practice. However, interferences occurred with aluminium chlorohydrate at normal and very high addition rates, and with other inorganic conditioners at very high addition levels (ie 50% above levels used in normal practice).

Flame atomic–absorption analysis was undertaken using a Perkin–Elmer Model 603, atomic absorption spectrophotometer equipped with deuterium background correction. The conditions for analysis were those recommended by the instrument manufacturer, with the exception of those for chromium for which slit 3 (0.2nm) was used to reduce spectral interferences.

Electrothermal analyses were undertaken using the same spectro-photometer in conjunction with a Perkin–Elmer HGA 76 heated graphite atomiser. The conditions and working ranges for electrothermal atomic absorption analysis are presented in Table 6.11. Aliquots of 20 or 50 μL were injected into the electrothermal atomiser with an Eppendorf micropipette. Analysis was performed by direct comparison with standards in 1% V/V nitric acid.

Table 6.11 Operating conditions for the atomic-absorption spectrophotometer and electrothermal atomiser

Conditions	Metal					
	Cd	Cr	Cu	Ni	Pb	Zn
Wavelength (nm)	228.8	357.9	324.8	232.0	283.3	307.6
Slit width (nm)	0.7	0.2	0.7	0.2	0.7	0.7
Drying						
temperature (°C)	100	100	100	100	100	100
Drying time(s)	30	30	30	30	30	30
Ashing						
temperature (°C)	250	1100	700	800	350	450
Ashing time(s)	40	30	30	30	40	35
Atomisation						
temperature (°C)	2100	2770	2770	2770	2300	2500
Atomisation time(s)	5	5	4	5	5	5*
Working range using						
20 µL/mg L^{-1}	0.005–0.03	0.02–0.2	0.05–0.4	0.10–1.0	0.025–0.4	0.02–2.0
Working range using						
50 µL/mg L^{-1}	0.002–0.02	0.01–0.1	0.02–0.2	0.05–0.4	0.01–0.2	0.05–1.0

Source: Royal Society of Chemistry [32]

Two sample pretreatment procedures were studied, homogenisation and nitric acid–hydrogen peroxide digestion.

Homogenisation
Approximately 250 mL of the sample, previously diluted 50-fold and acidified to 1% V/V with nitric acid, was homogenised in a 2 L tall-form borosilicate beaker with an Ultra Turrax T45N homogeniser for 5 min at 8000 rev/min^{-1}. To avoid contamination of the sample by chromium and nickel, the original stainless steel shaft, stator and rotor of the homogeniser were replaced with a replicate made of titanium. The suitability of this shaft for the homogenisation of samples to be analysed for cadmium, chromium, copper, nickel, lead and zinc is discussed above [27].

Nitric acid–hydrogen peroxide digestion [33]
A sample of 20 mL was digested on a thermostatic hot-plate with 30 mL of nitric acid until the volume was reduced to 10 mL. After cooling, 2 mL each of nitric acid and hydrogen peroxide were added repeatedly until the digestate turned a pale straw colour. For each of the pre-treatments (digestion and homogenisation of unconditioned and conditioned sludge) replicates and blanks were taken. The digested samples were analysed by flame atomic absorption spectrophotometry and the homogenised samples by electrothermal atomic absorption spectrophotometry.

Some typical results obtained by these procedures are quoted in Table 6.12 which illustrates that there are highly significant effects for all inorganic conditioners except zinc for which no significant differences were found. The results for cadmium, chromium and nickel in the presence of aluminium chloro–hydrate, at both rates of application, are statistically different from the others, in all instances lower results being obtained.

Thompson and Wagstaff [34] have described a simplified method for the determination of cadmium, chromium, nickel, lead and zinc in sewage sludge using atomic absorption spectrometry. These workers recommend a simple nitric acid digestion in calibrated glass tubes as an accurate safe and rapid technique for the routine atomic absorption determination of metals in sewage sludges. The technique can be used with aqua-regia, nitric acid/hydrogen peroxide and nitric acid/perchloric acid and involves gentle refluxing followed by evaporation to dryness. It requires less apparatus and bench space than the conventional beaker technique. Both wet and dried sludges can be digested and the nitric acid insoluble fractions were considered to be insignificant.

Thompson and Wagstaff [34] point out that the availability of metals from sewage sludge is a contentious subject [20, 35–39] and they thought it prudent to assume that all the metals not firmly bound in a siliceous matrix could ultimately become available for uptake by crops. In general, they expected that digestions with aqua regia, nitric acid–hydrogen peroxide or nitric acid–perchloric acid would release all metals from the sewage except those bound to the siliceous matrix while digestion with nitric acid hydrofluoric acid would release all metals including those bound in the siliceous matrix. In all, Thompson and Wagstaff [34] studied 13 methods of sludge digestion using the apparatus depicted in Fig. 6.1.

A Varian AA6 with automatic background correction and a Varian 1200 were used. Both instruments were equipped with the standard wide slot air-acetylene burner heads fitted with titanium burner grids.

The instrumental conditions used are listed in Table 6.13. Background correction was employed for cadmium, nickel and lead; the background absorption signals observed for chromium, copper and zinc were not considered significant. The typical range of background absorption signals observed expressed as milligrams per kilogram in the dried sludge are indicated in Table 6.13.

When necessary, the burner was rotated to increase the calibration range of the method; this minimised the number of dilutions required. This can result in a decrease in precision but was considered acceptable for routine sewage sludge analysis.

In the luminous air–acetylene flame chromium is especially prone to inter-element effects and the calibration graph has been reported to exhibit regions of negative slope [7, 40]. In order to minimise inter-element effects and obtain a virtually linear calibration graph, the burner

Table 6.12 Effect of different conditioners on the determination of cadmium, chromium, copper, nickel, lead and zinc in sewage sludge by electrothermal atomic-absorption spectrophotometry

Metal	Pre-treatment*	F–test level of significance	Mean concentration†/mg L^{-1}	RSD,‡ %
Cd	H$_2$O$_2$–HNO$_3$	0.01	0.26ᵃ	4.3
	Honog		0.25ᵃ	5.5
	CaO ± FeSO$_4$ (1)		0.26ᵃ	10.2
	CaO ± FeSO$_4$ (2)		0.24ᵃ	3.8
	CaO ± FeCl$_3$ (1)		0.24ᵃ	2.7
	CaO ± FeCl$_3$ (2)		0.24ᵃ	4.9
	AlCl$_3$ (1)		0.20ᵇ	6.5
	AlCl$_3$ (2)		0.18ᵇ	5.5
Cr	H$_2$O$_2$–HNO$_3$	0.01	3.4ᵃ	6.3
	Honog		3.2ᵃᶜ	5.9
	CaO ± FeSO$_4$ (1)		3.2ᵃᶜ	4.7
	CaO ± FeSO$_4$ (2)		3.4ᵃ	4.6
	CaO ± FeCl$_3$ (1)		3.2ᵃᶜ	3.5
	CaO ± FeCl$_3$ (2)		3.3ᵃ	5.1
	AlCl$_3$ (1)		2.9ᵇᶜ	4.0
	AlCl$_3$ (2)		2.6ᵇ	2.4
Cu	H$_2$O$_2$–HNO$_3$	0.01	16.9ᵃ	3.8
	Honog		17.2ᵃ	5.0
	CaO ± FeSO$_4$ (1)		17.5ᵃ	5.6
	CaO ± FeSO$_4$ (2)		21.0ᵇ	9.7
	CaO ± FeCl$_3$ (1)		17.7ᵃᶜ	5.9
	CaO ± FeCl$_3$ (2)		16.7ᵃ	7.6
	AlCl$_3$ (1)		12.4ᵈ	5.6
	AlCl$_3$ (2)		20.3ᵇᶜ	20.7
Ni	H$_2$O$_2$–HNO$_3$	0.01	6.5ᵃ	7.8
	Honog		6.2ᵃ	6.7
	CaO ± FeSO$_4$ (1)		6.4ᵃ	9.5
	CaO ± FeSO$_4$ (2)		6.5ᵃ	11.0
	CaO ± FeCl$_3$ (1)		6.2ᵃ	5.3
	CaO ± FeCl$_3$ (2)		5.7ᵃᵇ	9.4
	AlCl$_3$ (1)		5.1ᵇ	15.6
	AlCl$_3$ (2)		3.1ᶜ	16.7
Pb	H$_2$O$_2$–HNO$_3$	0.01	40ᵃ	7.8
	Honog		43ᵃ	5.9
	CaO ± FeSO$_4$ (1)		46ᵃ	8.4
	CaO ± FeSO$_4$ (2)		57ᵇ	9.4
	CaO ± FeCl$_3$ (1)		41ᵃ	5.9
	CaO ± FeCl$_3$ (2)		58ᵇ	9.1
	AlCl$_3$ (1)		42ᵃ	7.3
	AlCl$_3$ (2)		42ᵃ	6.7

Table 6.12 continued

Metal	Pre-treatment*	F–test level of significance	Mean concentration†/mg L^{-1}	RSD,‡ %
Zn	H_2O_2–HNO_3	0.01	35ᵃ	4.8
	Honog		34ᵃ	5.1
	CaO ± FeSO$_4$ (1)		31ᵃ	5.2
	CaO ± FeSO$_4$ (2)		33ᵃ	5.8
	CaO ± FeCl$_3$ (1)		33ᵃ	5.7
	CaO ± FeCl$_3$ (2)		35ᵃ	6.1
	AlCl$_3$ (1)		32ᵃ	6.1
	AlCl$_3$ (2)		34ᵃ	7.2

* H_2O_3–$HNO3$= hydrogen peroxide–nitric acid digestion; Homog = pre-treatment by homogenisation; CaO = lime; FeSO$_4$ = copper as; FeCl$_3$ = iron(III) chloride; AlCl$_3$ = aluminium chlorohydriage; (1) = rate 1; (2) = rate 2
† Means not followed by a common letter are statistically different at the 0.05 significance level
‡ RSD = relative standard deviation
§ NS = not significant at the 0.05 significance level

Source: Royal Society of Chemistry [32]

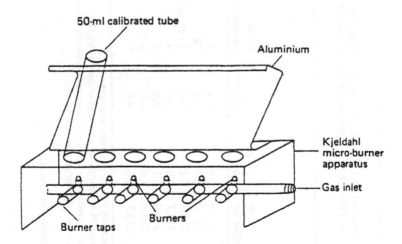

Fig. 6.1 Simplified digestion apparatus
Source: Royal Society of Chemistry [34]

Table 6.13 Instrumental operating conditions

The burner grid was set to 1 mm below grazing incidence for all metals except for chromium, for which a setting of 3.5 mm was used. Potassium (1000 μg mL^{-1}) was added to all samples, standards and blanks to suppress ionisation of chromium and lead when using the dinitrogen oxide–acetylene flame

Element	Wave-length nm	Spectral band pass/ nm	Flame*	Flame conditions†	Typical backbround absorption signal expressed as mg kg^{-1} of dried sludge	Upper concentration range of calibration graph with the burner parallel to the optical axis/mg kg^{-1}	Upper concentration of calibration graph with the burner rotated at 90° to the optical axis/ mg kg^{-1}
Cd	228.8	0.5	AA	S	1–2.5	200	2000
Cr	357.9	0.5	AA	FR	<5	1000	10000
Cr	357.9	0.2	NA	3 mm red feather‡	<5	2000	20000
Cu	324.7	0.5	AA	A	<5	750	7500
Ni	232.0	0.2	AA	S	5–12	1000	10000
Pb	217.0	1.0	AA	SFR	10–25	1000	10000
Zn	213.9	0.5	AA	S	<5	200	2000

*AA = air–acetylene; NA = dinitrigen oxide–acetylene

†S = stiocheiometric; FR = fuel-rich (on the verge of luminosity); SFR = slightly fuel-rich

‡Some measurements were also made in the dinitrogen oxide–acetylene flame for copper, lead and zinc. The burner grid was set at 3.55 mm below grazinb incidence and a flame with a 1 mm red feather was used

Source: Royal Society of Chemistry [34]

height was set 3.5 mm below grazing incidence and an air–acetylene flame on the verge of luminosity was used.

In this method a clean 10 mL wet sampling tube was vigorously shaken to remove any water and then rapidly immersed at an angle of approximately 45° to the vertical and adjusted to the vertical plane below the surface of the well shaken sludge. The tube was then immediately removed from the sludge and the outside was carefully washed down with de-ionised water from a washbottle into a sink. The contents of the tube were then carefully poured into a 50 mL calibrated glass tube and any remaining residues were washed from the sample tube into the 50 mL tube using a wash bottle with a very fine jet. The total washing volume was approximately 3–5 mL. Then the appropriate digestion reagents were added and the digestion was carried out. Digestion procedures are outlined below.

(i) Conventional digestion

One gram of the dried sludge was added to a 250 mL tall-form beaker wetted with 2 mL of de-ionised water and the appropriate digestion regents were added. The digestions were then carried out with a watch-glass covering each beaker. Finally, the contents of the beakers were quantitatively transferred into 100 mL flasks. Appropriate standards were carried through the digestion procedure.

(ii) Simplified digestion

An aliquot of the sludge (0.5 ± 0.002g of dried or 10 mL of wet) using the special sampling tube, was added to the 50 mL borosilicate calibrated glass tubes containing four or five anti-bumping granules. Dry samples were wetted with 1 mL of de-ionised water and then 6 mL of nitric acid (70% m/m) were carefully added from the automatic dispenser down the sides of each tube. The tubes were allowed to stand for about 1 min to allow any initial vigorous reaction to take place. A few drops of acid-washed kerosene were added to samples that exhibited excessive foaming, but this was rarely necessary. The liquid in the tubes was then heated using Kjeldahl multi-gas burner units. The apparatus is depicted in Fig. 6.1. The dried sludges were allowed to reflux gently for 15 min, whereas the liquid sludges were boiled at a fairly vigorous rate until the final volume was reduced to 7 ± 1 mL. When the digestion was complete the solutions were allowed to cool, diluted to 50 mL with de-ionised water, shaken and allowed to settle for at least 2 h or preferably overnight, prior to aspiration. Multi-element standards containing 10% V/V nitric acid were used for calibration.

The results in Table 6.14 show that the proposed simple nitric acid digestion gives comparable results to the nitric acid–hydrochloric acid,

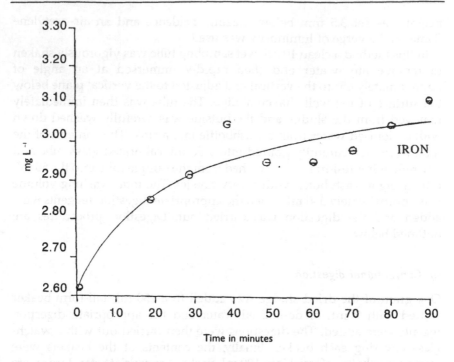

Fig. 6.2 Time required for complete microwave digestion of iron in sludges
Source: Own files

the nitric acid–hydrogen peroxide and the hydrochloric acid digestion methods in the analysis of digested sludge. The addition of hydrogen peroxide during the nitric acid digestion did not appear to be beneficial for sludge analysis. Table 6.15 demonstrates that other than for sludges with low (background) levels of nickel and chromium (ie sludges B, C and D) the nitric acid digestion results are not more than 10% below the nitric–perchloric acid digestion results. In fact, most of the results are within 5%. Table 6.16 shows that, except for chromium and nickel, the perchloric–nitric–hydrochloric acid digestions of eleven different sludges gave similar results to those obtained using the simple nitric acid digestion. Most of the copper and zinc results appeared to be slightly higher (3–5%) for the simple nitric acid digestion than the digestion involving hydrofluoric acid. This was attributed to small losses caused by foaming during the initial nitric acid digestion prior to the addition of the hydrofluoric acid and perchloric acids. This resulted in a small amount of the sludge creeping up the sides of the PTFE beakers. It was difficult to wash this back quantitatively into the digestion mixture. This was not considered to be a problem with the simple nitric acid digestion using the 50 mL Exelo calibrated tubes; in this instance digestion under reflux ensured that any residues on the tube walls were efficiently digested.

Table 6.14 Comparison of various digestion techniques

Sample size 0.5 g; sludge A2; digestion vessel 50 mL calibrated boiling-tube: final volume 50 mL; fame, air–acetylene. All analytical measurements were carried out on the same day

Digestion method	Parameter	Number of digestions	Mean concentration/mg kg^{-1}					
			Cadmium	Chromium	Copper	Nickel	Lead	Zinc
1	Mean	11	25.1	1114	3655	46.4	742	1705
	Standard deviation	11	0.44	12.1	23.0	3.78	4.83	16.9
2	Mean	3	24.8	1120	3647	56	740	1683
3	Mean	4	24.2	1098	3628	49	757	1672
4	Mean	4	25.0	1123	3690	53	777	1675

Method
1 Add 6 mL of nitric acid (70% m/m), simmer for 15 min
2 Add 4 mL of hydrochloric acid (36% m/m) and 4 mL of nitric acid (70% m/m), simmer for 15 min
3 Add 8 mL of nitric acid (70% m/m), simmer for 2 min over a 5-min period and carefully add 5 × 1 mL aliquots of hydrogen peroxide (50% m/m). Continue to simmer for 15 min
4 Add 8 mL of hydrochloric acid (36% m/m), simmer for 15 min

Source: Royal Society of Chemistry [34]

Table 6.15 Comparison of perchloric acid and simple nitric acid digestions

Sample size 0.5 g; digestion vessel 50 mL calibrated boiling-tube; final volume 50 mL; flame, air–acetylene with dinitrogen oxide–acetylene results in parentheses. All analytical measurements were carried out on the same day

Sludge designation	No of digestions	Mean concentration/mg kg^{-1}											
		Cadmium		Chromium		Copper		Nickel		Lead		Zinc	
		Nitric*	Per-chloric*	Nitric	Per-chloric	Nitric	Per-chloric	Nitric	Per-chloric	Nitric	Per-chloric	Nitric	Per-chloric
A Digested	†	24.6	25.2	—	—	3826	3751	51.6	52.7	741	760	1712	1722
B Digested	3	6.8	8.2	443 (495)	503 (487)	932	937	31.3	35	443	455	1810	1833
C Pressed	3	3.7	4.0	35.7 (36.8)	45 (48.7)	447	443	29.3	31.7	343	1010	1023	683
D Raw	6‡	5.1	4.75	28.7 (33.7)	40.3 (47)	450	435	32.3	34.5	671	655	685	683
E Digested	3	70	72	227 (265)	238 (268)	590	77	80	452	455	1263	1297	
F Digested	3	34.5	35	150 (178)	167 (187)	593	590	77	79.5	727	743	1137	1163

* Nitric = method 1; Perchloric = method 5
† The nitric acid results are for ten digestions, the perchloric acid for six digestions
‡ The chromium result is for only three digestions

Source: Royal Society of Chemistry [34]

Table 6.16 Comparison of standard digestion and nitric–perchloric–hydrofluoric acid digestion for dried samples

Sample size 0.5 g; digestion vessel 50 mL calibrated boiling-tube (nitric acid); final volume 50 mL; flame, air–acetylene with dinitrogen oxide–acetylene results in parentheses.

Sludge designation	No of nitric acid digestions*	Cadmium		Chromium		Copper		Nickel		Lead		Zinc		
		Hydro-fluoric†	Nitric†	Hydro-fluoric	Nitric	Hydro-fluoric	Nitric	Hydro-fluoric	Nitric	Hydro-fluoric	Nitric	Hydro-fluoric	Nitric	Per-Nitric
A Digested	15	23.0	24.5	(1225)	1180	3580 (3620)	3760	56	47	1110 (1095)	740	1690 (1630)		1750
B Digested	9	29	32.1	(990)	969	660 (665)	671	56	57	680 (640)	460	1840 (1775)		1903
C Digested	15	6.0	6.9	(495)	480 (495)	855 (890)				405 (410)		1665 (1660)		
D Pressed	15	4.2	4.0	(63)	34.7 (36.7)	400 (435)	437	42	28.5	323 (313)	322	945 (920)		979
E Raw	15	3.8	4.5	(70)	33.9 (33.7)	405 (415)	440	40	25.6	615 (612)	636	675 (645)		693
F Raw	15	7.6	8.4	(123)	89.6	698	735	46	32.5	537	588	925		939
Digested	9	194	206	(598)	576	510 (525)	550	657	738	315 (315)	327	1760 (1705)		1865
G Digested	9	74	76.8	(1395)	1378	1580 (1640)	1690	695	742	2305 (2335)	2334	7425 (7450)		7450
H Digested	9	9.7	10.2	(167)	163	260 (273)	278	65	54.4	318 (315)	317	835 (810)		862
I Digested	9	75.1	78.4	(2120)	2222	3255 (3200)	3355	1660	1657	840 (825)	828	5065		5250
J Digested	15	4.6	4.3	(65.5)	42.2	355 (345)	354	27.9	572	541 (570)	920	885		

* Hydrofluoric acid results are the mean of two digestions except for D2 (three digestions) and J (four digestions). † Hydrofluoric = method 6; Nitric = method 1.
Source: Royal Society of Chemistry [34]

Significantly higher results were observed with the hydrofluoric acid digestion for sludges that contained low background levels of nickel and chromium. This was thought to be caused by the release of these metals that had been firmly bound in the siliceous matrix. For the other four elements the contribution of the bound materials in the siliceous matrix was not considered significant for this type of analysis. In order to minimise matrix effects in the hydrofluoric acid digests, all chromium results were obtained using the dinitrogen oxide–acetylene flame. Multi-element standards carried through the procedure showed complete recovery (± 2%) for all of the metals except chromium. Some of the standards exhibited losses of chromium of up to 20%, while others exhibited no significant loss. The chromium results were calculated by ignoring these losses and it appears from the results, given in Table 6.16, for metal-contaminated sludges F and H that this was justified.

Thompson and Wagstaff [34] conclude that a simple nitric acid digestion in 50 mL calibrated glass tubes is an accurate safe and rapid dissolution technique for the routine atomic absorption determination of cadmium, chromium, copper, nickel, lead and zinc in typical sewage sludges. Wet or dried (105°C) sludge can be satisfactorily analysed by the proposed technique.

The calibrated tube technique exhibits significant advantages over the conventional beaker digestion technique with respect to the amount of volumetric apparatus required, the ease of digestion and the amount of bench space required. The technique can also be used successfully with other digestion reagents such as aqua regia, nitric acid–hydrogen peroxide and nitric acid–perchloric acid.

The cadmium, chromium, copper, nickel, lead and zinc in typical sludges appears to be efficiently extracted by refluxing with nitric acid. The nitric acid-insoluble fractions were not considered to be very significant when assessing the results in relation to the disposal of sewage sludge to agricultural land.

Jenniss et al. [41] used four procedures to prepare sewage sludge samples for the determination of cadmium and lead by atomic absorption spectrometry, namely muffle furnace ignition, digestion with nitric acid, high pressure digestion with nitric acid, and digestion with nitric acid and hydrogen peroxide. The results of analyses show increasing recoveries of the metals from primary settling tank sludge in progressing from samples prepared by muffle furnace ignition to nitric acid digestion to high pressure decomposition by nitric acid. Losses of metals greater than 10% occur with samples prepared by muffle furnace ignition or open digestion by nitric acid. The use of either high pressure nitric acid or nitric acid plus hydrogen peroxide digestion is recommended.

Katz et al. [42] prepared sludge samples from the primary settling tank and the first stage digester of a four-million-gallon-per-day sewage treatment plant for atomic absorption spectrometry by four methods: dry ashing, digestion with nitric acid, digestion with nitric acid and hydrogen peroxide, and digestion with nitric acid in pressure decomposition vessels. The prepared samples were analysed for cadmium, chromium, copper, iron, lead, nickel and zinc. These workers concluded that dry ashing is not an acceptable procedure to prepare sewage sludge samples for atomic absorption spectrometry. Digestion of such samples with nitric acid and hydrogen peroxide was concluded to be an efficient convenient and rapid procedure for such samples.

In this procedure 200 mg samples of each dried sludge was weighed into 125 mL Erlenmeyer flasks. The flasks containing the sludge samples and empty flasks serving as blanks were treated with 5 mL of redistilled nitric acid and 5 mL of high purity water. The flasks were heated gently until most of the acid was evaporated. After cooling, 5 mL more of redistilled nitric acid was added and the flasks were again heated gently until the volume was reduced to approximately 1 mL. The flasks were again allowed to cool, and they were treated with 1 mL of redistilled nitric acid, 1 mL of high purity water and 2 mL of 30% hydrogen peroxide. The flasks were alternately heated, cooled and treated with additional 1 mL increments of peroxide until no further changes were apparent in these samples. (This was accomplished with four additional 1 mL increments of peroxide). The contents of the flasks were treated with 5 mL of high purity water, filtered through Whatman No 42 paper into 50 mL volumetric flasks and brought to volume with high purity water. These solutions were diluted further for measurements by the metals by atomic absorption spectrometry.

In the nitric acid digestion procedure, 200 mg samples of the dried sludge was weighed into 120 mL Erlenmeyer flasks, treated with 5 mL of redistilled nitric acid and heated gently until the volume was reduced to approximately 1 mL. The flasks containing the samples as well as those included as blanks were cooled, treated with 5 mL of redistilled nitric acid and again heated gently. This process was repeated a total of four times to a point where additional acid produced no apparent changes in the samples. The contents of the flasks were treated with 5 mL of high purity water, filtered through Whatman No 42 paper into 50 mL volumetric flasks and brought to volume with high purity water. Further dilutions of these solutions were made as needed for the atomic absorption spectrometric measurements.

It will be noted in Table 6.17 that any ashing at 450°C (method 5) gives results which are perceptibly lower than those obtained by wet digestion with mineral acids. Indeed Katz et al. [42] comment that dry ashing appears to be the least efficient of the five sample procedures.

On the contrary, Ritter et al. [43] have found dry ashing to be superior to nitric acid and aqua regia digestion of sewage sludges and they conclude 'the dry ash digestion method is the best for preparing soil and sludge samples for atomic absorption analysis'. Hoenig et al. [44] have

Table 6.17 Levels of selected metals in primary settling tank sludge, mg/kg

	Method				
	1	2	3	4	5
Cd		76.7	74.1	64.7	61.0
		±1.7	±1.3	±2.2	±1.8
Cr	4440	19300	20200	21500	16400
	±140	±650	±865	±575	±1550
Cu	3190	3400	3400	3120	2580
	±210	±124	±115	±417	±92
Fe		15100	14800	13900	10600
		±2040	±353	±426	±656
Ni		67.0	75.9	90.6	42.5
		±8.5	±7.4	±16.7	±15.1
Pb		3410	3220	2970	2680
		±71	±73	±66	±124
Zn	8310	8370	8540	8200	7970
	±230	±316	±201	±186	±85

Method
1 Neutron activation analysis
2 Decomposition in nitric acid digestion bomb
3 Decomposition with nitric acid–hydrogen peroxide
4 Decomposition with nitric acid
5 Decomposition by dry ashing at 450°C by zirconium crucibles

Source: Gordon AC Breach, Nethrelands [42]

found that wet and dry ashing procedures can be applied successfully for the routine determination of trace elements in plant tissues. The US EPA [45] has considered a procedure for the determination of lead in sludges whereby the samples would be dry ashed for eight hours at 450° and then refluxed for 2 h with 50% (v/v) nitric acid prior to atomic absorption spectrometric measurement.

In further work Ritter [46] takes up the comment of Katz et al. [42] that there are 'conflicting reports on the efficiency of muffle furnace ignition compared to wet digestion and that lead and cadmium are lost from sewage sludge in dry ashing procedures'. Ritter [46] claims that, on the contrary, based on this earlier work [45] that the dry ashing method of preparing sewage sludge for atomic absorption spectrometry for lead, cadmium and other elements was better than wet ashing because precision is high (coefficient of variation ≤5%), the procedure is fast and values obtained from lead and cadmium on NBS standard 1571 orchard leaves are very close to those certified.

In his dry ashing procedure Ritter [46] ignited 1 g samples at 550°C in porcelain crucibles for 2.5 h, leached in 3N hydrochloric acid at 120°C for 2 h, filtered through a No 42 Whatman paper and diluted to 50 mL. In his

Table 6.18 Cadmium and lead levels (ppm) in Dayton, Ohio sewage sludge using two
preparation methods (12 samples prepared by each method)

	Dry-ash	High pressure vessel
Cadmium	338 ± 10.4	340 ± 7.7
Coefficient of variation	3.1%	2.3%
Lead	2391 ± 51	2612 ± 60
Coefficient of variation	2.1%	2.3%

Source: International Scientific Communications Inc, Shelton [46]

high pressure vessel procedure Ritter [46] placed 0.2 g of sludge in the
high pressure vessel with 2.5 mL of concentrated nitric acid and heated in
an 105°C oven overnight. The contents were filtered and diluted to
volume. The samples were analysed for cadmium and lead using an
acetylene–air flame in a Perkin–Elmer Model 603AA unit. The coefficient
of variation values obtained are given in Table 6.18.

Contrary to the results obtained by Jenniss et al. [41], Ritter [46]
concludes that no significant difference was found in the cadmium values
for the samples prepared by the two methods. On the other hand, there is
a significant difference in mean lead values. The values for samples
prepared by dry ashing are 8.5% lower than those digested in high
pressure vessels.

Moriyama et al. [47] have discussed the disadvantages of using atomic
absorption spectrometry for the determination of cadmium, lead, nickel
and copper in sewage sludge.

Legret et al. [48, 54] have discussed interferences by major elements in
the flameless atomic absorption spectrometric determination of lead,
copper, cadmium, chromium and nickel in sewage sludges. Lead,
cadmium and chromium were subject to most interferences.

Matrix interference in the determination of heavy metals in sludges by
electrothermal atomic absorption spectrometry was particularly severe
with regard to lead, cadmium and nickel determinations. Hydrofluoric
acid had to be eliminated by evaporation and perchloric acid was a
serious interferent. A technique for the reduction of chemical interference
in lead and nickel determinations was recommended, which consisted of
matrix modification by ammonium dihydrogen phosphate and ascorbic
acid. Rapid heating was recommended for cadmium determinations.

Nielsen and Hrudey [49] have described a rapid steam digestion
method for determining cadmium, chromium, copper, nickel and zinc in
sewage sludge. The method uses a domestic pressure cooker. The results
obtained for cadmium, chromium, copper, nickel and zinc are compared
with results obtained using open nitric acid digestion and high speed
homogenisation. All experimental results are within the reference range

except for the results for zinc. Details of the three digestion techniques are given below.

(i) Steam digestion

Sample for steam digestion were acidified (1% v/v) with concentrated nitric acid and then 20 mL aliquots were transferred to 50 mL Pyrex test tubes. The test tubes were covered with aluminium foil, placed upright in beakers in the pressure cooker and steam digested for 1 h at 210 k Pa (2 atm). The digested supernatant was decanted and analysed.

(ii) Open acid digestion

This was performed according to APHA [50] and USEPA [51] as follows:

Sample volumes of 20.0 mL were transferred to 100 mL Pyrex beakers, acidified with 3.0 mL of concentrated nitric acid and gently boiled to near dryness on a domestic electric hotplate. Further 3.0 mL volumes of concentrated nitric acid were added to the samples, and the beakers were covered with watch glasses and gently boiled with refluxing until the solution was light coloured and clear. Four 3.0 mL additions of acid were required for each sample, and the samples were boiled to near dryness after the last addition of concentrated acid. About 20 mL of dilute nitric acid (1% v/v) was then added to each sample and warmed. The samples were decanted into 50 mL volumetric flasks, the walls of the beakers were washed twice with 1% nitric acid and the rinsings were added to the volumetric flasks. The samples were cooled, made up to mark with 1% nitric acid and analysed.

(iii) High speed homogenisation

Samples to be digested were also acidified (1% v/v) with concentrated nitric acid. Samples (20 mL) were then transferred to Pyrex 50 mL test tubes and homogenised for 5 min at near maximum speed. The supernatant was decanted and analysed. Each type of sample was digested in triplicate by each method. Blank high purity water samples were concurrently treated and digested in the same manner as the sewage and sludge samples.

The operating conditions for the flameless atomic absorption metal determination are listed in Table 6.19. Pyrolytically coated graphite tubes were used, and triplicate injections to the furnace were made for all the samples. Where necessary, samples were diluted so that their absorbance values fell on the linear part of the respective calibration curves. Blanks and standards were analysed after about every third sample to correct for instrumental drift and graphite tube deterioration in the furnace.

The absence of matrix interferences was checked by the method of standard additions by using raw sewage and mixed liquor samples. Metal

Table 6.19 Operating conditions for the atomic absorption spectrophotometer and furnace

Conditions	Metal				
	Cd	*Cr*	*Cu*	*Ni*	*Zn*
Wavelength, nm	228.8	357.9	324.7	232.0	307.6
Slit width, nm	0.7(L)'	0.7(L)	0.7(L)	0.2(L)	0.7(L)
Integration time, s	5	10	5	5	5
Drying temp, °C	85	85	85	85	85
Drying time, s	40	40	40	40	40
Ashing time, °C	300	1100	650	800	300
Ashing time, s	35	35	40	35	35
Atomisation temp, °C	2100	2700	2700	2700	2400
Atomisation time, s	5	10	5	5	5

'(L) designates low slit height option

Source: American Chemical Society [49]

Table 6.20 Digestion results obtained by the steam, open HNO' and homogenisation methods with raw sewage, mixed liquor and primary sludge

Sample	Metal	Steam		Open HNO$_3$		Homogenisation	
Raw sewage	Cd	0.003	± 0.0004	0.0025	± 0.0002	0.0016	± 0.0003
	Cr	0.245	± 0.009	0.172	± 0.003	0.243	± 0.008
	Cu	0.090	± 0.911	0.081	± 0.021	0.069	± 0.010
	Ni	0.052	± 0.003	0.062	± 0.006	0.031	± 0.001
	Zn	0.31	± 0.01	0.29	± 0.01	0.30	± 0.03
Mixed liquor	Cd	0.0150	± 0.0001	0.0205	± 0.0030	0.0183	± 0.0010
	Cr	2.00	± 0.01	2.00	± 0.01	1.42	± 0.14
	Cu	0.375	± 0.025	0.433	± 0.194	0.167	± 0.029
	Ni	0.217	± 0.028	0.213	± 0.029	0.195	± 0.001
	Zn	1.4	± 0.2	2.3	± 0.2	1.6	± 0.03
Prkmary sludge	Cd	0.116	± 0.007	0.129	± 0.007	0.099	± 0.001
	Cr	12.7	± 0.4	10.5	± 0.3	10.4	± 0.17
	Cu	4.72	± 0.03	4.62	± 0.15	4.18	± 0.11
	Ni	1.50	± 0.03	1.20	± 0.01	1.40	± 0.03
	Zn	12.0	± 0.1	10.4	± 0.1	10.4	± 0.1

' Concentration presented as mean ± 1 standard deviation based on triplicate injections of each of three samples

Source: American Chemical Society [49]

recoveries from these samples were not significantly different ($p < 0.05$) from those using standard metal solutions prepared with 1% (v/v) nitric acid.

Table 6.20 shows total metal concentrations determined in raw sewage, mixed liquor and primary sludge samples from the City of Edmonton

Table 6.21 Metal concentrations determined in the US EPA reference municipal sludge after steam digestion

Metal	Measured concentration,[a] mg/kg	Reference concentration, mg/kg Mean	Range[b]	No. outside reference range
Cd	16.2, 17.9, 24.1	20.77	2.49 – 39.1	0
Cr	157, 145, 152	204.5	115 – 294	0
Cu	847, 905, 853	1095.3	831 – 1360	0
Ni	196, 209, 209	198.3	164 – 233	0
Zn	1165, 1149, 1251	1323.1	1190 – 1450	2

[a] Each value is the mean of three injections to the graphite furnace.
[b] Each range is mean ± (t0.95 × standard deviation)

Source: American Chemical Society [49]

Wastewater Treatment Plant which were digested by the steam, open nitric acid and homogenisation methods. Metal recoveries from samples which had been steam digested were equivalent or better than those obtained with either of the other digestion methods for all but one of the 45 cases. Only zinc recovery from mixed liquor was significantly ($p<0.05$) lower for steam digestion than for open nitric acid digestion. Furthermore, the precision (ie standard deviation) of the steam digestion method was generally as good as for either of the other two digestion methods.

The steam digestion method was further evaluated by analysing, in triplicate, a sample of reference municipal sludge from the US Environmental Protection Agency by this method. The results are shown in Table 6.21 together with the total metal concentration range supplied for the reference sludge. All experimental results were within the reference ranges except for the three zinc results.

A standard UK method has been described for the determination of cadmium, chromium, copper, lead, nickel and zinc in sewage sludge [52, 53].

The confused situation regarding condition for the digestion of sewage sludge samples prior to the determination of heavy metals by atomic absorption spectrometry is illustrated by a review of the literature presented by Christiansen *et al.* [58]. These workers state that based on a Nordic (26 laboratories) and a European (15 laboratories) interlaboratory comparison of metal determination in sewage sludge samples the most common digestion methods showed no significant differences in metal concentration levels [59, 60]. Muntau and Leschber state in their summary [61] of a recent European inter-laboratory comparison involving 39 laboratories, that aqua regia is suitable for digestion of sewage sludge samples. However, statistical tests (approximate *t*–test) of the observed average metal concentrations after exclusion of extreme observations support no preference of aqua regia to nitric acid. Both methods yielded higher metal concentrations of lead, nickel, copper and chromium – but not of cobalt, copper and zinc – than obtained with dry ashing of the sludge samples.

Ritter and co-workers [43] digested three different sludge samples by means of dry ashing, nitric + perchloric acid, nitric acid, aqua regia and hydrofluoric acid. They found that dry ashing, nitric + perchloric acid, and hydrofluoric acid in general resulted in the same metal concentrations, while nitric acid and aqua regia resulted in lower concentrations of the investigated metals (cadmium, lead, nickel, copper and zinc). Dry ashing was recommended due to speed of digestion and the high precision. Thompson and Wagstaff [34] investigated several digestion methods, including minor variants of the above mentioned five methods, with respect to cadmium, lead, nickel, copper and zinc. Dry ashing was abandoned because of low values for copper and methods involving sulphuric acid because of low values for lead. No major differences were found for the other methods: nitric acid, nitric acid + hydrogen peroxide, aqua regia, hydrochloric acid, nitric acid + perchloric acid, and hydrofluoric acid (no statistical test accomplished). The nitric acid digestion was recommended as an accurate safe and rapid digestion for sludge. The results of Delfino and Enderson [62] on comparison of five methods for sludge sample digestion are difficult to evaluate because of substantial uncertainty. However, dry ashing was considered to be unreliable with respect to chromium, nickel, magnesium and iron. Van Loon and co-workers [22, 63] found that aqua regia and perchloric acid resulted in the same metal concentrations of cadmium, chromium, lead, zinc and copper, but that perchloric acid resulted in higher concentrations of nickel and iron than aqua regia did. Andersson [64] found that dry ashing the residuals from nitric acid extraction of sludge increased the observed metal concentrations 14–19% for lead, cobalt, cadmium, nickel and chromium, but only 3–8% for copper, manganese and zinc. Lester and co-workers [65–67] found that dry ashing of sludge at 450°C resulted in lower concentrations of cadmium, chromium and copper compared to digestions with nitric acid + sulphuric acid or with nitric acid + hydrogen peroxide. No differences were observed for nickel and zinc, while the nitric acid + sulphuric acid digestion gave very low results for lead.

As indicated by the above summarised investigations, disagreement exists as to which method should be used for digestion of sludge samples prior to metal analysis by atomic absorption spectrophotometry. The apparent contradictory recommendations found in the literature are difficult to explain, although minor differences in equipment, concentrations of reagents, temperatures, reaction times, and biased habits may account for some of the discrepancies.

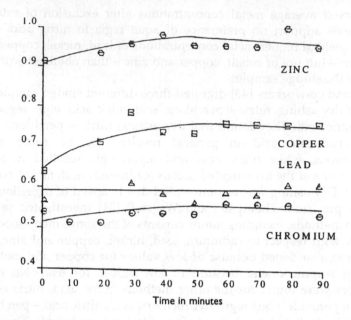

Fig 6.3 Time required for completion of microwave digestion of zinc, copper, lead and chromium in sludges
Source: Own files

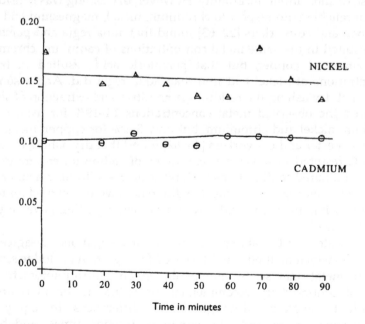

Fig. 6.4 Time required for complete microwave digestion of nickel and cadmium in sludges
Source: Own files

Table 6.22 Analysis of EPA standard sludge sample 5035 by microwave

Results in mg/kg(1)

Parameter	Microwave	V-microwave	EPA mean
Cd	20	19	19.1
Cr	185	193	193
Cu	1085	1086	1080
Fe	16745	15484	16500
Mn	200	200	202
Ni	186	184	19
Pb	578	629	5.6
Zn	1330	1373	1320

(1) Analyses performed by flame atomic absorption

Source: Own files

Davis and Carlton-Smith [75] of the Water Research Centre, UK, have organised an inter-laboratory comparison of results obtained in analyses of the full range of heavy metals in sewage sludges using the wide variety of sample digestion techniques employed by the 84 participating laboratories. The purpose of this exercise was to review the adequacy in terms of precision and bias of current methods of analysis as a preliminary stage in the development of recommended methods of analysis. The consensus of opinion now seems to be that digestion with nitric acid or mixtures of nitric acid and hydrogen peroxide or dry ashing at 450 to 550°C in zirconium crucibles are, overall, the best methods for the decomposition of sludge samples prior to the determination of heavy metals by atomic absorption spectrometry.

In recent years a further method has been evolved which involves sealed bomb digestion with acids in a microwave oven [55–57]. This procedure is claimed to have the advantages of reduced sample dissolution time (10–60 min compared to 4 to 48 h for open tube acid digestions), more complete digestion, lower blanks, higher sensitivity, and elimination of the need to use perchloric acid. Figures 6.2 to 6.4 show the times required to fully digest each of the heavy metals. It is seen that for most metals 60 min is adequate, while for iron 160 min suffices.

The good agreement in results obtained by this procedure on the EPA reference principal sludge sample 5035 are shown in Table 6.22.

The results obtained by microwave digestion on a sample of mixed sewage compared to those obtained by a conventional open tube wet acid digestion procedure are compared in Table 6.23 [57].

In this procedure [56] a Teflon lined 600 W microwave oven Model MDS81–D supplied by CEM Corporation [59] was used. A plastic vacuum desiccator was used to contain the samples and fumes. The knob on the

Table 6.23 Analysis of mixed sewage and industrial wastewaters

Results in mg/L(1)

| Parameter | Microwave | V-microwave | Wet ash(3) | Precision mg kg $^{-1}$ | |
				Microwave	V-microwave
Cd	0.02	0.02	0.02	0.6	1.0
Cr	0.53	0.51	0.56	7	2
Cu	0.46	0.43	0.47	31	10
Fe(2)	0.95	0.99	1.01	1433	201
Ni	0.09	0.11	0.09	1	2
Mn	0.08	0.08	0.09	3	1
Pb	0.09	0.12	0.10	39	80
Zn	1.45	1.40	1.50	42	23

(1) Analyses performed by flame atomic absorption
(2) Analyses performed on separate sample due to contamination
(3) The wet ash digestion procedure used 100 mL liquid samples and approximately 1 g samples of reference sludge. The liquid samples were 90 mL of D.I. water was added to each solid sample. The samples were placed in erlenmeyer flasks. 3 mL of nitric acid was added to each sample. The flasks were placed on a hot plate in a fume hood and coverd with a watch glass. The samples were slowly concentrated to 5 mL. 1 mL of hydrochloric acid was added and the samples refluxed for an additional five or 10 min. The samples were then filtered and brought to a volume of 100 mL with deionised water. This procedure usually reuiqred 3–5 h to complete.

Source: Own files

desiccator lid was removed and a flat piece of plastic cemented in its place. This allowed it to fit into the oven. The metal support plate that came with the desiccator was removed and a porcelain plate from a glass desiccator was installed. A square piece of plywood was fitted to the bottom of the oven to prevent the desiccator from interfering with the existing rotator.

The CEM Corporation provided 120 mL teflon digestion vessels which can withstand pressures up to 100 psi. They incorporate a pressure relief value for safety. An electric capping station tightens the vessels' caps and pressure relief valve to a repeatable torque. It also is used to open the vessels after digestion.

A Perkin–Elmer Model 5000 atomic absorption spectrophotometer was used to analyse the samples for metals concentrations. All analyses were performed using an air–acetylene flame. Data was collected by a model 7300 Perkin–Elmer computer.

Deionised water from an ion exchange system is used throughout. Sample 5035 of municipal digested sludge was obtained from the USEPA in Cincinnati. This material is ground, dried and ready for digestion as received. The EPA provides reference values and the 95% confidence level with the documentation. Samples that were digested by the desiccator microwave method were placed in 125 mL Erlenmeyer flasks with plastic screw caps. Fifty mL liquid samples and 1 g sludge samples were used. One mL of nitric acid and 1 mL of hydrochloric acid was added to each sample. The screw caps were tightened, then loosened by a quarter turn.

The flasks were placed in the vacuum desiccator. A vacuum line was attached and vacuum applied for 10–30 s. The desiccator was placed into the microwave oven. The oven was programmed as follows:

Power	Time
65%	3 min
10%	2 min
45%	3 min

This programme was run six times to complete the digestion. The low power setting in the middle step was used to prevent sample boil–over and loss. After digestion, the samples were filtered through Whatman No 4 paper and diluted to 50 mL. The microwave digestion methods produced recoveries from the municipal sludge sample that were well within the 95% confidence levels set by the EPA. The results were generally close to the mean values obtained by the EPA. Lead appears to have a significantly better recovery with the microwave digestions.

The microwave methods also show very similar results for liquid waste samples when compared to the wet ash procedure. The precision of the microwave methods were also comparable to those obtained by the wet ash procedure. The time required to prepare samples for analysis is reduced from 5 h using the wet ash method to 2 h for the microwave.

In further microwave digestions samples of standard EPA quality control sludge were digested on a hot plate with nitric acid. Duplicate samples were also digested with nitric acid in open vessels using microwave heaving. Both sets of samples were digested using equal volumes of nitric acid. The hot plate digestion time ranged from 8–24 h while microwave digestion time was 8.5 hours. Element recoveries for the two digestions, shown in Table 6.24, are comparable. Except for arsenic, both sets of data agree well with the average reference values for the sample. However, the range of reference data values for arsenic reported was 0–89 µg g $^{-1}$. The microwave digestion data falls well within this range.

Use of closed vessels in combination with microwave heating can speed up sludge digestion significantly. To demonstrate this, three sets of duplicate samples of this same EPA sludge sample were digested using different methods. The first set was microwave digested in closed vessels using 70% nitric acid and 30% hydrogen peroxide. The second set was also microwave digested in closed vessels but 1:1 nitric acid:water was used. The third set of samples was digested in glass beakers on a hot plate following EPA SW–846 procedures. The microwave digestions required 40 min for the nitric acid: hydrogen peroxide dissolution, and 60 min for the 1:1 nitric acid:water dissolution. The hot plate dissolution required 10 h. Agreement on element recoveries among the three digestion procedures is

Table 6.24 Element recovery for closed vessel microwave digestion versus open vessel hot plate digestion of EPA sludge sample (all values are µg/g)

Digestion	Ag	As	Cd	Cr	Cu	Fe	Pb	Ni	Se	Zn
Microwave										
Closed vessel	81	4	20	184	1139	17500	575	185	4	1308
10 mL HNO$_3$	81	3	20	184	1132	17450	576	187	4	1308
3 mL H$_2$O$_2$										
40 min										
Microwave										
Closed vessel	80	4	21	188	1136	18350	593	197	4	1319
20 mL 1:1	78	4	21	187	1125	18550	598	193	4	1325
HNO$_3$:H$_2$O										
60 min										
Open vessel										
Hot plate	84	3	21	183	1117	17700	601	191	4	1283
10 mL HNO$_3$	88	3	21	183	1128	17650	591	192	4	1287
4 mL H$_2$O$_2$										
10 h										
EPAX value	80.6	17.0*	19.1	193	1080	16500	526	194	not reported	1320

All values micrograms per gram
*Range of reported values was 0–89 micrograms per gram

Source: Own files

very good, and, except for arsenic again, they agree well with the average sample reference values.

Van de Wall *et al.* [119] invested the applicability of microwave digestion to the determination of heavy metals (also aluminium, mercury, magnesium, niobium and vanadium) in sewage sludge. Digestions were carried out using nitric acid or a mixture of nitric and hydrochloric acid. Repeatability of analyses was below 10% for all elements determined and conformity with known concentrations was acceptable for every element except aluminium and magnesium.

6.53.1.2 Heavy metals (chromium, iron, nickel, copper, cobalt, cadmium, zinc, lead and manganese) and sodium, potassium, calcium and magnesium

Christensen *et al.* [58] compared four methods for the digestion of sewage sludge samples for the analysis of the above metals by atomic absorption spectrometry. They concluded that based on results concerning observed metal concentrations, precision, recovery of digested standard solutions recovery of metal standards added to sludge samples, time consumption and laboratory safety, digestion of sewage sludge with nitric acid is considered to be the best method for routine measurements of metals by atomic absorption spectrophotometry.

The four methods examined were: (1) dry ashing followed by digestion with aqua regia, (2) digestion with aqua regia, (3) digestion with nitric acid, and (4) digestion with nitric acid followed by treatment of the residual by sulphuric acid and dry ashing.

The four digestion procedures are briefly summarised below.

(i) Dry Ashing
In a quartz crucible 2 g of sludge was ashed at 450°C for 1 h. The crucible was transferred to a hot plate (100°C), 10 mL of aqua regia (25% nitric acid + 75% hydrochloric acid) was added and the crucible was covered with a watch glass. After 30 min, another 10 mL of aqua regia was added. After another 60 min, the watch glass was removed and the sample evaporated to a low volume. The residual was dissolved in 1.4N nitric acid, filtered and diluted to volume with 1.4N nitric acid. The low ashing temperature should limit the loss of cadmium, and extraction of the residuals with aqua regia should recover copper better than extraction with nitric acid [68].

(ii) Aqua regia
In an Erlenmeyer flask 2 g of sludge was repeatedly treated at near boiling with 10 mL of aqua regia (25% nitric acid + 75% hydrochloric acid) until the organic matter was mineralised. During the digestion the Erlenmeyer flask was covered with a watch glass. The sample was evaporated to near dryness, redissolved in 1.4N nitric acid, filtered and diluted to volume with 1.4N nitric acid. The method is modified after Delfino and Enderson [62].

(iii) Nitric acid [64]
In an Erlenmeyer flask (supplied with a watch glass) 2 g of sludge was treated at 100°C with 20 mL of concentrated nitric acid. The digestion continued until brown nitrous oxide fumes ceased, but at least for 5 h. Approximately 10 mL of deionised distilled water was added, the sample was filtered and diluted to volume with 1.4N nitric acid.

(iv) Nitric acid + dry ashing [64]
After treatment of the sample as described in method (iii), the filter paper and residuals were transferred to a quartz crucible. 10 mL sulphuric acid was added and the crucible kept at 130°C overnight. This was followed by ashing at 450–500°C for 45 min. After cooling, 15 mL 4.7N nitric acid was added, the crucible covered with a watch glass and heated for 45 min at near boiling. After filtering, the sample was diluted to volume with 1.4N nitric acid.

Analytical operating conditions for flame atomic absorption spectrometry are summarised in Table 6.25.

Table 6.25 Analytical operation conditions for flame–AAS

Element	Line (nm)	Slit (nm)	Flame	Standards* (mg/L)
Co	240.7	0.2	air/C_2H_2	1 – 10
Ni	232.0	0.2	air/C_2H_2	1 – 10
Cr	357.9	0.7	air/C_2H_2	1 – 10
Cd	228.8	2.0	air/C_2H_2	0.05 – 0.75
Pb	283.3	0.7	air/C_2H_2	5 – 40
Cu	324.7	2.0	air/C_2H_2	1 – 25
Zn	213.9	0.7	air/C_2H_2	0.3 – 3
Mn	279.5	0.2	air/C_2H_2	0.5 – 3
Fe	248.3	0.2	air/C_2H_2	1 – 25
K[b]	248.3	0.2	air/C_2H_2	0.5 – 5
Na[c]	588.6	0.7	air/C_2H_2	0.1 – 2
Ca[d]	422.7	0.7	N_2O/C_2H_2	1 – 15[a]
Mg[d]	285.2	0.7	N_2O/C_2H_2	0.5 – 15[a]

[a] Standards adjusted to proper HNO_3 concentration
[b] 0.2 Na added to samples and standards
[c] 0.2 K added to samples and standards
[d] 6 La and 0.01% Na added to samples and standards

Source: Gordon AC Breach, Amsterdam [58]

Table 6.26 Observed metal concentrations in sewage sludge by four samples digestion methods (average of 3 replicates)

Method Element	Metal concentration (µg/g) with method:			
	i Dry ashing	ii Aqua regia	iii HNO_3	iv HNO_3 + ashing
Co	113.5 B	91.0 C	115.5 B	118.5 A
Ni	370 B	375 B	375 AB	385 A
Cr	1090 C	865 C	1265 B	1280 A
Cd	66.1 C	71.9 B	75.3 A	75.3 A
Pb	1210 C	1340 A	1275 B	1275 B
Cu	985 C	1080 B	1140 A	1150 A
Zn	2215 C	2210 A	2165 B	2165 B
Mn	485 B	520 A	495 B	495 B
Fe	16590 B	13180 D	15500 C	17070 A
K	1650 C	3020 A	1460 C	2170 B
Na	700 B	1040 A	805 B	910 A
Ca	49850 B	48350 C	54800 A	54900 A
Mg	3435 C	3915 A	3475 C	3715 B

A, B, C: Observations marked with A significantly higher than the observations (for the same element) marked with B, which furthermore are higher than those marked with C. An approximate t-test and a 95% significance level have been applied.

Source: Gordon AC Breach, Amsterdam [58]

Table 6.27 Relative coefficients of variation (CV) for observed metal concentrations[a] in sewage sludge by four sample digestion methods

	CV (%) for metal concentrations with method:			
Method	i	ii	iii	iv
Element	Dry ashing	Aqua regia	HNO$_3$	HNO$_3$ + ashing
Co	1.9	2.0	0.2	0.5
Ni	0.9	0.9	1.3	1.3
Cr	7.9[b]	20.2[b]	0.5	0.5
Cd	0.5	0.9	0.3	0.3
Pb	0.5	1.0	0.9	0.9
Cu	3.8[b]	1.4	1.2	1.2
Zn	1.1	1.0	0.9	0.9
Mn	3.4[b]	3.3[b]	1.4	1.4
Fe	1.4	9.3[b]	0.2	0.4
K	11.7[b]	2.4	2.0	1.4
Na	5.7[b]	11.0[b]	0.2	1.7
Ca	0.5	0.8	1.8	1.8
Mg	2.2	0.1	0.9	1.5

[a] Averages of observed concentrations in Table 6.26
[b] Coefficients of variation considered to be unsatisfactorily high. No statistical considerations attempted, since estimates are based on only three replicate samples

Source: Gordon AC Breach, Amsterdam [58]

The results of the triplicate digestions of a sewage sludge are shown in Table 6.26 in terms of average observed metal concentrations and in Table 6.27 in terms of coefficients of variation expressed relative to the observed averages. In Table 6.26, the observations marked with A are significantly higher than the observations (for the same element) marked with B, which furthermore are higher than the ones marked with C. An approximative *t*-test and a 95% significance level have been applied. It is seen that no single digestion method always yielded the highest metal concentrations, and hence that none of the investigated digestion methods was the most efficient for all elements. Defining an efficiency of the individual digestion method by expressing the observed metal concentration relative to the highest concentration found for a specific element by the four digestion methods helps generalising the results. In Table 6.28 the elements are grouped in heavy metals, manganese/iron and alkaline earth/alkali metals with and without potassium, and the average digestion efficiencies are calculated for these groups. Method (iv), nitric acid + dry ashing, showed the best overall (13 element) efficiency of 95.6%, while method (i), dry ashing, showed the lowest overall efficiency of 87.1%. For the heavy metals, method (iii), nitric acid, and method (iv), were better (98–99% efficiency) than the other two methods 90–92% efficiency). Method (i) was in particular inefficient for chromium,

Table 6.28 Average digestion efficiencies[a] of four sludge sample digestion methods for groups of investigated elements

Method	i	ii	iii	iv
Element	Dry ashing	Aqua regia	HNO₃	HNO₃ + ashing
Co, Ni, Cr, Cd,				
Pb, Cu, Zn	91.6	90.1	97.9	99.0
Mn, Fe	95.3	88.6	93.0	97.6
K, Na, Ca, Mg	75.1	97.0	78.6	88.6
Na, Ca, Mg	81.9	96.0	88.7	94.1
All 13 elements	87.1	92.0	91.2	95.6

Where column headers are HNO_3 and $HNO_3 + ashing$ for methods iii and iv.

[a] Digestion efficiency calculated for a specific element as observed metal concentration divided by the largest metal concentration observed for the four methods investigated

Source: Gordon AC Breach, Amsterdam [58]

Table 6.29 Recovery of a mixed metal standard added to blind reference digestion (average of duplicates)

Method	i	ii	iii	iv
Element	Dry ashing	Aqua regia	HNO₃	HNO₃ + ashing
Co	97.4	77.2[a]	94.1	94.1
Ni	99.0	97.4	104.9	104.9
Cr	97.0	87.3[a]	100.4	100.4
Cd	98.1	100.5	107.6	107.6
Pb	95.8	97.8	96.3	96.3
Cu	96.1	98.5	98.1	98.4
Zn	106.5	108.0	102.4	102.4
Mn	110.1a	91.0	113.2[a]	113.2[a]
Fe	99.7	87.8[a]	98.0	98.0
K	98.4	99.9	94.7	94.7
Na	97.0	102.6	99.4	99.4
Ca	103.5	97.1	105.9	105.9
Mg	97.3	104.5	100.7	100.7

Where the column headers for methods iii and iv are HNO_3 and $HNO_3 + ashing$ respectively, under the heading Recovery (%) of metal standard with method.

[a] The recovery is worse than 90–110% of the added metal

Source: Gordon AC Breach, Amsterdam [58]

cadmium and copper, while method (ii), aqua regia, was very inefficient for cobalt and chromium. All methods were better than 90% efficient for manganese and iron, except for aqua regia in the case of iron. For calcium and magnesium, all methods are considered acceptable with efficiencies better than 88%. For potassium and sodium, method (ii) showed the highest concentrations. Especially potassium showed highly variable

Table 6.30 Recovery of a mixed metal standard added to sewage sludge samples digested by four methods (average of duplicates)

	Recovery (%) of metal standard with method			
Method	*i*	*ii*	*iii*	*iv*
Element	Dry ashing	Aqua regia	HNO$_3$	HNO$_3$ + ashing
Co	96.6	132.3[a]	94.0	93.4
Ni	92.7	96.8	95.6	95.6
Cr	103.4	78.0[a]	93.3	94.8
Cd	88.0[a]	95.4	104.3	104.3
Pb	90.9	103.6	95.9	95.9
Cu	101.3	100.1	100.5	100.3
Zn	72.6[a]	89.7[a]	101.7	101.7
Mn	109.0	99.1	113.5[a]	113.5[a]
Fe	119.7[a]	–	103.0	98.2
K	84.5[a]	120.8[a]	107.9	101.1
Na	96.1	98.3	114.3[a]	108.0
Ca	103.9	103.0	112.6[a]	112.6[a]
Mg	106.7	122.7a	93.5	91.5

[a] The recovery is worse than 90–110% of the added metal

Source: Gordon AC Breach, Amsterdam [58]

results, most probably a result of a varying degree of destruction of clay minerals present in the sludge.

The coefficients of variation as shown in Table 6.27 were in general very low for methods (iii) and (iv): 0.2–2%. Both methods (i) and (ii) showed high coefficients of variation for five and four elements respectively. This indicates that nitric acid or nitric acid + dry ashing are the most precise digestion methods of the four methods investigated.

The recoveries of known amounts of metals are shown in Table 6.29 for addition to blind reference digestions and in Table 6.30 for addition to sewage sludge samples. Recoveries within 90–110% are considered acceptable. No statistical test can reasonably be used to identify the recoveries deviating from 100% since only duplicate digestions were accomplished. The results are blurred by significant uncertainty, in particular for digestion method (ii). The only substantial result from the standard addition experience is that methods (iii) and (iv) consistently showed good recoveries for the heavy metals: 94.1–107.6% recovery of additions to blind references, and 93.4–104.3% recovery of additions to sewage sludge.

Comparison of four methods for digestion of sewage sludge samples prior to determination of all the elements discussed above by atomic absorption spectrophotometry showed that digestion with nitric acid followed by dry ashing (method iv) gave the best results – in particular for the heavy metals, which are the elements of most environmental concern when sewage sludge is disposed of on land intended for agricultural production. This method resulted in the highest metal concentrations in general, in a low coefficient of variation, and in consistent and acceptable recoveries of standard additions. However, the method is very time- and manpower-consuming, and it is therefore to be recommended only to use nitric acid digestion (method iii), which is identical to the first part of method (iv), for routine determinations of metals in sewage sludge. Nitric acid digestion also resulted in consistent and acceptable recoveries of standard additions and in low coefficients of variation; concentrations were only 1–3% lower for the heavy metals and 5–12% lower for the other elements (except potassium) than obtained by nitric acid + dry ashing. Aqua regia (method ii) and dry ashing (method i) are considered to be less suitable because of low digestion efficiencies for a few heavy metals, and because of inferior precision.

Nitric acid (method iii) is thus considered to be a fairly efficient method with good accuracy and precision and without severe laboratory safety problems for digestion of sewage sludge prior to determination of metals with atomic absorption spectrophotometry.

6.53.1.3 Heavy metals (cadmium, lead and zinc) and arsenic, mercury, vanadium, silver and aluminium

Van Loon [69] has reviewed existing and newly developed methods for sewage sludge analysis designed to give information on the form of metals in waste samples. Sludges can be dry-ashed at 450°C without fear of loss of cadmium, lead or appreciable amounts of zinc, but dry-ashing should not be used when aluminium is to be determined. There is indication that incineration at temperatures usually employed will release large amounts of arsenic, cadmium, lead, mercury and zinc. Particular attention is given to non-flame atomic absorption methods for vanadium and silver, development of methods for metal speciation studies in sludges, and liquid chromatography atomic absorption studies.

Smith [70] has evaluated six techniques for the pretreatment of sewage sludges prior to the determination of heavy metals (chromium, cadmium, copper, iron, lead, manganese, nickel and zinc) also aluminium, calcium, mercury and arsenic by atomic absorption spectrometry. The use of hydrofluoric acid, hydrogen peroxide and perchloric acid was avoided for safety reasons. With the exception of aluminium, good recoveries were obtained using mineral acid digestion and/or extraction. The aqua regia extraction method is preferred for most metals for its safety, simplicity, speed and reliability. For accurate aluminium determination, fusion or hydrofluoric acid digestion is necessary, with appropriate safety precautions. Volatile trace metals, such as mercury or arsenic, require alternative pretreatment methods.

The six methods involved (i) dry ashing at 500°C followed by extraction with dilute hydrochloric acid; (ii) dry ashing at 500°C followed by extraction with aqua regia; (iii) nitric acid digestion followed by extraction with hydrochloric acid; (iv) extraction with aqua regia; (v) ashing with magnesium nitrate solution at 550°C followed by digestion with hydrochloric acid and extraction with nitric acid; and (vi) extraction with nitric acid.

Method (i) Dry ashing at 500°C followed by extraction with dilute hydrochloric acid
Approximately 0.5 g sample was accurately weighed into a porcelain basin and ignited at 500°C for 2 h in a muffle furnace, then cooled. The residue was then transferred to a 100 mL beaker and 25 mL of 3 mol L $^{-1}$ hydrochloric acid added. The mixture was gently boiled on a hotplate for 2 h, then cooled and filtered through Whatman No 42 paper into a 100 mL standard volumetric flask. The filtrate was diluted to 100 mL with deionised distilled water.

Method (ii): Dry ashing at 500°C followed by extraction with aqua regia
As for method (i) except that the residue from the ashing stage was boiled with 3 × 12 mL portions of aqua regia (three parts hydrochloric acid: one part nitric acid) for a total of 30 min prior to filtering.

Method (iii): Digestion with nitric acid, followed by extraction with hydrochloric acid
This is the technique recommended by the EPA for the pretreatment of sewage sludge samples for analysis. Approximately 0.5 g sample was accurately weighed into a 100 mL beaker and 3 mL of concentrated nitric acid were added. The beaker was placed on a hotplate and the mixture cautiously evaporated to dryness without allowing it to boil. After cooling, another 3 mL of nitric acid were added, the beaker covered with a watchglass and replaced on the hotplate. The temperature was increased until a gentle reflux action was taking place. Heating was continued until the digestion was complete (indicated by a light-coloured residue). Fifteen mL of 1:1 hydrochloric acid were added and the beaker again gently heated for about 15 min. The watchglass and beaker walls were then washed down with deionised distilled water and the mixture filtered through Whatman No 42 paper into a 100 mL standard volumetric flask. The filtrate was diluted to 100 mL with deionised distilled water.

Method (iv): Extraction with aqua regia
Approximately 0.5 g sample was accurately weighed into a 100 mL beaker and 12 mL of aqua regia added. The beaker was covered with a watchglass and the contents heated on the medium heat of a hotplate

until all bubbling had ceased (30 min minimum). The mixture was diluted with about 5 mL of deionised distilled water and filtered through Whatman No 42 paper into a 100 mL standard volumetric flask. The filtrate was diluted to 100 mL with deionised distilled water.

Method (v): Ashing with magnesium nitrate solution, digestion with hydrochloric acid and extraction with nitric acid

Approximately 0.5 g sample was accurately weighed into a silica crucible and 5 mL of a 950 g/mL solution of magnesium nitrate added. The mixture was stirred thoroughly with a glass rod, evaporated to dryness on a water bath, then ignited at 550°C for about 15 min. After cooling, 10 mL of 1:1 hydrochloric acid were added and the mixture was again evaporated to dryness. A further 5 mL of 1:1 hydrochloric acid were added and the mixture again evaporated to dryness. Another 5 mL of 1:1 hydrochloric acid were added and the mixture washed into a 250 mL beaker with 50 mL of deionised distilled water. Ten mL of 50 mL nitric acid solution were then added and the mixture heated on a steam bath for about 15 min, cooled and filtered through Whatman No 42 paper into a 250 mL standard volumetric flask, and the filtrate diluted to 250 mL with deionised distilled water.

Method (vi): Extraction with nitric acid, 6 h reflux

Approximately 0.5 g sample was accurately weighed into a 250 mL round-bottomed flask and 50 mL of concentrated nitric acid added. The mixture was then boiled under reflux for 6 h, after which 100 mL of deionised distilled water were added and the mixture was boiled for another hour. After cooling, the mixture was filtered through Whatman No 42 paper into a 250 mL standard volumetric flask and the filtrate diluted to 250 mL with deionised distilled water.

The air acetylene flame was used for the determination of all metals except aluminium and chromium; for these metals the nitrous oxide–acetylene flame was selected. Determination of chromium in the nitrous oxide–acetylene flame gave higher results than were obtained in the air–acetylene flame, even with the addition of interference suppressants and use of a reducing (fuel-rich) flame.

A caesium/lanthanum interference suppressant solution (0.5 g L^{-1} cerium and 2 g L^{-1} lanthanum) was added to each sample extract and standard solution in order to reduce chemical and ionisation interferences. Typical results obtained by this procedure on a municipal sludge are shown in Table 6.31.

Only a few mg/kg of cadmium were present. The results obtained by each method were almost identical with the exception of those from method (vi) (nitric acid extraction) which were slightly higher.

The differences among the results obtained for iron by the various pretreatment methods were not considered significant.

Table 6.31 Sample 2: (Pretoria Municipal dried sludge) – analysis results (mg/kg)

Method: Metal	i Dry ashing – 500°C Extraction with 3 mol/L HCl	ii Dry ashing – 500°C Extraction with aqua regia	iii HNO₃ digestion Extraction with HCl	iv Extraction with aqua regia	v Ashing with Mg(NO₃)₂ – 550°C HCl digestion Extraction with HNO₃	vi Extraction with HNO₃ (6 h reflux)	Mean value obtained by EPA method
Aluminium	18600	18000	12000	11000	14500	11400	10.133
Cadmium	6	6	6	6	6	8	5.8
Chromium	140	141	150	149	143	145	155
Copper	586	586	594	594	535	555	606
Iron	24600	24600	24000	23800	23900	24500	24.453
Lead	338	324	364	340	322	310	349
Manganese	452	460	462	464	450	460	460
Nickel	52	61	60	61	55	60	52
Zinc	1740	1740	1780	1790	1500	1850	1782

Source: Erudita Publications, Johannesburg [70]

Considerably higher aluminium results were obtained using pretreatment techniques incorporating an ashing stage (methods (i) and (ii)) than were obtained using direct acid digestion or extraction procedures, or both. For the more accurate determination of aluminium in sewage sludges, more drastic sample pretreatment procedures, such as fusion or hydrofluoric acid digestion, are necessary in order to dissolve the aluminium present in the form of silicates and possibly other relatively insoluble matrices.

Results for chromium, copper, lead, manganese, nickel and zinc showed that, in general, the highest recoveries were obtained using method (iii) (nitric acid digestion–hydrochloric acid extraction) and method (iv) (aqua regia extraction). The lowest values were obtained by methods involving an ashing step indicating that even at 500°C some volatilisation of these metals occurred. These findings are in contrast to those of Ritter et al. [43] who found that dry ashing at 500°C was the most suitable of five techniques investigated for the pretreatment of sludge samples for the determination of cadmium, copper, lead, nickel and zinc, but in agreement with those of Katz et al. [42], who obtained low recoveries with dry ashing pretreatment procedures in comparison with various acid digestion techniques. Thompson and Wagstaff [34] found nitric acid digestion to give the best results compared to six other pretreatment techniques including dry ashing at 420 and 600°C.

6.53.1.4 Heavy metals (cadmium and lead) and mercury

Kurfurst and Rues [71] have discussed the principles of Zeeman atomic absorption spectrometry and the manner in which the large degree of background interference associated with solid phase analyses can be overcome. The method was applied to the determination of lead, cadmium and mercury in standard sewage sludge samples which were not given any chemical pretreatment but were thoroughly homogenised and weighed directly into a graphite boat. Measured values based on peak heights agreed fairly well with the standard values, but were generally about 10% lower, indicating that matrix effects may have interfered with atomisation; peak area estimation and calibration with reference to solid phase standards are suggested as ways of overcoming this problem.

6.53.1.5 Aluminium, calcium, iron and magnesium

Carrondo et al. [72] have described a rapid flameless atomic absorption method for the determination of these metals in sewage sludge. The workers point out that many methods used for the determination of aluminium, calcium, iron and magnesium in sewage sludge are time-consuming. They compared a rapid flameless atomic absorption

Table 6.32 Conditions for flame atomic-absorption analysis

Metal	Wave-length nm	Spectral band-width nm	Flame type	Working range, µg/mL
Al	309.3	0.7	Nitrous oxide–acetylene Reducing (rich, red)	I – 50
Ca	422.7	0.7	Air–acetylene Oxidising (lean, blue)	0.05 – 5.0
Fe	248.3	0.2	Air–acetylene Oxidising (lean, blue)	0.05 – 5.0
Mg	285.2	0.7	Air–acetylene Oxidising (lean, blue)	0.005 – 0.5

Source: Elsevier Science, UK [72]

procedure, using low sensitivity lines utilising homogenisation of diluted samples as the only pretreatment, with wet and dry analytical methods followed by flame absorption spectrometry using high sensitivity lines in a statistically designed experiment. The flameless atomic absorption method is better than all other tested methods with the exception of the nitric-perchloric-hydrofluoric acid digestion procedure.

In this method [72] a Perkin–Elmer model 603 atomic absorption spectrophotometer equipped with deuterium background correction was used for flame analysis. The conditions for analysis and the working ranges used are presented in Table 6.32. In order to remove interferences or suppress ionisation, the samples and standards to be analysed for aluminium were made up to contain 2000 µg of potassium chloride per mL and those to be analysed for calcium and magnesium were made up to contain 0.5% w/v of lanthanum.

The same spectrophotometer and a Perkin–Elmer HGA 76 heated graphite atomiser were used for flameless analysis. The conditions and working ranges for flameless atomic absorption analysis are presented in Table 6.33. The atomisation programme consisted of drying at 100° for 30 s, two-stage thermal decomposition with temperature increase from 100° to 400° in 45 s (rate 2) followed by isothermal decomposition at 1200° for 30 s and atomisation at 2770° for 5 s, for all metals except aluminium (8 s). The first stage in the thermal decomposition avoided spattering of the sample which would have occurred if the temperature had been suddenly increased from 100° to 1200°.

Homogenisation
Approximately 250 mL of the sludge sample, previously diluted 50-fold and acidified with 1% of its volume of nitric acid, were homogenised in a two-litre tall-form Pyrex beaker with an Ultra Turrax T45N homogeniser

Table 6.33 Conditions for flameless atomic-absorption analysis

Metal	Wave-length, nm	Spectral band-width, nm	Sample volume, μl	Working range, μg/mL
Al	275.5	0.2	20	0.20 – 4.0
Ca	239.9	0.7	20	1.00 – 20.0
Fe	305.9	0.2	20	0.20 – 5.0
Mg	202.6	0.7	20	0.02 – 0.50

Source: Elsevier Science, UK [72]

(Scientific Instrument Co Ltd, London) for five minutes at 8000 rpm. Aliquots of 20 mL were injected into the flameless atomiser with an Eppendorf micropipette. Analysis was performed by direct comparison with standards prepared in 1% v/v nitric acid and checked by the method of standard additions.

Sulphuric–nitric acid digestion [19]
Heat was provided by an electric heating mantle.

To 50 mL of undiluted sludge in a 500 mL round-bottomed flask were added two glass anti–bumping granules (previously leached in 10% v/v nitric acid), 50 mL of concentrated nitric acid and 20 mL of sulphuric acid. The digestion was started at approximately 120°C, and when the volume was reduced to 10 mL, the mixture was allowed to cool before addition of a further 20 mL of nitric acid. Successive 20 mL portions of nitric acid were added until white fumes were evolved and the solution was pale straw in colour. At this point the digestion was deemed complete. The mixture was allowed to cool and 10 mL of distilled water were added. Insoluble matter was allowed to settle, and the solution was filtered through a Whatman GF/C glass fibre filter (previously leached with 10% v/v nitric acid) into a 100 mL standard flask.

The 500 mL flasks were leached twice by boiling 15 mL of 10% v/v hydrochloric acid in them for 20 min, the acid and two distilled water washes being added to the standard flask and the volume made up to 100 mL.

Nitric acid–hydrogen peroxide digestion [33]
20 mL of sample were digested instead of the 5 mL originally proposed by Geyer et al. [33], thus reducing the errors involved in the estimation of the volume of thick sludges. To 20 mL of sample were added 30 mL of concentrated nitric acid and two glass anti-bumping granules (previously leached overnight with 10% v/v nitric acid). When 30 mL had evaporated, a further 10 mL of acid were added and the digestion was continued until

the volume was reduced to approximately 5 mL. At this point 2 mL of 100 vol hydrogen peroxide and 2 mL of nitric acid were added, the addition being repeated until the solution was pale straw in colour.

The contents of the beaker were filtered into a 100 mL standard flask through a Whatman GF/C glass fibre filter (previously leached with 10% v/v nitric acid). The beaker was then leached twice by boiling 15 mL of 10% v/v hydrochloric acid in it for 10 min. The acid and two distilled water washes were added to the standard flask and the volume was made up to 100 mL.

Nitric-perchloric-hydrofluoric acid digestion [73]
To 10 mL of sludge in a 100 mL PTFE beaker 30 mL of concentrated nitric acid were added and the sample was evaporated nearly to dryness on a hotplate. The beaker was cooled and 5 mL of concentrated nitric acid, 2 mL of 60% perchloric acid and 6 mL of 40% hydrofluoric acid were added. The mixture was evaporated nearly to dryness on a hotplate at a temperature not exceeding 280°. A further 2 mL of nitric acid and 2 mL of perchloric acid were added and evaporated to ensure that silicon and fluoride were removed. The beaker was cooled and 20 mL of 5% v/v nitric acid were added to dissolve the salts. The solution was transferred to a 100 mL standard flask and made up to 100 mL with distilled water washings from the digestion beaker.

Destruction of organic matter by ashing
Samples (25 mL) were placed in 100 mL Pyrex beakers which were then covered with porcelain crucibles to prevent contamination by clay dust from the firebrick lining of the muffle furnace. Initially the samples were charred at 200° for 1 h, after which the temperature of the furnace was gradually increased over a period of 2 h to 450°. Ashing was continued for 14 h at this temperature. When it was completed, the beakers were removed and allowed to cool, after which 1 mL of concentrated nitric acid was added to each and the residues were then heated to dryness on a hotplate. The beakers were returned to the muffle furnace and the contents ashed at 450° for a further hour. The beakers were cooled, 10 mL of extraction acid were added to each (100 mL of concentrated hydrochloric acid, 650 mL of distilled water and 150 mL of concentrated nitric acid) and the contents heated almost at boiling point on a hotplate for 10 min. The extracted samples were filtered through a Whatman GF/C glass fibre fitler (previously leached with 10% v/v nitric acid) into 100 mL standard flasks. The beakers were rinsed twice with 15 mL of distilled water, the washings added to the contents of the volumetric flasks and the volumes made up to 100 mL with distilled water.

The mean values, within group relative standard deviation and the results of an analysis of variance by the F-test [74] obtained by flame and flameless atomic absorption spectrometry are reported in Table 6.34.

Table 6.34 Comparison of aluminium, calcium, iron and magnesium concentrations in sewage sludges, found by using sulphuric–nitric acid digestion, hydrogen peroxide–nitric acid digestion, perchloric–nitric–hydrofluoric acid digestion, and ashing at 450°C followed by flame atomic-absorption analysis, with those found for homogenised samples analysed by flameless atomic-absorption analysis

Metal	Pretreatment	Analytical method	F-test level of significance	Mean conc. µg/mL	RSD %
Al	H$_2$SO$_4$–HNO$_3$	F		140a*	6.2
	H$_2$O$_2$–HNO$_3$	F		120ac	6.4
	HClO$_4$–HF–HNO$_3$	F	0.01	223b	4.7
	Ashing (450°)	F		107c	4.0
	Homogenisation	FL		230b	7.8
Ca	H$_2$SO$_4$–HNO$_3$	F		764a	3.2
	H$_2$O$_2$–HNO$_3$	F		838b	2.3
	HClO$_4$–HF–HNO$_3$	F	0.01	828b	2.0
	Ashing (450°)	F		854b	2.4
	Homogenisation	FL		836b	2.0
Fe	H$_2$SO$_4$–HNO$_3$	F		113a	3.1
	H$_2$O$_2$–HNO$_3$	F		110a	5.7
	HClO$_4$–HF–HNO$_3$	F	NS	113a	3.7
	Ashing (450°)	F		108a	4.1
	Homogenisation	FL		111a	3.1
Mg	H$_2$SO$_4$–HNO$_3$	F		71ac	3.3
	H$_2$O$_2$–HNO$_3$	F		71a	2.1
	HClO$_4$–HF–HNO$_3$	F	0.01	83b	2.3
	Ashing (450°)	F		67c	2.7
	Homogenisation	FL		82b	2.3

* = means having a common following letter are not significantly different at the 0.05 probability level
NS = not significant at the 0.05 probability level
RSD = relative standard deviation
F = flame analysis
FL = flameless analysis
H$_2$SO$_4$–HNO$_3$ = sulphuric–nitric acid digestion
H$_2$O$_2$–HNO$_3$ = hydrogen peroxide–nitric acid digestion
HClO$_4$–HF–HNO3 = perchlorid–hydrofluoric–nitric acid digestion
Ashing (450°C) = ashing at 450°C for 16 h followed by dissolution
Homogenisation = pretreatment by homogenisation

Source: Elsevier Science, UK [72]

Tukey's test was used to identify which means were statistically different at the 0.05 significance level. The reproducibility of flameless analysis, based on 10 injections of the same diluted sample, is indicated in Table 6.35.

No significant differences were found between the treatments in the determination of iron, but for the determination of aluminium, calcium

Table 6.35 Reproducibility of flameless analysis of diluted sewage sludge samples (10 injections)

Metal		Mean	Concentration µg/mL SD	RDS %
Al	best	2.2	0.13	5.9
	worst	2.3	0.25	10.8
Ca		16.6	0.29	1.7
Fe		2.28	0.06	2.6
Mg		0.415	0.008	1.9

Source: Elsevier Science, UK [72]

and magnesium, highly significant differences were found between treatments. A comparison of the means by Tukey's test indicated that the sulphuric–nitric acid digestion procedure yielded lower results for calcium, but that there were no significant differences between the other pretreatments for this element.

For both aluminium and magnesium, Tukey's test indicated that no more than two of the pretreatments were in agreement for either element. It is perhaps significant that for both these elements the nitric–perchloric–hydrofluoric acid treatment in conjunction with flame atomic absorption spectrophotometry and homogenisation in conjunction with flameless atomic absorption spectrophotometry, are in agreement. Moreover these two treatments also yielded the highest recoveries for these two elements. It would be expected that the nitric–perchloric–hydrofluoric acid digestion procedure would give complete recovery, which further substantiates the suitability of the flameless atomic absorption method for the determination of total concentration of aluminium and magnesium.

The homogenisation procedure coupled with flameless atomic absorption analysis and direct comparison with aqueous standards proved to be a suitable method for the determination of aluminium, calcium, iron and magnesium in sewage sludges. This dispenses with the need for drastic pretreatments such as the nitric–perchloric–hydrofluoric acid digestion procedure and also the need to add different interference removal agents to permit analysis by flame atomic absorption spectrophotometry.

The rapid flameless procedure can be used advantageously for routine analysis. The time saved is considerable since homogenisation takes only 5 min as opposed to 3–6 h for digestion or approximately one day for ashing. This more than compensates for the additional time (2–3 min) required for flameless analysis as opposed to flame analysis. The method has a further advantage over flame atomic absorption in that it does not require interference removal agents to be added to samples and standards before analysis.

6.53.1.6 Antimony, arsenic, bismuth, tellurium, thallium and vanadium

Concentrations of arsenic [77, 78] bismuth [120] and antimony [77, 78] may be of the order of 30 µg/g, vanadium concentrations as high as 400 µg/g [20, 78], and tellurium and thallium, which have rarely been reported, occur at much lower concentrations [82].

Kempton *et al.* [76] have reported experiments undertaken to compare the applicability of flame, electrothermal and hydride generation atomic absorption spectrophotometry in conjunction with various sample pretreatments, for the determination of a number of metals in sewage sludge. The most suitable method appeared to be electrothermal atomic absorption spectrometry of homogenised samples by direct comparison with standard solutions. In the cases of vanadium and thallium, the standard addition method was required. For arsenic, antimony and tellurium, flame atomic absorption spectrometry was shown to be insufficiently sensitive for environmental concentrations and, although hydride generation is commonly used, it is subject to many interferences and high relative standard deviations. These effects can be reduced by using a high ratio of acid volume to digest volume (20 mL of 3% volume for volume hydrochloric acids to 20–100µL). The major cause of interference is reported to be incompletely oxidised organic matter.

A Perkin–Elmer model 5000 atomic absorption spectrophotometer fitted with standard burner heads was used for all flame atomic absorption determinations. The same spectrophotometer, fitted with a Perkin–Elmer HGA 500 heated graphite atomiser, and a model 603 atomic absorption spectrophotometer fitted with an HGA 400, were used for flameless determinations. A Perkin–Elmer MHS–1 hydride generation system was used in conjunction with the model 603 spectrophotometer for the determination of the hydride-forming elements (arsenic, antimony, bismuth and tellurium). Electrodeless discharge lamps were used for arsenic, bismuth, antimony, tellurium and thallium. A hollow cathode lamp was used for vanadium. Instrumental parameters are detailed in Table 6.36.

For flameless atomic absorption determinations of vanadium, pyrolytically coated graphite tubes were used. The coating was performed *in situ* by passage alternately of argon and of a methane-argon mixture (1:9 v/v) under temperature and gas flow conditions optimised to ensure uniform pyrolysis of the methane, thus producing an even coating. The conditions are summarised in Table 6.37. Pyrolytic coating extended the tube life, improved sensitivity and reduced memory effects, the last being the most pertinent in the determination of vanadium.

The sludge was diluted tenfold with 1% v/v nitric acid. An Ultra Turrax (Scientific Instrument Co, London) fitted with a titanium shaft was used to homogenise the samples [27].

Table 6.36 Parameters for atomic-absorption spectrophotometry

Element	Wave-length, nm	Slitwidth, nm	Flame AAS	Electrothermal AAS*			Hydride generation
				Step 1	Step 2	Step 3	
V	318.4	0.7	$N_2O-C_2H_2$ reducing (rich–red)	120°C† 30–60 sec	1000°C‡ 30–60 sec	2900°C‡ 7 sec	–
Bi	223.0	0.2	Air–C2H2 oxidising (lean–blue)	120°C 30–60 sec	600°C 30–60 sec	2300°C 5 sec	Programme I§
Tl	276.8	0.7	–	120°C 30–60 sec	700°C 30–60 sec	2700°C 5 sec	–
As	193.7	0.7	$Air-C_2H_2$ oxidising (lean–blue)	–	–		Programme I§
Sb	217.6	0.2	$Air-C_2H_2$ oxidising (lean–blue)	–	–		Programme I§
Te	214.3	0.2	Air-C2H2 oxidising (lean–blue)	–	–		Programme II¶

* Uncoated graphite tube used for Tl, Bi; pyrolytically coated tube for V (see text)
† Ramp time 20 s; held for 30–60 s according to volume injected (20–50 µL)
‡ Ramp time 0 s for V, 1 s for Bi,Tl
§ Argon purge 30 s; carrier-gas stream 40 s
¶ Argon purge 45 s; carrier-gas stream 40 s

Source: Elsevier Scientific UK [76]

Table 6.37 Conditions for *in situ* pyrolytic coating of graphite tubes

Step	1	2	3	4	5	6	7	8
Temperature °C	500	1950	20	500	1950	20	500	1950
Ramp time, sec	1	16	1	1	16	1	1	16
Hold time, sec	20	230	100	20	230	100	20	230
Internal flow, mL/min	300		300	300		300	300	
Internal alternative flow, mL/min		150			150			150

Source: Elsevier Science UK [76]

Decomposition procedures

Nitric acid–sulphuric acid digestion

A modified version of a recommended digestion [19, 72] was used. To undiluted sludge were added 30 mL of concentrated sulphuric acid and 50 mL of concentrated nitric acid; the mixture was heated at 120° and successive 20 mL portions of nitric acid were added until digestion was complete. The digest was filtered and then made up to volume in a 100 mL standard flask.

Nitric acid–hydrogen peroxide digestion

A modification of a previous method [33] was used. Preliminary oxidation with nitric acid ensured a controlled reaction [79]. An addition of 75 mL of nitric acid was made to 50 mL of sludge in a PTFE beaker; further 25 mL portions of nitric acid were added, followed by 10 mL additions of hydrogen peroxide until the digestion was complete. The digest was filtered and made up to volume in a 100 mL standard flask.

Nitric acid–perchloric acid digestion

The recommended method [27] was used. PTFE beakers each containing 50 mL of sludge were heated at 120°, with several 30 mL additions of nitric acid. Perchloric and nitric acid were added to the cooled reaction mixture, which was then heated to about 200°. Successive additions of perchloric and nitric acids were made until completion of the digestion. The digest was filtered before being made up to 100 mL.

Nitric–perchloric–hydrofluoric acid digestion

The method used was similar to that previously employed [73] and is an extension of the nitric–perchloric acid digestion described above, hydrofluoric acid being added as a final stage before filtration and making up to 100 mL.

Dry ashing
The ashing procedure adopted was as reported previously by Corrando [72]. The sludge (50 mL) was charred in covered Pyrex beakers at 200° for 1 h in a muffle furnace. The temperature was then increased to 450° and kept there for 14 h. After cooling and the addition of 5 mL of concentrated nitric acid, the ashing was continued for a further hour. The residue was boiled with extraction acids before filtration and making up to 100 mL.

Nitric acid–sulphuric acid–hydrogen peroxide digestion
The method was essentially that described by Arbab-Zavar *et al.* [80]. The sludge was charred with sulphuric acid; the digestion was continued with nitric acid and was completed by the addition of hydrogen peroxide. The digest was then filtered and made up to 100 mL.

Pre-reduction for tellurium, arsenic and antimony
Before hydride generation, reduction of the sample is sometimes required to ensure that the determinand is in the correct oxidation state [81]. To convert all tellurium into tellurium (IV) boiling for 2 min with concentrated hydrochloric acid [81] or boiling for a longer period with aqua regia [82] have been proposed. The latter method was used by Kempton *et al.* [76]: equal volumes of aqua regia (3:1 v/v hydrochloric acid–nitric acid) and digested sample were boiled together for 15 min. Standard solutions were treated in the same manner.

Similarly, the oxidation states of arsenic and antimony affect the sensitivity of the hydride-generation method [81]. Arsenic(V) gives lower sensitivity than arsenic(III) [81]. Potassium iodide [83] and sodium iodide [84] have been used to ensure that the arsenic is in oxidation state(III).

Antimony is also reduced with potassium iodide [81] or sodium iodide [85]. Antimony(V) gives only half the sensitivity given by antimony(III) [81]. As the reduction is instantaneous, the determination should be performed immediately; delay allows the formation of iodine from the acid solution, which may cause interference in hydride generation [81].

From the comparison of flame, electrothermal and hydride generation atomic absorption spectrometry for the determination of bismuth in sewage sludge, the most suitable method appeared to be electrothermal atomic absorption spectrometry of the homogenised samples by direct comparison with standard solutions.

For vanadium determination, electrothermal atomic absorption spectrometry was best, but the standard additions method was required. The same applies to determination of thallium.

For arsenic, antimony and tellurium, flame atomic absorption spectrometry has been shown to be insufficiently sensitive to detect environmental concentrations. Hydride generation is, therefore, the method of choice, but is subject to many interferences which may have contributed to the high relative standard deviations, eg for tellurium.

This effect was limited by using a large ratio of acid volume (20 mL of 3% v/v hydrochloric acid) to digest volume (20–100 mL). Since volatilisation of the tellurium hydride separates it from the majority of the matrix components, except other hydride species, interferences are then expected to be small. It has been reported that residual nitric acid and perchloric acid do not cause interference, the major cause being incompletely oxidised organic matter.

6.53.1.7 Silver, cobalt, manganese, molybdenum and tin

Sterritt [86] employed electrothermal atomic absorption to determine these elements in sewage sludge, and compared results obtained with those obtained by flame atomic absorption spectrometry on dry ashed samples and on acid digestions of samples. Sterritt [86] found that preliminary treatment of sludge by homogenisation was more rapid and reliable than other pre-treatment methods.

Sterritt [86] used a Perkin–Elmer Model 603 atomic absorption spectrophotometer fitted with standard burner heads for flame atomic absorption analyses. The same spectrophotometer fitted with a Perkin–Elmer HGA–76 heated graphite atomiser was used for all electrothermal atomic absorption analyses. The graphite atomiser was purged with high purity argon. Perkin–Elmer hollow cathode lamps were used as sources for silver, cobalt, manganese and molybdenum and a Perkin–Elmer electrodeless discharge lamp was the source for tin.

Aliquots of 50 µL were injected into the HGA–76 electrothermal atomiser with an Eppendorf micropipette, fitted with disposable polypropylene tips.

The analytical conditions used for electrothermal atomic absorption analysis were optimised in order to obtain the most effective destruction of the sample matrix without loss of the analyte during the ashing stage. The atomisation time for the molybdenum analysis was increased to 8 s in order to counteract memory effects encountered when an atomisation time of 5 s was used. The precision and sensitivity for the determination of tin were improved by using an ashing time of up to 150 s and aliquots of acidified homogenised sludge injected into the HGA–76 were neutralised by the injection of an equal volume of 50% v/v ammonia solution. Optimised conditions for electrothermal atomic absorption analyses are shown in Table 6.38.

A sample of sludge (25 mL), diluted to 250 mL and acidified by the addition of 2.5 mL of nitric acid, was homogenised in a two litre tall form beaker with an Ultra Turrax T45N homogeniser at 8000 rev/min $^{-1}$ for 10 min. The original stainless steel shaft of the Ultra Turrax was replaced with a replica machined from titanium.

Table 6.38 Conditions for flame and electrothermal atomic-absorption analysis

Metal	Wave-length nm	Spectral band width /nm	Flame atomic absorption		Electrothermal atomic absorption*				
			Flame type	Working mg L⁻¹	Ashing stage		Atomising stage		Working range used/ mg L⁻¹
					Temp./ °C	Time/ s	Temp./ °C	Time s	
Ag	328.1	0.7	Air–acetylene, oxidising (lean, blue)	0.5 – 4	400	75	2700	4	0.002 – 0.03
Co	240.7	0.2	Air–acetylene, oxidising (lean, blue)	0.05 – 5	1100	45	2700	4	0.02 – 0.2
Mn	279.5	0.2	Air–acetylene, oxidising (lean, blue)	0.03 – 5	1000	45	2700	5	0.005 – 0.04
Mo	313.3	0.7	Dinitrogen, oxide–acetylene, reducing (rich, red)	1 – 15	1750†	45	2770	8	0.005 – 0.03
Sn	286.3	0.2	Dinitrogen oxide–reducing (rich, red)	1 – 20	900	150	2700	5	0.08 – 0.8

* Samples (50 µL) were dried in the graphite furnace at 100°C for 60 s prior to ashing
† A temperature ramp of approximately 10°C s⁻¹ (Perkin–Elmer HGA–76, rate 2) was used when increasing the furnace temperature from 100 to 1750°C

Source: Royal Society of Chemistry [86]

Table 6.39 Comparison of silver, cobalt, manganese, molybdenum and tin concentrations in sewage sludge obtained by flame atomic-absorption analysis of sulphuric acid–nitric acid and nitric acid–hydrogen peroxide digestates and dry-ashed samples with electrothermal atomic-absorption analysis of homogenised samples

Metal	Pretreatment	Mode*	F-test level of significance†	Mean conc. µg/mL $^{-1}$‡	RSD %
Ag	H_2SO_4–HNO_3 digestion	F	0.05	0.78a	22.9
	HNO_3–H_2O_3 digestion	F		2.08b	29.4
	Dry ashing (450°C)	F		0.97a	22.2
	Homogenisation	E		1.39a	4.9
Co	H_2SO_4–HNO_3 digestion	F	0.05	1.80a	23.1
	HNO_3–H_2O_3 digestion	F		2.65b	24.1
	Dry ashing (450°C)	F		2.57b	8.0
	Homogenisation	E		3.09b	5.4
Mn	H_2SO_4–HNO_3 digestion	F	0.05	79.6a	9.1
	HNO_3–H_2O_3 digestion	F		106.2b	21.2
	Dry ashing (450°C)	F		94.1ab	1.4
	Homogenisation	E		87.9ab	9.2
Mo	H_2SO_4–HNO_3 digestion	F	NS	8.48a	8.0
	HNO_3–H_2O_3 digestion	F		8.40a	15.7
	Dry ashing (450°C)	F		9.08a	5.3
	Homogenisation	E		9.08a	6.3
Sn	H_2SO_4–HNO_3 digestion	F	0.05	4.32ab	33.3
	HNO_3–H_2O_3 digestion	F		3.95a	33.9
	Dry ashing (450°C)	F		6.07b	15.0
	Homogenisation	E		5.51ab	12.7

* F = flame atomic absorption analysis; E = electrothermal atomic-absorption analysis. Deuterium background correction was used wherever possible for flame atomic-absorption spectroscopy, although source-beam attenuation was often necessary. It was found that background correction was not necessary for electrothermal atomic-absorption spectroscopy, as it was possible to achieve effective destruction of the matrix with the ashing times and temperatures used, and it was therefore not employed.

† NS = not significant at 0.05 significance level

‡ Means not followed by a common letter are statistically different at the 0.05 significance level

Source: Royal Society of Chemistry [86]

Sulphuric acid–nitric acid digestion

A sample of undiluted sludge (100 mL) was digested with 50 mL of nitric acid and 20 mL of sulphuric acid in a 500 mL round-bottomed flask at 120°C until the volume was reduced to 10 mL. The digestion was continued by the successive addition of 20 mL aliquots of nitric acid until white fumes were evolved and the digestate was a pale straw colour. Insoluble particulate matter was removed by filtering the digestate through a Whatman GF/C glass fibre filter paper.

Nitric acid–hydrogen peroxide digestion
A 20 mL sample of undiluted sludge was heated with 30 mL of nitric acid, almost to dryness, on a hotplate at 120°C. The digestion was continued by the addition of a further 30 mL of nitric acid and the mixture again evaporated almost to dryness. Thereafter, 2 mL of nitric acid and 2 mL of hydrogen peroxide were added repeatedly until the digestion was complete.

Dry ashing
A 25 mL sample of sludge in a covered beaker was heated in a muffle furnace at 200°C for 1 h, after which time the temperature was increased to 450°C over a period of 2 h. After 14 h at 450°C the ashed sample was allowed to cool. Nitric acid (1 mL) was added, the residue heated to dryness and returned to the furnace for 1 h at 450°C. The residue was extracted with 10 mL of boiling nitric acid (25% v/v) and filtered.

In Table 6.39 are shown the relative standard deviations obtained in flame and electrothermal atomic absorption spectroscopic measurements of the five elements in sewage sludge samples prepared by acid digestion and dry ashing. The relative standard deviations observed for the flame atomic absorption analyses of acid-digested and dry-ashed samples were generally higher than those obtained for the electrothermal atomic absorption analyses of homogenised samples. The nitric acid–hydrogen peroxide digestion gave the highest scatter of results for all metals, the relative standard deviations ranging from 15.7% for molybdenum to 33.9% for tin, whereas homogenisation in conjunction with electrothermal atomisation gave relative standard deviations comparable to or better than those obtained from dry ashing, which was the best pre-treatment for flame atomic absorption. The highest relative standard deviation observed for homogenisation was 12.7% for the analysis of tin. The better precision obtained by using homogenisation as a pre-treatment may be due to the simplicity and reproducibility of the method compared with operator variation and variations in reaction conditions for more complex pre-treatments.

No significant differences were found between the treatments for the determination of molybdenum. However, for silver, cobalt, manganese and tin significant differences were obtained. The sulphuric acid–nitric acid digestion gave lower recoveries of cobalt and manganese than the other pre-treatments. The lowest recovery of tin was obtained for the nitric acid–hydrogen peroxide digestion; however, homogenisation yielded results that were comparable to the other pre-treatments for this element. The use of excess of nitric acid in digestions could possibly lead to some precipitation, but significant losses of tin may depend on its valency.

The recoveries of silver from the sulphuric acid–nitric acid digestion, dry ashing and homogenisation were all statistically lower than the values obtained for the nitric acid–hydrogen peroxide digestion. A possible explanation for the higher value obtained for the determination of silver after nitric acid–hydrogen peroxide digestion is that a component of the digestate caused enhancement effects during flame atomic absorption.

6.53.2 Inductively coupled plasma atomic emission spectrometry

6.53.2.1 Heavy metals (cadmium, chromium, copper, lead, nickel and zinc)

Hawke and Lloyd [87] used this technique to determine heavy metals in sewage sludge using a simplified acid digestion method in which the samples were mixed with nitric acid and digested in a sealed PTFE bomb at 110°C in an oven. The digest was diluted with water and insoluble residue allowed to settle. The supernatant was analysed by inductively coupled atomic emission spectrometry without filtration. Results were compared with those obtained using atomic absorption spectrometry after sample digestion with boiling nitric acid. The technique succeeded in cutting the time for each sample analysis to less than half that for the standard method. The procedure could be applied to liquid and solid sludges, including sludges conditioned with lime and iron salts.

6.53.2.2 Miscellaneous

Schramel et al. [88] examined the use of inductively coupled plasma atomic emission spectrometry for the determination of heavy metals and trace elements in a wide variety of sludge, sediment and soil samples. The particular features of this method of excitation (at temperatures of 10,000 K) are outlined and the necessary equipment and sample preparation techniques described. A comparison of aqua regia extraction and total ashing procedures for a range of sewage sludge samples is presented, the latter method giving considerably higher results. Calibration methods and the matrix effects encountered during the analysis of materials of different origin are discussed. A sixteen-channel ICP spectrometer was employed, permitting the simultaneous determination of sixteen elements by selection of the appropriate wavelengths.

6.53.3 Neutron activation analysis

6.53.3.1 Heavy metals (chromium, manganese, iron, cobalt, nickel, copper, zinc, cadmium and lead) and aluminium, scandium, titanium, vanadium, arsenic, antimony, bismuth, selenium, tellurium, magnesium, calcium, strontium, barium, sodium, potassium, caesium, rubidium, silver, gold, indium, thallium, lanthanum, cerium, neodynium, samerium, europium, gadolinium, terbium, dysprosium,

ytterbium, lutecium, tungsten, thorium, uranium, silicon, yttrium, hafnium, zirconium, molybdenum, tin, tantalum, mercury and iridium

This technique has been applied by various workers to the determination of heavy metals in sewage sludge [24, 42, 89, 91, 94, 95].

Katz [42] weighed 50 mg samples of each of the dried sludges into plastic irradiation vials which were heat sealed. Blanks and multi-element standards were similarly prepared. The samples, standards and blanks were irradiated for 10 min in a thermal flux of 5.0×10^{11} at the Dalhousie University Slowpoke Reactor facility. The irradiated materials were allowed to decay for 1 min and counted for ^{66}Cu activity using the 1039 keV photopeak. The samples, standards and blanks were then irradiated for 16 h in a flux of 5.0×10^{11} and allowed to decay for three weeks prior to counting. The levels of chromium, nickel and zinc were determined from the following photopeak activities: ^{51}Cr at 320 keV, ^{58}Co (from (n,p) reaction with ^{58}Ni) at 811 keV and ^{65}Zn at 1115 keV.

Dams *et al.* [91] in a pilot study applied neutron activation analysis to the determination of 41 elements in sewage sludge, sludge from an industrial water treatment plant and compost from a municipal compostation plant. These workers pointed out that thorough grinding and homogenisation of the samples was required before analysis.

Sewage sludge and industrial waste water plant samples of approximately 2 kg were taken and dried at 80°C in an oven for two days. The dry material was crushed and mixed in a ball-mill and sieved through a screen with 1 mm meshes. The samples were not sufficiently homogeneous for analysis on a 30 mg level. Therefore 2 g fractions were further homogenised in a microdismembrator II. In an egg-shaped hollow teflon holder a small teflon-lined steel ball was brought together with 2 g of sludge. This holder vibrated at a frequency of 50 Hz and with an amplitude of 1.5 cm. The holder was cooled by dipping it for 5 min in liquid nitrogen, and shaken three times for 1 min. This procedure was repeated three times, which resulted in a considerable reduction of the particle size and a very good homogeneity on the level required for analysis.

For the neutron irradiations in the reactor, 50 to 100 mg amounts of the sludges or the compost were weighed on clean 5.5 cm diameter Whatman 41 cellulose filters. After folding, the filters were pressed into pellets of 12 mm diameter and 3 mm thickness. Under these conditions at the 50 mg sample level inhomogeneity is less than 2.5%.

These pellets were irradiated in a Thetis reactor at a neutron flux of 2.5 $\cdot$ 10^{12} n $\cdot$ cm^{-2} $\cdot$ s^{-1}. A first irradiation of only 2 min together with a Ni neutron flux-monitor was followed by 2 Ge(Li) gamma-spectrometric measurements after decay times of 4 and 22 min, respectively. From this measurement, the following isotopes could be detected: ^{27}Mg, ^{28}Al, ^{49}Ca, ^{51}Ti, ^{52}V, ^{66}Cu, ^{116}In, ^{24}Na, ^{38}Cl, ^{56}Nm, ^{87m}Sr, ^{128}I, ^{139}Ba and ^{165}Dy. Standards have been previously irradiated of these elements together with the same Ni

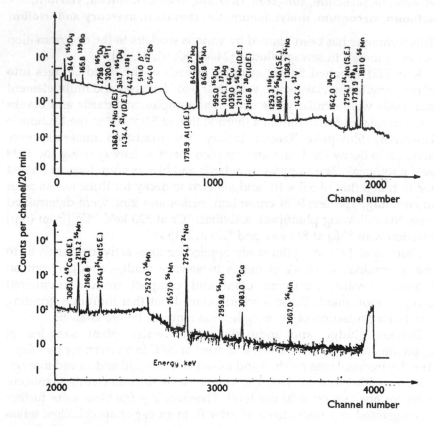

Fig. 6.5 Gamma-spectrum of sludge sample (Genk) obtained after a decay time of 20 min (t_{irr} = 2 min; t_d = 20 min; t_c = 20 min)
Source: Elsevier Sequoia SA, Lausanne [91]

flux-monitors. The same pellet or an additional one containing a slightly larger amount of sample was irradiated for a 2 h period at a neutron flux of $1.5 \cdot 10^{12}$n · cm^{-2} · s^{-1} together with standards. These standards were prepared by spotting solutions with mixtures of the elements to be determined on the cellulose filters, which were also pressed into pellets. Gamma-spectrometry was performed after decay times of one and 15 days. An additional counting after a decay period of approximately four days increases the sensitivity of the detection of a number of elements such as As, Br, Cd, U, etc. The length of these measurements varied between 20 min and 2 h. The following isotopes could be detected in most samples: ^{24}Na, ^{42}K, ^{64}Cu, ^{69m}Zn, ^{76}As, ^{82}Br, ^{115}Cd, ^{122}Sb, ^{140}La, ^{147}Nd, ^{153}Sm, ^{175}Yb, ^{187}W, ^{198}Au, ^{239}Np (daughter of ^{239}U) and ^{46}Sc, ^{51}Cr, ^{59}Fe, ^{60}Co, ^{58}Co (n, p reaction on Ni), ^{65}Zn, ^{75}Se, ^{110m}Ag, ^{124}Sb, ^{134}Cs, ^{131}Ba, ^{141}Ce, ^{152}Eu, ^{153}Gd, ^{160}Tb, ^{169}Yb, ^{177}Lu, ^{233}Pa (daughter of ^{233}Th). Figs. 6.5 and 6.6 illustrate two gamma-

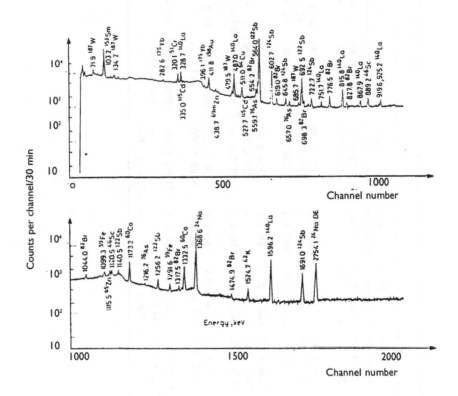

Fig. 6.6 Gamma-spectrum of sludge sample (Genk) obtained after a decay time of 4 days
(t_{irr} = 2 h; t_d = 4 d; t_c = 30 min)
Source: Elsevier Sequoia SA, Lausanna [91]

spectra obtained after decay times of 20 min and four days, respectively. Computerised data reduction of the gamma-spectra was possible after recording the data on a TU 10 magnetic tape unit and transfer to a VAX–11–780 computer. Net peak areas are calculated and after subtraction of blank, concentrations in the dry samples are calculated together with an estimation of the statistical uncertainties.

Dams *et al.* [91] obtained an independent check on the accuracy of their neutron activation analysis method.

Table 6.40 summarises results of analyses by X-ray fluorescence [92], atomic absorption [93] and ICP spectrometry, together with the corresponding mean values as obtained by instrumental neutron activation analysis. The agreement is satisfactory but not excellent.

The mean values as obtained are listed in Table 6.41. The standard deviations given are calculated from the reproducibility of the analyses. It

Table 6.40 Test for accuracy of neutron activation analysis. The standard deviations of the mean values are calculated from the reproducibility of the analyses

Sample	Technics	K, ppm	Ti, ppm	Cr, ppm	Mn, ppm	Fe, %	Co, ppm	Cu, ppm	Zn, ppm	Cd, ppm
Compost	NAA	7900 ± 360	2350 ± 300	55 ± 2	222 ± 15	0.90 ± 0.30	5.4 ± 0.5	275 ± 45	890 ± 50	<20
	XRF[a]	7340 ± 420	2108 ± 150	75 ± 5	299 ± 16	0.99 ± 0.07	–	382 ± 110	955 ± 70	<18
Sludge Genk	NAA	10300 ± 500	7500 ± 600	1190 ± 70	790 ± 20	3.5 ± 0.2	445 ± 20	400 ± 50	2980 ± 180	80 ± 10
	XRF[a]	7130 ± 400	8420 ± 600	1085 ± 80	800 ± 60	3.35 ± 0.25	–	500 ± 140	–	82 ± 6
	AAS[b]	–	–	721 ± 26	766 ± 5	3.46 ± 0.09	367 ± 5	268 ± 2	2860 ± 250	75 ± 1
Sludge Bocholt	NAA	8390 ± 300	1280 ± 310	56 ± 3	1290 ± 50	1.60 ± 0.10	60 ± 5	1300 ± 90	7 ± 2	<18
	XRF[a]	8395 ± 470	680 ± 50	62 ± 4	1540 ± 110	1.62 ± 0.12	–	700 ± 200	1320 ± 95	–
	AAS[b]	–	–	52 ± 2	1280 ± 45	1.58 ± 0.04	66 ± 2	414 ± 12	1290 ± 105	8.2 ± 0.2
Sludge Lommel	NAA	6810 ± 200	3700 ± 400	108 ± 10	350 ± 20	3.6 ± 0.4	36 ± 6	1200 ± 100	13500 ± 1500	130 ± 30
	AAS[b]	–	–	115 ± 2	322 ± 6	3.75 ± 0.05	48 ± 1	14200 ± 400	149 ± 2	–
Indust. Sludge	NAA	<1000	<900	117 ± 7	2700 ± 75	44.6 ± 1.7	34 ± 1	23500 ± 900	18000 ± 100	64 ± 4
	XRF[a]	–	167 ± 2	121 ± 2	3120 ± 220	41.9 ± 3.0	–	–	18400 ± 1300	–
	ICP[c]	–	–	–	–	44.6 ± 2.2	–	22200 ± 1200	–	–

[a] X-ray fluorescence
[b] Flame atomic absorption spectrophotometry
[c] Inductively coupled plasma emission spectrometry

Source: Elsevier Sequoia SA, Lausanne [91]

can be seen that concentration data were obtained for 41 elements in typical sludges from municipal waste water treatment plants. Owing to the extremely large concentrations of iron, zinc and copper in the industrial sludge, the sensitivity of the method was reduced, so that for six elements only upper limits could be given. In the product of a typical municipal compostation plant, the concentrations of 34 elements were determined and the concentrations of the seven other elements were below the detection limits of the method.

Chattopadhyay [90] also applied neutron activation analysis, also photon activation analysis (see section 6.53.4) to the determination of up to 50 elements in sewage sludges. They conducted an interlaboratory comparison of results obtained on sewage sludge samples. Sludge samples were irradiated separately for two different lengths of time at a flux of $1.0 \cdot 10^{12}$n· cm^{-2} · s^{-1} in the University of Toronto Slowpoke–1 Reactor. The third portion was irradiated at a flux of $1.5 \cdot 10^{13}$n · cm^{-2} ·s^{-1} in the McMaster University Nuclear Reactor. Following irradiations for 1 min at Slowpoke, concentrations of 10 elements, namely magnesium, aluminium, calcium, titanium, vanadium, copper, molybdenum and tin could be determined. Sludge samples were irradiated for 15 min at Slowpoke for the measurement of potassium, manganese, indium and barium. Up to 29 elements, namely sodium, scandium, chromium, iron, cobalt, nickel, zinc, arsenic, selenium, rubidium, strontium, zirconium, silver, antimony, caesium, lanthanum, cerium, neodymium, samerium, terbium, ytterbium, lutecium, hafnium, tantalum, tungsten, iridium, gold and mercury, could be determined after irradiations for 60 h at McMaster. The various nuclides and their gamma ray energies used in this study, and the best time for counting from the end of irradiations and typical analyses of raw sludge are given in Table 6.42.

In Table 6.43 are compared results obtained on a standard NBS coal fly ash sample SRM–1633 by neutron activation analysis. For most elements the standard deviation of the neutron activation results is between ±5% and ± 10% of the mean value.

6.53.4 Photon activation analysis

This is a sister technique to neutron activation analysis.

6.53.4.1 Heavy metals (chromium, manganese, iron, cobalt, nickel, zinc, cadmium and lead) and sodium, potassium, rubidium, calcium, barium, strontium, magnesium, arsenic, antimony, bismuth, scandium, titanium, vanadium, selenium, tellurium, yttrium, zirconium, molybdenum, silver, indium, tin, caesium, cerium, mercury, thallium and silicon

Segebade *et al.* [96] applied the photon activation analysis to determination of the distribution of lead, chromium, nickel and arsenic

Table 6.41 Results of analyses of siludges and compost. The concentrations are in ppm unless otherwise indicated

Element	Sludges from municipal plants				Industrial sludge	Compost
	Koersel	Lommel	Bocholt	Genk		
Na, %	0.15 ± 0.005	0.38 ± 0.01	0.46 ± 0.01	0.56 ± 0.2	0.10 ± 0.005	0.40 ± 0.02
Mg, %	0.47 ± 0.10	0.37 ± 0.13	0.43 ± 0.10	0.33 ± 0.10	<0.53	0.32 ± 0.10
Al, %	2.87 ± 0.1	3.44 ± 0.09	1.84 ± 0.06	3.78 ± 0.11	0.20 ± 0.01	1.92 ± 0.10
Cl	800 ± 60	1580 ± 70	1870 ± 80	590 ± 50	14800 ± 750	4030 ± 200
K, %	1.12 ± 0.5	0.68 ± 0.02	0.84 ± 0.03	1.03 ± 0.04	<0.1	0.79 ± 0.04
Ca, %	1.78 ± 0.17	1.31 ± 0.11	1.48 ± 0.12	1.46 ± 0.12	2.67 ± 0.27	2.29 ± 0.18
Sc	4.7 ± 0.2	5.7 ± 0.4	2.3 ± 0.1	5.4 ± 0.3	0.5 ± 0.2	2.0 ± 0.1
Ti, %	0.17 ± 0.05	0.37 ± 0.04	0.13 ± 0.03	0.75 ± 0.05	<0.09	0.23 ± 0.02
V	114 ± 5	62 ± 4	63 ± 3	80 ± 5	20 ± 2	26 ± 3
Cr	70 ± 10	103 ± 10	56 ± 3	1190 ± 70	117 ± 7	55 ± 2
Mn	890 ± 20	350 ± 20	1290 ± 50	790 ± 20	2700 ± 75	222 ± 15
Fe, %	8.8 ± 0.4	3.6 ± 0.4	1.6 ± 0.1	3.5 ± 0.2	44.6 ± 1.7	0.90 ± 0.30
Co	12.5 ± 0.9	36 ± 6	60 ± 5	445 ± 20	34 ± 1	5.4 ± 0.5
Ni	<900	<1800	150 ± 40	320 ± 80	<900	<500
Cu	160 ± 4	1170 ± 240	460 ± 50	450 ± 120	23500 ± 900	290 ± 80
Zn, %	0.17 ± 0.01	1.3 ± 0.1	0.13 ± 0.01	0.30 ± 0.02	1.87 ± 0.10	0.089 ± 0.005
As	43 ± 4	150 ± 20	12 ± 2	47 ± 3	27 ± 2	6.4 ± 0.9
Se	<15	<20	3 ± 1	10 ± 3	<15	<10
Br	40 ± 5	22 ± 3	72 ± 3	27 ± 4	17 ± 1	22 ± 2
Sr	285 ± 50	210 ± 60	<400	245 ± 65	<800	150 ± 30
Ag	13.8 ± 0.5	30 ± 4	4 ± 1	25 ± 3	3.2 ± 0.8	1.2 ± 0.2
Cd	15.3 ± 0.6	130 ± 30	7 ± 2	80 ± 10	64 ± 4	<20
In	0.17 ± 0.03	3.1 ± 0.2	<0.2	0.20 ± 0.03	<0.5	0.08 ± 0.02
Sb	4.45 ± 0.50	22 ± 3	2.85 ± 0.25	605 ± 30	1.5 ± 0.2	6.75 ± 0.50
I	31 ± 4	10 ± 2	53 ± 5	22 ± 3	<20	8.5 ± 3.0

Table 6.41 continued

Element	Koersel	Sludges from municipal plants Lommel	Bocholt	Genk	Industrial sludge	Compost
Cs	2.4 ± 0.4	<7	1.2 ± 0.2	2.4 ± 0.4	<5	1.0 ± 0.5
Ba	500 ± 40	1395 ± 100	385 ± 40	1740 ± 50	<400	1030 ±100
La	25 ± 4	74 ± 9	38 ± 2	54 ± 3	2.4 ± 0.3	6.7 ± 0.7
Ce	65 ± 5	142 ± 14	77 ± 5	106 ± 6	<10	14 ± 2
Nd	<7	28.1 ± 1.7	15 ± 2	17 ± 3	<5	<4
Sm	5.6 ± 0.6	18 ± —	8.9 ± 0.6	10.5 ± 0.3	0.61 ± 0.07	1.3 ± 0.1
Eu	1.2 ± 0.2	3.8 ± 0.3	2.1 ± 0.4	2.7 ± 0.3	<0.5	0.3 ± 0.1
Gd	1.5 ± 0.1	14 ± —	1.3 ± 0.1	1.5 ± 0.7	0.09 ± 0.1	<3
Tb	0.6 ± 0.1	1.5 ± 0.1	0.80 ± 0.07	1.0 ± 0.1	<0.4	<0.4
Dy	—	—	3.1 ± 0.1	3.4 ± 0.1	<1	—
Yb	0.90 ± 0.05	1.95 ± 0.15	1.15 ± 0.15	1.45 ± 0.25	< 0.6	0.24 ± 0.04
Lu	0.30 ± 0.01	0.7 ± 0.1	0.40 ± 0.02	0.58 ± 0.02	0.12 ± 0.03	0.08 ± 0.01
W	2.3 ± 0.4	4.5 ± 0.5	<16	6 ± —	2.0 ± 0.4	20 ± —
Au	0.56 ± 0.06	0.57 ± 0.006	0.145 ± 0.006	0.70 ± 0.06	0.026 ±0.006	0.10 ± 0.03
Th	3.7 ± 0.3	4.2 ± 0.4	1.9 ± 0.1	5.1 ± 0.8	<1	1.9± 0.3
U	4.8 ± 0.6	5.2 ± 0.6	4 ± 1	5.4 ± 0.5	3.1 ± 0.2	0.9 ± 0.4
Pb	395 ± 15[b]	660 ± 25[b]	235 ± 20[a]	690 ± 50[a]	7200 ± 520[a]	800 ± 60[a]
Vol %[c]	34.2 ± 0.2	55.3 ± 0.2	56.5 ± 0.2	39.6 ± 0.3	—	63.8 ± 0.3

[a] Obtained by XRF
[b] Obtained by AAS
[c] Vol: The % volatile material was determined by heating at 600°C

Source: Elsevier Sequoia SA, Lausanne [91]

Table 6.42 Nuclear data and range of element contents, sewage sludge nuetron activation analysis

Element	Activation reaction	Product half-life	Nuclear data Gamma-ray energy, keV	Decay time	Analysis (ppm, airdried basis) raw sludge
Na	^{23}Na (n, γ) ^{24}Na	15 h	1369	2 d	0.12–0.98
Mg	^{26}Mg (n, γ) ^{27}Mg	9.5 m	1014	7 m	0.25–1.05
Al	^{27}Al (n, γ) ^{28}Al	2.3 m	1779	7 m	1.1–2.63
Si					0.16–8.21
K	^{41}K (n, γ) ^{42}K	12.5 h	1525	2 h	0.14–1.18
Ca	^{45}Ca (n, γ) ^{49}Ca	8.8 m	3083	7 m	2.62–5.41
Sc	^{45}Sc (n, γ) ^{46}Sc	84 d	889	3 w	0.76–18.3
Ti	^{50}Ti (n, γ) ^{51}Ti	5.8 m	320	7 m	0.16–0.40
V	^{51}V (n, γ) ^{52}V	3.8 m	1434	7 m	9.51–38.3
Cr	^{50}Cr (n, γ) ^{51}Cr	27.8 d	320	3 w	0.06–0.57
Mn	^{53}Mn (n, γ) ^{56}Mn	2.58 h	847	2 h	0.021–0.1
Fe	^{58}Fe (n, γ) ^{59}Fe	45 d	1099	3 w	0.72–1.97
Co	^{59}Co (n, γ) ^{60}Co	5.26 y	1173	3 w	2.11–7.2
Ni	^{58}Ni (n, p) ^{58}Co	71.3 d	811	3 w	20.2–1800
Cu	^{65}Cu (n, γ) ^{66}Cu	5.1 m	1039	7 m	0.041–2.4
Zn	^{64}Zn (n, γ) ^{65}Zn	245 d	1115	3 w	0.023–0.28
As	^{78}As (n, γ) ^{76}As	26.3 h	559	2 d	1.51–16.3
Se	^{74}Se (n, γ) ^{75}Se	120 d	265	3 w	0.38–8.64
Rb	^{85}Rb (n, γ) ^{86}Rb	18.7 d	1077	3 w	10.3–28.0
Sr	^{84}Sr (n, γ) ^{85}Sr	64 d	514	3 w	5.50–31.6
Y					8.52–28.4
Zr	^{94}Zr (n, γ) ^{95}Zr–^{95}Nb	65.5–35.1 d	757	3 w	15.2–106
Mo	^{100}Mo (n, γ) ^{101}Mo	15 m	191	7 m	1.34–6.71
Ag	^{109}Ag (n, γ) ^{110m}Ag	253 d	658	3 w	3.65–15
Cd					5.2–65.7
In	^{115}In (n, γ) ^{116m}In	54 m	417	2 h	0.06–0.23
Sn	^{124}Sn (n, γ) ^{125m}Sn	9.7 m	332	7 m	4.0–98.6
Sb	^{121}Sb (n, γ) ^{122}Sb	2.8 d	564	2 d	10.2–95.4
Cs	^{133}Cs (n, γ) ^{123}Cs	2.05 y	605	3 w	0.25–6.42
Ba	^{138}Ba (n, γ) ^{139}Ba	83 m	166	2 h	0.01–0.10
La	^{139}La (n, γ) ^{140}La	40.2 h	1596	2 d	10.1–26.4
Ce	^{140}Ce (n, γ) ^{141}Ce	33 d	145	3 w	2.4–14.7
Nd	^{146}Nd (n, γ) ^{147}Nd	11.1 d	531	3 w	ND–10.3
Sm	^{152}Sm (n, γ) ^{153}Sm	47 h	103	2 e	ND–6.44
Tb	^{159}Tb (n, γ) ^{160}Tb	72.1 d	879	3 w	ND–1.20
Yb	^{168}Yb (n, γ) ^{169}Yb	32 d	177	3 w	ND–1.41
Lu	^{176}Lu (n, γ) ^{177}Lu	6.7 d	208	2 d	ND–2.60
Hf	^{180}Hf (n, γ) ^{181}Hf	42.5 d	482	3 w	ND–3.72
Ta	^{181}Ta (n, γ) ^{182}Ta	115 d	1221	3 w	ND–7.06
W	^{186}W (n, γ) ^{187}W	23.8 h	686	2 d	1.16–28.0
Ir	^{191}Ir (n, γ) ^{192}Ir	74 d	316	3 w	ND–3.57
Au	^{197}Au (n, γ) ^{198}Au	2.7 d	412	2 d	ND–1.26
Hg	^{202}Hg (n, γ) ^{203}Hg	46.6 d	279	3 w	2.62–61.4
Bi					ND–8.26
Pb					60.2–2900
Bi					ND–7.13

Source: Elsevier Sequoioa SA, Lausanna [90]

Table 6.43 Element contents (ppm/dry weight basis) of NBS SRM–1633 coal fly ash standard determined by neutron activation analysis

Element	NBS value		Neutron activation result		
Potassium		1.72	1.69	±	0.13
Vanadium	214	± 8	220	±	15
Chromium	131	± 2	135	±	6
Manganese	493	± 7	500	± 15	
Cobalt		38	42	±	1.6
Nickel	98	± 3	95	±	9
Copper	128	± 5	125	±	10
Zinc	210	± 20	200	±	20
Arsenic	61	± 6	59	±	4
Selenium	9.4	± 0.5	9.8	±	1
Rubidium		112	116	±	10
Strontium		1380	1500	±	180
Mercury	0.14	± 0.01	0.16	±	0.04
Lead	70	± 4	71	±	3

Source: Elsevier Sequoia SA, Lausanne [90]

and 24 other elements in sewage farm soils. The technique was not subject to matrix interferences and was shown to be capable of determining several elements which are not amenable to analysis by neutron activation analysis. Chattopadhyay [90] also applied photon activation analysis to the determination of 34 metallic elements in sewage sludges. In this method [90] portions of sludge samples were placed in a rotating sample vial holder assembly and irradiated at a maximum bremsstrahlung energy of 15, 20, 22, 35 and 44 MeV using the University of Toronto 45 MeV electron linear accelerator. The length of irradiation varied between 2 min (for product half-lives less than 10 min) and 6 h (for product half-lives greater than 12 days). The maximum bremsstrahlung energies used for activation, activation products and their gamma ray energies along with the best time for counting and typical analysis of raw sewage are given in Table 6.44.

In Table 6.45 are compared results obtained by this procedure with recommended values for a standard NBS coal fly ash sample SRM–1633. For most elements the standard deviation of the photon activation results is between ± 5% and ± 10% of the mean values.

Chattopadhyay [90] has pointed out that the selection of neutron activation analysis or photon activation analysis methods for determination of these elements in sewage sludges depends mainly on the availability of irradiation facilities. At present, access by analytical chemists to nuclear research reactors for neutron activation surpasses that to electron linear accelerators or betatrons for photon activation.

Table 6.44 Analysis conditions and range of element concentrations (ppm dry weight basis)

Activation reaction	E_{irr} MeV	Product half-life	Gamma-ray energy, keV	Decay time	Analysis of sewage sludge	Reference value[a]
²³Na (γ, n) ²⁴Na	20	15 h	1369	2 d	0.12 ± 0.98	0.17 ± 0.04
²⁶Mg (γ, p) ²⁷Mg	22	15 h	1369	1 d	0.601± 0.05	0.64 ± 0.10
³⁰Si (γ, n) ²⁹Al	20	6.5 m	1273	20 m	0.33 ± 0.03	
⁴¹K (γ, n) ⁴²K	20	7.7 m	2170	20 m	0.20 ± 0.02	0.2 ± 0.3
⁴⁵Ca (γ, p) ⁴⁹Ca	20	22 h	373	1 d	4.5 ± 0.20	4.53 ± 0.47
⁴⁵Sc (γ, n) ⁴⁶Sc	20	3.92 h	1156	2 h	0.75 ± 0.08	
⁵⁰Ti (γ, p) ⁴⁸Sc	22	1.83 d	1039	1 d	0.16 ± 0.02	
⁵¹V (γ, 3n) ⁵²V	44	16 d	983	3 w	26 ± 2	
⁵⁰Cr (γ, n) ⁵¹Cr	35	27.8 d	320	3 w	390 ± 30	391 ± 43
⁵³Mn (γ, n)) ⁵⁶Mn	20	303 d	835	3 w	400 ± 30	394 ± 30
⁵⁸Fe (γ, n) ⁵²Mn	35	5.6 d	1433	3 d	7.55 ± 0.69	7.65 ± 1.37
⁵⁹Co (γ, n) ⁶⁰Co	20	71.3 d	811	3 w	10.2 ± 1.1	
⁵⁸Ni (γ, n) ⁵⁷Ni	35	36 h	1378	12 h	25.6 ± 2	27.8 ± 3.9
⁶⁴Zn (γ, n) ⁶⁵Zn	35	245 d	1115	3 w	0.25 ± 0.02	0.27 ± 0.01
⁷⁸As (γ, n) ⁷⁶As	20	17.9 d	596	2 w	10.2 ± 0.9	
⁸²Se (γ, n) ⁸¹mSe	35	57 m	103	2 h	0.35 ± 0.1	
⁸⁵Rb (γ, n) ⁸⁶Rb	20	33 d	881	3 w	21 ± 2	
⁸⁴Sr (γ, n) ⁸⁵Sr	20	70 m	232	2 h	5.8 ± 0.4	
⁸⁹Y (γ, n) ⁸⁸Y	20	107 d	1836	3 w	25 ± 2	
⁹⁴Zr (γ, n) ⁹⁵Zr	22	78.4 h	909	1 d	8.5 ± 7	
¹⁰⁰Mo (γ, n) ¹⁰¹Mo	35	67 h	140	12 h	2.6 ± 0.3	
¹⁰⁹Ag (γ, n) ¹¹⁰mAg	20	8.4 d	451	2 w	65 ± 4	
¹¹¹Cd (γ, n) ¹¹¹mCd	15	49 m	247	2 h	16.5 ± 1.2	18.7 ± 1.8
¹¹⁵In (γ, γ) ¹¹⁶mIn	15	100 m	393	2 h	<0.1	
¹²⁴Sn (γ, n) ¹²⁵Sn	35	14 d	158	3 w	<8	

Table 6.44 continued

Activation reaction	E_{irr} MeV	Product half-life	Gamma-ray energy, keV	Decay time	Analysis of sewage sludge	Reference value[a]
^{121}Sb (γ, n) ^{122}Sb	15	2.8 d	564	1 d	7.4 ± 0.62	
^{130}Te (γ, n) ^{129}Te	20	69 m	459	2 h	<1	
^{133}Cs (γ, n) ^{123}Cs	15	6.5 d	668	3 d	2.3 ± 0.16	
^{138}Ba (n, γ) ^{139}Ba	35	28.7 h	268	1 d	620 ± 45	
^{140}Ce (γ, n) ^{141}Ce	35	138 d	166	3 w	8.2 ± 0.8	
^{119}Hg (γ, n) ^{197m}Hg	35	24 h	134	12 h	25 ± 2	26.5 ± 2.1
^{203}Tl (γ, n) ^{202}Tl	15	12 d	440	3 w	1.2 ± 0.1	
^{204}Pb (γ, γ) ^{204m}Pb	35	67 m	899	2 h	0.1 ± 0.05	0.10 ± 0.01
^{209}Bi (γ, 3n) ^{206}Bi	44	6.24 d	803	3 d	<1	

[a] Reference values obtained in Round Robin

Source: Elsevier Sequoioa SA, Lausanna [90]

Table 6.45 Element contents (ppm, dry weight basis) of NBS SRM 1633 fly ash standard determined by photon activation analysis

Element	NBS value			Neutron activation result		
Potassium		1.72		1.60	±	0.06
Vanadium	214	±	8	210	±	12
Chromium	131	±	2	131	±	6
Manganese	493	±	7	495	± 15	
Cobalt		38		40	±	2
Nickel	98	±	3	97	±	5
Zinc	210	± 20		215	±	20
Arsenic	61	±	6	60	±	2.6
Selenium	9.4	±	0.5	9.5	±	0.8
Rubidium		112		120	±	10
Strontium		1380		1370	±	120
Cadmium	1.45	±	0.06	1.5	±	0.1
Mercury	0.14	±	0.1	0.13	±	0.03
Thallium		4		3.7	±	0.4
Lead	71	±	3	70	±	4

Source: Elsevier Sequoia SA, Lausanne [90]

Therefore, it is not unexpected that neutron activation analysis can be more widely used than photon activation analysis for routine monitoring of elements that can be equally determined by both the methods. The neutron activation analysis method can also be used to advantage for measuring elements such as aluminium, copper and certain lanthanides (Table 6.42).

Reactions such as (n,p) and/or (n,d) can seriously interfere in the neutron activation analysis determinations of nuclides such as ^{27}Mg, ^{28}Al and others. A correction factor is also needed for the determination of mercury through 279 keV gamma ray of ^{203}Hg because of interference from the 280 keV gamma ray of ^{75}Se which has been detected in neutron irradiated sludge samples. A high resolution Ge(Li) detector is required to eliminate mutual interferences among 554 keV, 559 keV and 564 keV gamma rays of ^{82}Br, ^{76}As and ^{122}Sb respectively.

In photon activation analysis, a wide range of nuclides are produced by various photon-induced reactions such as (γ, γ), (γ, n), (γ, 2n), (γ, 3n), (γ, np), (γ, p), (γ, nα), etc. It is possible to use optimal photon activation products for measuring a number of elements in sewage sludges. The photon activation analysis method is particularly useful for the determination of silicon, yttrium, cadmium, tellurium, thallium, lead and bismuth in sludges. The neutron activation products of these elements are not very suitable for their instrumental measurement in a sludge matrix. Moreover, interferences from major elements such as sodium, magnesium, silicon, chlorine, potassium, calcium, titanium and manganese in sludge can be kept to a minimum by selecting appropriate irradiation energies. Mutual interferences observed in the neutron activation analysis determinations of ^{82}Br, ^{76}As and ^{122}Sb through closely lying photopeaks can be eliminated by choosing alternate photon activation products with well separated photopeaks. Measurements of short-lived nuclides such as ^{27}Mg, ^{49}Ca, ^{51}Ti, ^{52}V, ^{101}Mo, ^{125m}Sn and ^{128}I produced by neutron activation (Table 6.42) can be avoided by adopting the photon activation analysis method and assaying the long-lived photon activation products of these elements.

6.53.5 Spark source mass spectrometry

6.53.5.1 Miscellaneous elements

Spark source mass spectrometry (SSMS) has proved to be one of the most sensitive and comprehensive techniques for trace analysis of sludges [97]. It provides a sensitive method for the simultaneous determination of up to 82 chemical elements in such samples. Municipal sewage sludges from 16 US cities have been analysed (utilising SSMS, neutron activation analyses, emission spectrometry, voltammetry and fluorescence) to survey concentrations of 68 elements.

6.53.6 Differential pulse polarography

6.53.6.1 Heavy metals (chromium, manganese, zinc, iron, cadmium, lead and copper)

Cohen et al. [98] have developed a simple and rapid method to determine traces of chromium, manganese, zinc and iron in sludges produced during treatment of effluent from a large piggery. The samples were fused with sodium hydroxide at a temperature not exceeding 450°C, then cooled and dissolved in mannitol to give a solution equivalent to 0.5M mannitol and 3 M sodium hydroxide. Analysis was by differential pulse polarography with a mercury-drop electrode, a platinum counter-electrode and a silver/silver chloride reference electrode. The method has a detection limit in the ppb range using an initial potential of 300 mV, a final potential of 1700 mV, modulation amplitude 25 mV, sweep velocity 5 mW per s.

Cadmium, lead and copper can be simultaneously determined in sewage sludge by differential pulse anodic scanning voltammetry [99]. To avoid interference by organic constituents of the sample, 10 mL water and 5 mL 1 mL^{-1} sodium acetate/acetic acid are added to 0.5 mL of the sample. A computer controlled instrument (viz the Metrohm VA–Processor 646) is required which divides the sweep into three potential ranges viz –750 to –565V for cadmium, –562 to 350V for lead, and –350 to +100V for copper.

6.53.7 Potentiometric stripping analysis

6.53.7.1 Copper and lead

Pfeiffer Madsen *et al.* [100] applied potentiometric stripping analysis to determination of 7–70 µg g $^{-1}$ lead and 3–110 µg g $^{-1}$ copper in sludge. The precision obtained was in the range 2.5–4.1% for lead and 3.9–4.5% for copper. The accuracy of the method was checked on standard reference materials and found to be typically better than 7%.

A radiometer ISS–820 ion scanning system was employed in this procedure; this consists of an REA–120 ion scanning module plugged into an REC–80 servograph and TTA–80–IS titration cell. Three radiometer standard electrodes were used, ie glassy carbon F–3500, SCE K–4040 and platinum P–1312. The entire system has been thoroughly described [101, 102].

6.53.8 X-ray fluorescence spectroscopy

6.53.8.1 Miscellaneous metals

The application of this technique to the determination of metals in sludges has been discussed by Smits *et al.* [103] and Sleeman [104]. Workers at the Water Research Centre, UK have discussed the application of a wavelength dispersive X-ray fluorescence spectrometer to the determination of 40 elements in sludges. No sample digestion is required and the technique is sensitive and precise. After both spectral and matrix corrections have been made, the corrected intensity is converted into concentration by computer.

6.53.9 Miscellaneous

Corrondo *et al.* [65] compared electrothermal atomic absorption spectrometry with flame atomic absorption spectrometry for the determination of the metal content of sewage sludge following acid digestion and dry ashing. It was found that the precision of the rapid electrothermal atomic absorption procedure compares well with flame atomic absorption in conjunction with suitable pre-treatment.

Other workers who have studied the application of atomic absorption spectrometry to the determination of metals in sewage sludge include Martin [105], Stoveland *et al.* [67], Schwedt and Hockendorf [106].

Das [120] has described a method for determining trace metals in crude sewage which involves pre-treatment of the sewage samples with nitric acid and hydrochloric acid. The metal ions are converted into tartrate complexes which then adsorbed on the strongly basic anion exchange resin Dowex 2X–8 in columns. Elution is carried out and the metal ions present in the elutes are determined using atomic absorption spectrophotometry.

Gel filtration chromatographs [107], electrophoresis [108], atomic absorption spectrometry [107, 110], visible spectroscopy [109], elemental analysis [109], plasma emission spectrometry [110] and centrifugation [110] have all been used in studies of the speciation of metals in sludges.

Kingston and Walter [135] compared microwave digestion with conventional dissolution methods for the determination of metals in sludges.

Baham and Sposito [111] used proton titration to study the water soluble inorganic titratable groups extracted from sewage sludge. The titration curve represented the acid–base chemistry of a mixture of ligands which might be expected to occur in a soil solution shortly after incorporation of the sewage sludge into soil. Trace metal/water soluble extract complexes were chromatographed on a molecular size exclusion gel. Copper formed more stable complexes with the ligands in the water soluble extract than nickel or cadmium. Cadmium was present predominantly as the free metal ion or in inorganic complexes. The primary factor controlling copper solubility was the formation of soluble organo-copper complexes. A 'mixture model' was empirically derived and employed to model qualitatively the speciation of cadmium and copper in two metal-binding experiments.

Sterritt and Lester [112] determined conditional stability constants and complexation capacities for complexation of cadmium, copper and lead by activated sludge. Two methods were compared: direct determination of free metal ion using ion-selective electrodes, and separation of solids by membrane filtration followed by determination of free metal in the filtrate. The effects of several factors, including residual metal concentrations in the sample before filtration, interferences and disturbances of complex equilibria during sample processing, on the calculated values are considered.

Lun and Christensen [113] applied a procedure developed for determining cadmium species to leachates from sewage sludge. The characteristics of the leachates and the cadmium concentrations and species found are tabulated. Leachates from the sludge showed different fractions of cadmium. In most cases, the fraction of free divalent cadmium and the fraction of stable cadmium complexes did not exceed 10% each of the total cadmium content. Significant fractions of stable cadmium complexes (7–25%) were found in leachates containing stable organic matter, but none were found in leachates that were predominantly inorganic.

Angelis and Gibbs [114] used a sequential extraction technique to investigate the association of metals with the different sludge fractions. More than 88% of copper, lead and zinc, and more than 60% of cadmium and chromium, were associated with organic matter and sulphides. The acid-reducible phase was important only for iron (26–40%), and the dissolved metal phase was significant only for cadmium (3–25% of total cadmium). The fate of the metals depended mainly on physico-chemical changes in the oxidisable phase of the sludge, which was the major carrier of the metals.

Griepink *et al.* [115] determined a range of metals in various sewage sludges from different sources by a variety of methods and used the results as a basis for issuing certified values for the concentration of each metal in the sludges selected for use as reference materials. The certified values (after rejection of some unreliable results) and their confidence limits are presented in tabular form.

Van Baardwijk *et al.* [116] determined the caesium-137 content of sewage sludge in May 1986 and used it in a risk analysis. It was concluded that the sludge could continue to be used normally as an agricultural fertiliser. Subsequent measurements of the radioactive contamination at various sites and a model based on the distribution of radioactive caesium in sludges within a waste water treatment plant are presented. It was concluded that the decision to release sludge to agriculture was correct but that in future cases, an immediate and complete survey should be carried out before any decisions were made.

Imhoff *et al.* [117] showed that the radioactivity of digested sewage sludge at sewage works of the Ruhrverband had the additional activity resulting from the accident at the Chernobyl nuclear power plant in April 1986. Changes in the radioactivity of the soil, water and sediment in the Ruhr river, reservoir water and sediment, and aquatic plants are also tabulated.

Erlandsson and Mattsson [118] determined the concentration of gold-198 and thallium-201 medically used isotopes in sludge from a neighbouring sewage treatment plant. The sludge proved to be a very sensitive integrator of radioactive material released from an urban area.

6.53.10 Extraction of metals from sludges

This is discussed at the end of section 6.53.1.1. Regarding the heavy metals, three types of technique seem to have been evolved from the determination of these metals:

(i) open tube digestion with nitric acid or mixtures of nitric acid and hydrochloric acid or mixtures of nitric acid and hydrogen peroxide;
(ii) dry ignition at 450–550°C in zirconium crucibles;
(iii) microwave digestions with nitric acid, mixtures of nitric acid and hydrogen peroxide or mixtures of nitric acid and hydrochloric acid.

Each of these techniques does not necessarily work for every element present in the sample, for example bomb digestions in a microwave oven does not give reliable results for aluminium and magnesium. It is necessary, therefore, to carry out very carefully recovery checks on

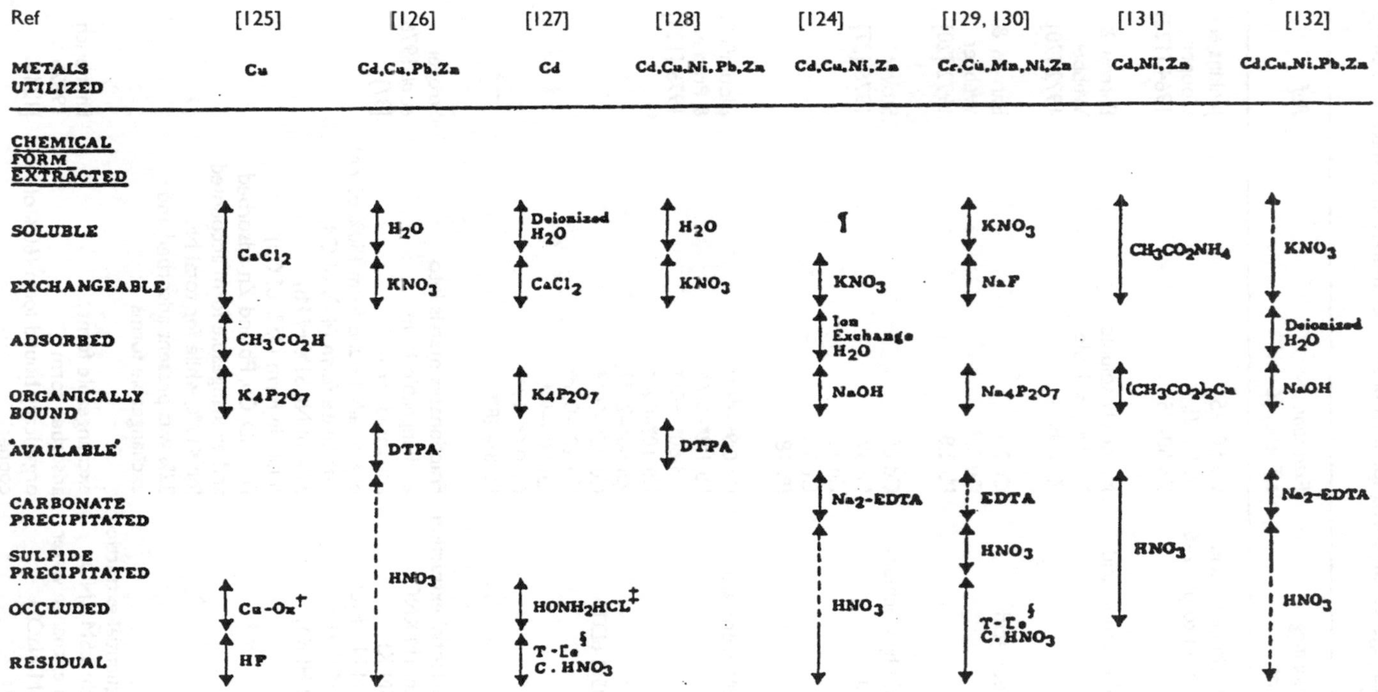

Fig. 6.7 Sequential extraction techniques employed to fractionate trace metals in soils and sewage sludge-amended soils.

Ref	[125]	[126]	[127]	[128]	[124]	[129, 130]	[131]	[132]
METALS UTILIZED	Cu	Cd,Cu,Pb,Zn	Cd	Cd,Cu,Ni,Pb,Zn	Cd,Cu,Ni,Zn	Cr,Cu,Mn,Ni,Zn	Cd,Ni,Zn	Cd,Cu,Ni,Pb,Zn
CHEMICAL FORM EXTRACTED								
SOLUBLE		H_2O	Deionized H_2O	H_2O				
EXCHANGEABLE	$CaCl_2$	KNO_3	$CaCl_2$	KNO_3	KNO_3	KNO_3	$CH_3CO_2NH_4$	KNO_3
ADSORBED	CH_3CO_2H				Ion Exchange H_2O	NaF		Deionized H_2O
ORGANICALLY BOUND	$K_4P_2O_7$		$K_4P_2O_7$		$NaOH$	$Na_4P_2O_7$	$(CH_3CO_2)_2Ca$	$NaOH$
'AVAILABLE'		DTPA		DTPA				
CARBONATE PRECIPITATED		HNO_3			Na_2-EDTA	EDTA	HNO_3	Na_2-EDTA
SULFIDE PRECIPITATED			$HONH_2HCL$‡			HNO_3		
OCCLUDED	Cu–Ox†		T-Ls§ C. HNO_3		HNO_3	T-Ls§ C. HNO_3		HNO_3
RESIDUAL	HF							

† Cu-Ox consists of oxalic acid ($CO_2H)_2$ and oxalate ($CO_2H)_2$; ‡Reagent used consisted of hydroxylamine hydrochloride (HONH2HCl) at pH 2 and sodium dithionate ($Na_2S_2O_4$) in sodium acetate buffer at pH 3.8; § Residual forms determined by subtracting sum of extracted forms from total metal concentration; ¶ Soluble metals separated from saturation extract cake.

Source: Own files

Table 6.46 Chemical agents employed for fractionation of metals in sewage or sewage amended soils

Sample	Percolating agent	Extraction efficiency %	Ref
Sewage sludge	0.095M citric acid buffered to pH 2–6	Cu 50–75 Ni 50–75 Zn 50–75	Jenkins & Cooper 1964 [122]
Sewage sludge	0.42M acetic acid	Mn highly soluble Ni highly soluble Zn >95	Barrow & Webber 1972 [20]
Sewage sludge	Acetic acid	Cr 3.1 Cu 6.9 Pb 2.8	Barrow & Webber 1972 [20]
Sewage sludge	0.5M hydrochloric acid	Cd 69 Ni 59 Zn 73 Cu 24 Pb 18	Stover 1976 [37]
Sewage sludge	0.5M acetic acid 0.05M EDTA	Cr 0.9*–46** Pb 1.2*–5.7* Cd 60*–50** Cu 10*–02** Zn 60*–20** Cd 75*–50** Cu 20*–0.5** Zn 70*–40* Cr 0.6*–17** Pb 8*–28**	Bloomfield & Pruden 1975 [123]
	Sequential extraction with: 1M KNO₃ 0.5M KF 0.1M Na4P₂O₇ 0.1M EDTA 1M HNO₃	Fractionates metals into exchangeable form adsorbed form organically bound form (50% of Zn) carbonate form (49% of Cd, 32% of Ni, 61% of Pb) sulphide form (35% of Cu) For CD, Cu, Pb and Zn adsorbed and exchangeable form accounted for <17%, while for total Ni, 22% was present adsorbed and exchangeable forms	Strover et al. 1976 [37]
	Sequential extraction with: 0.5M KNO₃ ion exchange water 0.5M NaOH	exchangeable form adsorbed form organically bound form (60% of copper)	Emmerich 1982b [124]

Table 6.46 continued

Sample	Percolating agent	Extraction efficiency %	Ref
	0.5M Na₂ EDTA	carbonate form (predominantly copper, nickel and zinc)	
	4M HNO₃	sulphite residual form	
	Sequential extraction with: CaCl₂ acetic acid Na₄ P₂O₇ oxalate	30% of zinc extracted	McLaren & Crawford [125]

* before incubation; ** after incubation

Source: Own files

standard samples of authenticated metal content by the method being considered and using standards similar in type to the sample being analysed.

Probably the most hopeful technique for the future and, certainly the most rapid, is bomb digestion in a microwave oven. Lake *et al.* [121] have reviewed the literature on methods for studying the distribution and speciation of metals in sludge and in soils treated with sludge. These methods include chemical extraction, elutriation and filtration for the solid phase, and chromatographic techniques for the liquid phase.

Some examples of the types of chemical agents that have been employed to fractionate metals in sewage or sewage-amended soils are listed in Table 6.46.

The diversity of reagents used to extract specific metal forms (Fig. 6.7) makes comparison of results difficult. Even when the same reagent is used, the rate and efficiency of leaching will be influenced by the type of sample, the size of particulate, and duration of extraction together with pH, temperature, strength of extractant and ratio of solid matter to volume of extractant. Unfortunately, the effect of varying such parameters has not been examined. Chemical reagents may themselves alter the indigenous speciation of a trace element, and in the choice of extractants, less powerful leaching solutions will probably be more selective for specific fractions than more severe reagents, which may attack other forms, although the overall efficiency may be lower.

Neuhauser [133] compared three solvents (2.5% acetic acid, 0.1N hydrochloric acid and 1.0N hydrochloric acid), commonly used in estimating the availability of heavy metals in soils and plants. The variability in the results obtained in terms of time required for maximum extraction of a range of metals leads the authors to propose that a standard extraction procedure be used by all workers to allow for comparisons of published data.

Rudd et al. [134] compared progressive chemical extraction, progressive acidification to pH 4.2 and 0.5, and repetitive extractions with 0.05 M calcium chloride for speciation of cadmium, copper, nickel, lead and zinc in seven sewage sludges (raw, activated or digested). Speciation in air-dried sludges was more reproducible than in liquid forms. Copper, nickel and zinc distributions were independent of sludge type but cadmium and lead retention depended on sludge physico-chemical properties. Sequential extraction indicated that predominant fractions of lead and zinc were organic and some insoluble inorganic forms. The largest cadmium and nickel fractions were carbonates and the copper fractions sulphides and organic phases. Progressive acidification mobilised significant levels of zinc at pH 4, cadmium and lead at pH 2 and nickel at all pH values studied. Copper was relatively immobile at all pH values. Copper, cadmium and lead forms were relatively stable. Repeated calcium chloride extractions led to a low degree of mobilisation and had little effect on speciation.

References

1 Webster, T. Water Pollution Control, 79, 405 (1980).
2 Gorsuch, T.T. Analyst (London), 84, 135 (1959).
3 HMSO, London, Methods for the Examination of Waters and Associated Materials (1987) 42 pp (40548). Selenium in waters: 1984 Selenium and arsenic in sludges, soils and related materials, 1985 A note on the use of hydride generator kits (1987).
4 Atsuya, I. and Akatsuya, K. Spectrochimica Acta, 36B, 747 (1981).
5 Pal, B.K., Kabiraj, U. and Ukiluddin, M. Analyst (London), 112, 171 (1987).
6 Esprit, M., Vandecasteele, C. and Hoste, S. Analytica Chimica Acta, 185, 307 (1986).
7 Thompson, K.C. and Wagstaff, K. Analyst (London), 104, 224 (1979).
8 Chakraborty, R., Das, A.K., Cervera, M.C. and De la Guardia, M. Journal of Atomic Spectroscopy, 10, 353 (1995).
9 Senesi, N. and Sposito, G. Water Air and Soil Pollution, 35, 147 (1987).
10 Senesi, N. and Sposito, G. Soil Science of America Journal, 48, 1247 (1984).
11 Jahns, G., Schunk, W. and Schwedt, G. Journal of Chromatography, 259, 195 (1983).
12 Magyar, B., Vonmont, H. and Cicciarelli, R. Mikrochimica Acta, 2, 407 (1982).
13 Watanabe, A., Arayashika, H., Mori, Y. and Moriyama, K. Journal of Japan Sewage Works Association, 20, 37 (1983).
14 HMSO, London, Methods for the Examination of Waters and Associated Materials (1982) Molybdenum, especially in sewage sludges and soils by spectrophotometry (1982).
15 Kunselman, G.C. and Huft, E.A. Atomic Absorption Newsletter, 15, 29 (1976).
16 HMSO, London, Standing Committee of Analysts, Methods for the Examination of Waters and Associated Materials, Phosphorus and silicon in effluents and sludges, 1992, 2nd edition (1992).
17 Legret, M. and Divet, L. Analytica Chimica Acta, 189, 313 (1986).

18   Kurochkina, N.I., Lyakh, V.I. and Perelyeva, G.L. Nauch Trudy Irkutsk, Gos Nauchno-issled Inst Redk Metall 1972 (2) 159. Reference Zhur Khim 19GD (13) Abstract No 13G148 (1972).

19   Government of the United Kingdom, Department of the Environment, Analysis of raw potable and waste water Her Majesty's Stationery Office, London 305 pp (1977).

20   Berrow, M.L. and Webber, J. *Journal of Science of Food and Agriculture*, **23**, 93 (1972).

21   American Public Health Association, American Water Works Association and Water Pollution Control Federation, Standard Methods for the Examination of Water and Waste Water, 13th edn (1971).

22   Van Loon, J.C., Lichwa, J., Rutton, D. and Kinnade, J. *Water Air and Soil Pollution*, **2**, 473 (1973).

23   Oliver, B.G. and Cosgrove, E.G. *Water Research*, **8**, 869 (1974).

24   Weaver, J.N., Hanson, A., McGaughey, J. and Steinkruger, F.J. *Water Air and Soil Pollution*, **3**, 327 (1974).

25   Boetlisz, J. *Vom Wasser*, **40**, 1 (1973).

26   Lester, J.N., Harrison, R.M. and Perry, R. *Science of the Total Environment*, **8**, 153 (1977).

27   Stoveland, S., Astruk, M., Perry, R. and Lester, J.N. *Science of the Total Environment*, **9**, 263 (1978).

28   HMSO, London, Methods for the Examination of Waters and Associated Materials 1987 39 pp (40493) Mercury in water, effluents, soils and sediments etc – additional methods (1985).

29   Mitchell, D.G., Mills, W.M., Ward, A.F. and Aldous, K.M. *Analytica Chimica Acta*, **90**, 275 (1977).

30   Kahl, M., Mitchell, D.G. and Aldous, K.M. *Analytica Chimica Acta*, **87**, 215 (1976).

31   Rees, T.D. and Hilton, J. *Laboratory Practice*, **27**, 291 (1978).

32   Corrondo, M.J.T., Perry, R. and Lester, J.N. *Analyst (London)*, **104**, 937 (1979).

33   Geyer, D., Martin, P. and Adrian, P. *Korrespondenz Abwasser*, **22**, 369 (1975).

34   Thompson, K. and Wagstaff, K. *Analyst (London)*, **105**, 883 (1980).

35   Furr, A.K., Kelly, W.C., Bache, C.A., Guterman, W.H. and Lisk, D.J. *Journal of Agriculture and Food Chemistry*, **24**, 889 (1976).

36   Sommers, L.E. *Journal of Environmental Quality*, **6**, 225 (1977).

37   Stover, R.C., Sommers, L.E. and Silviera, D.J. *Journal of Water Pollution Control Federation*, **48**, 2165 (1976).

38   Davis, R.D. and Beckett, P.H.T. *Water Pollution Control*, **77**, 193 (1978).

39   Gould, M.S. and Geneteili, E.J. *Water Research*, **12**, 889 (1978).

40   Thompson, K.C. *Analyst (London)*, **103**, 1258 (1978).

41   Jenniss, S.W., Katz, S.A. and Mount, T. American Laboratory August 18, 20 and 22–23 (1980).

42   Katz, S.A., Jenniss, S.W., Mount, T., Tout, R.E. and Chatt, A. *International Journal of Environmental Analytical Chemistry*, **9**, 209 (1981).

43   Ritter, C.J., Bergman, C.S., Cothern, C.R. and Zamierowski, E.E. *Atomic Absorption Newsletter*, **17**, 70 (1978).

44   Hoenig, M., Van Hoeyweghen, P. and Liboton, J. *Analusis*, **7**, 104 (1979).

45   Fisk, J.L. United States Environmental Protection Agency, Effluent guidelines, private communication (1979).

46   Ritter, C.J. *American Laboratory*, **14**, 72 (1982).

47   Moriyama, A., Wanatabe, A., Sugiura, S., Arayashiki, H. and Mori, Y. *Journal of Japan Sewage Works Association*, **19**, 68 (1982).

48   Legret, M., Demare, P., Marehandase, P. and Robbe, D. *Analytica Chimica Acta*, **149**, 107 (1983).

49   Nielsen, J.S. and Hrudey, S.E. *Environmental Science and Technology*, **18**, 130 (1984).

50 American Public Health Association, Standard Methods for the Examination of Water and Waste Water, 15th edn, American Public Health Association Washington DC Pt 302 (1980).

51 US Environmental Protection Agency, *Methods for Chemical Analysis of Water and Wastes*, US Environmental Protection Agency, Cincinnati, Ohio No EPA–625 16/74–003a p 81 (1976).

52 HMSO, London, Methods for the Examination of Waters and Associated Materials, Cadmium chromium copper lead nickel and zinc in sewage sludes by atomic absorption spectrometry (1981).

53 Department of the Environment (National Water Council Standing Technical Committee of Analysts), HMSO, London, Methods for the Examination of Waters and Associated Materials, 1983 (17 pp) (22BCENU) Determination of extractable metals in soils, sewage sludge treated soils and related materials 1982 (1983).

54 Legret, M., Divet, L. and Demare, D. *Analytica Chimica Acta*, **175**, 203 (1985).

55 National Bureau of Standards Special Publication No 719.

56 Kruse, D. Presented at Illinois Water Pollution Control Association Meeting, Naperville, Illinois, Microwave digestion of environmental samples for trace metal analysis, 15th May (1986).

57 CEM Corporation, Oxford Laboratories, High Wycombe, UK.

58 Christensen, T.H., Pedersen, L.R. and Tjell, J.C. *International Journal of Environmental Analytical Chemistry*, **12**, 41 (1982).

59 Tjell, J.C. and Knudsen, J. *Results of Nordic Intercalibration for Analysis of Trace Metals in Agricultural Samples*, Dept of Sanitary Engineering, Technical University of Denmark, Lyngby (1976).

60 Tjell, J.C. In *Treatment and Use of Sewage Sludge*. D. Alexander and H. Ott (eds). First European Symposium, Cadarache, France, 13–15 Feb (1979). Commission of the European Communities, Brussels, Belgium p 134 (1979).

61 Muntau, H. and Leschber, R. In *Characterization, Treatment and Use of Sewage Sludge*. P. L'Hermite, H. Ott (eds). Reidel Publishing Co, London, England, pp 235–250 (1981).

62 Delfino, J.J. and Enderson, R.E. *Water and Sewage Works*, **125**, 32 (1978).

63 Van Loon, J.C. and Lichwa, J. Environmental Letters, **44**, 1 (1973).

64 Andersson, A. *Swedish Journal of Agricultural Research*, **6**, 145 (1976).

65 Corrondo, M.J.T., Perry, R. and Lester, J.N. *Analytica Chimica Acta*, **106**, 309 (1979).

66 Corrondo, M.J.T., Perry, R. and Lester, J.N. *Science of the Total Environment*, **12**, 1 (1979).

67 Stoveland, S., Astrue, M., Perry, R. and Lester, I.N. *Science of the Total Environment*, **13**, 33 (1979).

68 Grabner, E., Hegi, R. and Guggenbühl, B. *ISWA Journal*, **30**, 15 (1980).

69 Van Loon, J.C. Environment Canada, Research Programme for Abatement of Municipal Pollution, Research Report No 51, 42 pp. Heavy metals in Agricultural Lands receiving Chemical Sewage Sludge (1976).

70 Smith, R. *Water, South Africa*, **9**, 31 (1983).

71 Kurfurst, K. and Rues, B. *Fresenius Zeitschrift für Analytische Chemie*, **308**, 1 (1981).

72 Corrondo, M.J.T., Lester, J.N. and Perry, R. *Talanta*, **26**, 929 (1979).

73 Agemian, H. and Chau, A.S. *Analytica Chimica Acta*, **80**, 61 (1975).

74 Bowker, A.H. and Liebermann, G.J. *Engineering Statistics*, Prentice Hall, New Jersey (1972).

75 Davis, R.D. and Carlton-Smith, C.H. *Water Pollution Control*, **82**, 3 (1983).

76 Kempton, S., Sterritt, R.M. and Lester, J.N. *Talanta*, **29**, 675 (1982).

77 Beckett, P.H.T. *Water Pollution Control*, **77**, 539 (1979).
78 Furr, A.K., Parkinson, T.F., Wachs, C.A., Bache, W.H., Gutermann, P., Wszolek, L.S., Pakkola, L.S. and Lisk, D.J. *Environmental Science and Technology*, **13**, 1503 (1979).
79 Krishamurty, K.V., Shpirt, E. and Reddy, M.M. *Atomic Absorption Newsletter*, **15**, 68 (1976).
80 Arbab-Zavar, M.H. and Howard, A.G. *Analyst (London)*, **105**, 744 (1980).
81 Sinemus, H.W., Mecker, M. and Welz, B. *Atomic Spectroscopy*, **2**, 81 (1981).
82 Thompson, K.C. and Thomerson, D.R. *Analyst (London)*, **99**, 595 (1974).
83 Wauchope, R.D. *Atomic Absorption Newsletter*, **15**, 64 (1976).
84 Fiorino, J.A., Jones, J.W. and Capar, S.G. *Analytical Chemistry*, **48**, 120 (1976).
85 Analytical Methods Committee, *Analyst (London)*, **105**, 66 (1980).
86 Sterritt, R.M. *Analyst (London)*, **105**, 616 (1980).
87 Hawke, D.J. and Lloyd, A. *Analyst (London)*, **113**, 413 (1988).
88 Schramel, P., Xu, L.Q., Wolf, A. and Hasse, S. *Fresenius Zeitschrift für Analytische Chemie*, **313**, 213 (1982).
89 Nadkari, R.A. and Morrison, G.H. *Environmental Letters*, **6**, 273 (1974).
90 Chattopadhyay, A. *Journal of Radioanalytical Chemistry*, **37**, 785 (1977).
91 Dams, R., Buysse, A.M. and Helsen, M. *Journal of Radioanalytical Chemistry*, **68**, 219 (1982).
92 Alluyn, F. Bepaling van zware metalen in slib en compost, door X-straal fluoriscentie. Thesis University of Ghent (1981).
93 Steegens, R. and Demuynck, M. Limbergs Centrum voor Toegepaste Ecologie, Diegenbeek, Belgium, private communication (1980).
94 Wiseman, B.H.F. and Bodri, G.M. *Journal of Radioanalytical Chemistry*, **24**, 313 (1975).
95 Bindra, K. MSc Thesis University of Toronto, Ontario (1975).
96 Segebade, C., Schmitt, B.F., Fusban, H.U. and Kuhl, M. *Fresenius Zeitschrift für Analytische Chemie*, **317**, 413 (1984).
97 Furr, A.K., Lawrence, A.W., Toug, S.S.C., Grandolfo, M.C., Hofstader, R.H., Bache, C.A., Gutenmann, W.H. and Lisk, D.J, *Environmental Science and Technology*, **10**, 683 (1976).
98 Cohen, E.S., Combes, R., Kammour, S. and Deville, D. *Comptes Rendus de l'Academie Bulgare des Sciences*, **36**, 1187 (1983).
99 Crompton, T.R. unpublished work.
100 Pheifler Madsen, P., Drobaek, I. and Sørensen., J. *Analytica Chimica Acta*, **151**, 479 (1983).
101 Graaback, A.M. and Jensen, O.S. *Industrial Research*, **21**, 124 (1979).
102 Labar, C. and Lamberts, L. *Analytica Chimica Acta*, **132**, 23 (1981).
103 Smits, J. and Van Grieken, R. *Analytica Chimica Acta*, **88**, 97 (1977).
104 Sleeman, P. *Water Services*, **86**, 431 (1982).
105 Martin, P. *Abwassertechnik*, **30**, 15 (1979).
106 Schwedt, G. and Hockendorf, A. *Zeitschrift für Wasser and Abwasser Forschung*, **19**, 72 (1986).
107 Lawson, P.S., Sterritt, R.M. and Lester, J.N. *Water Air and Soil Pollution*, **21**, 387 (1984).
108 Pupella, A., Campanella, L., Cardarelli, E., Ferri, T. and Petronio, B.M. *Science of the Total Environment*, **64**, 295 (1987).
109 Hernandez, T., Moreno, J.I. and Costa, F. *Biological Wastes*, **26**, 167 (1988).
110 Carre, J. and Welte, B. *Environmental Technology Letters*, **7**, 351 (1986).
111 Baham, J. and Sposito, G. *Journal of Environmental Quality*, **15**, 239 (1986).
112 Sterritt, R.M. and Lester, J.N. *Water Research*, **19**, 315 (1985).
113 Lun, X.Z. and Christensen, T.H. *Water Research*, **23**, 81 (1989).

114 Angelis, M. and Gibbs, R.J. *Water Research*, **23**, 29 (1989).
115 Griepink, B., Muntau, H. and Colinet, E. *Fresenius Zeitschrift für Analytische Chemie*, **318**, 490 (1984).
116 Van Baardwijk, F.A.N., de Vries, P.J.K. and Griffioen, S. *H20*, **20**, 528 (1987).
117 Imhoff, K.R., Koppe, P. amd Dietz, F. *Water Research*, **22**, 1059 (1988).
118 Erlandsson, B. and Mattsson, S. *Water Air and Soil Pollution*, **9**, 199 (1978).
119 Van de Wall, C.G.J., Van den Akker, A.H. and Stoks, P.G.M. *H2O*, **21**, 320 (1988).
120 Das, A.K. *Indian Journal of Environmental Health*, **22**, 130 (1980).
121 Lake, D.L., Kirk, P.W.W. and Lester, J.N. *Journal of Environmental Quality*, **13**, 175 (1984).
122 Jenkins, S.H. and Cooper, J.S. *International Journal Air Water Pollution*, **8**, 695 (1964).
123 Bloomfield, C. and Pruden, G. *Environmental Pollution*, **8**, 217 (1975).
124 Emmerich, W.E.L., Lund, J., Page, A.L. and Chang, A.G. *Journal of Environmental Quality*, **11**, 175 (1982).
125 McLaren, R.G. and Crawford, D.V. *Journal of Soil Science*, **24**, 172 (1973).
126 Silviera, D.J. and Sommers, L.E. *Journal of Environmental Quality*, **6**, 47 (1977).
127 Alloway, B.J.M., Gregson, S.K., Gregson, R. and Tills, A. In Management and Control of Heavy Metals in the Environment. International Conference, London, September, CEP Consultants Ltd, Edinburgh (1979).
128 Pettruzzelli, G., Lubrano, L. and Guidi, G. *Environmental Technology Letters*, **2**, 449 (1981).
129 Schalscha, E.B., Morales, M., Ahumada, I., Schirado, T. and Pratt, P.F. *Agrochimica*, **24**, 361 (1980).
130 Schalscha, E.B., Morales, M., Vergara, I. and Chang, A.C. *Journal of Water Pollution Control Federation*, **54**, 175 (1982).
131 Soon, Y.K. and Bates, T.E. *Journal of Soil Science*, **33**, 477 (1982).
132 Sposito, G.L., Lund, J. and Chang, A.C. *Soil Science Society of America Journal*, **46**, 260 (1982).
133 Neuhauser, E.F. *Journal of Environmental Quality*, **9**, 21 (1980).
134 Rudd, T., Lale, D.L., Mehrotra, I., Sterritt, R.M., Kirk, P.W.W., Campbell, J.A. and Lester, J.N. *Science of the Total Environment*, **74**, 149 (1988).
135 Kingston, H.M. and Walter, P.J. *Spectroscopy*, **7**, 20 (1992).

Chapter 7

Determination of anions in soil

7.1 Borate

7.1.1 Spectrophotometric methods

A method has been described [1] for the determination of borate in soils based on conversion of borate in a hot water extract to fluoroborate by the action of orthophosphoric acid and sodium fluoride.

The concentration of fluoroborate is measured spectrophotometrically as the blue complex formed with methylene blue which is extracted into 1,2 dichloroethane. Nitrates and nitrites interfere but these can be removed by reduction with zinc powder and orthophosphoric acid.

Aznarez et al. [2] used curcumin as a chromogenic reagent to estimate borate in soils. The borate is extracted from the sample with 2-methylpentane-2,4-diol into methylisobutyl ketone. The selectivity of the extraction of boric acid with 2-methylpentane-2,4-diol into methyliso-butyl ketone provides a pre-concentration method and the simultaneous elimination of numerous interferents. In this procedure 0.2–1g of finely ground soil is digested with 5 mL of concentrated nitric acid–perchloric acid (3 + 1) in a PTFE liquid pressure bomb at 150°C for 2 h. The solution is cooled and diluted and any residue is filtered off through Albet 242 filter paper. Acidity is neutralised with 6M sodium hydroxide and the solution diluted to 100 mL with hydrochloric acid (1 + 1) in a calibrated flask. A portion of this solution containing 10 to 100 μg boron is extracted three times with 10 mL portions of methylisobutyl ketone in order to eliminate iron interference then extracted with 10 mL of 20% v/v 2-methylpentane-2,4 diol in methylisobutyl ketone and the extract is shaken for 5 min and, finally, dried by the addition of 1g of anhydrous sodium sulphate. To carry out spectrophotometry 3 mL of the organic phase is transferred into a polyethylene test tube and 2 mL of a 0.1% m/v solution of curcumin in glacial acetic acid and 2 mL concentrated phosphoric acid added. The mixture is shaken and heated to 70 ± 3°C for 1 h. After cooling the absorbance of the solution is measured at 510nm against a reagent blank (Fig. 7.1). A calibration graph is prepared by adding various volumes of

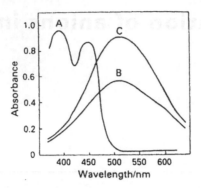

Fig. 7.1 Absorption spectra of A, reagent blank solution measured against IBMK as reference; B, boron–curcumin compound against reagent blank, 3 µg mL⁻¹ of boron; and C, as B, 5 µg mL⁻¹ of boron
Source: Royal Society of Chemistry [2]

standard borate solution containing 10–100 µg boron and to this adding an equal volume of hydrochloric acid (1 + 1). This solution is then extracted with 100 mL of 20% v/v methyl/pentane-2,4-diol and spectrophotometric measurements carried out as described above.

The calibration graph at 510nm is a straight line and Beer's law is obeyed from 0.5 to 5 µg mL⁻¹ of boron in the final measured solution (corresponding to 10–110 µg of boron in the aqueous phase). The molar absorptivity, calculated from the slope of the statistical working calibration graph at 510nm, was 2905 L mol⁻¹ cm⁻¹. The Sandell sensitivity was 0.011 µg cm² of boron. The precision of the method for 10 replicate determinations was 0.6%. The absorbance of the reagent blank solution at 510nm was 0.010 ± 0.003 for 10 replicate determinations. Therefore, the detection limit was 0.04 µg mL⁻¹ of boron in the final measured solution.

7.1.2 Fluorescence spectroscopy

Aznarez *et al.* [2] also applied molecular fluorescence spectroscopy to the determination of borate in soils. The soil extraction procedure is the same as described above in section 7.1.1. 3 mL of this organic extract is transferred into a polyethylene test tube with a hermetic cap and 2 mL 0.1% m/v dibenzoylmethane in methylisobutyl ketone and 2 mL concentrated phosphoric acid added. The solution is shaken for 2 min and then heated at 80 ± 3°C for 30 min. The selective fluorescence intensity of this is measured at 400nm, within 45 min, with excitation at 390nm and 0.05% m/v quinine sulphate in 0.1M sulphuric acid as reference solution. Aznarez *et al.* [2] employed a Pye–Unican SP 8100 spectrometer with special equipment for fluorescence measurements. The procedure is calibrated against portions of a standard borate solution

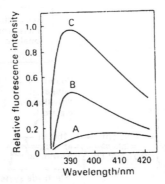

Fig. 7.2 Fluorescence excitation spectra against quinine sulphate solution as reference: A, reagent blank solution; B, boron–DBM, 50 µg L $^{-1}$ of boron; and C, boron–DBM, 100 µg L $^{-1}$ of boron
Source: Royal Society of Chemistry [2]

containing 0.5–5 µg of boron and to each of which is added an equal volume of hydrochloric acid (1 + 1). The fluorescent excitation spectrum of the boric acid – DBM compared in IBMK against quinine sulphate solution is shown in Fig. 7.2. The wavelength of the maximum excitation radiation was 390nm. The maximum relative fluorescence intensity was measured at 400nm or by using a Kodak 2B cut-off filter (400nm cut-off).

7.2 Bromide

Bromide occurs in soils mainly as a breakdown product of methyl bromide fumigant which is added to soils for crop protection.

7.2.1 Flow injection analysis

Van Staden [3] employed flow injection analysis coupled with a coated tubular solid-state selective electrode for the determination of bromide in soils. Soil extracted samples are injected into 10 mol L $^{-1}$ potassium nitrate carrier solution containing 100 mg L $^{-1}$ chloride as an ionic strength adjustment buffer. The sample buffer zone formed is transported through the bromide selective electrode onto the reference electrode. The method is applicable in the range 10–50000 mg L $^{-1}$ bromide. The coefficient of variation of this method is better than 1.6%.

Method

The basic design of the coated tubular flow-through solid-state bromide-selective membrane electrode used by Van Staden [3] was the same as that he used for the construction of a chloride-selective electrode [4]. The unit consisted of 0.025 mm thick silver metal foil wound around two pieces of

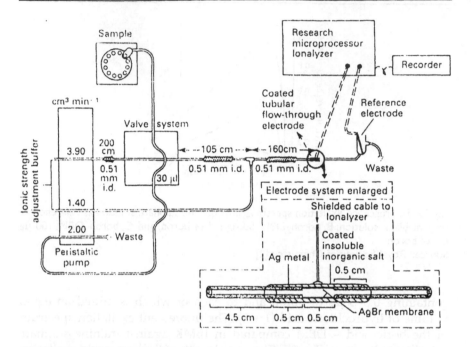

Fig. 7.3 Manifold and flow diagram of the FIA system. Valve loop size, 30 μL; sampling time, 45 s; wash time, 0 s; valve actuation at 43 s; sampling rate, 80 h⁻¹
Source: Royal Society of Chemistry [3]

Tygon tubing at both ends. An inner wire of a shielded cable was wound around the outside body of the silver metal cylinder to ensure electrical contact between the electrode and the Ionalyzer instrument. The whole unit was isolated with Araldite epoxy resin. The silver–silver bromide electrode was activated by anodic deposition of silver bromide as a fine membrane on the inner wall of the tubular silver cylinder. The coating was carried out at a current density of about 20 mA cm $^{-2}$ and 0.1 mol dm $^{-3}$ potassium bromide solution was circulated at a rate of 1.6 cm ^{3} min $^{-1}$ through the tubular cylinder.

A schematic diagram of the flow injection system used is illustrated in Fig. 7.3. A Carle microvolume two–position sampling valve (Carle No 2014) containing two identical sample loops was used. Each loop has a volume of 30 μL. A Cenco sampler unit was used to supply a series of samples to the sampling valve system. The timing of the sampler unit was 45 s for sampling with zero wash time and valve actuation at 45 s. A Cenco peristaltic pump operating at 10 rev min $^{-1}$ supplied the carrier and reagent streams to the manifold system; the sampling valve system was synchronised with a Cenco sampler unit. Tygon tubing (0.51 mm id) was used to construct the manifold; coils were wound round suitable lengths of glass tubing (15 mm od).

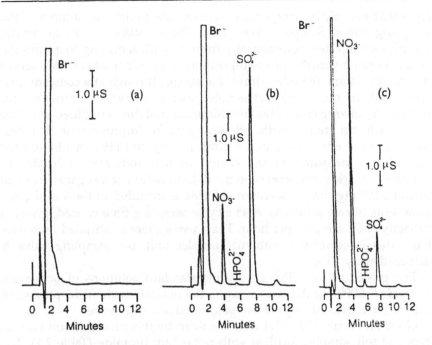

Fig. 7.4a–d Determination of anions in soil extracts: a. blank 10 mol L⁻¹ KCl; b. soil sample
A 10 mmol L⁻¹ KCl extract; c. soil sample B 10 mmol L⁻¹ KCl extract 1:500 dil.
AMPIC–NGI should also be used in series to remove humic acids
Source: Royal Society of Chemistry [3]

The tubular flow-through bromide-selective electrode was incorporated
into the conduits of the flow injection system as shown in Fig. 7.3. The
potentials were measured at room temperature with an Orion Research
(Model 901) micro-processor Ionalyzer. The detector output was recorded
with a two-channel Cenco recorder (Model 34195–041). The constructed
flow-through tubular indicator electrodes were used in conjunction with
an Orion 90–02 double-junction reference electrode with 10% m/V
potassium nitrate as the outer chamber filling solution.

Soil extraction
The samples were dried overnight at 105°C and ground to pass through a
2 mm sieve. The samples were transferred into screw-capped glass jars.
ISA solution (25 cm³), containing 0.1 mol dm⁻³ potassium nitrate and 100
mg dm⁻³ potassium chloride was added to each sample. The jars were
shaken vigorously on a reciprocating shaker for 39 min. The suspensions
were centrifuged, the supernatant liquid filtered (Whatman No 41 filter
paper) and bromide was determined in the filtrate.

The carrier stream (1 mol dm⁻³ potassium nitrate) is pumped at a
constant flow rate of 3.90 cm³ min⁻¹ (Fig. 7.4). A pulse suppressor coil (200

cm × 0.51 mm id) is incorporated between the peristaltic pump and the sampling valve. Samples taken from the turntable of an automatic sampler are injected automatically from a 30 μL sampling loop into the carrier stream by means of a two-position valve. Whereas one loop serves the carrier stream, the other draws the sample through at a constant flow rate of 2.0 cm^3 min^{-1}. Injected samples are mixed with the carrier stream in a 105 cm mixing coil. Potassium nitrate (1 mol dm^{-3}) is added at a flow rate of 1.40 cm^3 min^{-1} further downstream for improvement of hydrodynamic flow and mixed in a second mixing coil (160 cm) before the potential is measured in the coated tubular indicator electrode. To eliminate chloride interference in the determination of inorganic bromide in soils, 100 mg dm^{-3} potassium chloride is included in the 1 mol dm^{-3} potassium nitrate solutions. A 45 s cycle sampling time is used, giving a capacity of 80 samples per hour. The valve system is actuated on a time basis that is correlated with the sampler unit; the sampling valve is actuated every 43 s.

The procedure is calibrated against standard solutions of potassium bromide 5–5000 mg dm^{-3} made up in the potassium chloride–potassium nitrate ionic strength adjustment solution discussed above.

Good recoveries (90–98%) were obtained by this procedure on various types of soil samples fortified with potassium bromide (Table 7.1). The performance and reproducibility of the flow injection potentiometric method are shown in Table 7.2. In addition to a high sample throughput (80 h^{-1}) over a wide concentration range, the procedure is characterised by good reproducibility (< 1.6%). Direct flow injection potentiometric measurement of bromide ion in soil gave results fairly similar to those obtained by a standard iodimetric titration method (Table 7.2) with thiosulphate [5, 6].

7.2.2 Neutron activation analysis

Gladney and Perrin [7] used epithermal neutron activation analysis to determine down to 50 ppb bromine in US geological survey reference soils GXR–2, GXR–5 and GXR–6, and Canadian certified reference soils SO–1, SO–2, SO–3 and SO–4. The values reported in Table 7.3 indicate that good agreement was obtained between neutron activation analyses results and recommended values. The relative standard deviation was of the order of ± 10% over the concentration range 1–15 ppm bromine.

7.2.3 Gas chromatography

Roughan et al. [8] have described a gas chromatographic method for carrying out this analysis in which the soil is mixed with sodium hydroxide solution and then treated with ethanol prior to evaporation to dryness. After muffling, the residue is digested with sulphuric acid and to

Table 7.1 Recovery of inorganic bromide in soil samples fortified with potassium bromide

Bromide added/ μg g⁻¹	Bromide recovered							
	Sandy soil		Loamy sand soil		Sandy loam soil		Loam soil	
	Mass*/ $\mu g\ g^{-1}$	Mass* %	Mass*/ $\mu g\ g^{-1}$	Mass* %	Mass*/ $\mu g\ g^{-1}$	Mass* %	Mass*/ $\mu g\ g^{-1}$	Mass* %
0	<1	–	<1	–	<1	–	<1	–
3	3.1 ± 0.05	103	3.1 ± 0.05	103	2.9 ± 0.05	97	2.8 ± 0.04	93
5	5.2 ± 0.08	104	5.1 ± 0.08	102	4.9 ± 0.07	98	4.8 ± 0.09	96
10	9.6 ± 0.13	96	9.5 ± 0.14	95	9.5 ± 0.14	95	9.4 ± 0.15	94
20	19 ± 0.25	95	19 ± 0.27	95	19 ± 0.26	95	18 ± 0.26	90
60	59 ± 0.70	98	58 ± 0.74	97	57 ± 0.72	95	56 ± 0.74	93
100	96 ± 1.2	96	97 ± 1.1	97	94 ± 1.2	94	92 ± 1.4	92
250	246 ± 2.1	98	247 ± 2.0	99	246 ± 2.0	98	244 ± 2.0	98
1000	988 ± 4.6	99	989 ± 4.4	99	979 ± 4.6	98	974 ± 4.5	97

*Average of five soil replicate samples ± sandard error

Source: Royal Society of Chemistry [3]

Table 7.2 Performance and reproducibility of the flow injection method (FIA) for the determination of bromide in soil. Comparison of inorganic bromide concentrations (µg g^{-1}) recovered from soils using the proposed FIA method against a standard iodimetric titration method

Soil sample	Bromide concentration/µg g^{-1}		Coefficient of variation, *%
	Titration	FIA	
Sandy soil:			
1	3.0	3.1	1.59
2	9.5	9.6	1.48
3	58	59	1.27
4	247	246	0.82
5	987	988	0.43
Loamy sand soil:			
1	5.2	5.1	1.56
2	19	19	1.40
3	96	97	1.28
4	248	247	0.81
5	990	989	0.44
Sandy loam soil:			
1	2.9	2.9	1.58
2	19	19	1.41
3	56	57	1.28
4	95	94	1.09
5	247	246	0.81
Loam soil:			
1	4.7	4.8	1.57
2	9.5	9.4	1.48
3	57	56	1.28
4	245	244	0.81
5	977	974	0.44

* Mean result of 14 tests in each instance with relative standard deviation for the flow injection method

Source: Royal Society of Chemistry [3]

this solution are added acetonitrile and ethylene oxide:

$$H+ + BR^- + C_2H_4O \rightarrow HOCH_2 CH_2 Br$$

The 2-bromoethanol produced in this reaction is examined by gas chromatography using an electron capture detector. At the 10 mg kg^{-1} bromide level in soil a standard deviation of ±0.34 mg kg^{-1} was obtained, ie coefficient of variation was ±3%. Recoveries from soil were 81–94%.

Table 7.3 Bromine concentrations in Canadian certified reference soils

Soil ref. no.	Soil type	Bottle no.	Br, ppm, X ± α	Recommended value
SO–1	Regosolic	133	1.3 ± 0.1	1.4 ± 0.2
	clay	711	1.4 ± 0.2	
SO–2	Podzolic	97	14.8 ± 1.0	15 ± 2
	B horizon	903	15.2 ± 1.5	
SO–3	Calcareous	495	5.5 ± 0.7	5.2 ± 0.8
	C horizon	1023	4.8 ± 0.2	
SO–4	Chernozemic	103	5.5 ± 0.3	5.6 ± 0.5
	A horizon	441	5.8 ± 0.5	

Source: American Chemical Society [7]

7.3 Carbonate

7.3.1 Gasometric method

Collins [9] has described a gasometric method for the determination of carbonate in soil based on reaction with hydrochloric acid and subsequent measurement of the volume of carbon dioxide produced. This is also the basis of a standard HMSO method for the determination of carbonate in soils [10].

7.4 Chlorate

7.4.1 Spectrophotometric method

Banderis [11] has described a method for the determination of chlorate in water extracts of soil based on its conversion to free chlorine upon reaction with hydrochloric acid, followed by spectrophotometric evaluation of chlorine at 448nm by the spectrophotometric o–toluidine method. A correction is made for interference by iron(III), nitrite, free chloride derived from hypochlorites and strong oxidising agents by subtracting the absorbance of a modified blank, containing a lower concentration of hydrochloric acid, from that obtained in the test.

A standard UK method for the determination of chlorate is based on this procedure. Chlorate in water-extracts of soil has been determined by a spectrophotometric method based on the conversion of chlorate to 'free' chlorine by the addition of hydrochloric acid followed by reaction of chlorine with o-tolidine to produce a yellow colour which is evaluated spectrophotometrically at 448 nm [12].

7.5 Chloride

Chloride has been determined in soils by methods based on titration [13, 14], potentiometry [15, 16], ion selective electrodes [15, 17] and ion chromatography [18, 19].

7.5.1 Spectrophotometric method

In a method [13, 14] for determining chloride in a calcium sulphate extract of soil, the extract is acidified and the concentration of chloride determined by titration with 5 mmol L^{-1} mercuric chloride using diphenylcarbazone as indicator. Mercuric ion in the presence of chloride forms mercuric chloride which, although soluble, provides sufficient mercuric ion to form the mercuric diphenylcarbazone complex. When all the chloride has been so removed, addition of further mercuric ion produces a violet complex.

7.5.2 Potentiometric methods

Davey and Bembrick [15] have described a method for the determination of chloride in water extracts of soils based on measurement of the EMF developed between two silver–silver chloride electrodes in a cell with a liquid junction and suitable electrolyte.

McLeod et al. [17] carried out simultaneous measurement of chloride, pH and electrical conductivity in soil suspensions using a triple electrode system mounted on a single unit. A glass electrode and a silver–silver chloride electrode with a common reference electrode and two pH meters were used for the determination of pH and chloride respectively. The coefficient of variation obtained in determinations of chloride at the 50 mg L^{-1} level was approximately 7.3%.

7.5.3 Ion chromatography

Bradfield and Cooke [18] described an ion chromatographic method using a UV detector for the determination of chloride, nitrate, sulphate and phosphate in water extracts of soils. Soils are leached with water and Dowex 50–X4 resin added to the aqueous extract which is then passed through a Sep–Pak C_{18} cartridge and the eluate then passed through the ion chromatographic column. The best separation of these ions was obtained using a 5×10^{-1} mol L^{-1} potassium hydrogen phthalate solution in 2% methanol at pH 4.9. A reverse phase system was employed. Retention times were 5.5, 7.9, 12.6 and 18 min for chloride nitrate phosphate and sulphate respectively. Recoveries ranged from 84 to 108% with a mean of 97%. Figs. 7.4 a–c show chromatograms obtained in the ion chromatography of a soil extract [19]. See also section 7.13.2.

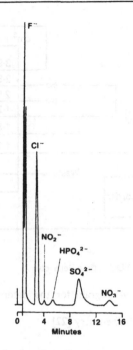

Fig. 7.5 Ion chromatography of mixed anions in soil extracts
Source: Royal Society of Chemistry [18]

7.6 Cyanide

7.6.1 Spectrophotometric method

Tecator Ltd produce an apparatus based on distillation and titration of spectrophotometry for the determination of free cyanides in soil [20].

7.7 Fluoride

Davis and Carlton Smith [21] have carried out an interlaboratory comparison of methods for the determination of fluoride in soils.

7.7.1 Ion chromatography

The ion chromatographic procedure [18] discussed in section 7.5.3 has been applied to the determination of fluoride in soils (see Fig. 7.5).

7.8 Iodide

7.8.1 Spectrophotometric method

Van Vleit *et al.* [22] have described a semi-automated spectrophotometric procedure using a Technicon Autonalyser for the determination of iodine

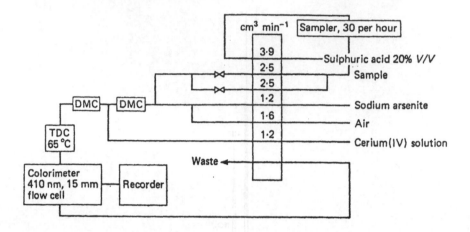

DMC = double mixing coil, TDC = time delay coil

Fig. 7.6 Autoanalyser flow diagram for the determination of iodine in soil and plant digests.
Source: Royal Society of Chemistry [22]

and iodide plus iodine in soil extracts with a coefficient of variation of 2.1% at the 8.6 mg L^{-1} to 6.1% at the 1.4 mg L^{-1} level.

The method is based on the catalytic action of iodine on the oxidation of arsenic(III) by cerium(IV).

In this method 1 g of air dried soil is boiled with 15 mL 2N sodium hydroxide for 45 min. After centrifuging off soil the supernatant solution 3 mL 20% sulphuric acid and 1 mL perchloric acid (72%)–nitric acid (55%) (2 + 1) are added and the solution heated at 265°C until clear. The solution is then diluted to 15 mL. The 15 mL volumes of solution were then transferred to the sampler units of the Autoanalyser.

Technicon Autoanalyser instruments were used as shown in Fig. 7.6. A photocell colorimeter equipped with a 15 mm tubular flow cell, a heating bath, a 26 min time delay and a Mark II proportionating pump were used. Between 96 and 97% recovery of added iodide spikes to soil was obtained by this method. The coefficient of variation was between 6.1% at the 1.4 ppm iodide level and 2.1% at the 8.6 ppm iodide level.

7.9 Molybdate

7.9.1 Ion chromatography

Mehra and Frankenberger [89] used ion chromatography to determine molybdate in soils.

7.10 Nitrate

Techniques used to determine nitrates in soils include titration [23], spectrophotometry [24–32, 35–37], flow injection analysis [26, 27], ion selective electrodes [33, 34] and ion chromatography [34, 38–49, 51].

7.10.1 Titration methods

A method [23] has been described for the determination of nitrate and nitrite nitrogen and ammonium ions in the 2M potassium chloride extracts of moist soils. Firstly an aliquot of the extract is made alkaline and the released ammonia determined using an ammonia selective probe or titrimetrically. The nitrate in the ammonia-free extract is then reduced to ammonia with Devadas alloy and the ammonia removed by distillation and determined titrimetrically. The concentration of nitrite in the extract is then determined spectrophotometrically as the red dye formed by coupling diazotised sulphanilic acid with N naphthylethylene diamine hydrochloride.

7.10.2 Spectrophotometric methods

Henrickson and Selmer-Olson [24] applied an autoanalyser to the determination of nitrate and nitrite in soil extracts. In an autoanalyser the water sample, buffered to pH 8.6 with aqueous ammonia–ammonium chloride, is passed through a copperised cadmium reductor column. The nitrite formed is reacted with sulphanilic acid and N-1-naphthylethylene diamine and the extinction of the azo dye is measured at 520nm. For soil extracts the range and standard deviation are, respectively, 0.5–1.0 and 0.007 mg L $^{-1}$.

Garcia Gutierrez [25] has described an azo coupling spectrophotometric method for the determination of nitrite and nitrate in soils. Nitrite is determined spectrophotometrically at 550nm after treatment with sulphanilic acid and N-1-naphthylethylene diamine to form an azo dye. In another portion of the sample, nitrate is reduced to nitrite by passing a pH 9.6 buffered solution through a cadmium reductor and proceeding as above. Soils were boiled with water and calcium carbonate, treated with freshly precipitated aluminium hydroxide and active carbon and filtered prior to analysis by the above procedure.

Tecator [26] have described a spectrophotometric method for the determination of nitrate and ammonium employing sulphanilamide and N-1- naphthyl ethylene diamine in 2M potassium chloride extracts of soil samples.

Lindau and Spalding [27] have studied the effects on nitrate and nitrite extractions from soil of 2M potassium chloride extractant ratios between 1:1 to 1:10 on nitrate recovery. Preliminary data indicated that concentrations of extractable nitrate and nitrogen isotopic values were influenced by the volume of extractant. The 1:1 extractions showed decreasing nitrogen isotope values with increasing nitrate levels, whereas in the 1:10 extractions these values were independent of each other. Incomplete extraction occurred at the 1:1 ratios. The ratio required for maximal recovery was not determined.

Elton Bott [28] and Osibanjo and Ajaya [29] determined nitrate in soil by a spectrophotometric method based on 3,4 xylenol. In one of these procedures [29], nitration of the 3,4 xylenol is carried out instantaneously at about 0°C in 80% sulphuric acid and the nitration product is extracted into toluene, the excess of the reagent remaining in the aqueous layer. The toluene layer is then treated with sodium hydroxide solution to form a coloured product (the sodium salt of the nitrophenol) in the aqueous layer, the absorbance of which is measured at 432nm. Interferences from common anions, including chloride and nitrate, was investigated.

In this method the soil extract and the blank and standards are evaporated nearly to dryness by gentle heating then cooled in an ice pack. 5 mL of 80% sulphuric acid and then 1 mL of 2% ethanolic solution of 3,4 xylenol are added. This solution is transferred to a separatory funnel with 80 mL ice-cooled distilled water. Toluene (10 mL) is added to the isolated toluene extract 5 mL of 1% sodium hydroxide is added to convert the phenol to the phenoxide. The lower aqueous phase is separated and evaluated spectrophotometrically at 432 nm using matched silica cells with distilled water in the reference cell.

Keay and Menage [30] carried out an automated determination of nitrate and ammonia in 2 M potassium chloride extracts of soils. The sample is reacted in an autoanalyser with a 0.25% suspension of magnesium oxide, the ammonia liberated from ammonium ion is absorbed in 0.1M hydrochloric acid and determined spectrophotometrically at 625nm by the indophenol blue method. The sum of ammonium and nitrate is determined similarly but with the addition of 4.5% titanous sulphate solution before distillation, thereby reducing nitrate but not nitrate to ammonia. The nitrate content of the soil can then be obtained by difference.

Hadjidemetriou [31] has carried out a comparative study of the determination of nitrates in calciferous soils by the phenoldisulphonic acid and the chromotropic acid spectrophotometric methods. He used 0.02N cupric sulphate as soil extractant. Silver sulphate was added to remove chlorides. Nitrites, if present, were eliminated by acidifying the extract with N sulphuric acid. The phenol disulphonic acid method is subject to interference by other ions. Details of the chromotropic acid method are given below.

In this method 50 mL of 0.02N copper(II) sulphate is added to 10 g of air-dried soil sample. A 3 mL volume of the filtrate is transformed to a 25 mL flask and cooled in ice. To this solution is added 1 mL 0.1% chromotropic

Table 7.4 Regression equations and correlation coefficients betwen the three methods for nitrate–nitrogen determination

Regression equation	Correlation coefficient
Chromotropic method = 1.92 + 0.99 (phenoldisulphonic acid method)	0.9998
Ion-selective electrode method = 1.58 + 0.96 (phenoldisulphonic acid method)	0.9998
Chromotropic method = 0.31 + 1.03 (ion selective electrode method)	0.9996

Source: Royal Society of Chemistry [31]

acid, sodium salt dissolved in concentrated sulphuric acid. Concentrated sulphuric acid (6 mL) is then added and the solution left for 45 min for colour to fully develop prior to spectrophotometric evaluation at 430nm. Standard solutions (0–35 mg L^{-1} NO$_3$) and blanks were subject to the same treatment.

Good agreement was obtained between results obtained on soil extracts by this method, the more lengthy phenolic disulphonic method and an ion-selective electrode method (see section 7.10.5) in the nitrate nitrogen range 3–200 mg kg^{-1}. The relationships between the three methods are shown in Table 7.4. There is a very close relationship between the methods: the correlation coefficients are almost unity, indicating that the phenoldisulphonic acid method could be replaced with the ion-selective electrode or the chromotropic acid method.

7.10.3 Flow injection analysis

Tecator [26] have described a flow injection system for the determination of nitrate and nitrite in 2 mol L^{-1} potassium chloride extracts of soil samples. Nitrate is reduced to nitrite with a copperised cadmium reductor and this nitrite is determined by a standard spectrophotometric procedure in which the soil sample extract containing nitrate is injected into a carrier stream. On the addition of acidic sulphanilamide a diazo compound is formed which then reacts with N-(1-Naphtyl)-ethylene-diamine dihydrochloride provided from a second merging stream. A purple azo dye is formed, the intensity of which is proportional to the sum of the nitrate and nitrite concentration. Nitrite in the original sample is determined by direct spectrophotometry of the soil extract without cadmium reduction.

7.10.4 Microdiffusion method

Waughman [32] has described a simple microdiffusion method for estimating nitrate (and ammonia) in soils. In this method nitrate in the soil extract is reduced to ammonia by titanous sulphate and the ammonia is then released from the solution, and diffused and absorbed onto a nylon square impregnated with dilute sulphuric acid. The nylon is then dipped into a solution of a chromogenic reagent for ammonia and the colour evaluated spectrophotometrically.

7.10.5 Nitrate selective electrode

The nitrate ion-selective electrode has been extensively used, even though there are interferences from other ions [39–41, 45, 48, 49]. The rapidity and the good accuracy achieved using this electrode [34, 38–48] have made it suitable for use in routine analysis and in soil agrochemical research [41, 45, 50].

Various workers [34, 41, 48] have used different extraction solutions in the ion-selective electrode method, depending on the soil being analysed. The most important are water [34, 39–42, 44], potassium sulphate [46], aluminium sulphate [43], copper(II) sulphate [40], calcium hydroxide [39] and copper sulphate(II) with aluminium and silver resins [48].

Hadjidemetriou [31] has described an ion selective electrode method for the determination of nitrates in calcareous soils as discussed in section 7.10.2. Hadjidemetriou [31] used an Orion Model 93–07 nitrate ion-selective electrode with a 1×2 sensing module construction, and an Orion Model 90–02 double junction reference electrode fitted on a pH meter. The outer chamber was filled with 0.04 M ammonium sulphate solution and the inner chamber with Orion 90–00–02 solution.

Goodman [34] has described an automated procedure for the determination of nitrate in soils. The apparatus automatically extracts and analyses batches of up to 60 soil samples. Analysis is performed electrochemically by means of an ion-selective electrode and reference electrode. Corning ion-selective electrodes were found to be superior to those produced by Orion in this application. Recoveries of nitrate in this method were between 94 and 95%. The calibration curve was linear down to 2.5 mg L $^{-1}$ nitrate. A plan of the general arrangement is shown in Fig. 7.7. It consists of a rail-mounted carriage, which carries rows of sample beakers past three 'stations' where each sample receives an aliquot of extractant, usually water, is thoroughly stirred and has an electrode or electrodes lowered into it. The electrical output from the electrode(s) is passed through an amplifier to a flat bed recorder. Control of the sequence of operations is completely automatic involving a system of three interlocking motor-driven cam timers, thereby ensuring that each sample receives identical treatment.

The apparatus used in this method consisted of a Corning liquid junction nitrate ion-selective electrode operating through a Pye Model 291 pH meter. This electrode has a flat end incorporating the sensing membrane. Also used was a Philips R44/2/–SD/1 double junction

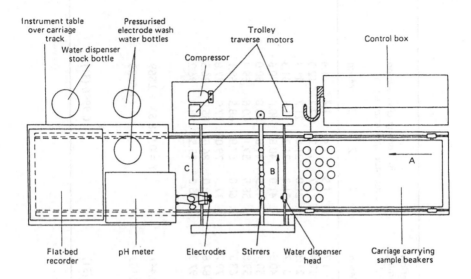

Fig. 7.7 Layout of apparatus. Arrows show direction of movement of carriage (A) and water and electrode trollies (B) and (C) respectively. (Not to scale)
Source: Royal Society of Chemistry [31]

reference electrode, containing 0.02M potassium chloride solution in the outer chamber.

In a series of experiments to test the Corning electrode with the apparatus, Goodman [34] added a range of standard nitrate solutions to weighed samples of a sandy loam soil. The nitrate contents of these modified samples were then determined by three different methods: (a) extracting 20 g of soil with 50 mL of water, filtering and analysing the filtrate by Kjeldahl distillation with alkaline titanium(III) sulphate; (b) preparing the filtrate as in (a) and determining the nitrate concentration in the extract by manually inserting the Corning electrode; and (c) extracting 10 g of soil with 25 mL of water on the apparatus and determining the nitrate content of the extract by automatic insertion of the Corning electrode into the soil suspension.

The results are given in Table 7.5. The slopes of the regression lines show that the electrode recorded 94–95% of the added nitrate, the response of the electrode in soil suspension being substantially linear down to 2.5 µg mL^{-1} of nitrate in the extract.

7.10.6 Ion chromatography

Bradfield and Cooke [18] give details of a procedure for the determination of nitrate (and chloride, phosphate and sulphate) in aqueous extracts of soil by an ion chromatographic technique with indirect ultraviolet

Table 7.5 Recovery of nitrate–nitrogen from soil by three techniques

NO3–N added/μg mL⁻¹	By shaking, filtration and distillation/μg mL⁻¹				By shaking, filtration and manually operating Corning electrode/μg mL⁻¹				By Corning electrode on automatic apparatus/μg mL⁻¹			
	1	2	3	Mean	1	2	3	Mean	1	2	3	Mean
0	0.7	0.4	0.9	0.7	1.8	1.7	1.7	1.7	1.5	1.4	1.4	1.4
10	10.0	10.9	9.7	10.2	11.7	13.7	11.7	12.4	11.7	12.2	12.2	12.0
20	20.8	19.8	20.7	20.4	21.0	20.8	22.0	21.3	21.0	21.6	21.7	21.4
30	30.7	29.9	29.7	30.1	32.5	30.5	32.0	31.7	31.3	31.2	31.5	31.3
40	41.3	40.4	40.4	40.7	40.8	38.8	42.0	40.5	41.0	40.8	40.0	40.6
50	49.0	50.7	50.0	49.9	51.5	45.8	53.0	50.1	49.5	49.8	50.8	50.0
60	60.9	60.0	61.5	60.8	61.0	58.0	60.5	59.8	58.5	59.1	60.5	59.4
70	71.3	70.8	71.0	71.0	70.0	67.5	57.6	68.3	68.0	67.8	68.5	68.1
80	79.1	81.9	77.2	79.4	79.0	77.0	83.0	79.7	77.5	79.0	77.0	77.8
90	90.4	91.8	93.3	91.8	90.0	85.0	88.5	87.8	88.0	86.0	85.0	86.3
100	96.6	100.3	992.	98.7	98.0	84.0	99.8	97.3	95.0	95.5	98.0	96.2
Regression of found (Y) on added (X) nitrate	Y = 0.996 0X + 0.554				Y = 0.952 2X + 2.447				Y = 0.938 5X + 2.589			
Standard error of slope	0.0069				0.0103				0.0050			
Correlation coefficient (degrees of freedom in parenthesis)	0.9992 (31)				0.9982 (31)				0.9996 (31)			

Source: Royal Society of Chemistry [34]

detection. Recoveries ranged from 84 to 108%. The technique is discussed further in section 7.5.3 (see Fig. 7.4, see also section 7.13.2).

7.11 Nitric oxide and nitrous oxide

7.11.1 Gas chromatography

Both indirect [53–57] and direct [54, 58–60] evidence indicate that gaseous forms of nitrogen can be lost from soil during the nitrification of ammonium or ammonium-forming fertilisers by soil micro-organisms. It appears that evolution of nitrogen, dinitrogen oxide and nitrogen oxide or its oxidative derivative, nitrogen dioxide, can occur, resulting in poor fertiliser efficiency.

Smith and Chalk [52] have described a simple method for determining nitrogen oxide and nitrogen dioxide, evolved from soils, in closed systems. These gases are absorbed by an acidic solution of potassium permanganate, and the resulting nitrate is determined by a steam distillation method. Excess of permanganate is reduced with iron (II) sulphate and neutralised with sodium hydroxide solution. Ammonium in solution is removed by distillation with magnesium oxide, and nitrate is determined by distillation after reduction to ammonium by Devarda's alloy. Nitrogen and dinitrogen oxide evolved from soils are measured using gas chromatography on a single 0.61 mm column of molecular sieve 5A, temperature programmed to 250°C at 39°C min^{-1} after an initial period of 1 min at 35°C. A complete analysis requires 19.5 min and 2 µg of nitrogen can be determined quantitatively for each gas.

Smith and Chalk [52] achieved a complete separation of oxygen, nitric oxide, nitrous oxide and carbon dioxide on a column packed with molecular sieve 5A (100–120 mesh), programmed between 35°C and 250°C at 39°C m^{-1}.

7.12 Nitrite

7.12.1 Spectrophotometric methods

Bhuchar and Amar [61] determined nitrites in soil by acidifying the sample to pH4 and adding mercapto acetic acid to produce a red coloured complex which is extracted into tributyl phosphate from a solution 2N in acid. The red colour is evaluated spectrophotometrically at 322nm. The method is applicable in the range 2-40 mg L^{-1} nitrite. Nitrate recoveries in water extracts of soil are in the range 95.6 to 102.0%.

Wu and Liu [62] have described a spectrophotometric method for the determination of micro amounts of nitrite in water extracts of soils. The chromogenic reagents were p-aminoacetophenone and resorcinol in sodium carbonate–sodium acetate medium at pH9 which form a golden coloured complex with nitrate at 435 nm:

$$CH_3CO-C_6H_4-HN_2^+ + NO_2^- + 2H^+ \rightarrow CH_3CO-C_6H_4 - N^+ \equiv N + 2H_2O$$

$$CH_3-CO-C_6H_4-N^+ \equiv N^+ + \overset{HO}{\underset{}{\diagdown}} \bigcirc-OH \xrightarrow{pH\,9} CH_3CO-C_6H_4-N=N- \bigcirc -OH$$

$$CH_3C-C_6H_4 - N = N - \bigcirc = O + H_2O$$
$$\underset{O^-}{\overset{KOH}{|}}$$

Foreign ions are masked with a composite EDTA-sodium hexametaphosphate reagent and interference by sulphide is overcome by the addition of mercuric chloride. Soil samples are digested with cold water containing mercuric chloride and the precipitated mercuric sulphide filtered off prior to the addition of chromogenic reagents and spectrophotometric evaluation. Recoveries of nitrite from red alluvial soil exceeded 97%.

Beer's Law is obeyed up to 20 µg nitrite in 60 ml of test solution. The effect of 20 foreign ions was examined. Most of the cations and common anions do not interfere. Sulphide, thiosulphate, sulphite, tetrathionate and iodide interfere but their effect is mitigated by the addition of mercuric chloride.

Chaube *et al.* [63] investigated the determination of ultra trace concentrations of nitrite in soil. In their method the nitrite is used to diazotize o-nitroaniline and the o-nitrophenyldiazonium chloride produced is coupled with N-naphthylethylene diamine hydrochloride. The red-violet dye produced is extracted into isoamylalcohol and evaluated spectrophotometrically at 545nm.

Beer's Law is obeyed in the range 0.1–0.6 mg L^{-1} nitrite in the original sample and 0.02–0.15 mg L^{-1} if the solvent extraction procedure is used. A wide range of foreign ions do not interfere in this procedure. Soil samples were acidified with sulphuric acid prior to filtration and analysis.

A method has been described [23] for the determination of 2M potassium chloride extractable nitrite, nitrate and ammonium ion in moist soils. In this method an aliquot of the extract is made alkaline and the released ammonia is determined either using a probe or, after removal by distillation, titrimetrically. The nitrate in the ammonia-free extract is then reduced to ammonia which is removed by distillation and determined titrimetrically. The concentration of nitrite in the extract is determined spectrophotometrically as the red dye formed by coupling diazotised sulphanilic acid with N-1-naphthylethylenediamine dihydrochloride (NEDD).

Henrickson and Selmer-Olson [24] have described an automated method for determining nitrite and nitrate in soil extracts. This method is discussed further in section 7.10.2.

7.12.2 Ion chromatography

The ion chromatography of nitrite is discussed in section 7.7.1 (Fig. 7.5), see also section 7.13.2.

7.13 Phosphate

7.13.1 Spectrophotometric methods

Spectrophotometric evaluation at 880nm of the phosphomolybdate complex has been used to determine phosphates in sodium bicarbonate extracts (pH 8.5) of soil [64, 65].

7.13.2 Ion chromatography

Bradfield and Cooke [18] have described an ion chromatographic method using a UV detector for the determination of phosphate chloride, nitrate and sulphate in water extracts of soils (see section 7.7.1). Soils are leached with water and Dowex 50–X4 ion exchange resin added to the aqueous extract which is then passed through a Sep–Pak C18 cartridge and the eluate then passed through the ion chromatographic column. The best separation of these anions was obtained using a 5×10^{-4} mol L^{-1} potassium hydrogen phthalate solution in 20% methanol at pH 4.9. A reverse phase system was employed. Detection times were 5.5, 7.9, 12.6 and 18 min for chloride, nitrate, phosphate and sulphate respectively. Recoveries ranged from 84 to 108% with a mean of 97%.

7.13.3 Miscellaneous

Bickford and Willett [68] have pointed out that the filtration of aqueous calcium chloride extracts of soils containing phosphate through Gelman 9A6 cellulose acetate membranes which contain a wetting agent caused low results in methods for determining phosphate due to contamination by some contaminant in the membrane. Gelman TCM–450 or Whatman No 42 membrane, on the other hand, does not interfere in the determination of phosphate.

7.14 Selenite

7.14.1 Ion chromatography

Karlson and Frankenberger [66] and Nieto and Frankenberger [67] developed a single column ion chromatographic method for the

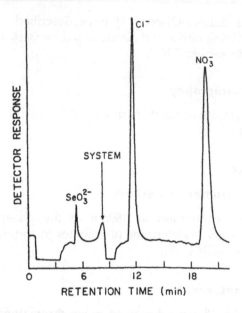

Fig. 7.8 Chromatogram of an aqueous extract of Panoche clay loam spiked with DeO_3^{2-}.
Peaks, 5.1 mg L^{-1} SeO_3^{2-}, system, 8.3 mg L^{-1} Cl$^-$, 6.5 mg L^{-1} NO_3^-
Source: American Chemical Society [66]

determination of selenite in soil extracts with the simultaneous determination of chloride, nitrite, nitrate and phosphate. Separation of the anions was conducted on a low capacity anion exchange column and anions were quantified by a conductiometric detection.

The element stream consisted of 1.5 m mol L^{-1} phthalic acid adjusted to pH 2.7 with formic acid. The method requires minimal sample treatment, allowing for precise measurements of trace levels of selenite in the presence of high background levels of chloride, nitrate and nitrite.

Interfering chloride anions were removed by reaction with a silver saturated cation exchange resin. The detection limit of selenite was 3 µg L^{-1} with a concentrator column. The relative standard deviation using a 500 µL loop was 2.0% with standards and 6.7% in soil extracts (0.5 mg L^{-1}). Selenite levels found in soil extracts ranged from 0.8 to 99.6 µg L^{-1}.

The apparatus used in this method has been described by Nieto and Frankenberger [67]. Soil extracts were prepared by adding 50 mL of deionised water to 10g soil, shaking for 1 h and filtering the suspensions through Whatman No 42 filter paper. The filtrate was then passed through a 0.22 µm Millipore GS membrane filter and introduced directly into the HPLC injector port. A typical chromatogram for an aqueous soil extract spiked with selenite is shown in Fig. 7.8. The solute sequence was selenite > chloride > nitrate system. The ions were separated into well-defined peaks with a time of analysis of 22 min. Table 7.6 indicates the

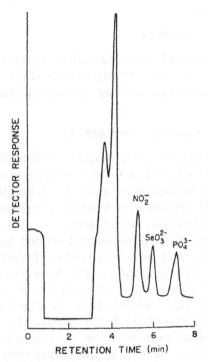

Fig. 7.9 Chromatogram of an aqueous extract of a subsurface sample of Panoche clay loam upon treatment with 40 mg mL^{-1} silver-saturated resin. Peaks, 50 µg L^{-1} NO$_2^-$, 99.6 µg L^{-1} SeO$_3^{2-}$, 4.2 µg L^{-1} PO$_4^{2-}$
Source: American Chemical Society [66]

Table 7.6 Single column ion chromatography analyses of the anionic content of aqueous soil extracts

Soil sample	Anion concentration[a]				
	SeO$_3^{2-}$	PO$_4^{3-}$	NO^{2-}	Cl$^-$	NO$_3^-$
Los Banos CL, surface	<6	10.9	47	6.4	2.2
Panoche CL, surface	<6	8.2	34	8.3	6.5
Panoche CL, subsurface	99.6	4.2	50	30.0	0.9

[a] Concentration units for SeO$_3^{2-}$, PO$_4^{3-}$ and NO$_2^-$ are micrograms per litre, while units for Cl$^-$ and NO$_3^-$ are milligrams per litre

Source: American Chemical Society [66]

inorganic anion composition of three soil extracts as determined by single column ion chromatography. Native selenite was detected in a subsurface sample of Panoche clay loam as illustrated in the chromatogram shown in Fig. 7.9, since chloride interference was eliminated by utilising the silver-saturated resin.

7.15 Sulphate

7.15.1 Titration method

A volumetric method based on addition of excess barium ions and back titration with potassium sulphate to the sodium rhodizonate end-point has been used to determine water soluble sulphates in soil [72].

7.15.2 Spectrophotometric methods

Ogner and Haugen [69] have described a technique for the automated determination of sulphate in water samples and soil extracts containing large amounts of humic compounds. This technique can be applied to the determination of sulphate in concentration ranges of 0–60 and 0–3000 mg L^{-1} (as sulphate) in the aqueous extract.

A turbidimetric method has been described for the determination of water soluble and acid soluble sulphate in hydrochloric acid extracts of soils [70]. Spectrophotometry has also been employed to carry out the determination of total water soluble sulphate in soil [71].

Landers et al. [73] have described a digestion procedure for the determination of phosphate buffer extractable sulphate in soils. The amount of sulphate extracted from a given substrate depends on the extraction procedure as well as the adsorptivity and solubility of the sulphate constituents. A phosphate buffer solution (2.23 gl^{-1} $Na_2H PO_4$ H_2O in double distilled water) will remove sulphate due to the higher affinity of phosphorus for anion exchange sites. Soil samples up to 5g are placed in 0.25 mL flasks and extracted in 100 mL of phosphate buffer solution by shaking vigorously for 1 h. The suspension is centrifuged to remove suspended particulates. The supernatant is then placed into the barrel of a 10 mL disposable syringe fitted with a filter adapter and the sample filtered through a GF/C (Whatman, 98% retention of 1.2 µm) filter. Filtrate (up to 2 mL) is added to the digestion flasks and the hydriodic acid reduction procedure followed to convert sulphate to sulphide. Sulphide is then estimated by the digestion–distillation procedure followed by spectrophotometry of the p-amino dimethylaniline–ferric ammonium sulphate complex.

7.15.3 Flow injection analysis

Krug et al. [74, 75] used flow injection turbidimetry to determine sulphate in natural waters and plant digests. They described an improved flow injection system with alternative streams of reagents. Samples were injected into an inert carrier comprising 0.3% EDTA disodium salt and 0.2 mol L^{-1} sodium hydroxide. The inert carrier is mixed with 5% barium chloride containing 0.05% polyvinyl alcohol to form a barium sulphate suspension. The range of the method can be extended to low concentrations by continuously adding sulphate to the sample carrier stream. System performance is improved by automatic pumping of the reagent stream and an alkaline EDTA solution at high flow rate. All operations were controlled by an electronically operated proportional injector commutator. No baseline drift was observed even after analysis of 3000 samples. The method is capable of analysing 120 samples per hour with a relative standard deviation of less than 1% for sulphate concentrations in the ranges 1–30 mg L $^{-1}$. Analytical recovery was 97–102%.

7.15.4 Atomic absorption spectrometry

In an indirect method for determining sulphate in soil extracts, Little *et al.* [76] precipitate sulphate as the lead salt in 40% ethanol medium. Unconsumed soluble lead is determined by atomic absorption spectrometry. The method is applicable to soil samples containing as little as 4 mg Kg $^{-1}$ sulphate.

7.15.5 Molecular emission cavity analysis

Molecular emission cavity analysis has been used to determine soluble sulphate in soil [77].

7.15.6 Ion chromatography

Bradfield and Cooke [18] give details of a procedure for the determination of sulphate, chloride, nitrate and phosphate and sulphate in aqueous extracts of plant materials and in soil solutions by an ion chromatographic technique with indirect ultra-violet detection. Recoveries ranged from 84 to 108% (see sections 7.5.3 and 7.7.1 (Fig. 7.5)).

7.15.7 Miscellaneous

Bolan *et al.* [78] used packed columns of soil with high and low sulphate adsorption capacity in laboratory studies of the movement of sulphate through the soil. Adsorption isotherms were obtained by batch experiments, using sulphur–35 as a tracer for the movement of the applied sulphur–32. The movement differences could be explained by reference to the adsorption isotherms. Breakthrough curves were obtained for varying concentrations of the applied sulphate solution. These were in good agreement with curves obtained by numerical solution of the dispersion–convection equation assuming a Freundlich absorption isotherm and instantaneous reversible adsorption. However, calculations based on these assumptions failed to account for sulphate distributions in the soil columns after leaching of a pulse of sulphate added to the surface soil. The reasons for this are discussed.

7.16 Sulphide

7.16.1 Spectrophotometric method

The widely used method for determining sulphide is based on the precipitation of the sulphide in hydrogen sulphide as zinc sulphide and subsequent determination by methylene blue formation or iodine titrimetry [80].

7.16.2 Atomic absorption spectrometry

Ramesh *et al.* [79] used an indirect rapid method based on atomic absorption spectrometry for the determination of sulphide in flooded acid sulphate soils.

Hydrogen sulphide, evolved during the anaerobic metabolism of sulphate, is readily converted into insoluble metal sulphides, chiefly iron(II) sulphide, in flooded acid sulphate soils that are especially rich in iron.

This method essentially involves the precipitation of zinc sulphide by the action of zinc on the hydrogen sulphide liberated on acidification of metal sulphides in flooded acid–sulphate soils, and then indirect determination of sulphide by determining the zinc in the precipitate and also the zinc remaining in solution, after the precipitation, by atomic-absorption spectrophotometry.

The assembly used for the conversion of metal sulphides into hydrogen sulphide consisted in a 250 mL Erlenmeyer flask, which was closed with a three-hole rubber bung, for holding a dropping funnel (for adding hydrochloric acid) and two glass tubes (an inlet for oxygen-free nitrogen and an outlet). The outlet was connected to two 100 mL Erlenmeyer flasks, in succession, containing an ammoniacal solution of zinc acetate to trap the hydrogen sulphide evolved. For standardisation of the method, a known amount of sodium sulphide was placed in a 250 mL flask and then treated with 100 mL of 1N hydrochloric acid. The hydrogen sulphide evolved was swept into the zinc acetate solution with oxygen-free nitrogen for 30 min to 1 h (until tests with lead acetate paper strips showed complete cessation of hydrogen sulphide evolution) in order to precipitate zinc sulphide. Zinc sulphide was accumulated in the first trap while the second trap was visibly free from any precipitate; this indicated that hydrogen sulphide was completely precipitated in the first trap.

After filtration, the precipitated zinc and soluble zinc remaining in the filtrate were assayed using a Varian Techtron atomic-absorption spectro-photometer, Model AA–1100. Sulphide equivalent to the zinc precipitated or to the decrease in the zinc content of the solution was then calculated. Care was taken to ensure that the molar concentration of sulphide was far exceeded by the molar concentration of zinc in order to provide a measurable excess of zinc in solution after complete precipitation of the sulphide.

Ramesh *et al.* [79] compared results obtained by this method with those obtained by the conventional iodimetric method [81], in which the zinc sulphide precipitated in the trap was reacted with excess of iodine plus 2.5 mL of concentrated hydrochloric acid and the unreacted iodine was titrated against standard thiosulphate solution.

Sulphide formed in two acid–sulphate soils, Pokkali (pH 5.0; organic carbon 2.28%; sulphate–S 0.056%; total sulphur 0.1%) and Kari (pH 3.9; organic carbon 4.65%; sulphate–S 0.039%; total sulphur 0.13%) under flooded conditions was determined by atomic absorption spectrophotometry and iodimetry. Soil samples (20g) were flooded with 25 mL of distilled water in test tubes (25 × 200 mm). After 40 d, the reduced soil samples were transferred into 250 mL flasks and treated with 100 mL of 1N hydrochloric acid in order to liberate hydrogen sulphide from the metal sulphides, chiefly iron(II) sulphide. Hydrogen sulphide was absorbed in an ammoniacal solution of zinc acetate with precipitation of zinc sulphide. The zinc in the precipitate and the filtrate was determined by atomic absorption spectrophotometry to give the result for the indirect determination of sulphide.

Care was taken to bubble oxygen-free nitrogen through the apparatus for 5 min prior to acidification of the soil samples to prevent the instantaneous oxidation of the hydrogen sulphide evolved in the flask. The sulphide was also determined by the conventional iodimetric method [81] by adding excess of iodine plus hydrochloric acid directly to the trap as described for the sulphide determination from sodium sulphide in the standardisation of the method. In a modification of this method, the precipitated zinc sulphide was first separated by filtration and then treated with excess of iodine plus hydrochloric acid to avoid any interferences from iodine consuming substances, if any, from the complex soil system.

The data in Table 7.7 showed that about 85% of sulphide was recovered from sodium sulphide standards by both the iodimetric and atomic absorption spectrophotometric methods. Also in the latter method, the sulphide values, obtained by determining the zinc either in the zinc sulphide precipitated or that remaining in solution in the filtrate, were almost identical. The atomic absorption spectrophotometric method was simple, rapid and reproducible with variations of less than 5% within replicates, while the iodimetric method, though equally sensitive, was somewhat tedious.

The determination of sulphide in two acid–sulphate soils after a 40 d flooding showed that the sulphide values from both soils were realistic and reproducible, ranging from 0.26 to 0.33 mg g^{-1} when determined by atomic absorption spectrophotometry (Table 7.8). As in the pure system when sodium sulphide was used, soil sulphide levels derived from the determination of zinc either in the precipitate or in the filtrate were identical. However, the iodimetric method in which iodine was added directly to the trap gave abnormally high sulphide values(> 1.36 mg g^{-1})

Table 7.7 Sulphide recovered from sodium sulphide standards

The results are for sulphide recovered

| Sulphide added/mg | Atomic-absorption spectrophotometry | | | | | | Iodimetry | | |
| | Result from determination of zinc in filtrate | | | Result from determination of zinc in precipitate | | | | | |
	Replicate/ mg	Mean/ mg	Recovery %	Replicate/ mg	Mean/ mg	Recovery %	Replicate/ mg	Mean/ mg	Recovery %
3.37	2.83 2.76 2.79	2.79	82.8	2.85 2.86 2.87	2.86	84.8	2.93 2.81 2.84	2.86	84.8
4.49	3.81 3.65 3.66	3.71	82.6	3.73 3.75 3.76	3.75	83.5	3.78 3.70 3.83	3.77	84.0

Source: Royal Society of Chemistry [79]

Table 7.8 Sulphide formed in acid–sulphate soils on 40 d flooding

The results are for sulphide formed in mg per gram of soil

Soil	Atomic-absorption spectrophotometry				Idiometry			
	Result from determination of zinc in filtrate		Result from determination of zinc in precipitate		Conventional*		Precipitate†	
	Replicate	Mean	Replicate	Mean	Replicate	Mean	Replicate	Mean
Pokkali	0.29 0.29 0.28	0.29	0.29 0.29 0.30	0.29	2.09 3.15 1.85	2.03	N.D.‡ N.D. N.D.	
Kari	0.29 0.30 0.26	0.28	0.29 0.29 0.29	1.68	1.70 1.65 1.56	0.31	0.31 0.32 0.33	

* Iodine was added directly to the trap
† Zinc sulphide precipitated was first separated by filtration and then treated with iodine
‡ N.D. not determined

Source: Royal Society of Chemistry [79]

for both soils; these levels are above the theoretical values for sulphide that could be generated from Pokkali soil with 0.056% sulphate–S and 0.1% of total sulphur and from Kari soil with 0.039% sulphate–S and 0.13% of total sulphur following flooding. However, the sulphide content of the Kari soil, obtained by the modified iodimetric method, in which the zinc sulphide was treated with iodine only after filtration, was realistic, reproducible and identical with that obtained by atomic absorption spectrophotometry. This would suggest that overestimation of soil sulphide, by the conventional method of adding iodine directly to the trap, was a result of interferences by some reduction products of the flooded soil system. For instance, reduced sulphur compounds such as sulphite, thiosulphate, tetrathionate and hydrosulphite may decompose during acidification leading to erratic results in the determination using iodimetry [81]. Likewise, the methylene blue method, widely used in determining sulphide, is not always reliable as not only hydrogen sulphide but other reduction products that commonly occur in anaerobic ecosystems can readily react with methylene blue to produce erroneous results. The method described by Ramesh et al. [79] of indirect determination of sulphide through the determination of zinc by atomic absorption spectrophotometry is simple, rapid and free from interference and thus has a definite advantage over the iodimetric and methylene blue methods, especially in a complex system such as waterlogged soil.

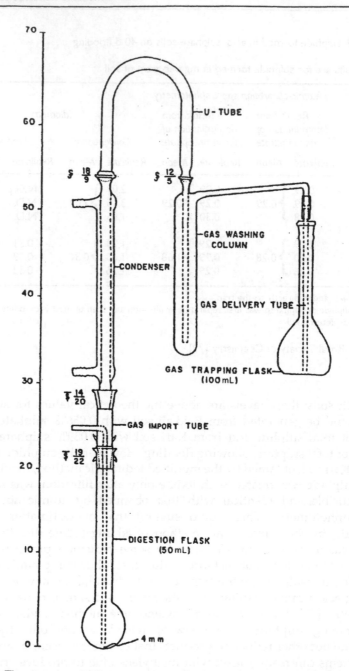

Fig. 7.10 The digestion–distillation apparatus. Tapered glass joints are sealed with a few drops of water and the ball and socket connections are very lightly coated with silicone stopcock preparation. Condensers held by adjustable clamps are mounted to a permanent frame. Gas washing columns are rinsed with distilled water after use. All other parts are interchangeable and are assembled ahead of each takedown to minimise down time
Source: Gordon and Breach UK [73]

7.16.3 Miscellaneous

Sulphur is an important component of both natural and anthropogenic processes. Due to its importance both in the formation of acidic precipitation and as a macronutrient required by all organisms, sulphur's role in atmospheric, aquatic and terrestrial systems has been investigated [82, 83]. Sulphur has a vast array of both inorganic and organic chemical species. The understanding of sulphur dynamics has been restricted due to lack of information on the role of specific sulphur constituents in affecting sulphur fluxes and transformations. For example, a knowledge of organic sulphur is very important in evaluating forest soils. Previous work on such substrates had generally ignored the organic sulphur constituents with most work focusing only on inorganic sulphate or sulphide.

Landers *et al.* [73] have combined and modified various analytical methods to determine the major sulphur constituents in soils. Independently, these methods are useful. However, in combination with the same digestion–distillation apparatus, they provide a reliable and convenient group of analytical methods which can be used in investigations of sulphur dynamics.

Landers *et al.* [73] have described a set of analytical methods for the determination of sulphide and sulphate in soils.

Landers *et al.* [73] used the digestion distillation flask shown in Fig. 7.10 to determine hydrochloric acid digestion, sulphur (method 1), zinc hydrochloric acid reducible sulphur (method 2) and hydroidic acid reducible sulphur (method 3) in soils.

Method 1: Hydrochloric acid digestion sulphur apparatus, (acid digestible inorganic sulphur) digestion–distillation [84] flask, see Fig. 7.10.

The following reagents are added to the trapping flask and the gas washing column respectively.

Acetate trapping solution

Stock solution: dissolve 50 g of zinc acetate in 12.5g of sodium acetate in double distilled water and dilute to one litre and filter.

Diluted solution: before each analysis mix 100 mL of stock solution with 200 mL of double distilled water and add 80 mL of this solution to each gas trapping flask.

Pyrogallol–sodium phosphate solution: Dissolve 10 g sodium dihydrogen phosphate ($NaH_2 PO_4 H_2O$) and 10 g of pyrogallol in 100 ml of double distilled water, bubble with nitrogen to dissolve. Prepare daily and discard when brown colouration develops. Add 10 mL of this solution to the gas washing column.

The soil sample (0.05–2g) is placed in the digestion flask and 10 mL of 1:1 hydrochloric acid added. All connections are closed quickly and the nitrogen flow commenced and the samples refluxed for 1 h. This treatment reduces various sulphur constituents to hydrogen sulphide which is moved by the stream of nitrogen into the trapping flask where it forms zinc sulphide. Colorimetric reagents, p-amino dimethylaniline sulphate [87] and 12.5% ferric ammonium sulphate in 2.5:97.5 sulphuric acid are added to the gas trapping flask and then acid digestible inorganic sulphur is estimated utilising the 670nm absorption maximum.

Method 2: Zinc hydrochloric and reducible sulphur (non-sulphate inorganic sulphur) [85]

The trapping solution is prepared as described under Method 1 and the same reagents are used in the trapping flask and the gas washing column. A wet sample (0.05–0.2g) is placed into a digestion flask containing about 2g of granulated zinc metal. The system is flushed with nitrogen, 10 mL of 1:1 hydrochloric acid is added and the solution is boiled for 1 h. The gas flow is continuous when adding the reagent to prevent liberated hydrogen gas from causing the sample to enter the gas import tubes. Extreme foaming has been a problem with some soil samples but the addition of 1.5 mL of an anti–foam spray (AH Thomas Co) has solved the problem and no interference has been found. Sodium thiosulphate is used as a standard.

The hydrogen sulphide produced is estimated by the p-amino dimethylaniline–ferric ammonium sulphate spectrophotometric method [87], utilising the 670nm absorption maximum. Reagent blanks are run. These are low (<0.03 μM s).

Method 3: Hydriodic acid acid reducible sulphur [86] (non-carbon bonded sulphur)

The trapping system is prepared as described in Method 1 and the same reagents are used in the trapping flask and the gas washing column. Prepare the hydriodic acid reducing agent as follows: combine 300 mL of hydriodic acid, 75 mL hypophosphorus acid (50%) and 150 mL of 88% formic acid. Boil gently with a nitrogen gas stream for 10 min after reaching 115°C. During the 10 min temperature is kept between 115° and 117°C. Upon completion the reagent may appear bright yellow or brown, apparently depending on the quality of the reagents used. The mixed reagent has a shelf life of about two weeks.

A wet sample (0.01–0.1 g) is added to the digestion flask and 4 mL of mixed reagent are added. Gas flow is started and the sample is refluxed for 1 h. Potassium sulphate is used for a standard. The hydrogen sulphide produced is estimated by the p-amino dimethylaniline-ferric ammonium sulphate spectrophotometric method [87] utilising the 670nm absorption maximum. Reagent blanks are run. These are low (< 0.03 μM s).

Examples of soil analyses carried out by Landers *et al.* [73] for acid digestible inorganic sulphur (HCl–S), non-sulphate inorganic sulphur

Table 7.9 Sulphur constituents in soils in mole g $^{-1}$

	Total S	HCl–S	Zn–HCl–S	HI–S	SO$_4^{2-}$–S	C–S	C–O–SO$_3$
Forest soil 1	50 ± 9.3	n.d.	0.62 ± 1.1	8.25 ± 0.93	0.48 ± 0.08	41.9	7.16
Forest soil 2	16.5 ± 1.6	n.d.	0.69 ± 0.03	5.00 ± 0.62	0.72 ± 0.13	11.5	3.59

n.d. Not detectable

Gordon and Breach, Amsterdam [73]

(Zn–HCl–S) and non-carbon bonded sulphur (HI–S), also total sulphur, sulphate, carbon-bonded sulphate (CS) and ester sulphate (C–O–SO$_3$) are shown in Table 7.9. Ester sulphate and carbon-bonded sulphur are the main sulphur constituents of these soils.

Clark and Lesarge [87] have described a method for the determination of elemental sulphur in soils using gas chromatography with flame photometric detection after the sulphur is reacted to form Ph$_3$ PS.

7.17 Tungstate

7.17.1 Ion chromatography

Mehra and Frankenberger [88, 89] used ion chromatography to determine tungstate in soil.

7.18 Vanadate

7.18.1 Spectrophotometric method

Abbasi [86] determined metavanadate in soil by a method based on the formation of a violet colour with vanadium V on addition of a chloroform solution of N–(p–NN dimethylanilino-3-methoxy-2-naphtho)–hydroxamic acid to the acidified (4–6 mol L $^{-1}$ hydrochloric acid) sample. This solution was evaluated spectrophotometrically at 570nm. The detection limit was 0.05 µg vanadium at a dilution ratio of 1:10. Very few interferences occur in this procedure.

References

1 Ducret, L. Analytica Chimica Acta, 17, 213 (1957).
2 Aznarez, J., Bonilla, A. and Vidal, J.C. Analyst (London), 108, 368 (1983).
3 Van Staden, J.F. Analyst (London), 112, 595 (1987).
4 Van Staden, J.F. Analytica Chimica Acta, 179, 407 (1986).

5 Kolthoff, I.M. and Belcher, R. *Volumetric Analysis, Volume 3 Titration Methods,* Interscience, New York, p 25 (1957).
6 Kempton, R.J. and Maw, G.A. *Ann Applied Biology,* **72,** 71 (1972).
7 Gladney, E.S. and Perrin, D.R. *Analytical Chemistry,* **51,** 2015 (1979).
8 Roughan, J.A., Roughan, P.A. and Wilkins, J.P.G. *Analyst (London),* **108,** 742 (1983).
9 Collins, S.H.J. *Society of Chemical Industries (London),* **25,** 518 (1906).
10 HMSO, London. *The Analysis of Agricultural Materials* RB427, 2nd edn (ISBN 0112403522), Method 15, Carbonate in Soil (1979).
11 Banderis, J. *Journal of Science of Food and Agriculture,* **16,** 558 (1965).
12 HMSO, London. *The Analysis of Agricultural Materials* RB427, 2nd edn (ISBN 0112403522), Method 16, Chlorate in Soil (1979).
13 Clarke, F.E. *Analytical Chemistry,* **22,** 553 (1950).
14 HMSO, London. *The Analysis of Agricultural Materials* RB427, 2nd edn (ISBN 0112403522), Method 19, Chloride in Soil (1979).
15 Davey, B.G. and Bembruck, M.J. *Proceedings of the Soil Science Society of America,* **33,** 385 (1969).
16 Johnson, C.M. and Nishita, H. *Analytical Chemistry,* **24,** 736 (1952).
17 McLeod, S., Stace, H.T.C., Tucker, B.M. and Bakker, D. *Analyst (London),* **99,** 193 (1974).
18 Bradfield, E.G. and Cooke, D.T. *Analyst (London),* **110,** 1409 (1985).
19 Dionex Corporation, *Environmental Monitoring with Dionex Ion Chromatography* Ref LPN 325794/85 (1985).
20 Tecator Ltd, Box 70, S–26321 Hoganes, Sweden, Application Notes AN89/87 and 86/87. Cyanides in waste waters, soils and sludges using the 1026 distilling unit (1987).
21 Davis, R.D. and Carlton-Smith, C.H. *Water Pollution Control,* **82,** 290 (1983).
22 Van Vleit, H., Basson, W.D. and Bohmer, R.G. *Analyst (London),* **100,** 405 (1975).
23 HMSO, London. *The Analysis of Agricultural Materials* RB427, 2nd edn (ISBN 0112403522), Method 60, Ammonium, nitrate and nitrite nitrogen, potassium chloride extractable in soil (1979).
24 Henrickson, A. and Selmer-Olson, A.R. *Analyst (London),* **95,** 514 (1970).
25 Garcia Gutierrez, G. *Infeion Quin analet Pura apl Ind,* **27,** 171 (1973).
26 Tecator Ltd, Box 70, S–26321 Hoganes, Sweden, Application Note AN65/83. Determination of Nitrate and Ammonia in Soil Samples, extractable with 2 M potassium chloride (1983) and Application Note ASN65–31/83, Determination of Nitrate in Soil Samples, extractable with 2 M potassium chloride using flow injection analysis (1983).
27 Lindau, C.W. and Spalding, R.F. *Groundwater,* **22,** 273 (1984).
28 Elton-Bott, R.R. *Progress Technology,* **8,** 215 (1977).
29 Osibanjo, D. and Ajayi, S.O. *Analyst (London),* **105,** 908 (1980).
30 Keay, J. and Menage, P.M.A. *Analyst (London),* **95,** 379 (1970).
31 Hadjidemetriou, D.G. *Analyst (London),* **107,** 25 (1982).
32 Waughman, A. *Environmental Research,* **26,** 529 (1981).
33 Bremner, J.M., Bundy, L.G. and Agarwal, A.S. *Analytical Letters (London),* **1,** 837 (1968).
34 Goodman, D. *Analyst (London),* **101,** 943 (1976).
35 Jackson, M.L. ed, *Soil Chemical Analysis,* Constable, London p 197 (1962).
36 Bremner, J.M., In Black, C.A. ed, *Methods of Soil Analysis, American Society of Agronomy,* Madison, Wisconsin, Part 2, p 1191 (1965).
37 Sims, J.R. and Jackson, G.D. *Soil Science of America Society Proceedings,* **35,** 603 (1971).
38 Myers, R.J.K. and Paul, E.A. *Canadian Journal of Soil Science,* **48,** 369 (1968).

39 Mahendrappa, M.K. *Soil Science*, **108**, 132 (1969).
40 Qien, A. and Selmer-Olsen, A.R. *Analyst (London)*, **94**, 888 (1969).
41 Fiskell, J.G.A. and Breland, H.A. *Soil Crop Science, Society Fla Proceedings*, **29**, 63 (1969).
42 Milham, P.J., Awad, A.S., Paull, A.S. and Bull, J.H. *Analyst (London)*, **95**, 751 (1970).
43 Smith, G.R. *Analytical Letters, London*, **8**, 503 (1975).
44 Krupsky, N.K., Alexandrova, A.M., Gubareva, D.N. and Varenik, V.A. *Agrokhimya*, **10**, 133 (1978).
45 Revek, A. *Soil Science*, **116**, 388 (1973).
46 Tchagina, E.G., Dubinina, R.I., Golovin, V.A., Materova, E.A. and Grekovitch, A.A. *Agrokhimiya*, **5**, 134 (1980).
47 Houba, V.J.G., van Schowenburg, J.C. eds *Soil Analysis, 11: Methods of analysis for soils*, Agricultural University, Wageningen, The Netherlands, pp 43 & 83 (1971).
48 Bound, G.F. *Journal of Science of Food and Agriculture*, **28**, 501 (1977).
49 Instruction Manual for Nitrate Ion Electrode, Model 93–07, Orion Research, Cambridge, Massachusetts (1978).
50 Nasko, B.S., Alexandrova, A.M., Gubareva, A.M. and Razday, V.C. *Agrokhimya*, **4**, 131 (1980).
51 HMSO, London. *The Analysis of Agricultural Materials* RB427, 2nd edn (ISBN 0112403522), Method 59, Nitrate nitrogen, calcium sulphate, extractable in soil (1979).
52 Smith, C.J. and Chalk, P.M. *Analyst (London)*, **104**, 538 (1979).
53 Gerretsen, F.C. and de Hoop, H. *Canadian Journal of Microbiology*, **3**, 359 (1957).
54 Wagner, G.H. and Smith, G.E. *Soil Science*, **85**, 125 (1958).
55 Soulides, D.A. and Clark, F.E. *Proceedings of Soil Science Society of America*, **22**, 308 (1958).
56 Clark, F.E., Beard, W.E. and Smith, D.H. *Proceedings of Soil Science Society of America*, **24**, 50 (1960).
57 Khan, M.F.A. and Moore, A.W. *Soil Science*, **106**, 232 (1968).
58 Schwartzbeck, R.A., MacGregor, J.M. and Schmidt, E.L. *Proceedings of Soil Science Society of America*, **25**, 186 (1961).
59 Meek, B.D. and MacKenzie, A.J. *Proceedings of Soil Science Society of America*, **29**, 176 (1965) .
60 Steen, W.C. and Stojanovic, B.J. *Proceedings of Soil Science Society of America*, **35**, 277 (1971) .
61 Bhuchar, V.M. and Amar, U.K. *Indian Journal of Technology*, **10**, 433 (1972).
62 Wu, Q.F. and Liu, P.F. *Talanta*, **30**, 374 (1983).
63 Chaube, A., Bajeva, A.K. and Gupta, U.K. *Talanta*, **11**, 391 (1984).
64 Murphy, J., Riley, J.P. *Analytica Chimica Acta*, **37**, 31 (1962).
65 HMSO, London. *The Analysis of Agricultural Materials* RB427, 2nd edn (ISBN 0112403522), Method 65, Phosphorus, extractable in soil (1979).
66 Karlson, U. and Frankenberger, W.T. *Analytical Chemistry*, **58**, 2704 (1986).
67 Nieto, K.F. and Frankenberger, W.T. *Soil Science of America Journal*, **49**, 587 (1985).
68 Bickford, G.P. and Willett, I.R. *Water Research*, **15**, 511 (1981).
69 Ogner, G. and Haugen, A. *Analyst (London)*, **102**, 453 (1977).
70 HMSO, London. *The Analysis of Agricultural Materials* RB427, 2nd edn (ISBN 0112403522), Method 87, Acid soluble and water soluble sulphate sulphur in soil being considered for tile draining (1979).
71 HMSO, London. *The Analysis of Agricultural Materials* RB427, 2nd edn (ISBN 0112403522), Method 76, Sulphate sulphur, total water soluble, in soil (1979).
72 Chauhan, P.P.S. and Chauhan, C.P.S. *Soil Science*, **128**, 193 (1979).

73 Landers, D.R., David, M.B. and Mitchell, M.J. *International Journal of Environmental Analytical Chemistry*, **14**, 245 (1981).
74 Krug, F.J., Zagatto, E.A.G., Reis, B.F., Bahia, F.O.O., Jacintho, A.O. and Jorgensen, S.S. *Analytica Chimica Acta*, **145**, 179 (1983).
75 Krug, F.J., Bergamin, F.O.H., Zagatto, E.A.G. and Jorgensen, S.S. *Analyst (London)*, **102**, 503 (1977).
76 Little, L.P., Reeve, R., Proud, G.M. and Luchan, A.J. *Science of Food and Agriculture*, **20**, 673 (1969).
77 Al-Ghabsha, T.S., Bogdanski, S.L. and Townshend, A. *Analytica Chimica Acta*, **120**, 383 (1980).
78 Bolan, N.S., Scotter, D.R., Syers, J.K. and Tillman, R.W. *Soil Science Society of America Journal*, **50**, 1419 (1986).
79 Ramesh, C., Ray, P.K., Nayer, A.K., Misra, O. and Sethunathan, N. *Analyst (London)*, **105**, 984 (1980).
80 American Public Health Association and Water Pollution Control Federation, *Standard methods for the examination of water and waste water*, 13th edn. American Public Health Association, New York, p 551 (1971).
81 Nriagu, J.O. and Hem, J.D. In *Sulphur in the Environment*, Wiley International, New York, Part 2, pp 211–270 (1978).
82 Shriner, D.S. and Henderson, G.S.J. *Environmental Quality*, **7**, 392 (1978).
83 Smittenberg, J., Harinsen, G.W., Quispel, A. and Otzen, D. *Plant and Soil*, **3**, 353 (1951).
84 Aspiras, R.B., Keeney, D.R. and Chesters, G. *Analytical Letters*, **5**, 425 (1972).
85 Johnson, C.M. and Ulrich, A. *California Agricultural Experimental Station Bulletin No 766* (1959).
86 Abbasi, S.A. *International Journal of Environmental Studies*, **18**, 51 (1981).
87 Clark, P.D. and Lesage, K.L. *Journal of Chromatographic Science*, **27**, 259 (1989).
88 Mehra, H.C. and Frankenberger, W.T. *Analyst (London)*, **144**, 707 (1989).
89 Mehra, H.C. and Frankenberger, W.T. *Analytica Chimica Acta*, **217**, 383 (1989).

Chapter 8

Determination of anions in non-saline sediments

Published work on the determination of anions in sediments is limited to methods for the determination of nitrate and sulphate. Should it be required to determine anions other than these in sediments then consideration should be given to applying soil analysis methods as discussed in Chapter 7. In some, but probably not all, cases soil analysis methods will be applicable with little or no modification to sediment analysis.

8.1 Nitrate

8.1.1 Spectrophotometric method

Most of the methods for the determination of nitrate such as spectrophotometry using phenoldisulphonic acid or brucine, formation of azo dyes are subject to severe interferences and are not very sensitive.

The nitration of phenolic-type compounds has been investigated following the report by Holler and Huch [1] that these compounds are specific and sensitive spectrophotometric reagents for nitrate determination. It is surprising that most of the studies carried out so far have involved the use of 2,4–xylenol[2, 3] and 2,6–xylenol [4–6] and not 3,4–xylenol, which is claimed to be the most suitable isomer [1]. Problems such as non–stoicheiometric nitration products and interferences from chloride and nitrite have also been reported with these reagents.

However, Elton-Bott [7] has reported a sensitive spectrophotometric method for the determination of nitrate with 3,4–xylenol reagent, involving the distillation of the nitration product into sodium hydroxide solution and measurement of the absorbance of the coloured solution obtained at 432nm. Based on these observations, Osibanjo and Ajay [8] has described a method for the determination of nitrate with 3,4–xylenol. The method, which is also relatively rapid, involves extraction of the nitration product with toluene and is relatively free from interferences. It is highly sensitive and 96–108% recoveries were achieved with 5 µg levels of nitrate.

Table 8.1 Effects of various ions on the relative absorbances of nitrate ion

Values given are percentage changes in absorbance

	NO$_3^-$ to ion ratio			
Ion	*1:1*	*1:10*	*1:100*	*1:1000*
F$^-$	0.77	4.26	4.26	6.60
Cl$^-$	1.16	6.59	2.32	80.62
Br$^-$	4.65	9.30	29.45	75.96
I$^-$	4.65	2.71	3.88	39.92
CO$_3^{2-}$	4.65	4.65	9.30	2.30
PO$_4^{3-}$	0.77	1.16	3.87	3.10
NO$_2^-$	0.38	10.79	11.24	11.62
SO$_3^{2-}$	0.77	5.42	6.98	9.30
NH$_4^+$	0.77	2.71	9.30	9.30

Source: Royal Society of Chemistry [8]

Osibanjo and Ajay [8] investigated the effects of various interferences on a sample containing 40 µg of nitrate. For each interfering ion, the ratios of nitrate to ion investigated were 1:1, 1:10, 1:100 and 1:1000. The results are given in Table 8.1.

The coefficient of variation for 40 µg of nitrate (10 replicate readings) was 5.0%. An ion is considered to interfere in the nitration of the reagent if the percentage change in absorbance (Table 8.1) is greater than twice the coefficient of variation. Hence there are no serious interferences from most of the anions studied, including nitrite and chloride, to which most methods for nitrate determination are normally intolerant. However, bromide interferes at the 1:100 level and above, and nitrite interferes slightly at the 1:10 level and above. These two interferents can be removed, however, by the addition of silver sulphate or mercury sulphate and sulphamic acid, respectively, to the sample solution prior to extraction with toluene.

8.1.2 Chemiluminescence method

Because of the sensitive nature of the chemiluminescence detector it is possible to analyse samples containing nanomolar concentrations or nanogram amounts of nitrite and nitrate ions. The chemiluminescence analysis method then is of importance in environmental analyses, chemical oceanography, and other applications where trace nitrite and nitrate data are needed. Parts-per-billion concentrations can be analysed with milliliter sample volumes while parts-per-million and higher concentrations can also be determined by using microlitre range sample sizes or by dilution.

The earlier methods used an acetic acid–potassium iodide mixture at room temperature for nitrite reduction to nitric oxide. Ferrous ammonium sulphate with ammonium molybdate in hot approximately 50% sulphuric acid, was used for reduction of nitrate plus nitrite. Nitrate was determined as the difference between analyses of the same sample by the two methods. Extensive trapping of the analyte carrier gas was needed to prevent introduction of acidic gases into the NO_x detector.

Braman and Hendrix [10] studied the use of vanadium(III) as a reductant for nitrate and found that it has substantial advantages. It is more reactive as a reducing agent than iron(II)–molybdate and can be used at far lower acidities. As was found, sequential multiple large volume water samples (10–1000 mL) can be analysed by using the same reduction solution, a feature not achieved by the iron(II)–molybdate reduction method which requires fresh blanked reduction solution for each sample of substantial volume. Addition of large water samples reduces the acidity of the iron(II)–molybdate reagent to the point that it no longer reduces nitrates.

In this method [10] nitrate in saline or non-saline sediment samples is reduced at room temperature to nitric oxide in acidic medium containing vanadium(III). Nitrate is also rapidly reduced after heating to 80–90°C. Nitric oxide is removed from the reaction solution by scrubbing with helium carrier gas and is detected by means of a chemiluminescence NO_x analyser. Nanogram detection limits are obtained. The method has the advantage of not requiring highly acidic solutions for nitrate reduction.

In this method solutions being analysed for nitrites or nitrates by reduction reactions producing nitric oxide were degassed by using helium which was then passed into a Bendix Model 8101 chemiluminescence analyser. The analyser inboard flow rate was controlled by a micrometering valve set to approximately 200 mL/min. The helium degassing flow was set at approximately 120 mL/min, while oxygen make-up gas to the 'T' was set at 100 mL/min. A single bubbler containing 1–2M sodium hydroxide at room temperature was used to remove any acidic gases from the inboard flow into the detector. The 'T' system avoids the problem of matching the flow of the analysis stream from the reaction chamber to the inboard flow demanded by the detector.

Cold trapping is not necessary in this apparatus arrangement because the mix of dry make-up gas with the saturated carrier gas produces a relative humidity near 60%. 50–100 mL reaction flasks were used. The detector was used in the NO mode. The reducing reagent (20–50 mL) comprising solutions of vanadium(III) approximately 0.10M, also 1–2M in hydrochloric acid, were produced by reduction of acidic 0.10M solutions of vanadyl sulphate using a Jones reductor. Generally, the solution developed a pink–purple colour during this step indicating the presence of vanadium(II). The vanadium(II) was converted to vanadium(III) by bubbling air or oxygen through the solution.

8.2 Sulphate

8.2.1 Ion chromatography

Ion chromatography has been used to determine sulphate in fresh water sediments [9].

8.3 Sulphide

8.3.1 Gas chromatography

Sulphur enters into numerous biogeochemical reactions in the aquatic environment. In anoxic waters, the sulphate/sulphide redox couple is thought to control the free electron activity, particularly in the marine environment. Dissolved sulphide produced from microbial sulphate reduction in anoxic waters affects the solubility of trace elements such as iron. In anoxic sediments these insoluble metal sulphides can accumulate, or be remobilised through sedimentary diagenetic reactions. Iron sulphide minerals including pyrite (FeS_2) and mackinawite (FeS) are the most common metal sulphides in sediments and represent the major form of sedimentary sulphur. Thus, determinations of dissolved sulphide and sedimentary sulphur speciation are important not only to investigations of sulphur itself but also to studies of trace elements in the aquatic environment.

Cutter and Oatts [11] determined sedimentary sulphur speciation using gas chromatography with photoionisation detection. The method employs selective generation of hydrogen sulphide, liquid nitrogen-cooled trapping, and subsequent gas chromatographic separation/photoionisation detection. Hydrogen sulphide is generated from sedimentary acid volatile sulphides (AVS) via acidification, from greigite using sodium borohydride and potassium iodide, and from pyrite using acidic chromium (II). The detection limit for these sulphur species is 6.1 µg of S/g, with the precision not exceeding 7% (relative standard deviation). This method is rapid and free of chemical interference, and field determinations are possible. Numerous sediment samples have been analysed by using the described procedures.

The detector is an HNU Systems photoionisation unit and electrometer (Model PI–52) equipped with a 10.2 eV lamp. The detector output is processed with a digital plotter/integrator (Hewlett–Packard 3392A). An ultramicro balance (Cahn 29) is used for sample weighings. The following operating parameters are utilised: helium stripping/carrier gas, 60 cm^3 min $^{-1}$; detector temperature, 50°C; Porapak column temperature, 50°C; PID lamp intensity setting, 4. Total sedimentary sulphur is determined with a Carlo Erba ANA 1500 NCS analyser.

8.3.2 Miscellaneous

Dissolved sulphide is typically determined by spectrophotometric procedures which utilise multiple reagents and sample manipulations. The lowest detection limit reported for a spectrophotometric sulphide method is approximately 0.1 mol/L. In sediments, metal monosulphides (acid volatile sulphides or AVS) are determined via acidification and collection of the evolved hydrogen sulphide, which is then quantified by using spectrophotometric or titrimetric procedures. Two principal methods are used to determine pyrite. One method uses selective chemical leaching to remove all nonpyritic iron, followed by a nitric acid digestion to solubilise pyrite-bound iron; the resulting solution is subjected to atomic absorption analysis. This method requires substantial sample preparation efforts, and the removal of all nonpyritic iron is crucial to the method's accuracy. Pyrite can also be determined by using acidic Cr(II) reduction of pyritic sulphur to hydrogen sulphide, which is then determined via AVS methods. Pyrite determinations via Cr(I) reduction method are direct, but the available procedures are very time-consuming.

References

1 Holler, A.C. and Huch, R.V. *Analytical Chemistry*, **21**, 1385 (1949).
2 Norwitz, G. and Gordon, H. *Analytica Chimica Acta*, **89**, 177 (1977).
3 Andrews, D.W.W. *Analyst (London)*, **89**, 730 (1964).
4 Hartley, A.M. and Asai, R.J. *Analytical Chemistry*, **35**, 1207 (1963).
5 Yulin, L.T. *Analytica Chimica Acta*, **91**, 373 (1977).
6 Hajos, P. and Inczédy, J. *Hungarian Scientific Instrumentation*, **34**, 25 (1975).
7 Elton-Bott, R.R. *Analytica Chimica Acta*, **90**, 215 (1977).
8 Osibanjo, O. and Ajay, S.O. *Analyst (London)*, **105**, 908 (1980).
9 Hardijk, C.A. and Capponburg, T.E. *Journal of Microbiological Methods*, **3**, 205 (1985).
10 Braman, R.S. and Hendrix, S.A. *Analytical Chemistry*, **61**, 2715 (1989).
11 Cutter, G.A. and Oatts, T.J. *Analytical Chemistry*, **59**, 717 (1987).

Chapter 9

Determination of anions in sludges

9.1 Borate

9.1.1 Spectrophotometric methods

Borate has been determined [1] in amounts down to 0.02 mg L^{-1} in sewage effluents spectrophotometrically by reaction with phenol and carminic acid in concentrated sulphuric acid to produce a coloured compound with an absorption maximum at 610nm.

The standard curcumin method [2, 3] has been found to be suitable for the determination of borate in industrial effluents and sludges.

9.2 Chloride

9.2.1 Ion selective electrodes

Hindin [4] has shown that chloride in concentrations in sewage can be determined by the ion specific electrode method as an alternative to the standard mercuric nitrate method, providing two precautions are taken – the addition of an ionic strength adjusting solution to overcome any effect the ionic strength of the sample or standard may have, and the removal of sulphide ions by a cadmium ion precipitating solution.

9.3 Cyanide

9.3.1 Spectrophotometric methods

Tecator [5] produce apparatus based on distillation and titration or spectrophotometry for the determination of cyanides in sludges.

Kodura and Lada [6] determined cyanide in sewage spectrophotometrically using ferron. Iron, copper sulphide, acrylonitrile, phenol, methanol, formaldehyde, urea, thiourea, caprolactam or hexamine did not interfere at concentrations up to 1 g L^{-1}.

9.3.2 Atomic absorption spectrometry

In an indirect atomic absorption procedure [7] for determining down to 20 μmol L^{-1} cyanide, the sample is treated with sodium carbonate and a known excess of cupric sulphate to precipitate cupric cyanide. Excess copper in the filtrate is then determined by atomic absorption spectrometry and hence the cyanide content of the sample calculated. Iron(III) does not interfere except at low cyanide concentrations. Quantitative recovery of cyanide from sewage was obtained by this procedure.

9.4 Fluoride

9.4.1 Spectrophotometric method

Devine and Partington [8] have shown that errors in the determination of fluoride in sewage by the SPADNS colorimetric method are due to sulphate carried over during the preliminary distillation step. It is suggested that the colorimetric method should be replaced by the fluoride ion electrode method following distillation.

9.4.2 Ion selective electrodes

Rea [9] used an ion selective electrode to determine fluoride in sewage sludge – a trisodium nitrate buffer was used. Calibration was achieved by the standard addition procedure.

9.5 Nitrate

9.5.1 Spectrophotometric method

Workers at the Water Research Centre, UK [10] have described detailed procedures based on the use of the Technicon Autoanalyser AA11 for the determination of nitrate and nitrite in sewage and sewage effluents. Measurements of nitrate in sewage at the 10–50 mg L^{-1} level were made with a within-batch standard deviation of 0.1 mg L^{-1} nitrate. Nitrate recoveries at the 12 mg L^{-1} level were in the range 100–100.7%. Standard deviations of nitrite determinations were in the range 0.003 (at 0.2 mg L^{-1} nitrite) to 0.01 (at 1 mg L^{-1} nitrite). Nitrite recoveries in the 0.4 mg L^{-1} region were between 98 and 102%.

9.5.2 Ion selective electrodes

Nitrate levels in sewage in amounts down to 1 mg L^{-1} have been determined by specific ion electrodes [11, 12]. Petts [10] used a non-porous plastic membrane nitrate selective electrode and compared results obtained with this electrode and those obtained by a standard

Table 9.1 Nitrate analysis of effluents of sewage treatment plants[a]

Sample no.	Potentiometric NO$_3$ N, mg L^{-1}	Spectrophotometric[b] NO$_3$ N$_3$ mg L^{-1}
1	28	28
2	43	44
3	52	41
4	43	37
5	34	39

[a] The data represent a single analysis
[b] As measured independently

Source: Water Research Centre, Stevenage [10]

spectrophotometric method on a sewage works effluent. The results indicate that the selective ion electrode is suitable both for laboratory and plant monitoring purposes. Only chloride and, to a lesser extent, nitrite and bicarbonate interfere in these measurements of nitrate.

Table 9.1 gives a survey of results obtained by both methods on sewage effluents with a high nitrate content.

9.6 Nitrite

9.6.1 Spectrophotometric method

See section 9.5.1.

9.7 Phosphate

9.7.1 Spectrophotometric method

The Department of the Environment, UK [13] has issued details of spectrophotometric methods for the determination of orthophosphate in sewage effluents.

These methods are based on reaction with acid molybdate reagent to form a phosphomolybdenum blue complex which is determined at 882nm. The first method has a range of 0–0.40 mg L^{-1} and the second is mainly designed for oligotrophic waters with phosphorus contents in the range 0–25 µg L^{-1}. In addition, the report discusses various methods for converting other forms of phosphorus to orthophosphate, and the elimination of interference due to arsenic, based on the reduction of arsenate to arsenite.

Bretscher [14] discusses reduction reagents for the spectrophotometric determination of phosphate in sewage as the phosphomolybdenum blue complex. He points out that the disadvantage of aqueous stannous

Table 9.2 Condition for the determinatino of phosphorus molybdohetcropoly yellow with HPLC

Packing material	Lichrosorb RP–18
	(mean particle size: 5 μm)
Column size	4 mm ∅ × 150 mm
Eluent	30% H_2O in CH_3CN
Flow rate	0.9 ml min (70 kg cm $^{-2}$)
Temperature	ambient
Detector	UV 251 nm
Range	0.04 AUFS
Chart speed	5 mm min $^{-1}$
Injecting sample solution	5 μL

Source: Springer Verlag Chemie GmbH [15]

chloride reduction reagent is that it oxidises rapidly and a fresh solution must be prepared daily.

Solutions of stannous chloride in glycerol were found to be stable for at least six months.

9.7.2 High performance liquid chromatography

Sakurai *et al.* [15] have described a high performance liquid chromatographic procedure for the determination of down to 0.5 mg L $^{-1}$ phosphate in waste water sewage effluents. The method is based on the solvent extraction of molybdoheteropoly yellow with methyl propionate. The corresponding silicon compound is not extracted into this solvent. Thus the interference by silicon is excluded, even at concentrations as high as 10 L $^{-1}$.

The instrument used was a Hitachi Model 635 high speed liquid chromatograph equipped with a UV detector and a Hitachi Model 100–50 double beam spectrophotometer with a 5 cm glass cell.

For determination of dissolved inorganic phosphate, 20 mL of the sample is transferred to a 50 ml beaker and 1 mL of 4.4N nitric acid added. The solution is heated in a water bath at about 50–60°C, and 1 ml of 3% ammonium molybdate solution added, and the mixture allowed to stand for a further 5 min in the water bath with occasional stirring. After standing for a further 5 min in running water, the solution is transferred into a separatory funnel (50 mL) with 10 mL of water containing 3 mL of methyl propionate. The complex is extracted into the organic phase by shaking for 2 min, and centrifuged for 2 min (3000 rpm) to remove water suspended in the organic phase. Molybdoheteropoly yellow in the organic extract is determined by a high performance liquid chromatograph with UV spectrophotometric detector at 251 nm and under the working conditions shown in Table 9.2.

For determination of total phosphorus 20 mL of the sample solution is transferred into a 50 mL beaker and heated with 5 mL each of concentrated nitric acid and perchloric acids on the hot plate until white smoke is produced. After cooling to ambient temperature, 10 mL of water and 1 mL of 4.4N nitric acid are added to the beaker and the residue dissolved by warming. The solution obtained in a water bath is heated at about 50–60°C and the analysis continued as above.

A wavelength of 251nm at the UV detector gave an absorption maximum for phosphorus molybdoheteropoly yellow in methyl propionate.

9.8 Sulphide

9.8.1 Ion selective electrodes

Glaister *et al.* [16] studied three sulphide ion selective electrodes in cascade flow and flow-through modes to investigate carrier stream, sample size and flow rate parameters in the analysis of sulphide in sewage effluents. Results were compared with those obtained by direct potentiometry. The electrodes were successfully used as detectors of sulphide during flow injection analysis and the presence of ascorbic acid in the standard antioxidant buffer minimised deleterious effects of hydrogen peroxide in the samples. Both the cascade flow and flow-through modes of electrodes yielded sulphide concentrations similar to those obtained by colorimetric methods.

9.8.2 Draeger tube methods

Ballinger and Lloyd [17] have described a detailed procedure for the rapid determination of down to 0.06 mg L^{-1} sulphides in sewage samples. Hydrogen sulphide is brought to solution-vapour equilibrium in a closed flask under controlled conditions. The concentration of hydrogen sulphide vapour is determined by means of Draeger tubes, and related to the concentration in solution by means of a calibration graph.

In aqueous solution, hydrogen sulphide dissociates as a weak diprotic acid:

$$H_2S + H_2O \Leftrightarrow HS^- + H_2O^+$$
$$HS^- + H_2O \Leftrightarrow S^{2-} + H_3O^+$$

At a given pH, the degree of dissociation depends upon the temperature and activity of the sulphide species. At pH 5.0, dissolved sulphides are present almost entirely as undissociated hydrogen sulphide. It follows from Henry's law that the vapour pressure of hydrogen sulphide above its aqueous solution is a function of the mole fraction of dissolved hydrogen sulphide. In this method, hydrogen sulphide is brought to solution-vapour equilibrium in a closed flask, and the concentration of hydrogen sulphide vapour in the air space is determined by a

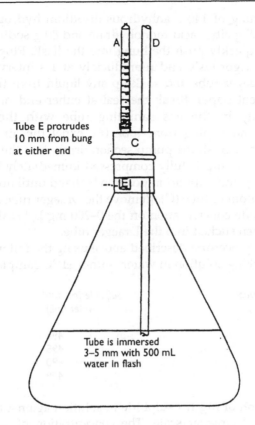

Tube E protrudes
10 mm from bung
at either end

Tube is immersed
3–5 mm with 500 mL
water in flash

Fig. 9.1 Diagram of apparatus for determination of sulphide
Source: Bureau of National Affairs, Washington [17]

conventional gas analyser. The pH of the solution is adjusted to 5.0 by means of a buffer solution and the ionic strengths of samples and standards are maintained at a constant level by the strong electrolytes incorporated into this solution.

The apparatus, depicted in Fig. 9.1, consists of a Draeger multi-gas detector pump Model 21/31. Draeger hydrogen sulphide tubes 0–200 mg L^{-1}, catalogue reference 1/C (A). 1 L Pyrex conical quickfit flask, socket 34/35 (B). 35 mm silicone rubber bung, drilled in two places to give a tight fit to a 9 mm rod (C). 300 mm soda glass tube 9 mm OD, 5 mm id (D). Small rubber bung to plug (E). 60 mm PVC tube 10 mm od, 6 mm id (F). Thread the tube through one of the holes in (C). Ensure that reagents and water for dilution are at the same temperature ± 1 °C. Without breaking the seal at either end of the Draeger tube, push it firmly into the gas sampling tube (F). Quickly transfer a volume (× ml) of sample to the flask containing 5 L water. Add samples by pipette, with the tip dipping below the surface of the dilution water. Immediately add 25 mL of pH 5 buffer solution (consisting of 140 g anhydrous disodium hydrogen phosphate (Na_2HPO_4), 106.5 g citric acid monohydrate and 80 g sodium chloride in 2 L water) and quickly push the bung into the flask. Plug tube (D) and shake the flask vigorously and reproducibly at 1 s intervals for 1 min. Remove the Draeger tube and soak up any liquid from tube (F) with a twist of absorbent paper. Break the seal at either end of the tube and replace it firmly in the gas sampling tube with the white band downwards. Remove plug from tube (F), expel the air from the gas detector pump and push the pump section firmly into the Draeger tube, ensuring that the pump is fully compressed immediately before release. Release the pump and support in position by hand until no more bubbles emerge from bottom of tube (C). Remove the Draeger tube and record the hydrogen sulphide concentration on the 0–200 mg L^{-1} scale. Ensure that no liquid has been sucked into the Draeger tube.

Carry out the procedure described above using the following dilutions of sulphide working solution in water, within ±1 °C sample temperature.

Sulphide working solution (mL) (50 mg L^{-1} S^{-1})	Sulphide free tap water (mL)
2.5	498
5.0	495
10.0	490
12.5	488

Construct a graph of mg L^{-1} sulphide in solution against the reading on the 0.200 mg L^{-1} Draeger scale. The concentration of sulphide in the reagents and dilution water are negligible and a blank correction is not normally necessary.

The above method determines undissociated hydrogen sulphide, and hydrosulphide and sulphide ions. To determine total sulphide, including above plus other compounds liberating hydrogen sulphide from a cold solution containing 0.5 mol L^{-1} hydrochloric acid, follow the above procedure using 50 mL acid chloride solution instead of buffer solution.

The effect of various compounds likely to occur in sewage on results obtained by this method is shown in Table 9.3. The comparative freedom of interference from other sulphur-containing anions is of particular note. Sulphite, metabisulphite and thiosulphate do not interfere significantly at 100-fold excess; detergents depress the recovery of sulphide at 100-fold excess but do not interfere significantly at 10-fold excess.

9.9 Tungstate

9.9.1 Ion chromatography

Mehra and Frankenberger [12] used ion chromatography to determine tungstate in sludges.

Table 9.3 Effect of constituents and contaminants of sewage on a dissolved mg L^{-1} sulphide solution

Compound	Apparent % change in sulphide concentration		
	10 mg L^{-1}	100 mg L^{-1}	1000 mg L^{-1}
Nitrite N as sodium nitrate		<5%	−95%
Ammn.N as NH$_4$Cl		<5%	<5%
Sulphite SO$_3$ as Na$_2$SO$_3$5H$_2$O		<5%	<5%
Thiosulphate S$_2$O as Na$_2$S$_2$O$_3$		<5%	+9.5%
Metabisulphite S$_2$O$_3$ as Na$_2$S$_2$O$_3$		<5%	−20%
Anionic detergents, as Manoxol–OT	<5%	−24%	
Acetone			<5%
Mineral oil			<5%
Methylated spirit			<5%
Chloroform			<5%
Effect of sewage organic strength (simulated by a dispersion of raw sausage meat in water)		Approx COD 6090 mg L^{-1} oil and grease 300 mg L^{-1} SS 300 mg L^{-1}	Approx COD 6000 mg L^{-1} oil and grease 3000 mg L^{-1} SS 3000 mg L^{-1}
		−7%	−10%

* In order to maintain pH 5.0 ± 0.2 it was necessary to add 100 mL buffer solution to overcome the alkalinity or sodium sulphite

Source: Bureau of National Affairs Inc Washington [17]

References

1 Lionnel, L.J. *Analyst (London)*, **95**, 194 (1970).
2 Burton, N.G. and Tait, B.H. *Journal of the American Water Works Association*, **61**, 357 (1969).
3 American Public Health Authority (APHA), *Standard methods for the analysis of waters and waste waters*, 12th edn, New York (1965).
4 Hindin, E. *Water and Sewage Works*, **122**, 60 (1975).
5 Tecator Ltd, Box 70, S–26321 Hoganes, Sweden, Application Notes AN 89/87 and 86/87. Cyanides in waste waters, soils and sludges using the 1026 distilling unit (1987).
6 Kodura, I. and Lada, Z. *Chemia Analit*, **17**, 871 (1972).
7 Manahan, S.E. and Kunkil, R. *Analytical Letters (London)*, **6**, 547 (1973).
8 Devine, R.F. and Partington, G.L. *Environmental Science and Technology*, **9**, 678 (1975).
9 Rea, R.E. *Water Pollution Control*, **78**, 139 (1979).
10 Petts, K.W., Water Research Centre, Stevenage, Herts, UK. Technical Memorandum No 108, Determination of nitrogen compounds by Technicon Autoanalyser AA11 (1975).
11 Longmuir, D. and Jacobson, R.L. *Environmental Science and Technology*, **4**, 834 (1970).
12 Mehra, H.C. and Frankenberger, W.T. *Analytica Chimica Acta*, **217**, 383 (1989).
13 Department of the Environment/National Water Council Standing Committee of Analysts, HMSO, London, Method for the Examination of

Water and Associated Materials: phosophorus in waters, effuents and sewages 1980 (1981).

14 Bretscher, U. and Gas, U. *Abwasserfach (Wasser, Abwasser)*, **117**, 31 (1976).

15 Sakuri, N., Kadohata, K. and Ishinose, N. *Fresenius Zeitschrift für Analytische Chemie*, **314**, 634 (1983).

16 Glaiser, M.G., Moody, G.J. and Thomas, J.D.R. *Analyst (London)*, **110**, 113 (1985).

17 Ballinger, D. and Lloyd, A. *Water Pollution Control*, **80**, 648 (1981).

Chapter 10

Sampling procedures

10.1 Introduction

Sampling procedures are extremely important in the analysis of soils, sediments and sludges. It is essential to ensure that the composition of the portion of the sample being analysed is representative of the material being analysed. This fact is even more evident when it is conceded that the size of the portion of sample being analysed is in many modern methods of analysis extremely small. It is therefore essential to ensure before the analysis is commenced that correct statistically validated sampling procedures are used to ensure as far as is possible that the portion of the sample being analysed is representative of the bulk of material from which the sample was taken.

The collection and handling of samples prior to analysis has been discussed by various workers and organisations, including Smith and James [1], Kratochvil et al. [2,3], Gy [4], Woodget and Cooper [5], Harrison [6], Walton and Hoffman [7], Laitinen [8,9], Ingamells and Pitard [10], Kratochvil and Taylor [11], Kratochvil [12], Wallace and Kratochvil [13], Ministry of Agriculture, Fisheries and Food [14] and HMSO [15]. Other bodies who have discussed sampling procedures include the American Society for Testing Materials, the US Environmental Protection Agency, the American Public Health Association, the British Standards Institution, [16] etc.

The principal step of the sampling process is the taking of the sample. Here we intend to deal only with the risk of contaminating the sample during its collection, storage and processing, since any subsequent separation is applied only after the sample has been brought into solution.

By knowing the history of the sample it is possible to act correctly during all of the sampling steps, in order to avoid contamination of the material either from the utensils used to collect the sample, or from the reagents, the laboratory atmosphere and even the laboratory personnel. The contamination risk is greatest in analysis for trace components. Trace analysis requires the use of specially acclimatised, sometimes over-pressurised laboratories, of very pure reagents and sensitive instruments, and of specialised personnel, who possess a broad range of knowledge both in basic and analytical chemistry. In such cases, the 'method-man-instrument' correlation has to be correctly applied, since any failure of the system would result in unreliable analytical information being obtained.

In analysis of solid materials [17,18], the determination of trace elements requires knowledge of the exact manner in which these traces were introduced or pre-existed in the samples. The most complex problem occurs whenever the trace component to be determined has properties which are very similar to those of a major component of the sample to be analysed. To set up an efficient extraction process when using these techniques, it is necessary to resort once again to a knowledge of basic chemistry. For selecting the separation methods, besides knowing the history of the sample, a decision must be made about the choice of separation methods, a decision which will involve knowledge of the solvents and the ion-exchangers used, the kinetics of the processes and the nature and mechanism of the extraction equilibria and the ion-exchange. Once these problems have been solved, the analytical separation process can be set up.

The 'art' of the analytical chemist consists of knowing the history of the sample, and of choosing the simplest possible analytical procedure.

Ristenpart et al. [42] and Houba [43] have evaluated various sediment samplers. A sediment shovel proved highly practical but was limited because small particles tend to be lost when the shovel is lifted. A cryogenic sediment sampler was less convenient to use, but allowed the collection of nearly undisturbed samples. Houba described a different device for the automatic subsampling of sediments for proficiency testing. Thoms [44] showed that freeze-sampling collects representative sediment samples, whereas grab-sampling introduces a bias in the textural composition of the 120 mesh fraction, due to washout and elutriation of the finer fractions.

Rubio and Ure [51] have discussed the risks of the contamination of sediment samples using inappropriate materials, containers and tools as well as possible analyte loss during sample handling.

Meriwether et al. [29] have devised a coring sampler made from standard plumbing parts. It was found especially useful for sampling and maintaining the depth profile of the soft sediments underlying relatively shallow waters.

Wehrens et al. [78] have discussed a decision support system for the sampling of aquatic sediments in lakes.

Vernet et al. [79] used three methods of fluvial sediment sampling to determine the validity and representativeness of the geochemical information obtained by each technique. Bottom sediments and trap sediments showed similar results for metals.

Fortunati et al. [45] have reviewed problems associated with techniques and strategies of soil sampling.

Meriwether *et al.* [46] has suggested a new type of pedologically based soil sampling technique based on the soil horizon rather than incrementally with depth for the assessment of radionucleides in soil. He gives an example where classical sampling approaches would lead to erroneous conclusions about anthropogenic contamination.

Rasemann *et al.* [47] have demonstrated in a non-uniformly contaminated soil site that mercury concentrations depend on the method of handling soil samples between sampling and chemical analysis. Sample pretreatment contributed substantially to the variance in results and was of the same order as contributions from the sample inhomogeneity.

Davies [48] has reported a systematic approach to interpreting the results from surveys of soils contaminated with trace metals. He uses lead as an example.

Kimbrough and Wakakuwa [49] have reported an interlaboratory study involving 160 accredited hazardous materials laboratories. In this study, each laboratory performed a mineral acid digestion on five soils spiked with As, Cd, Mo, Se and Tl. Instrumental detection methods were inductively coupled plasma atomic emission spectrometry, inductively coupled plasma mass spectrometry, flame atomic absorption spectrometry, electrothermal atomic absorption spectrometry and hydride generation atomic absorption spectrometry. At most concentrations, inductively coupled plasma atomic emission spectrometry exhibited higher precision and accuracy than the other techniques, but also the highest rate of false positives and negatives.

Einax *et al.* [50] have used chemometric techniques to investigate the representativity of soil sampling.

Rubio and Ure [51] have discussed the risks of the contamination of soil samples using inappropriate materials, containers and tools as well as possible analyte loss during sample handling. Factors affecting the realism of the collected sample have been discussed by Burton [52]. Different sampling designs are needed, depending upon whether the soil contamination is expected to be 'spread' over the whole area or exists in localised 'hot spots' [53].

Lame showed that the fundamental sampling error for soil only affects the analytical variance when sample sizes are less than 10 g [54]. For larger samples, the variance is determined by the segregation error. A sampling board method for estimation of the segregation error was described. Skalski showed that a two-way compositing strategy could be used to attribute detected contamination in composited samples directly to constituent samples without further analyses [55].

Droppo *et al.* [80] studied the effects of concentrating suspended sediment samples on the primary grain size distribution. The initial and resuspended size distributions were not significantly different using either cellulose or polycarbonate filters.

Truckenbrodt and Einex [81] have shown in an analysis of overall analytical error that, independent of grain size, sampling was the main source of variance in the determination of metals in river sediments. Two approaches are described to determine the number of samples required for representative sampling.

Ruiz *et al.* [82] have described a method for the determination of the distribution of metallic constituents in sediment particles in torrential rivers according to particle size. A rapid sampling method using passive sampling devices for soil contaminant characterisation has been shown to provide a more thorough site assessment [30].

10.2 Sample homogeneity

The problem of the homogeneity of the sample is closely related to the problem of the history of the sample.

Whereas the literature reports many specific sampling situations, there are few papers which consider the fundamental aspects of the sampling process and its implications for the general analytical process.

The aspects which relate to the homogeneity of the sample have to be considered in the context of the nature of the analytical process. Also the nature of the analytical process is determined by the characteristics of the sample to be analysed. Hence the two groups of analytical methods, destructive and non-destructive, should be considered separately. Although both groups have wide applicability, the analytical chemist has a tendency to prefer the non-destructive methods. Since such methods act directly upon the sample, they have the advantage of partly – and with some precautions, totally – eliminating the risk of contamination of the sample.

The choice of one or other of the methods depends on the nature of the sample. Generally, the non-destructive methods are applied to samples with relatively simple composition. However, for more complex samples or when the determination of major and minor components is required, preliminary separation of the components and concentration of the minor or trace components are necessary, before the actual determination may be performed. It follows therefore, that the analytical chemist must resort in many cases, willingly or not, to destructive methods of analysis.

The homogeneity of the sample depends on the physical state of the material. Because of natural diffusion processes, liquid and gaseous samples are much more homogeneous than solid ones. Solid samples are often heterogeneous, and have first to be homogenised by mechanical means (grinding, ball-milling, etc.) before specimens are selected. The lack of homogeneity of solid samples is the main factor which renders their processing difficult.

10.3 Destructive analysis

'Destroying' a sample means to bring it into a homogeneous form as a solution, normally in an aqueous or a partially non-aqueous medium. There are two means of bringing solid samples such as soils, sediments and sludges into solution, either by dissolving them, or by decomposing or disintegrating them in dry form by means of fluxes. These supplementary operations not only increase the duration of the analysis proper, but also introduce the risk of contamination of the samples by reagents and working techniques.

The dissolution agents used for soil and sediment samples are very diverse, and the analytical chemist must understand thoroughly the chemistry underlying the dissolution process when using a particular reagent. Although the most common dissolution agent is water, there are many situations where water may be unsatisfactory.

To dissolve certain solid samples, acids or mixtures of acids may be used. However, besides dissolving the sample, the acids may interfere with the subsequent analysis either by converting some components of the sample into extremely stable complexes or by creating volatile components which may be lost partly or even totally during the dissolution process. Hence, before dissolving a sample, it is necessary to acquire some knowledge of the nature of its components and their relative proportions, i.e. some knowledge of the origin of the sample, and its history.

Some wet dissolution/decomposition reagents such as hydrofluoric or hydrochloric acid may have strong competing action. Very often, the complex formed may prevent the determination proper from being performed, because it is kinetically or thermodynamically very stable. In many cases the dissolution/decomposition reagents are used to destroy an organic substrate. For example, the use of nitric–perchloric or nitric–sulphuric–perchloric acid mixtures is well known.

The art of the analytical chemist consists in choosing the most suitable dissolution/decomposition system for a given sample, so that the resulting solution contains the components in a form directly usable in the subsequent concentration and separation processes.

When the aim of the analytical chemist is to determine trace components after a wet decomposition with water or acids, care must be taken to use clean vessels and pure reagents in order to minimise contamination risks. Trace analysis presupposes an appropriate sampling procedure and dedicated high-purity reagents. The water to be used must be purified by ion-exchange and then distilled, and stored in polyethylene vessels. The acids and other reagents used for decomposing samples must be of suitable purity, e.g. so-called 'electronic' or 'semiconductor' grade.

An interesting example of the contamination risks which may be caused by a laboratory vessel is that of boron. Determination of very low boron concentrations, involves a prior separation by distillation and subsequent analysis by spectrometry, with a suitable reagent such a curcumin or carminic acid. The use of laboratory vessels made of borosilicate glass (such as Duran or Pyrex) could lead to very large errors in the boron content found due to sample contamination from the boron present in the glassware.

For samples which are virtually insoluble in water or acids, either cold or hot, so-called dry decomposition may be used. This system is more tedious than wet decomposition since it involves two independent operations, the decomposition proper and the succeeding dissolution of the product. For dry decomposition various fluxes can be used, such as $Na_2CO_3+K_2CO_3$, Na_2CO_3+borax, or Na_2CO_3+S (Freiberger decomposition). To transfer the sample completely into solution, it must first be perfectly homogenised with the decomposition agents. These decomposition systems normally require the use of expensive laboratory vessels, for instance platinum crucibles. However, if one is unaware of the history of the sample, such vessels might be damaged or even destroyed (e.g. whenever the samples contain sulphur, phosphorus, arsenic, antimony, etc.).

The fact that 'classical' systems of dry decomposition still persist in today's analytical chemistry is due to the traditional thinking of the analytical chemist, who still believes that the most favourable agent for speeding up the process is temperature. For this reason, dry decomposition has now become the greatest drawback in sampling; on the one hand, it has led to the lengthening of the analysis time and, on the other, to increased contamination risks due to the decomposition agents used. Unlike acids, which can now be obtained in a high degree of purity, solid reagents are often of insufficient purity for trace analysis. It is this aspect of trace analysis which has led to the development of some non-contaminant decomposition systems. The simplest way of achieving faster (and non-contaminating) decomposition has been to resort to an additional physical parameter, namely pressure, coupled with an adequate decomposition temperature. As discussed later, the use of high-pressure decomposition vessels requires much lower temperatures for decomposing a sample, than those necessary for dry decomposition at atmospheric pressure. The appearance of the high-pressure decomposition vessels (bombs) is a direct result of the availability of a chemically inert plastic, namely Teflon. Teflon exhibits good thermal stability and offers minimal contamination risks. High-pressure decompositions involve the use of some decomposition agents which can be prepared easily in a high degree of purity (e.g. hydrochloric acid). The great advantage offered by these disintegration systems is that they make use of relatively cheap laboratory apparatus and avoid expensive materials such as platinum. These high-pressure decomposition systems have now become commonplace in the laboratory.

In analysing solid samples, regardless of the chosen decomposition system, a preliminary and extremely important step is granulometry, which plays a decisive role when preparing solid samples for chemical analysis. Although many studies have been written on granulometry, these studies could also be considered as part of the general sampling process. Wolfson and Belyaev [19] reviewed current work in this field and discussed the role and importance of granulometry for the general sampling process. This work underlined the necessity of granulometric control of the composition of a sample, during its preparation before chemical analysis. Vulfson and Belyaev [19] examined the modern methods of fine grinding and granulometric analysis and attention was given to problems of the influence of the granulometric composition of the dispersed substance on the chemical analysis results and sampling errors.

A great number of separation processes are based on solvent extraction, especially since this is also a concentration technique. For these reasons, solvent extraction will be considered, both from the point of view of the sampling process and from that of the general analytical process. Solvent extraction is ultimately a process of partitioning between two immiscible solvents, and for its optimisation it is necessary to know first of all the operational parameters of the system.

The technique of solvent extraction has long been used in organic chemistry for concentrating and purifying some substances. In the case of organic compounds, the separation process is simple, in many cases being based only on differences in the solubility of the compounds in different solvents.

Many attempts at classifying solvent extraction systems have been made. Thus Diamond and Tuck [20] have described a classification of the solutes that can be separated by solvent extraction.

A number of conclusions may be drawn from this discussion of the destructive analysis of samples.

1 Destructive analysis will continue to be necessary, for many types of samples, owing to the nature of certain samples. The complexity of the composition of some samples often imposes use of a separation step prior to the analysis itself. Furthermore, the classical techniques are usually the only ones suitable for validation of the major components of reference materials.

2 Destructive analysis is also necessary when there is a need to concentrate the components of a given sample, the components being present at very low concentrations (or in traces), often in a very complex matrix.

3 The techniques of destructive analysis involve taking some special precautions concerning the sampling in general, and the dissolution–decomposition process in particular, in order to avoid the risk of contamination.

4   The techniques of destructive analysis may lead to reliable analytical results only if the analytical process is correctly planned, by taking into account the interdependence of all the operational parameters of the process.

5   The concentration–separation process is a necessary step used in most destructive analyses. Regardless of the actual concentration–separation procedures used, this process belongs to the general sampling process. For this reason, terms such as chromatographic analysis, perpetuated by long routine use, should be replaced by terms such as 'chromatographic separation'. In the framework of the chromatographic separation systems, the detector would constitute an independent, well-defined entity of the chromatograph, the 'analyser' proper.

6   Owing to its complexity, in most cases destructive analysis, including the separation methods, resorts not only to the theoretical and basic knowledge of the analytical chemist, but also to the analyst's 'art' of optimising the analytical process step-by-step, taking account of the whole series of factors which may interfere with the sampling process. In this connection much experimental work might be saved by using simplex optimisation [41] in the exploratory research.

Considering these conclusions, it is apparent that, destructive analysis still has a place in the analysis of soils and sediments – and for this very reason, we should correlate the necessary knowledge so as to simplify as much as possible the analytical process. Such an action is necessary in order to shorten the analysis time.

10.4 Analysis of soils and sediments

The analytical chemist is frequently requested to determine trace elements in solid samples.

The sample when it arrives in the laboratory is usually in a form unsuitable for analysis, e.g. a river sediment or sewage sludge suspended in water. Table 10.1 shows steps that may then be required to convert the sample into a form suitable for analysis.

Table 10.1 Conversion of sample into form suitable for analysis

Sediments, sludges
Comminute sieve Particle-size measurement possibly Digest sample with acid

Source: Own files

Table 10.2 Laboratory homogenisers and comminution equipment supplied by Fritsch

Homogenisers		Description	Sample type
Comminution			
A	Vibrating cup mill	Pulverisette 9	A gate graining to 20µm of dry or wet sediments, fish, crustacea or plant material
B	Laboratory desk mill	Pulverisette 13	Grinding 0.1mm of dried sediments, soils, sewage sludge, hydrological sediments and drilling cores
C	Mortar grinder	Pulverisette 2	Grinding to 10µm of sediments and soils
D	Centrifugal mill	Pulverisette 6	Grinding to 1µm of sediments and soils
E	Planetary mill	Pulverisette 5	Grinding to 0.1µm of sediments and soils
F	Sieving devices Vibratory sieve Shaker for micro-precision sieving	Analysette 3	Dry and wet sieving (25mm to 20µm) or micro-precision sieving (5–100µm)
G	Rotary sieve shaker	Analysette 18	Satisfied ASTM E–11–190 BS 410 1969; AFNOR NFX 11–501 and DIN 53477/1 separation of coarse grain material

Source: Own files

Note if determinations of certain volatile elements such as mercury or selenium are required it is necessary to carry out these analyses on the wet sample as received (to avoid loss of element by drying at 105°C). The dry weight of material in the sample is obtained by determining moisture in a separate position of the sample and applying a correction to the sample weight used in metals determination.

As an alternative to drying at 105°C microwave drying has been used to remove moisture from aqueous slurries [21].

10.4.1 Comminution of samples

Various comminution devices (Table 10.2 A–E) are available for handling these types of samples.

Grinding elements are offered in various non-contaminating materials such as corundum (Al_2O_3), agate (SiO_2), or zirconium oxide (ZrO_2).

10.4.2 Sieving analysis of samples

Having comminuted the sample it may now be required to carry out a sieving analysis in order to obtain different size fractions for chemical analysis. Fritsch supply a range of devices for sieving analysers (Table 10.2).

A practical scheme for sieving a soil sample is discussed below.

10.4.2.1 Field or moist soil

Remove the sample from its container and either chop the soil with a knife or rub the soil through a 5.6mm mesh wire sieve. If an additive has been used to prevent nitrification during transit, gloves should be worn. Thoroughly mix the chopped or sieved sample and commence the analysis without delay.

10.4.2.2 Air-dried soil

Transfer the soil sample to a suitable metal tray to form a thin layer and, as far as possible, remove any stones present. With very heavy soils it is necessary to break any clods between the fingers. Dry the soil by placing the tray in a current of air at a temperature not exceeding 30°C. With large numbers of samples it is convenient to place the trays on a series of metal racks over which air may be blown from thermostatically controlled fan heaters. Continue the process until the soil feels quite dry. If the soil appears to contain moisture after grinding, return it to the drying rack.

10.4.3 Grinding of samples

Grind the air-dried soil until the whole of the sample, excluding stones, any fibrous material from roots etc. passes through a 2mm mesh sieve. There is a limited range of apparatus available for grinding soil but the Rukuhia-type soil grinding machine is suitable. (This is obtainable from D. Mackay, 85 East Road, Cambridge, CBI 1BY.) The apparatus consists of a number of cylinders into which the samples and metal pestles are placed. The cylinders, which have walls of 2mm mesh perforated steel, are rotated horizontally by means of electrically driven rollers. As the cylinders rotate, the soil is ground by the pestle and falls through the mesh into a tray below. When grinding samples containing soft rock, the action of the pestle should be cushioned by encasing it in a nylon or polythene tube. With very heavy soils it may be better to grind the sample while still slightly moist and to complete the air-drying after grinding.

Houba et al. [56] studied the influence of grinding procedures and demonstrated that the availability of some analytes is significantly influenced by the grinding of some soils.

10.4.4 Particle-size distribution measurement

A complete particle-size analysis can require the use of various analysis technologies. A microscopic examination may be performed before the sieve analysis, which in turn can be followed by a sedimentation analysis or the recording and the evaluation of a diffraction pattern.

The working ranges of the analysis methods overlap and can be subdivided as shown in Table 10.3, which also details equipment suppliers.

Table 10.3 Suppliers and working ranges of particle-size distribution methods

Method	Particle-size range	Equipment supplier	Model
Dry sieving	63um–63mm	Fritsch	Analysette 3, (20μm–25mm)
Wet sieving	20μm–200μm	Fritsch	Analysette 18
Microsieving	5μm–100μm	Fritsch	
Sedimentation in gravitational field	0.5μm–500μm	Fritsch	Analysette 20
Laser diffraction	0.1μm–1100μm	Fritsch	Analysette 22
Electrical zone sensing	0.4μm–1200μm	Coulter	Model ZM, Coulter multisizer
Electron microscopy	0.5μ m–100μm	–	–
Photocorrelation spectroscopy	0.5μm–5μm	–	–
Sedimentation in centrifugal field	0.5μm–10μm	Fritsch	Analysette 21 (Anderson Pipette centrifuge)
Diffraction spectroscopy	1μm–1mm		
Optical microscale	0.5μm–1mm		
Projection microscopy	0.05μm–1mm		
Image analysis systems	0.8μm–150μm down to 0.5μn	Joyce–Leebl Leitz, Karl Zeiss, Cambridge Instruments	Magiscan and Magiscan P Autoscope P Videoplan II Quantimet 520

Source: Own files

10.4.4.1 Sieving methods (5μm to 63mm)

Sieving methods have been discussed in section 10.4.2.

10.4.4.2 Gravitational sedimentation, 0.5–500μm

An optical measuring system is used in sedimentation analysis, whereby a concentrated beam of light is deflected horizontally through the lower section of a measuring vessel onto a photoelectric cell. The amount of light absorbed by the sedimenting particles decreases with time as the number of particles passing the measuring beam increases. The increase in the photoelectric current as a function of time is then a measure of the particle size.

A major step on the road to reducing the measuring time is provided by the 'Analysette 20' scanning photo sedimentograph (Table 10.2). Using this device, the measuring time is considerably reduced by a continuous movement of the light beam towards the direction of fall of the particle.

10.4.4.3 Centrifugal sedimentation (0.05–10µm)

The Andereasen pipette (Fritsch Analysette 21) (Table 10.3) is extremely well suited to this type of analysis. The measuring radius is determined by six rotating capillaries of equal length in a centrifuge drum. At certain predetermined times, samples are drawn from this radius using a pipette and the solid content of these samples mathematically evaluated to determine the particle-size distribution for the whole sample. The volume of material remaining in the centrifuge is reduced with each sampling and the distance between the surface of the sample liquid and the measuring plane is also reduced, thus reducing the sedimentation time for the smallest particles without the accuracy of the measurement being affected.

10.4.4.4 Laser diffraction (0.1–1100µm)

This is a universally applicable instrument for determining particle-size distributions of all kinds of solids which can be analysed either in suspension in a measuring cell or dry by feeding through a solid particle feeder. In the Fritsch Analysette 22 laser diffraction apparatus (Table 10.3) the measured particle-size distribution is displayed on the monitor in various forms, either as a frequency distribution, as a summary curve or in tabular form and can be subsequently recorded on a plotter, stored on hard disk or transferred to a central computer via an interface. The time required for one measurement is approximately 2min.

10.4.4.5 Electrical zone sensing (0.4–1200µm)

This is the classical method of carrying out particle-size analysis. Coulter supply two instruments – the Model ZM (video display optical) and the top-of-the-range multisiser – the latter having built-in video display of results.

The Coulter method of sizing and counting is based on measurable changes in electrical resistance produced by non-conductive particles suspended in an electrolyte.

By means of the Coulter channeliser 256 module an optional extra on the model ZM but built-in on the multisiser, enables biological cell-size distributions to be measured. This provides an ability to measure suspension concentration and distributions of populations against size with a choice of 64-, 128- or 256-channel resolution over a range approximately 3:1 diameter. Size differences as small as $0.05µm^3$ (fL) are detected.

A data management system is also available for the model ZM.

10.4.5 Digestion of solid samples preparatory to chemical analysis

Having as necessary, dried, homogenised or comminuted the samples, they must now be digested in a suitable reagent to extract elements in a suitable form for chemical analysis. In many organisations we have reached the point where the analyses pass from the hands of the person who took the sample to those of the analytical chemist. In the author's experience, however, it must be emphasised that to ensure best-quality results the whole procedure from, for example, statistically sampling a sediment to the final chemical analysis, should be handled by the same person.

10.4.5.1 Wet ashing

Digestion of the sample with hydrochloric acid, hydrofluoric and (if silicaceous material present) nitric acid and aqua regia have all been used. Aqua regia will dissolve most metals. Nitric acid provides an oxidising attack for organic materials which are usually present at very high concentrations in soil, sediment and biological specimens. Perchloric acid is a very strong oxidising agent, especially when used in conjunction with nitric acid, but its use is not favoured by all chemists and certainly it must not be used in the pressure dissolution technique discussed below.

10.4.5.2 Fusion

Fusion with a flux such as sodium hydroxide, potassium bifluoride potassium pyrosulphate has been used extensively in the water industry.

10.4.5.3 Dry ashing

This is often used to remove organic material from the sample. The sample is weighed into a suitable container such as a ceramic or metal crucible, heated in a muffle furnace and the residue dissolved in an appropriate acid. It is not suitable for the analysis of volatile elements such as mercury and arsenic, since they may volatise during the ashing process. Magnesium nitrate has been used as an ashing agent to prevent volatilisation or arsenic during dry ashing. Dry ashing has been used in the analysis of municipal waste [22,23].

10.4.5.4 Pressure dissolution

Pressure dissolution and digestion bombs have been used to dissolve samples for which wet digestion is unsuitable. In this technique the sample is placed in a pressure dissolution vessel with a suitable mixture of acids and the combination of temperature and pressure effects dissolution of the sample. This technique is particularly useful for the analysis of volatile elements which may be lost in an open digestion [24].

10.4.5.5 Microwave dissolution

More recently, microwave ovens have been used for sample dissolution. The sample is sealed in a Teflon bottle or a specially designed microwave digestion vessel with a mixture of suitable acids. The high-frequency microwave, temperature (ca. 100–250°C) and increased pressure have a role to play in the success of this technique. An added advantage is the significant reduction in sample dissolution time [25,26].

(a) Digestion of soils

Kingston and Walter [57] compared microwave digestion with conventional dissolution methods for the determination of metals in soils.

Reynolds [58] has reviewed microwave digestion procedures for the analysis of metal contaminated soils.

Real et al. [59] showed that optimising the microwave heating procedure would optimise results obtained in sequential extraction procedures.

Torres et al. [60] found that a microwave assisted robotic method for trace metals in soil decreased sample digestion times from 2 h to 3 min.

Sturgeon et al. [61] have demonstrated that a continuous flow microwave assisted digestion of soil samples gave an average of 90% recovery of trace elements with good precision.

An appreciable amount of work has been carried out on the application of microwave digestion techniques to the determination of heavy metals, arsenic and uranium in soils and sediments [61–68].

Lo and Fung [62] studied the recovery of heavy metals from soils duping acid digestion with different acid mixtures by a block heater and by microwave heating.

Chakraborty et al. [63] determined chromium in soils by microwave assisted sample digestion followed by atomic absorption spectrometry without the use of a chemical modifier.

Feng and Barrett [64] showed that microwave dissolution of soil and dust samples with nitric–hydrofluoric acid gave recoveries of cadmium (and lead) of over 90% in 30 min digestion.

Kratchvil and Mamba [65] showed that all the zinc and copper were released from soils within 7 min using a commercial microwave oven.

The microwave extraction of cadmium from a soil reference sample gave results comparable to those found after using conventional extraction procedures [66].

Two methods involving dissolution in hydrogen chloride gas and microwave dissolution have been compared for the remote dissolution of uranium in soil [67].

Comparative studies using different digestion procedures have been performed for the determination of heavy metals and arsenic in the fine grain particle size fraction of suspended particulate matter [68]. The highest metal concentrations were found for microwave heating in a closed system using an acid mixture of nitric and hydrofluoric acids.

A continuous flow microwave assisted digestion of environmental soil samples was also found to be an effective approach as trace element recoveries averaged 90% with good precision [70].

(b) Digestion of non-saline sediments

Various workers have applied microwave digestion to the determination of metals in non-saline sediments [57, 61, 63, 65, 69–77].

Mahan et al. [75] used a microwave digestion technique in the sequential extraction of calcium, iron, chromium, manganese, lead in zinc in non-saline sediments.

The sequential extraction scheme of Tessler partitions metals in sediments into exchangeable carbonate bound iron–manganese oxide, bound organic bound and residual binding fractions. Extraction rate experiments using conventional and microwave heating showed that microwave heating procedures results comparable to the conventional procedure. Sequential microwave extraction procedures were established from the results of the extraction rate experiments. Recoveries of total metals from NBS SRM 1645 ranged from 76% to 120% for the conventional procedure, and 62% to 120% for the microwave procedure. Recoveries of total metals using the microwave and conventional techniques were reasonably comparable except for iron (62% by microwave vs 76% by conventional). Substitution of an aqua regia/ hydrofluoric acid extraction for total/residual metals results in essentially complete recovery of metals. Precision obtained from 31 replicate samples of the California Gulch, Colorado, sediment yielded about an average 11% relative standard deviation excluding the exchangeable fraction which was more variable.

Millward and Kluckner [70] have demonstrated that metals showing the poorest precision in conventional digestion showed the greatest improvements when microwave digestion was used to study a standard reference sediment.

Kammin and Brandt [69] showed that microwave digestion had considerable promise as a high speed alternative to the Environmental Protection Agency digestion method 3050 for the determination of trace metals in non-saline sediments.

Kratchvil and Mamba [65] showed that all the zinc and copper were released from non-saline sediments within 7 min using a commercial microwave oven.

Nieuwenholze et al. [76] analysed six reference sediments after microwave aqua regia extraction. The results obtained showed close agreement with the reference values, and microwave extraction gave the same or slightly higher results than those obtained by conventional reflux extraction methods for seven metals tested in 30 samples.

Kingston and Walter [57] compared microwave digestion with convent- ional dissolution methods for the determination of metals in sediments.

Table 10.4 Temperature pressure data for acids heated in a 120ml closed vessel

Acid (wt%)		Temperature (°C)	Pressure (kg cm⁻²)
HNO_3	70	200	8.5
HCl	37	153	8.5
HNO_3	70	193	7.1
HCl	37	30	7.1

Source: Own files

Elwaer and Belzile [77] have compared the use of a closed vessel microwave assisted dissolution method and conventional hotplate digestion for the determination of selenium in lake sediments. A mixture of hydrochloric acid, nitric acid and hydrofluoric acid with microwave digestion resulted in the best recoveries of selenium. Poor recoveries were obtained by hotplate digestion.

Chakraborty et al. [63] determined chromium in non-saline sediments by microwave assisted sample digestion followed by atomic absorption spectrometry without the use of any chemical modifier.

Sturgeon et al. [61] have demonstrated that a continuous flow microwave assisted digestion of sediment samples gave an average recovery of trace elements of 90% with good recovery.

10.4.5.6 Equipment for sample digestions

Pressure dissolution acid digestion bombs

Inorganic and organic materials can be dissolved rapidly in Parr acid digestion bombs with Teflon liners and using strong mineral acids, usually nitric and/or aqua regia and, occasionally, hydrofluoric acid. Perchloric acid must not be used in these bombs due to the high risk of explosion.

Table 10.4 contains temperature and pressure data obtained while using microwave heating with a single closed vessel for two different acids. For nitric acid, 200°C (80°C over the atmospheric boiling point) and 7 kg cm⁻² was achieved in 12 min and for hydrochloric acid 153°C (43°C over the atmospheric boiling point) and 7kg cm⁻² was obtained in 5 min.

At such elevated temperatures these and other acids become more corrosive. Materials that digest slowly or will not digest at the atmospheric boiling points of the acids become more soluble so dissolution times are greatly reduced. The aggressive digestion action produced at the higher temperatures and pressures generated in these bombs result in remarkably short digestion times, with many materials requiring less than 1 min to obtain a complete dissolution, ie considerably quicker than open-tube wet-ashing or acid-digestion procedures (Table 10.5).

Table 10.5 Single-vessel dissolution of inorganic sample using HF:HNO$_3$:H$_2$O

Sample size	Acid volume	Microwave digestion time	Hot-plate digestion time	Time saved
Ig	36mL [1]	Ih	5h	4h

[1]HF:HNO$_3$:H$_2$O, 1:1:1
HF–48wt %
HNO$_3$–70wt %

Source: Own files

Table 10.6 Pressure digestion bombs

Supplier	Oven part no.	Bomb part no.	Comments
Acid digester types			
Parr Instruments	Not supplied	4781	See Table 10.7 for metals
		4782	determination in organic material
CEM Corporation	MD581D	Solid	See Table 10.7 for metals
		PTFE	determination in organic material
Prolabo	Microdigest 300	Solid	See Table 10.7 for metals
	Microdigest A300	PTFE	determination in organic material
Oxygen combustion types			
Parr instruments	Not supplied	1108	For sulphur, chlorine, etc. determination

Source: Own files

Several manufacturers supply microwave ovens and digestion bombs (Tables 10.6 and 10.7(b)). CFM Corporation state that their solid PTFE bombs are suitable for the digestion of soils and sediments.

10.4.5.7 Oxygen combustion bombs (Tables 10.6 and 10.7)

Combustion with oxygen in a sealed Parr bomb has been accepted for many years as a standard method for converting solid and liquid combustible samples into soluble forms for chemical analysis. It is a reliable method whose effectiveness stems from its ability to treat samples quickly and conveniently within a closed system without losing any of the sample or its combustion products. Sulphur compounds are converted to soluble forms and absorbed in a small amount of water placed in the bomb. Organic chlorine compounds are converted to hydrochloric acid or chlorides. Any mineral constituents remain as ash but other inorganic elements such as arsenic, boron, mercury, phosphorus and nitrogen and

Table 10.7 Pressure digestion bombs supplied by Parr Instruments

Catalogue No.	4781	4782
Maximum charge of:		
inorganic sample	1.0	1.0
organic sample	0.1	0.2
Maximum temperature (°C)	250	250
Cup seal	Teflon-o-ring	Teflon-o-ring
Overpressure protection	Compressible relief disc	Compressible relief disc
Closure style	Band tighten	Band tighten
Bomb dimensions, cm:		
height overall	112	14.3
max. o.d.	7.8	7.8
Cup dimensions, cm:		
Inner diameter	3.1	3.1
Inner depth	3.0	6.1
weight, g	515	625

Source: Own files

all of the halogens are recovered with the bomb washings. In recent years the list of applications has been expanded to include metals such as chromium, iron, nickel, manganese, beryllium, cadmium, copper, lead, vanadium and zinc by using a quartz liner to eliminate interference from trace amounts of heavy metals leached from the bomb walls and electrodes [27, 28].

10.4.6 Elemental analysis of sample digests

Once the sample is in solution in the acid and the digest made up to a standard volume the determination of metals is completed by standard procedures such as atomic absorption spectrometry or inductively coupled plasma optical emission spectrometry.

If the sample matrix is complex, it may be necessary to determine if there are any interference effects from the matrix, on the analyte response. This is usually done by spiking the sample with a known amount of analyte. Two equal portions of sample are taken and an appropriate quantity of analyte is added to one to effectively double the absorbance. A similar quantity of analyte is added to water to make a 'spike-alone' solution. Readings are taken for sample, sample-plus-spike and spike-alone solutions and the amount of interference calculated as a percentage enhancement or suppression of the response. The interference can then be corrected or preferably removed by use of a separation technique.

It is advisable to include in the sample run standard materials of a type similar to the samples being examined. Standard biological

Table 10.8 SRM 1645 river sediment microwave digested in 1:1 $HNO_3:H_2O$

Element	(a) in 1:1 $HNO_3:H_2O_2$ Amount recovered (%)	(b) in 5:3 $HNO_3:H_2O_2$ Amount recovered (%)	Certified value (%)
			0.0066
As	0.0060, 0.0060	0.0076 0,0070	0.0012±0.00015
Cd	0.0012, 0.0012	0.0011, 0.0012	2.96±0.28
Cr	3.00, 2.98	3.04, 2.96	0.0109±0.0019
Cu	0.0122, 0.0113	0.0118, 0.0119	0.74±0.02
Mg	0.72, 0.72	0.70, 0.70	0.0785±0.0097
Mn	0.0790, 0.0780	0.0720, 0.0725	0.00458±0.00029
Ni	0.0050, 0.0050	0.0044, 0.0055	0.0714±0.0028
Pb	0.0736, 0.0737	0.0736, 0.0733	(0.00015)
Se	0.0001, 0.0001	0.0001, 0.0001	0.0001 0.172±0.017
Zn	0.170, 0.168	0.160, 0.160	

Source: Own file

materials and river sediments are available from the National Bureau of Standards USA.

Table 10.8 shows results obtained in the digestion in closed vessels of 1 g samples of NBS SRM 1645 river sediment samples, digested (a) in 20 mL of 1:1 nitric acid water and (b) in 5 mL concentrated nitric acid and 3 mL 30% hydrogen peroxide. In the former, at a power input of 450W, the temperature and pressure rose to 180°C and 7kg cm^{-2}. At that point, microwave power was reduced to maintain the temperature and pressure at those values for an additional 50 min. In the latter case, 1 g samples were open-vessel digested in 1:1 nitric acid:water for 10 min at 180W. After cooling to room temperature, 5 ml of concentrated nitric acid and 3ml of 30% hydrogen peroxide were added to each. The vessels were then sealed and power was applied for 15 min at 180W followed by 15 min at 300W power. The temperature rose to 150–160°C at 2.8 kg cm^{-2} after the final 15 min of heating. With both reagent systems element recoveries are in good agreement with the certified values obtained using a hot plate total sample digestion technique which typically requires 4–6 h.

Table 10.9 demonstrates the fact that in the case of sewage sludge the use of closed vessels in combination with microwave heating can speed up sludge digestion significantly. To demonstrate this, three sets of duplicate samples of the same standard EPA sludge sample were digested using different methods. The first set was microwave digested in closed vessels using 70% nitric acid and 30% hydrogen peroxide. The second set was also microwave digested in closed vessels but 1:1 nitric acid:water was used. The third set of samples was digested in glass beakers on a hot

Table 10.9 Element recovery for closed vessel microwave digestion versus open vessel hot plate digestion of EPA sludge sample (all values are µg g⁻¹)

Digestion	Ag	As	Cd	Cr	Cu	Fe	Pb	Ni	Se	Zn
Microwave closed vessel										
10 mL HNO₃	81	4	20	184	1139	17,500	575	185	4	1308
3 mL H₂O₂	81	3	20	184	1132	17,450	576	187	4	1308
40 min										
Microwave closed vessel										
20 mL 1:1 HNO₃:H₂O	80	4	21	188	1136	18,350	593	197	4	1319
60 min	78	4	21	187	1125	18,550	598	193	4	1325
Open vessel hot plate										
10 mL HNO₃	84	3	21	183	1117	17,700	601	191	4	1283
4 mL H₂O₂	88	3	21	183	1128	17,650	591	192	4	1287
10 h										
EPA X value	80.6	17.0[1]	19.1	193	1080	16,500	526	194	not reported	1320

[1] Range of reported values was 0–89 µg g⁻¹

Source: Own files

plate following EPA SW–846 procedures. The microwave digestions required 40min for the nitric acid:hydrogen peroxide dissolution and 60min for the 1:1 nitric acid:water dissolution. The hot-plate dissolution required 10h. Agreement on element recoveries among the three digestion procedures was very good for selenium and other metals and, except for arsenic, they agree well with the average sample reference values.

10.4.7 Sampling equipment

Meriwether reported development of a coring sampler made from standard plumbing parts [29]. It was found especially useful for sampling and maintaining the depth profile of the soft sediments underlying relatively shallow waters. A rapid sampling method using passive sampling devices for soil contaminant characterisation can provide a more thorough site assessment [30]. Analysis of the overall analytical error showed that, independent of the grain size fraction, sampling was the main source of variance for the determination of metals in river sediments [31]. Two approaches were described to determine the number of samples required for representative sampling. Hewitt found that volatile organic compounds are readily lost from soil samples unless care is taken to limit surface area exposure and to ensure subsample isolation [32]. Volatile organic carbon losses were found to be most abundant during field collection and storage. Hewitt reported that fortified soils held in sealed glass ampoules at 4°C, or dispersed in methanol and held at 22°C, showed no significant losses over 20 and 98 days, respectively [33].

Hunt [34] has described a simple method of filtering soil extracts that eliminates the need for filter-funnels and receivers. It therefore reduces the risk of contamination and speeds up the procedure. It also offers a convenient means of obtaining filtrates in the field for subsequent analysis.

After shaking the soil suspension in the extraction bottle, a tube of filter-paper folded about the centre to form a V with the open ends uppermost is inserted into the bottle. Clear filtrate collects inside the paper tube and aliquots are removed with a pipette.

Fig. 10.1 shows two types of tube that were satisfactory. Type B is preferred because it is easier to produce. Type A is formed from a piece of filter-paper of dimensions 92 × 85mm with two edges glued together with clear impact adhesive. Type B is made from a piece of filter-paper of dimensions 200 × 60mm glued along the long edge with a 4 mm overlap (shown folded). The adhesive did not produce any contamination in soil extracts.

Bates et al. [35] collected suspended particulate matter from river water and wastewater effluents using high speed continuous flow centrifugation, and analysed the isolated solids for hydrocarbons. The results were compared with those obtained on samples obtained by glass filter

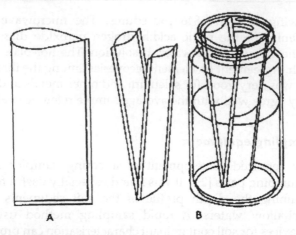

Fig. 10.1 Filtration apparatus
Source: Royal Society of Chemistry [34]

filtration. It was concluded that the use of a continuance flow centrifuge allows the concentration of organic associated with suspended particulate matter to be estimated more accurately.

10.4.8 Slurry sampling technique

A relatively new technique for analysing sediment samples is to slurry the sample ultrasonically prior to introduction into a graphite furnace atomic absorption spectrometer [83–85]. Good agreement was obtained by this technique on certified reference sediments, and of course, sample preparation time is considerably reduced.

Epstein *et al.* [84] have described a method for automated slurry sample introduction for use in the analysis by graphite furnace atomic absorption spectrometry with Zeeman-effect background correction of river sediments. This automated slurry sampling system uses a retractable ultrasonic probe for mixing the slurry solutions prior to sampling and deposition into a graphite furnace. Metals were determined in a standard reference river sediment (SRM 2704). Different methods of slurry preparation are tested, optimum analysis parameters are determined, and sources of variability in the graphite furnace atomic absorption spectrometry measurements are characterised. Measurement variability is found to increase in proportion to the percent of analyte not extracted into the aqueous phase of the slurry solution and is highly dependent on the homogeneity of analyte distribution in the sample. Analytical results for the four elements determined in SRM 2704 are in good agreement with certified values and confirm the utility of slurry sample introduction combined with graphite furnace atomic absorption spectrometry for analysis of a complex matrix.

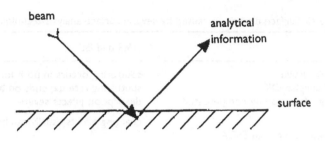

Fig. 10.2 Schemative view of beam analysis
Source: Own files

Klemm and Bombach [85] have reported a considerable simplification of sample preparation for the determination of trace elements in sediments by using ultrasonic slurry sampling and graphite furnace atomic absorption spectrometry. Before analysis, the samples were ground to a grain size of < 63 μm.

10.5 Non-destructive analysis of solid samples

10.5.1 Introduction

There are methods which are capable of showing the distribution of elements on the surface of solid samples such as soils and sediments. As such they enable one to ascertain the homogeneity of distribution of elements on the surface of and, presumably, within the portion of sample analysed.

We have at our disposal a large number of methods for analysing solid materials without altering the sample in any way all of which enable us to characterise them qualitatively, quantitatively and sometimes structurally, by the direct action of a 'reagent' upon a previously prepared surface of the sample. Although there are a number of techniques which involve the destruction of at least a fraction of the sample, either because of the width of the 'reagent' beam used or the sensitivity of the determination (eg laser-source emission spectrometry or spark-source mass spectrometry). In the following discussion we will refer mainly to those techniques for analysis of solid materials which maintain the sample almost intact after impact of the 'reagent'. These are the so-called surface-analysis techniques or, more correctly, beam-analysis techniques. The general scheme of beam analysis is given in Fig. 10.2, from which it may be concluded that beam-analysis systems can be practically unlimited, and even multibeam systems, such as electron spectrometry for chemical analysis (ESCA) are available: these techniques provide various data concerning the sample analysed.

Table 10.10 Surface disruption caused by several surface analysis techniques

ESCA	SIMS and ISS
least destructive	sputtering occurs in both for all materials
95% of samples OK	sputtering rate depends on beam current
charging effects can be compensated	desorption effects severe
AES	charging effects usually can be compensated
more destructive than ESCA	
particularly bad for organic materials	
desorption and diffusion altered by electrons	
charging effects usually can be compensated	
all effects: insulators >> >> semiconductors	
> conductors	

Source: American Chemical Society [36]

As mentioned already, many surface-analysis techniques are available nowadays. In the opinion of some specialists in this field [36,37], four of these are greater in importance: X-ray photoelectron spectrometry (ESCA), Auger electron spectrometry (AES), secondary-ion mass spectrometry (SIMS), and low-energy ion scattering spectrometry (ISS).

The importance of these surface-analysis techniques has resulted in the development of a range of highly automated instruments. In the effort to obtain multiple analytical data, a trend has occurred during the last ten years to build combined instruments, that is apparatus which will permit measurements by several techniques, in a single vacuum system. In this way, greater utilisation of the complex instrumentation involved and a more economic use of the functional parameters of the instruments are ensured.

There is no such thing as a completely non-destructive analysis. Upon interaction of the beam with the sample, a series of surface-disruption phenomena can occur. This fact is illustrated in Table 10.10 for the four major surface analysis techniques mentioned above. The least surface disruption occurs in ESCA and the measurements are characteristic of the surface. However, when charged particles such as electrons and ions are used, this is not necessarily true, as ions may cause considerable sputtering of any material. Although they induce less sputtering, electrons are chemically active and thus may cause chemical effects, which are usually most severe in insulators and least severe in conductors. ESCA measures electrons photo-ejected from a surface by soft X-rays. The technique is applicable to all elements with an atomic number greater than two. The information obtained by this technique is valuable – for instance we can gather data on oxidation states, structural effects, etc.

Table 10.11 Types of X-ray spectrometers used for X-ray fluorescence analysis

Category	Type	Typical source	No. of elements measured	Speed	Cost (×$1000)
Wavelength-dispersive	Single-channel	High-powered X-ray tube, 2–4kW	One at a time	Slow	60–100
	Multi-channel	High-powered X-ray tube, 3–4kW	20–30	Very fast	120–180
Energy-dispersive	Bremsstrahlung source	Low-powered (pulsed) X-ray tube, 0.5–1kW transmission target	In principle, all	Fast	30–70
	Secondary fluorescer	High-powered bremsstrahlung source, 2–4kW, used to generate monochromatic radiation from a selected fluorescer	Groups of 1–10	Relatively fast	60–90

Source: American Chemical Society [38]

10.5.2 X-ray fluorescence spectroscopy

This technique is extremely useful for determining the surface concentration and distribution of elements eg chlorine, arsenic, etc. in soils and sediments.

X-ray fluorescence spectrometry was the first non-destructive technique for analysing surfaces and produced some remarkable results. The Water Research Association, UK, has been investigating the application of X-ray fluorescence spectroscopy to solid samples. Some advantages of non-destructive methods are no risk of loss of elements during sample handling operations, the absence of contamination from reagents, etc. and the avoidance of capital outlay on expensive instruments and highly trained staff.

A wide variety of X-ray fluorescence spectrometers may be used, depending on the nature and complexity of the sample, and on the number of samples to be analysed. To prove this and to indicate the substantial influence which the sample has on the choice of measuring instrument, let us consider some of the main characteristics of some X-ray fluorescence instruments used today [38]. These are shown in Table 10.11.

10.5.3 Electron probe X-ray microanalysis

Even though X-ray fluorescence is now widely used to analyse a large variety of samples, it does have some drawbacks. For instance the X-ray beam used is wide, and this is of no great use for analysing tiny inclusions present in samples, and also does not allow point-by-point analysis on surfaces ('scanning analysis'). The first electron probe X-ray primary emission spectrometer was built in 1949. No doubt this encouraged the use of surface analyses, by allowing samples of very small dimensions to be studied. This was possible because the electron beam had a diameter of only about 1 μm. Although the small size of this beam permitted the analysis of some micro-inclusions in samples, and also multiple analyses by scanning, the main problem, which still remains unsolved, is that of the microhomogeneity and microtopography of the samples. Thus, whereas polishing the solid samples with a 30–100 μm grade abrasive is usually satisfactory for X-ray fluorescence spectrometry, a 0.25 μm grade abrasive or finer may be required for electron-probe microanalysis.

In principle, the difference between X-ray fluorescence spectrometry and electron-probe microanalysis lies in the fact that the analytical information is provided, in the first case, by secondary, fluorescence X-rays, and in the second by primary X-rays, emitted as a result of the impact of the electron beam on the sample's electrons.

Owing to the small size of the electron beam on the one hand, and to the high sensitivity of the method on the other (a sensitivity which can go down to detection of 10^{-16}g), electron-probe microanalysis has found applications in many fields.

Some of the disadvantages of the electron-probe method may be overcome, as in other methods, by the use of complementary techniques. Such techniques can complete the results obtained by electron microprobe. For instance, the introduction of a proton microprobe [39], which is much more sensitive (by two orders of magnitude) than the electron microprobe, and may be used with very good results in geochemical and cosmo-chemical studies.

10.5.4 Auger electron spectrometry

Auger electron spectrometry (AES), reported by Auger in 1923 [40], is also a valuable technique for analysing surfaces. The technique is somewhat similar to ESCA, measuring electrons emitted from a surface as a result of electron bombardment. In both cases, the sampling depth is ca. 20A. Coupling this technique with scanning electron microscopy (SEM) produced a tandem (AES–SEM) technique which has proved extremely productive.

10.5.5 Secondary ion mass spectrometry

In secondary ion mass spectrometry (SIMS), a primary ion beam bombards the surface and a mass spectrometer analyses the ions sputtered from the surface by the primary bombardment. This extremely sensitive technique provides both elemental and structural information.

10.5.6 Ion scattering spectrometry

Ion scattering spectrometry (ISS) is also a technique which is sensitive for all elements with an atomic number greater than 2, and measures the energy change of the bombarding ions, caused by elastic collisions with surface atoms. Like SIMS, it has limited spatial capabilities.

References

1 Smith, R. and James, G.V. *The Sampling of Bulk Materials*, The Royal Society of Chemistry, London (1981).
2 Kratochvil, B., Wallace, D. and Taylor, J.K. *Analytical Chemistry*, **56**, 113R (1984).
3 Kratochvil, B. and Taylor, J.K. *A Survey of Recent Literature on Sampling for Chemical Analysis*, NBS. Technical Note 1153. US Department of Commerce, Washington, DC, January (1982).
4 Gy, P.M. *Sampling of Particulate Mixtures: Theory and Practice*, Elsevier, New York (1979).
5 Woodget, B.W. and Cooper, D. *Samples and Standards*, Wiley, Basingstoke (1987).
6 Harrison. T.S. *Handbook of Control of Iron and Steel Production*, Harward, Chichester (1979).
7 Walton, W.W. and Hoffman, J.I. in *Treatise on Analytical Chemistry*, Part I, vol. 1, 1st edn (eds I.M. Kolthoff and P.J. Elving), Wiley-Interscience, New York, pp.67–97 (1969).
8 Laitnen, H.A. *Chemical Analyses*, 1st edn. McGraw-Hill, New York (1960).
9 Laitnen H.A. and Harris, W.E. *Chemical Analysis*, 2nd edn., McGraw-Hill, New York (1975).
10 Ingamells, C.O. and Pitard, F.F. *Applied Geochemical Analysis*, Wiley-Interscience, New York (1986).
11 Kratochvil, B. and Taylor, J.K. *Analytical Chemistry*, **58**, 924A (1981).
12 Kratochvil, B. *Samples for Microanalysis: Theories and Strategies.* Paper presented at the 11th International Symposium of Microchemical Techniques, Wiesbaden, 28th August–1st September 1989 (1989).
13 Wallace, D. and Kratochvil, B. *Analytical Chemistry*, 59, 226 (1987).
14 Ministry of Agriculture, Fisheries and Food. *The Analysis of Agricultural Materials* R.B. 427, HMSO, London (1979).
15 HMSO *Sampling and Initial Preparation of Sewage and Waterworks Sludges, Soils, Sediments and Plant Materials*, London (1977).
16 British Standards Institution *Sampling Procedures*, London (1971).
17 Veillon, C. *Analytical Chemistry*, **58**, 851A (1986).
18 Versieck, J., Barbier, F., Gornelis, R. and Hoste, J. *Talanta*, **29**, 973 (1982).
19 Wulfson, E.K. and Belyaev, Yu I. *Zhur Analit Khim*, **40**, 1364 (1985.
20 Diamond, R.M. and Tuck, D.G. in *Progress in Inorganic Chemistry*, vol. 2. (ed. F.A. Cotton), Wiley-Interscience, New York, pp.109–192 (1960).
21 Kuchn, D.G., Brandvig, R.L., Lunden, D.C. and Jefferson, R.H. *International Laboratory*, 82, September (1986).
22 Haynes, W. *Perkin–Elmer Atomic Absorption Newsletter*, **17**, 49 (1978).
23 Dalton, E.F. and Melanoski, A.J. *Journal of Association of Official Analytical Chemists*, **52**, 1035 (1969).
24 Adrian, W.A. *Perkin–Elmer Atomic Absorption Newsletter*, **10**, 96 (1971).
25 Reverz, R. and Hasty, E. *Recovery study using an elevated pressure temperature microwave dissolution technique.* Paper presented at the Pittsberg Conference and Exposition on Analytical Chemistry and Applied Spectroscopy, March 1987 (1987).

26 Nadkarni, R.A. *Analytical Chemistry*, **56**, 2233 (1984).
27 Nadkarni, R.A. *American Laboratory*, **13**, 2 August (1981).
28 Parr Manual 207M Parr Instrument Co., 211 53 Rd St. Moine, Illinois 61265 (1974).
29 Meriwether, J.R., Shew, W.J., Hardway, C. and Beck, J.N. *Microchem. Journal*, **53**, 201 (1996).
30 Johnson, K.A., Naddy, R.B. and Weisskopf, C.P. *Toxicology and Environmental Chemistry*, **51**, 31 (1995).
31 Truckenbrodt, D. and Einax, J. *Fresenius Journal of Analytical Chemistry*, **352**, 437 (1995).
32 Hewitt, A.D. *ASTM Special Publication*, **1261**, 170. (Volatile Organic Compounds in the Environment) (1996).
33 Hewitt, A.D. *Special Technical Publication*, **1261**, 181. (Volatile Organic Compounds in the Environment) (1996).
34 Hunt, J. *Analyst (London)*, **106**, 374 (1981).
35 Bates, T.S., Hamilton, S.E. and Cline, J.D. *Estuarine, Coastal and Shelf Science*, **16**, 107 (1983).
36 Hercules, D.M. *Analytical Chemistry*, **50**, 734A (1978).
37 Hercules, D.M. *Analytical Chemistry*, **58**, 1177A (1986).
38 Jenkins, R. *Analytical Chemistry*, **56**, 1099A (1984).
39 Bosch, F., El Goresy, A. Martin B. *et al. Science*, **199**, 765 (1978).
40 Auger, P. *Compt. Rend.*, **177**, 169 (1923).
41 Betteridge, D., Wade, A.P. and Howard, A.G. *Talanta*, **32**, 709, 723 (1985).
42 Ristenpart, E., Gitzel, R. and Uhl, M. *Water Science and Technology*, **25**, 63 (1992).
43 Houba, V.J.G., *Fresenius Journal of Analytical Chemistry*, **345**, 156 (1993).
44 Thoms, M.C. *Journal of Geochemistry and Exploration*, **51**, 131 (1994).
45 Fortunati, G.L., Banfi, C. and Pasturenzi, M. *Fresenius Journal of Analytical Chemistry*, **348**, 86 (1994).
46 Meriwether, J.R., Burns, S.F., Thompson, R.H. and Beck, J.N. *Health Physics*, **69**, 406 (1995).
47 Rasemann, W., Seltmann, U. and Hempel, M. *Fresenius Journal of Analytical Chemistry*, **351**, 632 (1995).
48 Davies, B.E. *Environmental Geochemistry Health*, **11**, 137 (1989).
49 Kimbrough, D.E. and Wakakuwa, J. *Analyst (London)*, **119**, 383 (1994).
50 Einax, J., Machelett, B., Geiss, B. and Danzer, K. *Fresenius Journal of Analytical Chemistry*, **342**, 267 (1992).
51 Rubio, R. and Ure, A. *International Journal of Environmental Analytical Chemistry*, **51**, 205 (1993).
52 Burton, G.A. In *Sediment Toxicity Assessment*, Burton, G.A. ed, Lewis, Boca Raton, Florida pp 37–66 (1992).
53 McBratney, A.B., van Hoof, P., Buydens, L., Kateman, G., Vossen, M., Mulder, W.H. and Bakker, T. *Analytica Chimica Acta*, **1**, 435 (1993).
54 Lame, F.P.J. and Defize, P.R. *Environmental Science and Technology*, **27**, 2035 (1993).
55 Skalski, J.R. and Ward, J.Q. *Environmental Toxicology and Chemistry*, **13**, 15 (1994).
56 Houba, V.J.G., Charden, W.J. and Roelse, K. *Communications in Soil Science and Plant Analysis*, **24**, 1591 (1993).
57 Kingston, H.M. and Walter, P.J. *Spectroscopy*, **7**, 20 (1992).
58 Reynolds, A.R. In *Engineering Aspects of Metal–Waste Management*, Iskander, I.K. and Selim H.M. eds, Lewis, Boca Raton, Florida, pp 49–61 (1992).
59 Real, C., Barreiro, R. and Carballeira, A. *Science of the Total Environment*, **152**, 135 (1994).

60 Torres, P., Ballesteros, E. and Luque de Castro, M.D. *Analytica Chimica Acta*, **308**, 371 (1995).

61 Sturgeon, R., Willie, S.N., Methven, B.A., Lam, J.W.H. and Matusiewicz, H. *Journal of Atomic Spectroscopy*, **10**, 981 (1995).

62 Lo, C.K. and Fung, Y.S. *International Journal of Environmental Analytical Chemistry*, **46**, 277 (1992).

63 Chakraborty, R., Das, A.K., Cervera, M.L. and De la Guardia, M. *Journal of Analytical Atomic Spectroscopy*, **10**, 353 (1995).

64 Feng, Y. and Barratt, R.S. *Science of the Total Environment*, **143**, 157 (1992).

65 Kratchvil, B. and Mamba, S. *Canadian Journal of Chemistry*, **68**, 360 (1990).

66 Krishnamurti, G.S.R., Huang, P.M., Van Rees, K.C.J., Kozak, L.M. and Rostad, H.D.W. *Communications Soil Science and Plant Analysis*, **25**, 615 (1994).

67 D'Silva, A.P., Bajie, S.J. and Zamzow, D. *Analytical Chemistry*, **65**, 3174 (1993).

68 Stachel, B., Elsholz, O. and Reincke, H. *Fresenius Journal of Analytical Chemistry*, **353**, 21 (1995).

69 Kammin, W.R. and Brandt, M. *Journal of Spectroscopy*, **4**, 49 52 (1989).

70 Millward, C.G. and Kluckner, P.D. *Journal of Analytical Atomic Spectroscopy*, **4**, 709 (1989).

71 Li, M., Barban, R., Zucchi, B. and Martinotti, W. *Water Air and Soil Pollution*, **57–58**, 495 (1991).

72 Hewitt, A.D. and Reynolds, C.M. *Atomic Spectroscopy*, **11**, 187 (1990).

73 Paudyn, A.M. and Smith, R.G. *Canadian Journal of Applied Spectroscopy*, **37**, 94 (1992).

74 Hewitt, A.D. and Reynolds, C.M. Report CRREL–SP–90–19 CETHA–TS–CR– 90052 Order No AD–A–226367 (Avail NTIS) (1990).

75 Mahan, K.I., Foderaro, T.H., Garza, T.L., Martinez, M., Maroney, G.A., Trivisonno, M.R. and Willging, E.M. *Analytical Chemistry*, **59**, 938 (1987).

76 Nieuwenholze, J., Poley–Vos, C.H., Van den Akker, A.H. and Van Delft, W.I. *Analyst (London)*, **116**, 347 (1991).

77 Elwaer, N. and Belzile, N. *International Journal of Environmental Analytical Chemistry*, **61**, 189 (1995).

78 Wehrens, R., Van Hoof, D., Buydens, L., Kateman, G., Vossen, M., Mulder, W.H. and Bakker, T. *Analytica Chimica Acta*, **271**, 11 (1992).

79 Vernet, J.P., Favarger, P.Y., Span, D. and Martin, C. *Trace Metals Environ (1 Heavy Metals, Environment)*, **1**, 397 (1991).

80 Droppo, I.G., Krishnappan, B.G. and Onley, E.D. *Environmental Science and Technology*, **26**, 1655 (1992).

81 Truckenbrodt, D. and Einax, J. *Fresenius Journal of Analytical Chemistry*, **352**, 437 (1995).

82 Ruiz, E., Echeandia, A. and Romero, F. *Fresenius Journal of Analytical Chemistry*, **340**, 223 (1991).

83 Hoenig, M., Regnier, P. and Wallast, R. *Journal of Analytical Atomic Spectroscopy*, **4**, 631 (1989).

84 Epstein, M.S., Carnrick, G.R., Slavin, W. and Miller-Ihil, N.J. *Analytical Chemistry*, **61**, 1414 (1989).

85 Klemm, W. and Bombach, G. *Fresenius Journal of Analytical Chemistry*, **353**, 12 (1995).

Chapter 11

Accumulation processes in sediments

It has been observed that sediments in rivers and the oceans have the property of adsorbing some types of dissolved substances present in the overlying water so that the concentration in the sediment (in mg kg^{-1}) is appreciably greater than that in the water (in µg L^{-1}) with which the solid is in contact.

A convenient method of expressing this phenomena is by calculating a concentration factor expressed by:

$$\frac{\text{Concentration of substance in sediment (µg kg}^{-1})}{\text{Concentration of substance in water (µg L}^{-1})}$$

Observed concentration factors for a range of metal ions in different types of water are tabulated in Table 11.1. Where the concentration factor is appreciably greater than unity, the dissolved phase shows a tendency to be adsorbed by the sediment. Examination of the data in Table 11.1 indicates that all the inorganic metal ions listed are strongly adsorbed onto sediments (see Figure 11.1.). This data indicates the magnitude of accumulation of metals in sediments that can occur ranging from values of the order of 100 for lead and antimony to several million in the use of tin.

Both sources of pollution, ie dissolved or sedimentary, are capable of entering living creatures with possible adverse effects. The concentration of toxicants present in sediments is a measure of its concentration in the water over a period of time and is therefore a measure of the risk to creatures. In the case of bottom feeding creatures there is the additional risk of direct ingestion of contaminated sediments in the gills and mouth with consequent adverse effects.

When a creature is exposed to toxicants in the water or sediments in which it lives then the concentration of those toxicants in its tissues gradually increases as a function of exposure time and the concentration of toxicant in the water until the concentration in the tissues of the creature is many times that present in the water. This phenomenon is known as bioaccumulation, which is not to be confused with bioamplification, a process whereby an increase in toxicant levels occurs

Table 11.1 Concentration factors from metals between sediments and liquid phases in water

	In water µg L^{-1}	In sediment µg kg^{-1}		Factor = sediment µg kg^{-1} / water (µg L^{-1})		
				minimum	maximum	mean
Mercury	0.009–13.0 (mean 6.5)	910–46,800 (mean 23,850)	river	3600	101,110	3669
Lead	0.02–200 (mean 100)	23,000–38,200 (mean 30,600)	coastal	191	1.15 × 10^6	306
Tin	<0.0001	1,000–20,000 (mean 10,500)	coastal	<10^7	>2 × 10^8	71.05 × 10^9
Arsenic	1.00–1.04 (mean 1.02)	1,600–117,000 (mean 59,300)	coastal	1,600	112,500	58,140
Copper	0.069–9.7 (mean 4.85)	5,400–84,800 (mean 45,100)	coastal	8,742	78,260	9,298
Nickel	0.2–15.0 (mean 7.6)	30,000–57,000 (mean 43,500)	coastal	3,800	1.5 × 10^5	5,723
Selenium	<0.01–0.08 (mean 0.04)	1,500–9,000 (mean 5,250)	coastal	112,500	150,000	131,250
Antimony	0.30–0.82 (mean 0.56)	6,200–13,400 (mean 9,800)	coastal	20,666	163,414	17,500
Antimony	0.08–0.42 (mean 0.25)	10–2,900 (mean 1,455)	river	125	6,904	5,820
Manganese	0.35–250 (mean 125)	21,800–750,000 (mean 386,000)	coastal	3,000	62,285	3,088

Source: Own files

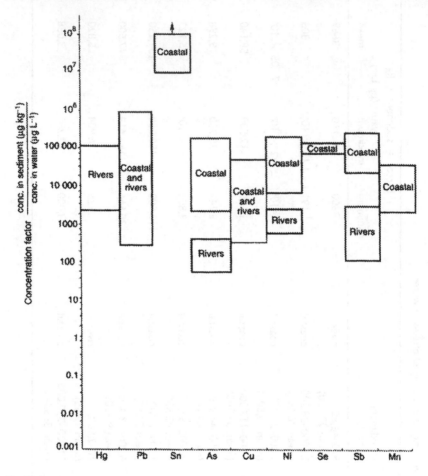

Fig. 11.1
Source: Own files

Table 11.2 Accumulation of bismuth in sediments from Narragensett Bay and North Pacific Ocean

	Naragensett Bay			North Pacific		
	Max	Min	Mean	Max	Min	Mean
Sediment µg kg^{-1}	640	270	455	120	100	110
Water µg L^{-1}	0.00063	0.000083	0.000356	0.00005	0.00005	0.00005
Accumulation factor $\left(\dfrac{\text{sed µg kg}^{-1}}{\text{water µg L}^{-1}}\right)$	7.71×10^6	4.28×10^5	1.28×10^6	2.4×10^6	2.4×10^6	2.2×10^6

Source: Own files

along a food chain, eg plants – minute creatures – fish, as occurs for example in the case of chlorinated insecticides.

Lee [1] using a hydride generation atomic absorption spectrometric method has investigated the accumulation of bismuth on marine sediment samples collected in Narragansett Bay and the North Pacific ocean.

The concentrations of bismuth found in the sediment and in the overlying sea water are tabulated in Table 11.2.

It is seen that the weak concentrations of bismuth in the sediment (in µg kg $^{-1}$) are some one to two million times greater than they are in the overlying water (in µg L $^{-1}$).

Reference

1 Lee, D.S. *Analytical Chemistry*, **54**, 1682 (1982).

Chapter 12

Disposal of wastes to land

12.1 Introduction

In all developed countries, large quantities of agricultural, industrial and municipal wastes are generated. As authorities move to protect the environment by regulating waste disposal practices, environmentally sound methods of waste disposal are being sought. In particular, land application of wastes as a means of disposal, also nutrient recycling and water conservation are becoming increasingly popular [1]. Before discussing the question of disposal of wastes to land, other means of waste disposal currently in use will be briefly discussed below.

At present land filling and disposal to water are the most common means of disposing municipal and industrial waste in the US and most OECD countries. In Europe about 60–70% of municipal and industrial wastes are disposed of this way [2]. Other methods which are growing in importance are incineration and discharge to land as discussed below.

12.2 Disposal of waste by landfilling

Table 12.1 shows that the leachates percolating out of landfill sites contain significant concentrations of anions, alkali, alkaline earth and heavy metals. A major attraction of landfill disposal has been its low cost. However, the increasingly stringent regulatory requirements have resulted in sharp increases in the cost of landfilling in recent years.

12.3 Disposal of waste by incineration

This process is not really suitable for non-combustible wastes including heavy metals. Inorganic pollutants emitted to the atmosphere include mercury, cadmium and lead.

Table 12.1 Typical concentrations of leachates (gm³) from landfill sites

Species	UK	New Zealand
Chloride	3400	200–2000
Sulphate	340	200
Sodium	2185	200–2000
Potassium	888	100–1000
Magnesium	214	50–400
Calcium	88	100–1000
Chromium	0.05	0.05–0.5
Manganese	0.5	5–50
Iron	10	50–400
Nickel	0.04	0.1–1
Copper	0.09	0.05–0.5
Zinc	0.16	1–100
Cadmium	0.02	0.01–6
Lead	0.1	0.05–0.5

Source: Own files

12.4 Disposal of wastes to the oceans

Discharge of sewage waters, sludges and other wastes (eg dredged soils, hazardous wastes) into the marine environment (rivers, lakes and sea) is practised in many countries [2]. Heavy metals such as mercury have been found at elevated concentrations in marine organisms and sediments, particularly near effluent discharge sites [3–7]. World, regional and national organisations have imposed, or are imposing, increasingly strict regulations on the discharge of wastes to sea. The 1972 London Dumping Convention specifies the ban of sea dumping of certain hazardous wastes unless it is proven that the hazardous substance is in trace amounts and would be harmless in the sea [2]. In the EU, discharge of untreated sewage to sea will be phased out in the next few years, except in special circumstances.

12.5 Disposal of waste to land

The management of waste application on land is a challenging task and requires rigorous scientific input. Sludges and effluents contain significant concentrations of metals, also plant nutrients, particularly nitrogen, phosphorus and organic matter. Their application on land has been shown, in many cases, to result in significant increases in plant yields and improvements in soil physical conditions and chemical fertility. The constraints with some wastes, particularly those of industrial and municipal origin, are that they contain undesirable constituents, eg heavy metals, toxic organics, pathogens and salts, or have extremely high or low pH. High concentrations of nitrate and phosphate derived from wastes are also of concern for ground and surface water contamination. The processes that control the fate of wastes in the soil are complex and many of them are poorly understood, eg rate of release of nutrients and other chemicals; leaching of nutrients, metals and organics through macropores and as suspended solids; emission of greenhouse gases; impact of solvents, surfactants and sludge organic matter on the sorption, degradation and leaching of hydrophobic organics; and the long-term bioavailability and fate of metals and organics fixed by soil organic matter.

Land application of wastes is becoming more widespread as regulatory authorities move to protect water quality by restricting waste disposal into rivers, lakes and the marine environment. It is not clear, however, that soil is in fact an appropriate dumping ground for all our wastes.

The pressure to dispose of wastes onto land rather than into water often results in engineers being forced to design land treatment systems with little rigorous scientific information to guide them. One of the main problems is that there is such a wide range of waste materials with different physical, chemical and biological characteristics that it is inappropriate and indeed risky to transfer guidelines from one waste disposal system to another.

In addition to the phasing out of waste disposal into waterways and the ocean, the renewed interest in land application of wastes in the past 20 years is partly because of the need to conserve water and nutrient resources and to use them efficiently and because of the high costs of incineration and landfilling.

In many parts of the world, land application of organic waste is not only an economic imperative, but also a management necessity in order to stem the degradation and erosion of soil.

Although the nutrient content of wastes makes them attractive as fertilisers, the application of many industrial wastes and sewage is constrained by the presence of heavy metals, hazardous organic chemicals, salts and extreme pH values. An example of the type of wastes that might be applied to land is shown in Table 12.2.

While the nutrients (eg nitrogen and phosphorus) contained in some types of wastes applied to land make them attractive as fertilisers, their application on land may be constrained by the presence of toxic metals, (also toxic organics) excessive concentration of salt and high pH. Unfortunately, sewage sludge can contain a range of metals. Actual sludge composition varies with time, and between treatment plants, depending on the type of sewage and waste water received and the nature of the treatment process. Many urban sewage treatment works receive a mixture of domestic and industrial sewage as well as urban run-off from roads and other sealed surfaces. This makes actual prediction of sewage sludge composition extremely difficult.

Table 12.2 Composition of sludges or effluents from tannery and pulp and paper sludges

	Tannery secondary effluent g m³	Pulp and paper sludges mg kg⁻¹	Sewage effluents	
			wet weight basis mg kg⁻¹	dry weight basis mg kg⁻¹
Sodium	2,700	4,586		
Potassium	–	2,905		
Calcium	340	17,000		
Magnesium	36	2,000		
Iron		2,842		380–51,000
Manganese		483		120–1,540
Aluminium		18,000		184–37,800
Tin		15		3.5–66
Arsenic		0.17		0.85–1.50
Cadmium		4.5	3.6–77	3.2–338
Chromium		20	36–630	7–20,000
Copper		206	245–716	7.4–3,650
Lead		42	178–341	10–3,500
Nickel		35	26.3–146	26–940
Mercury		0.3		1.4–26
Zinc		513	603–2,633	100–14,200
Chloride	3,430			
Sulphate	2,410			

Source: Own file

Cameron *et al.* [1] have reviewed the effects of land application of wastes on soils and discuss particularly metals and leaching of metals by plants. Application of sewage sludge to land could result in these metals entering the food chain.

The fate of metals applied to soil in this way is controlled by several processes including rainwater leaching, sorption onto soil particles and pick-up by grass and crops

Sorption refers to the processes by which metals in the soil sorb onto soil solid surfaces or penetrate into the solid matrices. Health hazards to humans and grazing animals result from metal contamination of food for human consumption either by direct ingestion of grass or by contamination of animal feeds, and contamination of crops, fruit and vegetables.

Also, of course, migration of metals from soil by rainwater into the aquifer and subsequent migration to rivers which might be an intake for water treatment plants is another possible source of health hazards to humans. In this connection, there is still variation between and uncertainty over guidelines on permissible concentrations of metals in drinking water supplies.

Table 12.3 Metal pollutant levels found in dry soil and dry sewage sludge

Heavy metals	Found in soil mg kg⁻¹	Acceptable level in soil mg kg⁻¹	Found in sewage sludge mg kg⁻¹
Zinc	11–600	300	100–14,200
Copper	1–240	100	7.4–3,650
Nickel	3–74	100	26–940
Cobalt	0.2–21	50	2–445
Cadmium	0.04–5.1	5	3.2–335
Lead	3–710	100	10–3,500
Chromium	9.2–171	100	7.0–20,000
Total heavy metals (excluding manganese and iron)	28–1,820	755	222–43,073
Major constituents			
Manganese	188–2,750	400	120–1,540
Iron	400–77,500		380–51,000
Arsenic	6–1,375		0.85–150
Titanum	0.5–8,600		0.1–8,420
Aluminium	9,990–85,000		184–37,800
Barium	500–700		110–1,740
Molybdenum	0.2–437		1.26–3.7
Vandium	4–200		11.3–114
Selenium	0.01–111		0.14–10
Minor constituents			
Antimony	0.6–66		2.8–605
Bismuth	0.0–4.0		0.01–4.3
Mercury	0.03–3.3		1.4–26
Beryllium	2.7		–
Silver	0.05–0.5		1.7–104
Uranium	1.2–4.0		4.1–5.4
Tin	1.5–7.1		3.5–66
Tungsten	1.2–18		1.96–16
Hafnium	0.6–4.1		0.01–2
Lanthanum	15–47		8.1–74
Tellurium	–		0.01–2.8
Cerium	–		1.6–142

Source: Own files

12.5.1 Effect of sewage sludge applied to land on the metal content of soil

The terrestrial ecosystem considered includes soil, the crops grown on it and foodstuffs derived from these crops and farm animals. The consequences of applying sewage sludge to land are also considered as pollutants in soil have a bearing on the health of plants and animals, and consequently on humans who eat them. The range of concentrations of various metallic pollutants that have been found in a wide variety of soil samples are reported in Table 12.3. These results show that total heavy metal contents in soil (excluding iron and manganese) can range from as low as 28 mg kg $^{-1}$ in a relatively unpolluted soil to as high as 1820 mg kg $^{-1}$ in heavily polluted soils. This maximum value of 1820 mg kg $^{-1}$ is about twice the maximum acceptable level for total heavy metals (lc 755 mg kg $^{-1}$) in soil which ensures the toxicological acceptability of crops for human and animal consumption.

Zinc, copper, lead and chromium at their highest levels found in soil, *viz* 600, 240, 710 and 171 mg kg $^{-1}$, all exceed maximum acceptable levels for these individual elements, *viz* 300, 100, 100 and 100 mg kg $^{-1}$ respectively. It is noteworthy that metals other than heavy metals that can occur at concentrations exceeding 100 mg kg $^{-1}$ in soil include arsenic, titanium, aluminium, barium, molybdenum, vanadium and selenium.

Some 45% of the 1.24 million tonne of dry sewage sludge produced annually in the UK and Wales is used as an agricultural fertilizer, the remainder being disposed of to rivers or oceans or dumped at sea. This sludge, if it originates in a sewage works that also handles industrial effluents, can contain high levels of metals. The disposal of such sludge to land can be harmful to plants and/or animals and consequently, indirectly, to man who consumes farm crops and animals.

Typical concentrations of metals present in sewage sludge are quoted in Table 12.3. At the maximum observed concentrations appreciably higher concentrations of all the heavy metals also barium, antimony, mercury, silver and uranium can occur in the sludges than in native soil, because contamination of the soil by these metals will be a consequence of the application of sewage sludge to land.

Pollution of good quality farmland soil by heavy metals, originating in sewage sludge additions is illustrated in Table 12.4 which shows the effect of sewage treatment on Scottish arable soil and Canadian prairie soil. In all cases the maximum metal levels in sewage treated soil exceed acceptable levels for crop growing, ie maximum total heavy metal content of treated soil of up to 5690 mg kg $^{-1}$ against a maximum acceptable value of 755 mg kg.

Clearly for a sewage sludge amended to meet the maximum recommended metal concentration limits quoted in Table 12.3, controls must be exerted on both the composition of the sewage and its application rate and application frequency to land.

Consider the case of one hectare of land (100 m × 100 m) of density of cm ^{3} and thickness to which sewage is applied t_m. If the density is 1.5 g cm ^{3} and the soil thickness is 0.2 m then the top 0.2 m of land weighs 10^4 × 0.2 × 1.5 = 3000 tonne. The application of 100 tonne of sewage to the top 0.2 m of soil represents a 3.3% addition by weight. If, for example, the cadmium content of a sewage sludge were 0.08 mg kg $^{-1}$ (Table 12.3) then the cadmium content of the top 0.2 m of sewage treated land would increase by 3.3% of 0.08, ie 0.00264 mg kg $^{-1}$.

Table 12.4 Metal contents (mg kg $^{-1}$) of sewage located land compared to metal contents of untreated good quality farmland

	Untreated Scottish arable soil	Untreated Canadian soil	Sewage treated soil	Acceptable value for crop growing
Cadmium	0.07–0.39	0.07–0.38	1–130	5
Chromium	18–60	9.2–171	20–1,190	100
Copper	3.1–26	4.3–56	1–1,170	100
Lead	8–44	4.2–9.8	3–1,400	100
Nickel	3.8–26	<3–74	2–320	100
Zinc	16.3–121	52–157	0.1–1,420	300
Cobalt	–	–	12–60	50
Total heavy metals	49–277	70–458	38–5,690	755
Manganese	–	–	12–43	400

Source: Own files

Similarly, if the initial cadmium content of the soil was 3 mg kg $^{-1}$ and the cadmium content of the sewage was 100 mg kg $^{-1}$, then application of 100 tonne per hectare of sewage would increase the cadmium content of the top 0.2 m of soil by 3.35% of 100 mg kg $^{-1}$ = 3.3 mg kg $^{-1}$, ie the cadmium content of the top 0.2 m of soil sewage mixture would increase from 3 mg kg $^{-1}$ to 6.3 mg kg $^{-1}$ and would then exceed the 5 mg kg $^{-1}$ limit for soil acceptable for crop growing.

It is interesting to note that Davis *et al.* [8] have noted in their study of the distribution of metals in grassland soils following surface applications of sewage sludge that all the metals studied migrated to a depth of 10 cm (compared to the 200 cm used in the above calculations) but an average of 87% of each metal stayed in the upper 5 cm of soil. They concluded that in the case of cadmium, chromium, copper, molybdenum, nickel, lead and zinc, at least sampling to a depth of 5–7.5 cm would be most suitable for long term monitoring of grassland treated with surface applications of sewage sludge.

We discuss first the simple case in which it is assumed that once sewage has been applied to land, the metallic contaminants present in the sewage remain in the land for prolonged periods of time, ie the metal salt has low water solubility and there is practically no loss with time due to rainwater leaching. This is probably seldom if ever realised in practice. We then proceed to the second case in which the contaminant content of the land due to sewage application gradually increases with repeated applications of sewage, but this effect is offset by a gradual decrease in the metal contaminant content of the land caused by leaching of the contaminant from the soil by rainwater.

12.5.1.1 Simple case in which once sewage is applied to land no subsequent losses of metals occur

We consider the case of applying sewage containing cadmium to soil.
If cadmium content of a soil prior to addition of sewage = Mmg kg^{-1}
Cadmium content of sewage = Smg kg^{-1}
Application rate of sewage to soil = T tonne per ha
Thickness of soil layer to which sewage added = tm
Density of soil to which sewage added = d g cm^3
Weight (W tonne) of one hectare of untreated soil = $10^4 \times t \times d$
Weight (mg) of cadmium in $10^4 \times t \times d$ – T tonne untreated soil = 1000M
$(10^4 \times t \times d - T)$
Weight (mg) of cadmium in T tonne sewage = 1000 STmg
Weight (mg) of cadmium in $(10^4 \times t \times d - T) + T = 10^4 \times t \times d = 1000M$
$(10^4 \times t \times d - T) + 1000$ TS cadmium per $10^4 \times t \times$ dtonne of mixture
Tonne of soil – sewage mixture immediately following addition of sewage
to soil
∴ mg kg^{-1} of cadmium in soil – sewage mixture

$$I_I \text{ (mg kg}^{-1}) = \frac{M (10^4 \times t \times d - T) + ST}{10^4 \times t \times d} \tag{1}$$

To simplify this equation, it is assumed that t = 0.2m, d = 1.5 gcm^3 and M = 0.01 mg kg^{-1}
Then: cadmium content of sludge treated soil (mg kg^{-1})

$$= \frac{30 + T (S - 0.001)}{3000} \tag{2}$$

In Table 12.5 are reported values of cadmium contents of treated soil that would be obtained by applying between 100 and 500 tonnes per hectare of sewage containing between 5 and 50 mg kg^{-1} of cadmium. It is seen that depending on the cadmium content of the sewage and its application rate the sewage has imported to the soil an additional 0.176 to 8.34 mg kg^{-1} cadmium over the 0.01 mg kg^{-1} assumed present in the untreated soil. These data apply to a single application of sewage. Higher levels would result if sewage applications were made yearly as is often the case.

Rearranging equation (1) it is possible to calculate the maximum permissible application rate of sewage of known cadmium content to land (ie T$_{max}$ tonne per hectare) that will permit the cadmium content of the soil sewage mixture immediately after mixing to meet advised cadmium levels.

$$T_{max} = \frac{(I_I - M) \times 10^4 \times t \times d}{S - M} \text{ tonne per hectare} \tag{3}$$

It is also possible by rearranging equation 1 to calculate the maximum allowed cadmium of sewage (S max mg kg^{-1}) that is permitted for the

Table 12.5 Cadmium contents of sewage treated soil $= \dfrac{30 + T\ (S - 0.01)}{3000}$ see equation (2)

Cadmium content of sewage (S), mg kg^{-1}	5			10			50		
Application rate of sewage, T tonnes per ha	100	300	500	100	300	500	100	300	500
Cadmium content of treated soil mg kg	0.176	0.343	1.68	0.51	1.01	5.01	0.84	1.67	8.34

Cadmium content of untreated soil M 0.01 mg kg^{-1} soil density of 1.5 g cm^3, soil thickness to ?? sewage applied t 0.2 m

Source: Own files

cadmium content of the treated soil (I mg kg^{-1}) not to exceed an advised limit.

$$S_{max}\ mg\ kg^{-1} = \frac{(10^4 \times t \times d)\,I_I - M(10^4 \times t \times d - T)}{T} \qquad (4)$$

Thus in equation (3) if the advised maximum concentration of cadmium in treated soil were 0.05 mg kg (I_I), the cadmium content of the sewage (S mg kg^{-1}) was 0.5 mg kg^{-1} and the initial cadmium content of the soil (M mg kg^{-1}) was 0.01 mg kg^{-1}, then the maximum permitted application range of sewage to land would be 245 tonne per hectare on the top 0.2 m of land. Thus after the application of 5 tonne of sewage per hectare of land of a sewage containing between 10 and 50 mg kg^{-1} of cadmium, the cadmium content of the soil sewage mixture (5–8 mg kg^{-1}) approaches or exceeds the maximum recommended cadmium value of 5 mg kg^{-1} for crop growing.

12.5.1.2 Case in which losses of metal from soil by rainwater elution run in parallel with gains in metal content of soil caused by sewage addition

When a metal is introduced into soil it may to varying degrees become bonded to soil particles or may not become so bonded to soil particles. The factors governing the rate of loss of metals from soil by rain elution are complex. Depending on the degree of bonding, the metal will to varying degrees be eluted from the soil by rain and will drain off to water courses, ie the concentration of metal in the soil will decrease as a function of time and the amount of rain.

Of course, some metals form such tight bonds with soil particles that they are never eluted by rain whilst other metals elute from the soil very quickly. As shown in equation 1, the concentration of metals in a sewage treated soil immediately following application of sewage is given by:

$$I_I \, mg \, kg^{-1} = \frac{M \, (10^4 \times t \times d) + ST}{10^4 \times t \times d} \qquad (1)$$

where
M mg kg^{-1} = concentration of metal in soil prior to sewage application
S mg kg^{-1} = metal content of sewage
T tonne per hectare = application rate of sewage
t m = thickness of soil layer to which sewage applied
d gcm^3 = density of soil
If P% represents the annual loss of metals caused by rainwater leaching etc then:

$$\frac{I_n}{I_I} = \left(1 - \frac{P}{100} \right)^n \qquad (5)$$

where
I_n mg kg^{-1} = concentrated metal in sewage treated soil at end of year
I_I mg kg^{-1} = concentration of metal compound in treated soil immediately following application of sewage
Then combining equations (1(and (5),

$$I_n = I_I \left(1 - \frac{P}{100} \right)^n = \left[\frac{M \, (10^4 \times t \times d - T) + ST}{10^4 \times t \times d} \right] \left(1 - \frac{P}{100} \right)^n \qquad (6)$$

I_n gives the net concentration of metal left in the soil at the end of time n years, corrected for losses by leaching etc.
When $I_n = 0.5 \, I_I$
Then

$$\frac{I_n}{I_I} = 0.5 \quad \text{and} \quad \log \frac{I_n}{I_I} = \log 0.5 = n \log \left(1 - \frac{P}{100} \right)$$

$$\text{Therefore } n = \text{half life} = \frac{\log 0.5}{\log \left(1 - \frac{P}{100} \right)} = \frac{-0.301}{\log \left(1 - \frac{P}{100} \right)} \qquad (7)$$

The results in Table 12.6 show that if, immediately following addition of sewage contaminated with a metal to soil, the metal content of the soil is 0.05 mg kg^{-1} and if the annual percentage of loss (P) of metal from the sewage treated soil is for example 30% then at the end of three years, the metal content of the treated soil reduces by 66% to 0.017 mg kg^{-1}, ie the

Table 12.6 Concentrations of metal in sewage treated soil as a function of time

Residual metal content retained in sewage treated soil at end of stipulated year $n = 1, 2, 3$ years etc. $I_n = I_i(1 - P/100)^n$ See equation (4)

P, annual % of metal lost from sewage treated soil by rain elution volatilisation etc	Initial ie metal content of sewage treated soil immediately after addition of sewage I_i mg kg^{-1}	Metal content of sewage treated soil at end of year 1 $= I_n$ (mg kg^{-1}) $= I_i\left(1 - \dfrac{P}{100}\right)$	Metal content of sewage treated soil at end of year 2 $= I_n$ (mg kg^{-1}) $= I_i\left(1 - \dfrac{P}{100}\right)^2$	Metal content of sewage treated soil at end of year 3 $= I_n$ (mg kg^{-1}) $= I_i\left(1 - \dfrac{P}{100}\right)^3$	Half life (n½) of metal in years $n = \dfrac{-0.301}{\log\left(1 - \dfrac{P}{100}\right)}$ (see equation (5))
0	0.050	0.050	0.050	0.050	–
10	0.050	0.045	0.040	0.036	6.54
20	0.050	0.040	0.032	0.025	3.10
30	0.050	0.035	0.024	0.017	1.94
50	0.050	0.025	0.0125	0.0062	1.01
70	0.050	0.015	0.0045	0.0013	0.57
90	0.050	0.005	0.0005	0.00005	0.30
0	0.500	0.500	0.500	0.500	–
10	0.500	0.450	0.400	0.360	8.54
20	0.500	0.400	0.320	0.230	3.10
30	0.500	0.350	0.245	0.170	1.94
50	0.500	0.250	0.125	0.062	1.01
70	0.500	0.150	0.045	0.013	0.57
90	0.500	0.050	0.005	0.0005	0.30

Source: Own files

half-life of the metal is 1.9 × years, ie the initial metal content of 0.05 mg kg $^{-1}$ reduces to 0.025 mg kg $^{-1}$ in 1.94 years. Correspondingly, under similar circumstances an initial metal addition of 0.5 mg kg $^{-1}$ reduces to 0.17 mg kg $^{-1}$ in three years assuming 30% of the initial addition is lost per annum.

In actual fact the regime of adding sewage to land is usually an annual event, ie a certain weight of sewage is incorporated per hectare into the soil annually. So in parallel with the steady decrease in metal content of the soil with time due to losses by elution etc with each further addition of sewage there occurs a further increase in metal content. Estimates of the net metal content of soil at any given time after the applications of sewage as a consequence of these two opposing mechanisms are possible by the following treatment.

If the initial concentration of added metal in the soil at the start of year 1 is I_I mg kg $^{-1}$ then by equation 6 the concentration of metal (I_n mg kg $^{-1}$)

at the end of years 1, 2, 3 are respectively given by $I_I \left(1 - \dfrac{P}{100}\right)$ mg kg $^{-1}$

at the end of year 1, $I_I \left(1 - \dfrac{P}{100}\right)^2$ at the end of year 2, and $I_I\left(1 - \dfrac{P}{100}\right)^3$ at the end of year 3.

If at the end of years 1, 2 and 3, a further addition of sewage to soil adds a further I_I mg kg $^{-1}$ of metal, then the net concentrations in the treated soil will be

$$I_I + I_I \left(1 - \frac{P}{100}\right) = I_I \left[1 + \left(1 - \frac{P}{100}\right)\right] \text{ mg kg}^{-1}$$

at the end of year 1,

$$I_I + I_I \left(1 - \frac{P}{100}\right)^2 + I_I \left(1 - \frac{P}{100}\right) = I_I \left[\left(1 - \frac{P}{100}\right)^2 + \left(1 - \frac{P}{100}\right) + 1\right] \text{ mg kg}^{-1}$$

at the end of year 2 and

$$I_I + I_I \left(1 - \frac{P}{100}\right)^3 + I_I \left(1 - \frac{P}{100}\right)^2 + I_I\left(1 - \frac{P}{100}\right) + 1 =$$

$$I_I \left[\left(1 - \frac{P}{100}\right)^3 + \left(1 - \frac{P}{100}\right)^2 + \left(1 - \frac{P}{100}\right) + 1\right] \text{ mg kg}^{-1}$$

at the end of year 3.

If at the start of year 1, an amount of sewage is added to the soil which imparts 0.05 mg kg $^{-1}$ ($=I_I$) metal to the mixture and if at the end of that year and subsequent years a further amount of sewage is added which imparts an additional 0.05 mg kg $^{-1}$ of metal then the levels of metal to be expected in the soil assuming various P value (ie percentage annual loss of metal due to rain leaching etc) are illustrated in Table 12.7.

Table 12.7 Net effect of sewage additions and leaching on losses on metal content of sewage treated soil

P annual % of metal lost from sewage treated soil by rain elution, volatilisation etc	Metal content of treated soil (I_n mg kg)			
	Start of year 1 $I_1 = 0.05$ mg kg	End of year 1 $I_1\left[1-\left(1-\dfrac{P}{100}\right)\right]$	End of year 2 $I_1\left[\left(\left(1-\dfrac{P}{100}\right)^2+\left(1-\dfrac{P}{100}\right)\right)+1\right]$	End of year 3 $I_1\left[\left(\left(1-\dfrac{P}{100}\right)^3+\left(1-\dfrac{P}{100}\right)^2+\left(1-\dfrac{P}{100}\right)\right)+1\right]$
10		0.095	0.135	0.171
30		0.085	0.109	0.126
70		0.065	0.069	0.071
90		0.055	0.0555	0.0555

Source: Own files

It is seen in Table 12.7 that if only 10% (P) of the metal is lost annually by rain leaching then after three years of sewage treatment the metal content of the soil has doubled. If, however, 70–90% (P) is lost annually, by rain water elution then the metal content of the treated soil remains practically constant despite annual sludge additions.

Calculations of the kind discussed in this section, while not providing precise values, are very useful in that they provide a feel for what happens to toxic metals incorporated into a soil by sewage addition and effects of rainfall etc on these values over a period of time.

It must be understood that although a particular metal salt might be very soluble in rain and would be expected to leach quickly from the soil, if the seasonal rainfall is low then losses by leaching will be low leading to increased levels in crops. Also, when heavy rainfall does occur a sudden surge of toxicant leaching will occur to the watercourse leading possibly to fish kills in adjacent rivers.

References

1 Cameron, K.C., Di, H.J. and McLaren, R.G. *Australian Journal of Soil Research*, **35**, 995 (1997).
2 UNEP Organic Contaminant in the Environmental Pathways and Effects. Ed K.C. Jones, pp 275–289, Elsevier London (1993).
3 Smith, G. *Water Wastes New Zealand*, **86**, 50 (1995).
4 Brechin, J. and McDonald, O. *Australian Journal of Experimental Agriculture*, **34**, 505 (1994).
5 Quin, B.F.C. and Woods, P.H. *New Zealand Journal of Agricultural Research*, **21**, 419 (1978).
6 Smith, K.A., Unwin, R.J. and Williams, J.H. Experiments on the Fertiliser Value of Animal Waste Slurries. In *Long-term effects of sewage sludge and farm slurries applications*. Eds J.H. Williams, G. Guidian and P.L. Hermite, pp 124–135, Elsevier London (1985).
7 Carnus, J.M. and Mason, I. *Land Treatment of Tannery Wastes, Land Treatment Collective Review*, **10**, 31–39 (1994).
8 Davis, R.D., Carlton-Smith, C.H., Stark, J.H. and Campbell, J.A. *Environmental Pollution*, **49**, 99 (1988).

Chapter 13

Relationship between metal contents of soil and crops grown on soil

Information on the levels of various metals present in farmland is given in the first column of Table 13.1. Depending on the history of the land, various studies have shown that the toxic metal levels can vary over a wide range, for example lead concentrations between 3 and 710 mg kg^{-1} have been found against a maximum acceptable level for farmland of 100 mg kg^{-1}. While some of this metal is naturally occurring, i.e. may have a geological origin, the remainder represents manmade pollution such as airborne pollutants and the use of agrochemicals or sewage sludge as a fertiliser.

A proportion of the metals in the soil enter grass and crops grown on the soil and this represents a potential hazard to farm animals and man, who eats the animals and also crops grown on the land.

It would be expected that there exists a relationship between the metal content of soils and the metal content of the crop. In Table 13.2 is given data on the maximum metal contents observed in soils (taken from Table 13.1) and the maximum determined metal contents of various crops including corn, wheat and rice flours, apples, potatoes, broccoli and kale. A plot of maximum metal contents (mg kg^{-1}) in soil and crops, respectively, shows the relationship between these parameters (Fig. 13.1). Metal contents in crops in the range 001–1000 mg kg^{-1} increase with increases in metal contents of the soil in the range 1–100,000 mg kg^{-1}.

This relationship is, understandably, not precise as would be expected due to the many variables involved. It does, however, provide benchmark data.

The relationship is:

$$\log C_s = \log C_C + 2.1 \pm 0.8$$
$$\text{ie } \log C_s = \log C_s - 2.1 \pm 0.8$$
$$\text{ie } \log \frac{C_C}{C_S} = -2.1 \pm 0.8$$

where C_s = concentration mg kg^{-1} of metal in soil
and C_C = concentration mg kg^{-1} of metal in crop

Table 13.1 Metal pollutant levels in soil

	Found in soil mg kg^{-1}	Acceptable level mg kg^{-1}
Heavy metals[a]		
Zinc	11–600	300
Copper	1–240	100
Nickel	3–74	100
Cobalt	0.2–21	50
Cadmium	0.04–5.1	5
Lead	3–710	100
Chromium	9.2–171	100
Total heavy metals	28–1,820	755
Major constituents		
Heavy metals		
Manganese[a]	188–2,750	400
Iron[b]	400–77,500	–
Others		
Arsenic	6–1,375	
Titanium	0.5–8,600	
Aluminium	9,990–85,000	
Barium	500–700	
Minor constituents		
Antimony	0.6–66	
Bismuth	0.1–4.0	
Mercury	0.03–3.3	
Vanadium	4–200	
Beryllium	2.7	
Selenium	0.01–111	
Silver	0.05–0.5	
Uranium	1.2–4.0	
Tin	1.5–7.1	
Molybdenum	0.2–437	
Tungsten	1.2–18	
Hafnium	0.6–4.1	
Lanthanum	15–47	

[a] Heavy metals for which acceptable levels exist
[b] Heavy metals for which acceptable levels do not exist

Source: Own files

This is in fair agreement with the ratio proposed by O'Connor [1], admittedly for organic compounds, of:

$$\frac{C_C}{C_s} = <0.01 \quad ie \quad \log\frac{C_C}{C_s} = -2$$

Table 13.2 Relationship between maximum metal contents of soils and crops

Element	Soils		Crops							
			Cornflour		Wheatflour		Rice flour		Apples	
	mg kg⁻¹ (See Table 13.1)	log mg kg⁻¹	mg kg⁻¹	log mg kg⁻¹	mg kg⁻¹	log mg kg⁻¹	mg kg⁻¹	log mg kg⁻¹	mg kg⁻¹	log mg kg⁻¹
Zinc	600	2.78	37	1.57	10.6	1.02	5.2 / 19.4	0.72 / 1.29	0.71	−0.15
Copper	240	2.38	7.5	0.87	2.0	0.30	—	—	0.3	0.52
Nickel	74	1.87	3.6	0.56	0.06	−1.22	—	—	—	—
Cobalt	21	1.32	—	—	0.013	−1.89	—	—	0.007	−2.15
Cadmium	5.1	0.71	0.10	−1.00	0.04	−1.40	—	—	—	—
Lead	710	2.85	0.61	−0.22	0.016	−0.80	—	—	0.10	−1.00
Chromium	171	2.23	2.5	0.40	0.04	−1.4	—	—	0.66	−0.18
Manganese	2,750	3.44	—	—	8.5	0.93	20	1.30	43	0.63
Iron	77,500	4.89	—	—	20	1.30	9	0.95	—	—
Antimony	66	1.82	—	—	<0.002	<−0.27	—	—	—	—
Bismuth	40	1.60	—	—	1.0	0.00	—	—	—	—
Mercury	3.3	0.52	—	—	—	—	—	—	—	—
Vanadium	200	2.30	—	—	—	—	—	—	0.04	−1.40
Selenium	111	2.04	0.53	−0.28	1.9	0.28	0.30 / 1.87	−0.52 / 0.27	0.004	−2.40
Silver	5.0	0.70	—	—	0.02	−1.7	—	—	—	—
Tin	7.1	0.85	—	—	<0.02	<−1.7	—	—	—	—
Arsenic	1,375	3.14	—	—	1.0	0.00	0.45	−0.35	—	—
Molybdenum	434	2.64	—	—	—	—	1.6	0.20	—	—

Table 13.2 continued

Element	Potatoes		Crops Spinach		Kale		Overall crops	
	mg kg⁻¹	log mg kg⁻¹	mg kg⁻¹	log mg kg⁻¹	mg kg⁻¹	log mg kg⁻¹	mg kg⁻¹	log mg kg⁻¹
Zinc	10.9	1.04	56.2	1.75	35	1.54	56.2	1.75
Copper	3.1	0.49	13.9	1.14	6	0.78	13.9	1.14
Nickel	–	–	7.5	0.87	1.1	0.04	7.5	0.87
Cobalt	0.06	–1.22	1.67	0.22	0.06	–1.22	1.67	0.22
Cadmium	–	–	2.5	0.40	1.1	0.04	2.50	0.40
Lead	1.20	0.08	3.2	0.50	2.9	0.46	3.2	0.50
Chromium	0.06	–1.22	3.86	0.59	0.35	–0.46	3.86	0.59
Manganese	5.7	0.76	173	2.24	16.9	1.23	1.73	2.24
Iron	24	1.38	586	2.77	126	2.10	586	2.77
Antimony	–	–	0.04	–1.40	–	–	0.04	–1.40
Bismuth	–	–	<0.008	<–2.10	–	–	<0.008	<–2.10
Mercury	0.15	–1.82	–	–	–	–	0.15	–0.82
Vanadium	–	–	–	–	0.4	–0.40	0.40	–0.40
Selenium	–	–	0.78	–0.11	0.14	–0.85	1.9	0.28
Silver	0.03	–2.52	0.15	–0.82	0.03	–2.52	0.15	–0.82
Tin	–	–	<0.02	<–1.70	–	–	<0.02	–1.70
Arsenic	–	–	0.17	–0.77	–	–	1.00	0.00
Molybdenum	–	–	–	–	–	–	1.6	0.20

Source: Own files

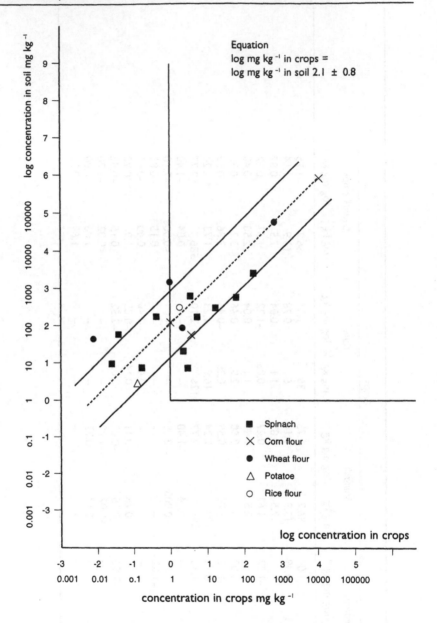

Fig. 13.1 Relationship between metal contents of soil and metal contents of crops grown in soil

Thus if a soil contained 1000 mg kg^{-1} of a metal then the concentration in the crop would be 8 mg kg^{-1} with a range of 1.3 to 50.2 mg kg^{-1}. The corresponding median values in crops if the levels of metal in soil are 10,000 mg kg^{-1} and 100,000 mg kg^{-1} are, respectively, 80 and 800 mg kg^{-1}

levels which would definitely be of environmental concern (see maximum levels in crops of manganese (2750 mg kg $^{-1}$), iron (77500 mg kg $^{-1}$), arsenic (1375 mg kg $^{-1}$), titanium (8600 mg kg $^{-1}$) and aluminium (85000 mg kg $^{-1}$) reported in Table 13.1).

Reference

1 O'Connor, G.A., Chaney, R.L. and Ryan, J.A. *Review Environmental Contamination and Toxicology*, **121**, 129 (1991).

Appendix 1

Instrument suppliers

(Cross-referenced with Chapter 1)

Section 1.1.1 Polarography, voltammetry

646 VA Processor/647 VA stand/675 VA sample changer and 6654 decimat
506 Polarecord
626 Polarecord

UK agents
VA Howe & Co Ltd
12–14 St Anns Crescent
London SW18 2LS
UK

Metrohm Ltd
CH 9100 Herisau
Switzerland

ECP 100 Differential Pulse Polarograph
ECP 120 ASW Programmer
ECP 140 anodic stripping analyser

EDT Analytical Ltd
14 Trading Estate Road
London NW10 7LD
UK

ECP CYSY–1 Computer controlled electro-analytical system
PDV–2000 portable trace metal analyser
CVA–2000 on-line voltammetric analyser
ECP–100 differential pulse polarograph

EDT Analytical Ltd
14 Trading Estate Road
London NW10 7LD
UK

Cypesso Systems Inc
PO Box 3931
Lawrence KS 66046
USA

PDV–2000 on-line voltametric analyser

Chemtronics Ltd
Bentley
Western Australia

Section 1.1.2.1, 1.2.2.1 Visible ultraviolet and near infrared spectrometers

UV, visible and near infrared
PU 8620 basic instrument
Optional PU 8700 scanner for colour graphics
PU 8800 research applications

Pye Unicam Ltd
York Street
Cambridge CB1 2PX
UK

Phillips Electronic Instruments
85 McKee Drive
Mahway NJ 07430
USA

Phillips Nederland BV
HSD Analysetechniken VB3
Postbus 90050
5600 PB Eindhoven
Netherlands

Cecil Instruments Ltd
CE 2343 Visible range spectrophotometer
CE 243D Visible range spectrophotometer
CE 2393 Digital visible grating spectrophotometer
CE 2292 Digital ultraviolet spectrophotometer
CE 2303 Grating spectrophotometer
CE 2202 Ultraviolet spectrophotometer
CE 2373 Linear read-out grating spectrophotometer
CE 2272 Linear read-out ultraviolet spectrophotometer
CE 594 Ultraviolet and visible double-beam spectrophotometer
CE 6000 Ultraviolet visible double-beam spectrophotometer with
CE 6606 Superscan graphic plotter

Cecil Instruments Ltd
Milton Industrial Estate
Cambridge CB4 4AZ
UK

Kontron Instruments
Unikon 860 Ultraviolet visible double-beam spectrophotometer
Unikon 930 Ultraviolet – visible graphics

Kontron Instruments AG
Bernerstrasse SUD 169
8010 Zurich
Switzerland

Perkin–Elmer Ltd
Lambda 2 Ultraviolet – visible double-beam spectrophotometer
Lambda 3 Ultraviolet – visible double-beam spectrophotometer
Lambda 5 and Lambda 7 Ultraviolet visible spectrophotometers
Lambda 9 Ultraviolet visible – near infrared spectrophotometer
Lambda Array 3430 Spectrophotometer

Perkin–Elmer Ltd
Post Office Lane
Beaconsfield
Buckinghamshire HP9 1QA, UK

Perkin–Elmer Corporation
Analytical Instruments Division
761 Main Avenue
Norwalk CT 06859, USA

Bodenseewerk Perkin–Elmer P 60 GmbH
Postfach 1120
D 770 Uberlingen
Germany

Section I.1.2.2, I.2.2.2 Luminescence instruments and spectrofluorimeters

Luminescence instrument LS–3B
Luminescence instrument LS–5B
Accessories: low flow cell, cell holders, bioluminescence spectroscopy, fluorescence spectroscopy, recorders/printers, low temperature luminescence, fluorescence plate reader, polarization accessory, microfilm fluorimeter LS–2B

Perkin–Elmer Corporation
Analytical Instruments Division
761 Main Avenue
Norwalk CT 06859–0012
USA

Perkin–Elmer Ltd
Post Office Lane
Beaconsfield
Buckinghamshire HP9 1QA
UK

SFM–25 spectrofluorimeter

Kontron Instruments
Kontron AG
Bernerstrasse Sud 169
8010 Zurich
Switzerland

Chemi and bioluminescence; lumicon, luminescence instrument

Hamilton Co
PO Box 1003
Reno
Nevada 89520
USA

Hamilton Bonaduz AG
PO Box 26
CH 7402 Bonaduz
Switzerland

Section I.1.2.3, I.2.2.2 Flow injection analysis

FIA star 5020
FIA star 5032
FIA star 5025 ion selective electrode meter
FIA star 5010
The Aquatec system

UK Supplier
EDT Analytical Ltd
14 Trading Estate Road
London NW10 7LD
UK

Tecator AB
Box 70
S 26301 Hanagas
Sweden

FIA System LGCI

Advanced Medical Supplies Ltd
Caker Stream Road
Mill Lane Industrial Estate
Alton, Hampshire GU34 2PL, UK

Chemlab FIA system

Chemlab Instruments Ltd
Hornminster House
129 Upminster Road
Hornchurch
Essex RM11 3XJ
UK

FIA

Skalpar BV
Spinvald
LL 4815 HS Breda
PO Box 3237
NL 4800 DE Breda
Netherlands

Fialite 600
Fiatrode 400
Fiatrode 410

Fiatron Laboratory Systems
5105 South Wortington Street
Oconomowoc WI 53066
USA

Fiazyme 500 Series Automated Carbohydrate Analysis Models

Sections 1.1.3.1(a), 1.2.4 Flame and graphite furnace atomic absorption spectrometry

IL and Video Series:

Thermoelectron Ltd
(formerly Allied Analytical Systems)
830 Birchwood Boulevard
Birchwood
Warrington
Cheshire WA3 7QT
UK

Thermoelectron Ltd
590 Lincoln Street
Waltham
MA 02254
USA

Perkin–Elmer 2280, 2380, 1100 and 2100

Perkin–Elmer Ltd
Post Office Lane
Beaconsfield
Buckinghamshire HP9 1QA
UK

Perkin–Elmer Corporation
Analytical Instruments Division
761 Main Avenue
Norwalk CT 06856
USA

Varian Associates Spectr AA 30/40 and Spectr AA 10/20

Varian Associates Ltd
29 Manor Road
Walton on Thames
Surrey KT12 2QF
UK

Varian Techtron Pty Ltd
679 Springvale Road
Mulgrove
Victoria
Australia 3170

Varian Instruments Division
611 Hansen Way
Palo Alto
California 94303
USA

Varian AG
Steinlauserstrasse
CH 6300 Zug
Switzerland

GBC 903 and 902

GBC Scientific Equipment Pty Ltd
22 Brooklyn Avenue
Dandenong
Victoria
Australia 3175

UK Agent
Techmation Ltd
58 Edgware Road
Edgware
Middlesex HA9 8JP
UK

Shimadzu AA 670 and AA 670G

Shimadzu Corporation
International Marketing Division
Shinjuki Mitsui Building 1–1
Nishe–Shinjuku 2–chome
Shinsuku–ku
Tokyo 163
Japan

UK Agent
VA Howe & Co Ltd
12–14 St Anns Crescent
London SW18 2LS
UK

Autosamplers
20.020 20 Position autosampler
20.080 80 Position autosampler and fraction collection atomic absorption model comprising:
PSA 20.080 SS Stainless steel probe
PSA 20.080 PP Polypropylene probe
PSA 20.080 CC Complete automatic control from computers
PSA 20.080 OA On-line dilution probe
PSA 20.080 FC Triple-probe fraction collector
PSA 20.080 VP Vial piercing option
PSA 20.080 PH PH Electrode assembly
PSA 20.080 TM Turrax mixer assembly and interface requirements:
TTL logic
RS 232 Random access
RS 232 Other requirements

PS Analytical Ltd
Arthur House
Cray Avenue
Orpington
Kent BR5 3TR
UK

Gilson 300 position programmable autosampler

Gilson International
Box 27
300 W Beltine
Middleton
Wisconsin 53562
USA

Gilson Medical Electronics (France) SA
BP 45
F 95400 Villiers le Bel
France

Section 1.1.3.1(b), 1.2.4 Zeeman atomic absorption spectrometry

Perkin–Elmer Zeeman 3030 and Zeeman 5000

Perkin–Elmer Ltd
Post Office Lane
Beaconsfield
Buckinghamshire HP9 1QA
UK

Perkin–Elmer Corporation
Analytical Instruments Division
761 Main Avenue
Norwalk CT 06856
USA

Varian Associates Spectr AA 30/40 and Spectr AA 300/400

Varian Associates Ltd
29 Manor Road
Walton on Thames
Surrey KT12 2QF
UK

Varian Techtron Pty Ltd
679 Springvale Road
Mulgrove
Victoria
Australia 3170

Varian Instrument Division
611 Hansen Way
Palo Alto
California 94303
USA

Varian AG
Steinlauserstrasse
CH 6300 Zug
Switzerland

Section 1.1.3.1(c), 1.2.4 Hydride generator assemblies

PS Analytical Ltd Model PSA 10.002
 Dr P B Stockwell
 PS Analytical Ltd
 Arthur House
 Far North Building
 Cray Avenue, Orpington
 Kent BR5 2TR
 UK

Varian AG VGA–76

Varian Associates Ltd
28 Manor Road
Walton on Thames
Surrey KT12 2QF
UK

Varian Techtron Pty Ltd
679 Springvale Road
Mulgrove
Victoria
Australia 3170

Varian Instrument Division
611 Hansen Way
Palo Alto
California 94303
USA

Varian AG
Steinlauserstrasse
CH 6300 Zug
Switzerland

Section 1.1.3.2 Inductively coupled plasma optical emission spectrometers

Spectroflame

Spectro Analytical UK Ltd
Fountain House
Great Cornbew
Halesowen
West Midlands B63 3BL
UK

Spectro Inc
160 Authority Drive
Fitchbury
MA 01420
USA

PU 7450 and PV 8050

Phillips Analytical Agents
Pye Unicam Ltd
York Street
Cambridge CB1 2PX
UK

Phillips Electronic Instruments
85 McKee Drive
Mahway NJ 07430
USA

Section 1.1.3.3 Inductively coupled plasma mass spectrometers

Plasmaquad

VG Isotopes Ltd
Ion Path, Road Three
Winsford
Cheshire
UK

Elan 500

Perkin–Elmer Ltd
Post Office Lane
Beaconsfield
Buckinghamshire HP9 1QA
UK

Auto samplers
Intelligent autosampler 84100
R 5232 C serial
Communications interface

AB Labtam Ltd
43 Malcomb Road
Braeside
Victoria
Australia 3195

Section 1.1.3.6 Spark source mass spectrometry

Model 251 Isotope ratio mass spectrometer
Model 281 UF6 mass spectrometer
Models 271 and 271/45 Hydrocarbon group type mass spectrometer
Thermionic quadrupole mass spectrometer
Model 261 Magnetic sector mass spectrometer
Delta and Delta E stable isotope ratio mass spectrometers
Delta isotope ratio mass spectrometer

Finnigan MAT
355 River Oaks Parkway
San Jose
CA 95134–1991
USA

Finnigan MAT Ltd
Paradise
Hemel Hempstead
Herts HP2 4TQ
UK

Isochrom II
Gas chromatograph–infrared–mass spectrometer for isotope ratio mass spectrometry

VG Isogas Ltd
Cheshire
UK

Section 1.1.6.1 Energy dispersive and total reflection X-ray fluorescence spectrometers

Energy-dispersive types:
Link Analytical XR 200/300

Link Analytical Ltd
Halifax Road
High Wycombe
Buckinghamshire HP12 3SE
UK

Link Analytical Ltd
240 Twin Dolphin Drive
Suite B
Redwood City
CA 94065
USA

Phillips PW 1404

Pye Unicam Ltd
York Street
Cambridge CB1 2PX
UK

Phillips Electronic Instruments
85 McKee Drive
Mahway NJ 07430
USA

Phillips Nederland BV
Afd Analysetechniken VB3
Postbus 90050
5600 PB Eindhoven
Netherlands

Energy dispersive and total reflection types:
Siefert Extra 2

Richard Siefert & Co GmbH & Co KG
Bogenstrasse 41
Postfach 1280
D 2070 Ahrenberg
Germany

Section 1.1.8.1, 1.2.6.1 Gas chromatography

There are numerous suppliers of gas chromatography equipment, a selection of which are given below

Carlo Erba Instruments
Strada Rivoltana
20090 Rodano
Milan
Italy

Models 8100, 8200, 8400, 8500 and 8700 sigma 2000 range

Perkin–Elmer Corporation
Analytical Instruments
761 Main Avenue
Norwalk CT 06856
USA

Perkin–Elmer Ltd
Post Office Lane
Beaconsfield
Buckinghamshire HP9 1QA
UK

GC 14A, GC 15A, GC 16A, GC 8A

Shimadzu Corporation
International Marketing Division
Shinjuki Mitsui Building
1–1 Nishi Shinjuku 2–chome
Shinjuku–ku
Tokyo 163
Japan

Dyson Instruments Ltd
Hetton Lyons Industrial Estate
Hetton
Houghton le Spring
Tyne and Wear DH5 3RH
UK

Micromat HRGC 412

Nordion Instruments Co Ltd
PO Box 1
SF 003171 Helsinki
Finland

Silchromat 1–4 and Silchromat 2–8

Siemens AG
Instrumentation and Control Division
E 687
Postfach 211262
D 7500 Karlsruhe 21
Germany

Siemens Ltd
VE6
Siemens House
Eaton Bank
Congleton
Cheshire CW12 1PH
UK

Section 1.1.10 Radioactivity measurements

Low-level $\alpha\beta\gamma$ counting system 2401 and 2401F (manual system), 2400 and 2400F (automatic system), series 95 multi–channel analyser, series 10 plus portable multi-channel analysis system, germanium detectors, well type, extended range type, coaxial type, low-energy type, reverse electrode coaxial type and planar type, lead shield 747 cryostats 7500, 7500 SL, 7600 and 7913–30 types, portable cryostats

Canberra Packard Ltd
Brook House
14 Station Road
Pangbourne
Berkshire RG8 7DF
UK

Segmented γScanner (234U and 239U) assays in waste.
Whole body scanners 2250, 2260, 2270, Abacus II

Canberra Instruments Ltd
One State Street
Meriden
CT 06450
USA

Series 10 plus portable gamma multi-channel analyser for field use 7404 Quad
Alpha, alpha spectroscopy system

Canberra Industries Ltd
45 Gracey Street
Meriden
CT 06450
USA

Gammamatic I/II automatic gamma counting system

Kontron AG
Bernerstrasse Sud 169
8010 Zurich
Switzerland

Betamatic IV–V liquid scintillation counters

Kontron Instruments Ltd
Blackmoor Lane
Croxley Centre
Watford
Hertfordshire WD1 8XQ
UK

Tricarb/LL series, low level liquid, scintillation analysers, Tricarb 2050 CA, Tricarb
1550, Tricarb 1050, Tricarb 2250, Tricarb 100 (low cost), Tricarb 1500 sample
changer, Tricarb 1900 CA computer, Tricarb 2200 CA colour graphics, Autogamma 5550/5530 gamma spectrometer, Cobra 5010/5005 autogamma automatic
multi-detector gamma counter

Packard Instrument Co
2200 Warrenville Road
Downers Grove
IL 60515
USA

Canberra Packard International SA
Renggerstrasse 2
CH 8038 Zurich
Switzerland

Labted scale/rate meter alpha-beta scintillators analyser

Bicron Corporation
12345 Kingsman Road
Newbury
OH 44065
USA

Bicron Corporation
PO Box 271
2410 AG Bodegraven
Netherlands

92X Spectrum master gamma spectroscopy workstation

EG&G Ortec
100 Midlands Road
Oak Ridge
TN 37831–0895
USA

EG&G Instruments
Division of EG&G Ltd
Bracknell
Berkshire
UK

Lead castles, encapsulated liquid scintillators, flow cells, crystal scintillators, Na (TI), CgI (Na or TI), Ca F; Manual liquid scintillation counter (tritium and 14 C)

Nuclear Enterprises Ltd
Bath Road
Beenham
Reading
Berkshire RG7 5PR
UK

Integral sodium iodide detectors for gamma spectrometry, Planchette alpha-beta counting and analysis systems, PSR 8 Portable scaler, ST7 scaler timer, SR8 scaler ratemeter

NE Technology Ltd
Bankhead
Medway
Sighthill
Edinburgh EH11 4BY
Scotland

Radon measuring systems; 20 MCA Radon counting system (gamma)

Canberra Industries Inc
One State Street
Meriden
CT 06450
USA

Ortec Airguard Radon Monitoring System (gamma)

EG&G Ortec
100 Midlands Road
Oak Ridge
TN 37831–0895
USA

EG&G Instruments
Division of EG&G Ltd
Bracknell
Berkshire
UK

Picorad Radon Analysis system with Tricarb 1900 CA, Tricarb 1500 or Tricarb 2200 CA liquid scintillation analysers

Packard Instruments Co
2200 Warrenville Road
Downers Grove
IL 60515
USA

Tritium analysis – radioactive gas monitor RG M1/1

Nuclear Enterprises
Bath Road, Beenham
Reading
Berkshire RG7 5PR
UK

Personal radiation monitoring probes
Alpha probes AP2,AP3 and AP4, alpha beta probes DP3 and DP2, beta probes
BP5, BD5 and BP4, high energy gamma probes BP3/4A, beta X-ray probe BP1/4A,
gamma X-ray probes GP6, GP7 and GP9/4A

Personal gamma monitoring system PPDI,Si, Obex contamination monitor (125 I,
beta and X-rays), portable beta gamma dosimeter PPDMI, gamma inspection
monitor PDR4–SV, portable gamma survey monitor PDR3–SV, and PDR2–SV,
low-level gamma radiation monitor PDR ISV, portable beta-gamma
contamination meter PCM 5, radiation gamma doserate meter RDM 1, ratemeter
RH6 for alpha, beta and gamma probes, portable scaler ratemeter PSR 8

Nuclear Enterprises Ltd
Bath Road, Beenham
Reading
Berkshire RG7 5PR
UK

'Analyst' Portable analyser, 'Microanalyst' µR gamma, ion chamber, survey meter,
R7050 and RG050 beta, gamma, X-ray ion chamber survey meters

Bicron Corporation
12345 Kingsman Road
Newbury
OH 44065
USA

Bicron Corporation
PO Box 271
2410 AG Bodegraven
Netherlands

Section I.2.6.2 High performance liquid chromatography

2000 series, 2500 series, 5000 series, 5500 series, 9060 diode array detector, 9060 LC
autosampler

Varian
220 Humboldt Court
Sunnyvale
CA 94069
USA

Varian Associates Ltd
28 Manor Road
Walton on Thames
Surrey KT12 2QF
UK

Series 10 chromatography LC–95 variable wavelength UV/visible detector, LC–90
variable wavelength UV detector, LC–135 and LC–235 diode array detectors, LC
1–100 computing integrator, ISS–100 intelligent sampling system, Series 410 LC
pump

Perkin–Elmer Corporation
Analytical Instruments Division
761 Main Avenue
Norwalk CT 06859
USA

Perkin–Elmer Ltd
Post Office Lane
Beaconsfield
Buckinghamshire HP9 1QA
UK

System 400, comprising 420 and 414 pumps, 460 autosampler, 430 and 432 detectors, 450 data system, 480 column oven, 425 gradient former, Anacomp 220 data management, 306 autosampler, MSI 66 autosampler, 740 LC variable wavelength detector, 720 LC digital variable wavelength detector, 735 LCC variable wavelength detector

Kontron Instruments Ltd
Blackmoor Lane
Croxley Centre
Watford
Hertfordshire WD1 8XQ
UK

Series 4500i

Dionex Corporation
PO Box 3063
Sunnyvale
California
USA

Dionex (UK) Ltd
Albany Park
Camberley
Surrey GU15 2PL
UK

2350 pump, 2360 gradient programmer, 2351 gradient controller V4, UA5 and 228 wavelength detectors, FL–2 fluorescence detector, Chem Research data management/gradient control system, 015A recorder, autoinjector, Foxy, Retriever II and Cygnet fraction collectors, Peak collection instrument

Isco
4700 Superior
Lincoln
NE 67504
USA

2150 pump, 2249 gradient pump, 2152 controller, 2157 autosampler, 2154 injector, 2510 UV detector, 2151 variable wavelength detector, 2140 rapid spectral detector, 2142 refractive index detector, 2143 electrochemical detector, 2221 integrator, 2145 data system, 2210 and 2240 recorders, 2134, 2133, 2135, 2134, 2131 column

Pharmacia LKB
Bjorkgaten 30
75182 Uppsala
Sweden

Pharmacia Ltd
Pharmacea LKB Technology Division
Midsummer Boulevard
Central Milton Keynes
Buckinghamshire MK9 3HP
UK

LC 6A system comprising SCL–6A controller, SPD–6A and SPD–6AV spectrophotometric detectors, CTO–6A column oven, LC–6A pump, SIL–6A auto-injector

LC–8A preparative HPLC comprising LC–8A pump, FCV–100AL reservoir switching valve, FCV–130AL valve/pump box, FCV–120AL recycle value, SCL–8A system controller, SPD–6A UV detector, SPD 6AV UV-visible detector, SIL–8A autoinjector, 7125 manual injector, FCV 100B fraction collector, C–R4A data processor, PC–11L 3–pump interface, PC–30L pump interface, PC–24L interface for reservoir switching valve FCV 110 AL and recycle valve FCV 120 AL, PC–16N interface for fraction collector FCV 100B, PC–14N interface for data processor C–R4A, LC–7A bicompatible system comprising LC–7A pump, LC–7A gradient system, 7125/T sample injector, SPD–7A UV detector, SPD 7AV UV/ visible detector

Shimadzu Corporation
International Marketing Division
Shinjuki Mitsui Buildings
1–1 Nishi Shinjuku 2–chome
Shinjuku–ku
Tokyo 163
Japan

Dyson Instruments Ltd
Hetton Lyons Industrial Estate
Hetton
Houghton le Spring
Tyne and Wear DH5 3RH
UK

HP 1050 Series comprising programmable variable-wavelength detector, multiple-wavelength detector, pumping system, autosampler

Hewlett Packard
PO Box 10301
Palo Alto
CA 94303–0890
USA

HPLC columns only

HPLC Technology
Wellington House
Waterloo St West
Macclesfield
Cheshire SK11 6PJ
UK

Spectroflow 400 system
HPLC pump only

Kratos Analytical Instruments
170 Williams Drive
Ramsey
NJ 07446
USA

Chromo-A-Scope comprising UV visible detector and data processing system only

Barspec Ltd
PO Box 560
Rehovot 76103
Israel

Barspec Ltd
PO Box 430
Mansfield
MA 02048
USA

Roth Scientific Co Ltd
Alpha House
Alexandra Road
Farnborough
Hampshire GU14 6BU
UK

Hichrome Ltd
6 Chiltern Enterprise Centre
Station Road
Theale, Reading
Berkshire RG7 4AA
UK

Series 100 comprising CE 1100 pump, CE 1200 variable wavelength monitor, CE 1300 gradient programmer, CE 1400 refractive index detector, CE 1500 electrochemical detector, CE 1700 computing integrator, CE 1710 recorder and

1720 recorder, CE 1800 sample injector, 1200, 2000 column monitoring panel and sample valve

Cecil Instruments Ltd
Milton Technical Centre
Cambridge CB4 4AZ
UK

Model 5100 A Coulochem electrochemical detector only

ESA Inc	Severn Analytical
45 Wiggins Avenue	30 Brunswick Road
Bedford	Gloucester GL1 1JJ
MA 01730	UK
USA	

Model RR/066 351 and 352 pumps: models 750/16 variable-wavelength UV monitor detector 750/11 variable filter UV detector, MPD 880S multiwave plasma detector, 750/14 mass detector, 750/350/06 electrochemical detector refractive index detector; HPLC columns; column heaters, autosamplers, precolumns derivatization systems, solvent degassers, preparative HPLC systems

Applied Chromatography Systems Ltd
The Arsenal
Heapy Street
Macclesfield
Cheshire SK11 7JB
UK

LCA 15 system; LCA 16 system

EDT Ltd
EDT Research
14 Trading Estate Road
London NW10 7LD
UK

Aspec Automatic sample preparation system; Asted automated sequence trace diazylate enricher; 231/401 HPLC autosampling injector; 232/401 automatic sample processor and injector

Gibson Medical Electronics	Gilson Medical Electronics Inc
(France) SA	Box 27
72 rue Gambetta	300 010 Beltine Highway
BP 45	Middleton
F 95400 Villiers le Bel	Wisconsin 53562
France	USA

Isoflo HPLC radioactivity monitor, iso mix interface between HPLC column and Isoflo HPLC radioactivity monitor:

Nuclear Enterprises Ltd
Bath Road
Beenham
Reading
Berkshire RG7 5PR
UK

Advanced automated sample processor:

Varian AG
Steinhauserstrasse
CH 6300 Zug
Switzerland

Varian Associates
28 Manor Road
Walton on Thames
Surrey KT12 2QF
UK

Section 1.2.6.3 Ion chromatography

Wescam System single-channel ion chromatograph, dual-channel ion chromatograph and dual-channel automated ion chromatograph

UK agents
Alltech Associates Applied Science Ltd
6–7 Kellet Road Industrial Estate
Carnforth
Lancashire LA5 9XP
UK

Wescan Instruments Inc
2051 Waukegau Road
Deerfield
IL 60015
USA

6200 Ion analyser, 6210 pump

Tecator AB
Box 70
S 26301 Hanagas
Sweden

Tecator Ltd
Cooper Road
Thornbury
Bristol BS12 2UW
UK

UK Agent
EDT Analytical Ltd
14 Trading Estate Road
London NW10 7LU
UK

HIC 6A High performance ion chromatographer

UK Agent
Dyson Instruments Ltd
Hetton Lyons Industrial Estate
Hetton
Houghton le Spring
Tyne and Wear DH5 3RH
UK

Shimadzu Corporation
International Marketing Division
3 Kanda
Nisi–kicho 1–chome
Chiyoda–ku
Tokyo 101
Japan

Ion chem system

UK Agent
Severn Analytical
30 Brunswick Road
Gloucester GL1 1JJ
UK

ESA Inc
45 Wiggins Avenue
Bedford MA 01730
USA

Monitor III ion chromatography system

LCD Milton Royal
PO Box 10235
Riviera Beach
FL 33404
USA

LDC UK
Milton Roy House
52 High Street
Stone
Staffordshire ST15 8AR
UK

HPLC Technology Ltd
Wellington House
Waterloo Street West
Macclesfield
Cheshire SK11 6PS
UK

690 Ion chromatograph

Metrohm Ltd
CH 9100 Herisau
Switzerland

UK Agents
VA Howe & Co Ltd
12–14 St Anns Crescent
London SW18 2LS
UK

Dionex Autoion 400 series 2000i, 4000i, 4500i ion chromatographs

Dionex UK Ltd
Selmoor Road
Farnborough
Hampshire
UK

Dionex Corporation
PO Box 3603
Sunnyvale
CA
USA

ILC Series ion chromatograph

Waters Division of Millipore
Millipore UK Ltd
Waters Chromatography Division
11–15 Peterborough Road
Harrow
Middlesex HA1 2YH
UK

Section 1.2.7 Ion-selective electrode equipment

EA 940
EA 920
SA 720
SA 720
Orion 960 Autochemistry System with optional 960SC sample changer

Orion Research UK
Freshfield House
Lewes Road
Forest Row
East Sussex RH18 5ES
UK

Orion Research Incorporated
Laboratory Products Group
The Schraft Centre
529 Main Street
Boston MA 02129
USA

Ion-selective electrodes

Ingold Electrodes Ltd
261 Ballandvale Street
Wilmington
MA 01887
USA

Ion-selective electrodes

EDT Analytical Ltd
14 Trading Estate Road
London NW10 7LU
UK

Section I.3 Water purification units

Autostill range

Jencons Scientific Lab
Cherrycourt Way Industrial Estate
Stanbridge Road
Leighton Buzzard
Bedfordshire LU7 8UA
UK

Waters Division of Millipore
Millipore UK Ltd
Waters Chromatography Division
11–15 Peterborough Road
Harrow
Middlesex HA1 2YH
UK

Millipore Intertech
PO Box 255
Bedford
MA 01730
USA

Water 1 column unit

Gelman Sciences
10 Horrowden Road
Brackmills
Northampton NN0 0EB
UK

L4 stainless steel water cell

Manestry Machines
Speke
Liverpool L24 9LQ
UK

48C/24C Deioniser

Houseman (Burnham) Ltd
UK Industrial Division
Waterslade House
53–57 High Street
Maidenhead
Berks SL6 1JU
UK

Still

Hamilton Laboratory Glassware
Europa House
Sandwich Industrial Estate
Sandwich
Kent CT13 9LR
UK

Aquatron Stills

J Bibby Science Products Ltd
Stone
Staffordshire ST15 0SA
UK

Nanopure ultra pure water system and Cyclon advanced ultrapure still and RO60 reverse osmosis unit

Fistreem Water Purification
Belton Road West
Loughborough
Leicestershire LE11 0TR
UK

Elgastat Spectrum
Elgastat UHP
Elgastat UHQ
Elgastat Prima

Elga Ltd
Lane End
High Wycombe
Buckinghamshire HP14 3JH
UK

Milli Q System

Milli RO System
Water Chromatography Division
Millipore (UK) Ltd
11–15 Peterborough Road
Harrow
Middlesex HA1 2YH
UK

Section 7.4 Laboratory homogenisers

Planetary micromill Pulverisette 7, vibration micro pulverizer, Pulverisette 0, rotary speed mill, Pulverisette 14

Fritsch GmbH
Laborgeraetebau
Industriesstrasse 8
D 6580 Idar Oberstein
Germany

Laboratory comminuters: vibrating cup mill, Pulverisette 9, laboratory disk mill, Pulverisette 13, mortal-grinders, Pulverisette 2, centrifugal mill, Pulverisette 6, planetary mill, Pulverisette 5

Laboratory sieving devices: vibratory sieve shaken for micro-precision sieving, Analysette 3, rotary sieve shaker, Analysette 18

Christison Scientific Equipment Ltd
Albany Road
Gateshead
Tyne & Wear NH8 3AT
UK

Particle size distribution measurement:
(a) Sedimentation in gravitational field, Analysette 20
(b) Laser diffraction, Analysette 22
(c) Sedimentation in centrifugal field, Analysette 21

Fritsch GmbH
Laborgeraetebau
Industriesstrasse 8
D 6580 Idar Oberstein
Germany

Christison Scientific Equipment Ltd
Albany Road
Gateshead
Tyne & Wear NH8 3AT
UK

Appendix 2

Standard official methods of analysis for cations and anions

Her Majesty's Stationery Office, London

Methods for the examination of waters and associated materials

Cations

Arsenic
Arsenic and selenium in sludges, soils and related materials
Arsenic in sludges, soils and related materials (1985). A note on the use of hydride generator kits (1987)

Cadmium
Cadmium in sewage sludge by nitric acid/atomic absorption spectrophotometry (1981)

Calcium
Calcium in water and sewage effluents by atomic absorption spectrophotometry (1977)

Chromium
Chromium in sewage sludge by nitric acid/atomic absorption spectrophotometry (1981)
Chromium in raw and potable waters and sewage effluent (1980)

Copper
Copper in sewage sludge by nitric acid/atomic absorption spectrophotometry (1981)

Lead
Lead in sewage sludge by nitric acid/atomic absorption spectrophotometry (1981)

Mercury
Mercury in waters, effluents and sludges by flameless atomic absorption spectrophotometry (1978)
Mercury in waters, effluents, soils and sediments (1985)
Mercury in waters, effluents and sediments etc. Additional methods 40453 29 pp (1987)

Molybdenum
Molybdenum, especially in sewage sludges and soils by spectrophotometry (1982)

Nickel
Nickel in sewage sludge by nitric acid/atomic absorption spectrophotometry (1981)

Phosphorus
Phosphorus in water effluents and sewages 1980 (1981)
Phosphorus and silicon in effluents and sludges, 2nd edition (1982)

Selenium
Arsenic and selenium in sludges, soils and related materials. A note on the use of hydride generator kits (1987)

Silicon
Phosphorus and silicon in effluents and sludges, 2nd edition (1982)
Phosphorus and silicon in waters, effluents and sludges (1992)

Tin
The determination of organic, inorganic, total and specific tin compounds in water, sediments and biota (1992)

Zinc
Zinc in sewage sludge by nitric acid/atomic absorption spectrophotometry (1981)

Miscellaneous cations

Government of the United Kingdom Department of the Environment Analysis of Raw Potable and Waste Water 305 pp (1972)

A survey of multi-element and related method for analyses of waters, sediments and other materials of interest to the water industry (1980)

Extractable metals in soils, sewage, sludge, treated soils and related materials (1982)

Determination of extractable metals in soils, sewage sludge, treated soils and related materials 1982 (1983)

Information on Concentration and Determination Procedures in Atomic Spectroscopy (1992)

Emission spectrophotometric multi-element methods of analysis for waters, sediments and other materials of interest to the water industry (1980)

Anions

Boron	Boron in waters, effluents, sewage and some solids (1980)
Bicarbonate and carbonate	Titrimetric determination of total and bicarbonate alkalinity and volatile fatty acids in sewage sludge 1980/89 (1989)
Chloride	Chloride in waters, sewage and effluents (1981)
Fluoride	Fluoride in waters, effluents, sludges, plants and soils (1982)
Sulphate	Sulphate in waters, effluents and solids (1979)
	Sulphates in waters, effluents and solids (1988)

Miscellaneous anions/cations

The determination of anions and cations, transition metals, other complex ions and organic acids and bases by chromatography (1990)

Miscellaneous

Suspended solids	Suspended, settleable and total solids in waters and effluents (1980)
Sampling sludges soils etc	Sampling and initial preparation of sewage and waterworks sludges, soils, sediments and plant materials prior to analysis (1977)
α β activity	Measurement of α β activity of water and sludge samples (1985/6)
pH	Determination of pH value of sludge, soil, mud and sediment and the lime requirements of soil (1992)

From: The Analysis of Agricultural Materials, Agricultural Development and Advisory Service (ADAS), 2nd edition, RB427, London HMSO (1979) ISBN 011240352.2

Method
No
2	Preparation of samples of soil, incl drying, grinding
8	Boron, water soluble in soil
10	Cadmium, extractable in soil
11	Cadmium, nitric–perchloric soluble, in soil
15	Carbonate in soil
16	Chlorate in soil

19	Chloride in soil
22	Cobalt, extractable in soil
23	Cobalt, nitric–perchloric soluble, in soil
24	Conductivity and density of soil
26	Copper, EDTA extractable in soil
27	Copper, nitric–perchloric soluble, in soil
29	Divalent-ion deficit in soil
34	pH and lime requirements of mineral soil
43	Lead, extractable in soil
44	Lead, nitric–perchloric soluble in soil
46	Magnesium, extractable in soil
48	Manganese, exchangeable and easily reducible, in soil
50	Molybdenum, extractable in soil
51	Molybdenum, total in soil
53	Nickel, extractable in soil
54	Nickel, nitric–perchloric soluble in soil
59	Nitrate nitrogen, calcium sulphate extractable in soil
60	Ammonium, nitrate and nitrite nitrogen, potassium chloride, extractable in moist soil
63	Particle size distribution, clay, coarse sand, coarse silt, fine sand, fine silt, medium sand
65	Phosphorus, extractable in soil
68	Potassium, extractable in soil
72	Sodium, extractable in soil
84	Ammonium acetate extractable cations (incl calcium) in soil
87	Acid soluble sulphate in soils for tile draining
68	Ammonium acetate extractable potassium in soil
84	Calcium and magnesium, potassium and sodium – lithium acetate extractable and cation exchange capacity of soil
86	Mercury in soil and plant material
84	Manganese, in soil exchangeable
76	Sulphate–sulphur in soil, water soluble
82	Zinc, in soil, extractable
83	Zinc, in soil, nitric–perchloric soluble

USA & Canada

Cutter, G.A. Electric Power Research Institute, Palo Alto, California, Dept EPRI: EA, 4641 Vol 1. Speciation of selenium and arsenic in natural waters and sediments, arsenic speciation (1986)

Bishop, J.N., Taylor, L.A. and Neary, B.P. The determination of mercury in environmental samples. Ministry of the Environment, Canada (1973)

Methods for the analysis of water and wastes, US Environmental Protection Agency, Cincinnati, Ohio, p 134 (1974)

American Public Health Association, American Water Works Association and Water Pollution Control Federation. Standard methods for the examination of water and waste water, 15th edition (1980)

Fisk., J.L., United States Environmental Protection Agency. Effluent guidelines (1979)

US Environmental Protection Agency, *Methods for chemical analysis of water and wastes.* US EPA Cincinnati, Ohio NO EPA–625 16/74–0039 (1976)

Van Loon, J.C., Environment Canada, Research programme for abatement of municipal pollution, Research Report No 51, 42 pp, Heavy metals in agricultural lands receiving chemical sewage sludges (1976)

Chromium

Dissolved and particulate chromium in natural waters, Water Research Centre (UK) Report TR 215

Suspended solids

Suspended solids and ash, accuracy of determination, Water Research Centre (UK) Report 217

Appendix 3

Typical metal contents of sediments

A detailed of results obtained for fresh waters is found in Table A3. In Tables A1 and A2 are summarised concentrations of metallic elements that have been found in fresh water and sea water sediments.

List 1 in Table A1 shows the results obtained for the toxic elements that have been discussed in various EU directives. List 2 covers the major naturally occurring elements, and List 3 the minor elements, many of which are naturally occurring, most of which are of little toxicological concern.

References

1 Sakata, M. and Shimodo, O. *Water Research*, **16**, 231 (1982).
2 Panko, J.F., Leta, D.P., Lin, J.W., Ohl, S.E., Shim, W.P. and Januer, G.E. *Science of the Total Environment*, **7**, 17 (1977).
3 Pillay, K.K.S., Thomas, C.C., Sondel, J.A. and Hycke, C.M. *Analytical Chemistry*, **43**, 1419 (1971).
4 Fresnet Robin, M. and Ottman, F. *Estuarine Marine and Coastal Science*, **7**, 425 (1978).
5 Lum, K.R. and Edgar, D.G. *Analyst (London)*, **108**, 918 (1983).
6 Goulden, P.D., Anthony, D.H.J. and Austen, K.D. *Analytical Chemistry*, **53**, 2027 (1981).
7 Sandhu, A.S. *Analyst (London)*, **106**, 311 (1981).
8 Zink Neilson, I. *Vatten*, **1**, 14 (1977).
9 Agemian, H. and Chau, A.S.Y. *Archives of Environmental Contamination and Toxicology*, **6**, 69 (1977).
10 Breder, R. *Fresenius Zeitschrift für Analytische Chemie*, **313**, 395 (1982).
11 Lichfusse, R. and Brummer, G. *Chemical Geology*, **21**, 51 (1978).
12 Malo, B.A. *Environmental Science and Technology*, **11**, 277 (1977).
13 Agemian, H., and Chau, A.S.Y. *Analyst (London)*, **101**, 761 (1976).
14 Nadkarni, R.A. and Morrison, G.H. *Analytica Chimica Acta*, **99**, 133 (1978).
15 Aspila, K.I., Agemian, H., and Chau, A.S.Y. *Analyst (London)*, **101**, 187 (1976).
16 Minagarva, K., Takizawa, Y. and Fifure, I. *Analytica Chimica Acta*, **115**, 103 (1980).
17 Jirka, A.M. and Carter, M.J. *Analytical Chemistry*, **50**, 91 (1978).
18 Hodge, W.F., Seidel, S.L. and Goldberg, D. *Analytical Chemistry*, **51**, 1256 (1979).
19 Edenfield, H. and Greaves, M.J. in Wong, C.S. (ed.), *Trace Metals in Seawater*, Proceedings of a NATO Advanced Research Institute on Trace Metals in Seawater, 30 March to 3 April 1981, Sicily, Italy, Wiley, New York (1981).

Table A.1 Elements in freshwater sediments

	Concentration (mg kg^{-1})	
Element	Rivers	Lakes
(1) Elements covered in EU directives		
Al	9,890–46,200	26,200–63,800
As	0.22–7.1	1.9–26
Sb		0.01–2.9
Ba		163–2,700
Cd	0.06–27.5	3.5–40
Cr	0.48–1,143	16–110
Co	1.8–53	3.9–200
Cu	0.07–244	50
Pb	0.11–5,060	20–180
Hg	0.91–46.8	1.95–6.8
Ni	1.4–238	1–218
Se	0.09–0.93	0.03–1.0
Ag	1–5.53	0.1–8.05
Ti		800–3,800
U		0.78–4.3
V		26–68
Zn	0.31–9,040	10–450
(2) Naturally occurring elements		
Br		23–96
C		12,300–40,000
Cl		20–609
Fe	16.9–31,000	14,700–30,600
Li		50
Mg		5,900–16,800
Mn	0.34–9,640	214–4500
P	675–1,870	
Na		3,000–9,200
Sr		10–242
(3) Minor elements (few or no toxicity data)		
Ru		19–49
Cs		0.5–14
Au		0.25–19
Th		4.0–9.4
Hf		1.7–12
Zr		55–488
In		5.3–19
Ru		45–500
Sc		3.3–9.2
Ta		0.14–1.4
Tm		0.19–7.4
Ce		53–160
Yb		2.3–9.3
Dy		5.3–15
Gd		6.4–22

Table A.1 continued

	Concentration (mg kg^{-1})	
Element	Rivers	Lakes
La		28–73
Tb		0.95–2.4
Nd		15–137
Sm		7.9–28
Ir		0.5–48
Os		1–4.5
Pt		0.3–8.1

Source: Own files

Table A.2 Metals in marine sediments

Element	Location	Concentration (mg kg^{-1})	Ref
Bismuth	Narragonsett Bay, USA	Surface 0.40	[3]
		49–54 mm core 0.27	
	Pacific	0.1	[3]
Mercury		Sand < 0.1–1.4	[17]
		Clay <0.1–0.8	
	River Loire estuary, salinity 20–35%	13.2	[4]
	River Loire 0–10 km upstream of estuary	28.0	[4]
	River Loire 10–15 km upstream of estuary	22.9	[4]
	River Loire 15–30 km upstream of estuary	46.8	[4]
Tin	Narragonsett Bay, USA	1 cm core 20	[18]
		80 cm core 1	[18]
Lanthanum	Deep-sea sediments	65.1	[19]
Cerium		91.0	
Neodynium		92.5	
Samerium		22.9	
Europium		5.7	
Gadolinium		25.2	
Dysprosium		23.0	
Erbium		13.4	
Ytterbium		13.1	

Source: Own files

Table A.3 Metals in fresh waters, mg kg⁻¹ dry weight

	River			Lake/pond		
	Location	Concentration	Ref.	Location	Concentration	Ref.
Aluminium	–	46,200	[9]	–	26,200–63,800	[14]
	–	Total: 9,890–11,500 acid extractable 522–19,200	[12] [12]	Lake Ontario, Canada	43,000	[13]
Arsenic	River Edisto, USA	0.22–0.63	[7]		1.9–26	[14]
	–	1.9–7.1	[6]			
Antimony				–	0.01–2.9	[14]
Barium				Lake Ontario, Canada	163–175	[14]
Bromine					2,700	[13]
Cadmium	–	0.08–1.22	[1]	Lake Ontario, Canada	23–96	[14]
	River Arno, Italy	Tota: 1.01–9.6	[10]		3.5–8.0	–
		0.05–27.5		Lake Ontario, Canada	40.0	[13]
		Acid extractable 0.1–15.4	[12]			
Caesium				–	0.5–14.0	[14]
Calcium				–	12,300–40,000	[14]
Cerium				–	53–160	[14]
Chlorine				–	20–609	[14]
Chromium	–	0.48–0.49	[8]	Lake Ontario, Canada	16–50	[14]
	River Susquehanna, USA	31.4–1,143	[2]		110	[13]
		108	[9]			

Table A.3 continued

	River			Lake/pond		
	Location	Concentration	Ref.	Location	Concentration	Ref.
	River Arno, Italy	450	[10]			
		Total: 3–368	[12]			
		Acid extractable 1.3–128				
Cobalt	River Arno, Italy	21.9	[10]	–	3.9–16.0	[14]
	–	57	[9]	Lake Ontario, canada	200	[13]
		Total: 2.2–5.3	[12]			
		Acid extractable 1.8–48.9				
Copper	–	0.07	[12]			
	River Rideau, Canada	4.2	[8]	Lake Ontario, Canada	50	[13]
	River Arno, Italy	59.5–244	[9]			
	–	1.9–226	[10]			
		Total: 1–148	[11]			
		Acid extractable 6.6–74	[12]			
Dysporsium				–	5.4–74.0	[14]
Europium				–	0.77–194	[14]
Gadolinium				–	6.4–22	[14]
Gold				–	0.25–19	[14]
Hafnium				–	1.7–12	[14]
Indium				–	5.3–19.0	[14]
Iridium				–	0.5–48	[14]

Table A.3 continued

	River			Lake/pond		
	Location	Concentration	Ref.	Location	Concentration	Ref.
Iron	—	16.9–18.4	[8]	—	14,700–30,600	[14]
	—	31,000	[9]	Lake Ontario, Canada	30,000	[13]
		Total: 6,960–15,700	[12]			
		Acid extractable	[12]			
		1,600–79,800				
Lanthanum				—	28–83	[14]
Lead	River Arno, Italy	0.11–0.13	[8]		20–180	[13]
		60.7–170	[10]			
	—	84	[9]	Lake Ontario, Canada	100	[15]
		17–59	[1]			
		Total: 51–5,060	[12]			
		Acid extractable				
		5–5,160				
Lithium				Lake Ontario, Canada	50	[13]
Lutecium				—	0.52–1.20	[14]
Magnesium				—	5,900–16,800	[14]
				Lake Ontario, Canada	16,000	[13]
Manganese	—	0.34	[8]	—	214–4,500	[14]
	River Arno, Italy	553–704	[10]	Lake Ontario, Canada	4,500	[13]
	—	582	[9]			
	—	5–3225	[11]			
	—	Total: 113–9640	[12]			
		Acid extractable				
		37–9,600				

Table A.3 continued

	River			Lake/pond		
	Location	Concentration	Ref.	Location	Concentration	Ref.
Mercury	River Arno, Italy	0.91–4.4	[10]	Lake Erie, Canada	1.95–6.79	[3]
	River Loire, France	13.2–46.8	[4]			
	River inorganic	6.5–9.0	[16]			
		Total: 12–21.0				
Neodymium				—	15–137	[14]
Nickel	River Arno, Italy	60.0–79.0	[10]	—	1–218	[14]
	—	72	[9]	Lake Ontario, Canada	200	[13]
	—	Total: 7–238	[12]			
		Acid extractable 1.4–67.6	[12]			
Osmium	—					
Phosphorus	—	Total: 675–1,870	[15]	—	1–4.5	[14]
Platinum				—	0.3–8.1	[14]
Potassium				—	5,600–22,900	[14]
Rubidium				—	19–49	[14]
Ruthenium				—	45–500	[14]
Samarium				—	7.9–28.0	[14]
Scandium				—	3.3–9.2	[14]
Selenium	—	0.09–0.93	[6]	—	0.03–1.0	[14]
Silver	—	1–5.53	[5]	—	0.1–1.0	[14]
				Lake Moira, Canada	1.0–8.05	[5]
Sodium	River Arno, Italy	9.3	[10]	—	3,000–9,200	[14]
Strontium				—	10–242	[14]
Tamerium				—	0.19–0.74	[14]

Table A.3 continued

	River			Lake/pond		
	Location	Concentration	Ref.	Location	Concentration	Ref.
Tantalum				—	0.4–1.4	[14]
Terbium				—	0.95–2.4	[14]
Thorium				—	4.9–9.4	[14]
Titanium				—	800–3800	[14]
Uranium				—	0.78–4.3	[14]
Vanadium				—	28–68	[14]
Ytterbium				—	2.34–9.34	[14]
Zirconium				—	54–488	[14]

Source: Own files

Index

Non-destructive analysis of solids 653–657
Non-saline sediments accumulation of metals 630–633; aqueous extraction 366–397; digestion 326, 327, 339; microwave extraction 366, 397–401, 644–648; particle size distribution 640–642; PIXE analysis 365; sequential extraction 366, 401–408

Particle size distribution; solid samples 640–642
Photon activation analysis, determination of aluminium 151, 276; antimony 276, 417, 458–460, 481, 561–569; arsenic 172, 276, 420, 458–460, 487, 561–569; barium 422, 458–460, 483, 561–569; bismuth 483, 561–569; cadmium 484, 561–569; caesium 484; calcium 426, 458–460, 485, 561–569; cerium 485, 561–569; chromium 189, 458–460, 487, 561–569; cobalt 191, 427, 458–460, 488, 561–569; indium 490, 561–569; iron 195, 276, 429, 458–460, 561–569; lead 199, 276, 430, 491, 458–460, 561–569; magnesium 200, 276, 431, 458–460, 561–569; manganese 492, 561–569; mercury 493; molybdenum 494, 561–569; nickel 216, 276, 433, 458–460, 495, 561–569; potassium 434, 458–460, 495, 561–569; rubidium 495, 561–569; selenium 496, 561–569; scandium 496, 561–569; silicon 496, 561–569;